VOLUME SEVENTY FOUR

Advances in
ORGANOMETALLIC
CHEMISTRY

VOLUME SEVENTY FOUR

Advances in ORGANOMETALLIC CHEMISTRY

Edited by

PEDRO J. PÉREZ
Laboratorio de Catálisis Homogénea
CIQSO-Centro de Investigación en Química Sostenible and
Departamento de Química
Universidad de Huelva - Huelva
Spain

Founding Editors

F. GORDON A. STONE

ROBERT WEST

Academic Press is an imprint of Elsevier
50 Hampshire Street, 5th Floor, Cambridge, MA 02139, United States
525 B Street, Suite 1650, San Diego, CA 92101, United States
The Boulevard, Langford Lane, Kidlington, Oxford OX5 1GB, United Kingdom
125 London Wall, London, EC2Y 5AS, United Kingdom

First edition 2020

Copyright © 2020 Elsevier Inc. All rights reserved

No part of this publication may be reproduced or transmitted in any form or by any means, electronic or mechanical, including photocopying, recording, or any information storage and retrieval system, without permission in writing from the publisher. Details on how to seek permission, further information about the Publisher's permissions policies and our arrangements with organizations such as the Copyright Clearance Center and the Copyright Licensing Agency, can be found at our website: www.elsevier.com/permissions.

This book and the individual contributions contained in it are protected under copyright by the Publisher (other than as may be noted herein).

Notices
Knowledge and best practice in this field are constantly changing. As new research and experience broaden our understanding, changes in research methods, professional practices, or medical treatment may become necessary.

Practitioners and researchers must always rely on their own experience and knowledge in evaluating and using any information, methods, compounds, or experiments described herein. In using such information or methods they should be mindful of their own safety and the safety of others, including parties for whom they have a professional responsibility.

To the fullest extent of the law, neither the Publisher nor the authors, contributors, or editors, assume any liability for any injury and/or damage to persons or property as a matter of products liability, negligence or otherwise, or from any use or operation of any methods, products, instructions, or ideas contained in the material herein.

ISBN: 978-0-12-820692-8
ISSN: 0065-3055

For information on all Academic Press publications
visit our website at https://www.elsevier.com/books-and-journals

Publisher: Zoe Kruze
Acquisitions Editor: Sam Mahfoudh
Editorial Project Manager: Shellie Bryant
Production Project Manager: Denny Mansingh
Cover Designer: Alan Studholme

Typeset by SPi Global, India

Contents

Contributors   ix
Preface   xi

1. **Metallodendrimers as a promising tool in the biomedical field: An overview**   1
   Natalia Sanz del Olmo, Riccardo Carloni, Paula Ortega,
   Sandra García-Gallego, and F. Javier de la Mata

   1. Introduction   1
   2. Applications of metallodendrimers in biomedicine   5
   3. Conclusions   44
   Acknowledgments   45
   References   45

2. **Sigma-bond activation reactions induced by unsaturated Os(IV)-hydride complexes**   53
   Miguel A. Esteruelas, Montserrat Oliván, and Enrique Oñate

   1. Introduction   54
   2. H—H bond activation   56
   3. B—H bond activation   58
   4. C—H bond activation   61
   5. C—C bond activation   73
   6. C—O bond activation   76
   7. C-halogen bond activation   77
   8. Si—H bond activation   78
   9. Sn—H bond activation   79
   10. N—H bond activation   83
   11. O—H bond activation   87
   12. O—N bond activation: Azavinylidene compounds   89
   13. Rupture of the oxygen molecule   97
   14. Cl—H bond activation   98
   15. Conclusions   98
   Acknowledgments   100
   References   100

3. **N-heterocyclic germylenes and stannylenes: Synthesis, reactivity and catalytic application in a nutshell**   105
Rajarshi Dasgupta and Shabana Khan

  1. Introduction   106
  2. Synthesis of heavier tetrelylenes [germylenes and stannylenes]   107
  3. Reactivity of heavier tetrelylenes   127
  4. N-heterocyclic germylenes and stannylenes as organocatalyst for basic organic transformations   137
  5. Summary at a glance   145
  References   145

4. **Pincer ligands incorporating pyrrolyl units: Versatile platforms for organometallic chemistry and catalysis**   153
C. Vance Thompson and Zachary J. Tonzetich

  1. Introduction   153
  2. Diaminopyrrolyl pincers (^{R}NNN)   154
  3. Diiminopyrrolyl pincers (^{R}DIP)   160
  4. Dipyridylpyrrolyl pincers (DPP)   167
  5. Bis(pyrazolyl)pyrrolyl pincers (BPP)   170
  6. Bis(oxazolinyl)pyrrolyl pincers (pyrrbox and pyrrmebox)   176
  7. Diphosphinopyrrolyl pincers (^{R}PNP)   194
  8. Pyridinedipyrrolyl pincers (^{R}PDP$^{R'}$)   210
  9. Other pyrrolyl-containing pincers   219
  10. Conclusions and outlook   232
  Acknowledgment   232
  References   232

5. **Low-coordinate M(0) complexes of group 10 stabilized by phosphorus(III) ligands and N-heterocyclic carbenes**   241
Raquel J. Rama, M. Trinidad Martín, Riccardo Peloso, and M. Carmen Nicasio

  1. Introduction   243
  2. Three-coordinate complexes   244
  3. Two-coordinate complexes   289
  4. Conclusions and outlook   305
  Acknowledgments   305
  References   306

6. Recent advances in Pd-catalyzed asymmetric addition reactions — 325
Wenbo Li and Junliang Zhang

1. Introduction — 325
2. Palladium-catalyzed asymmetric conjugate addition — 326
3. Palladium-catalyzed asymmetric 1,2-addition — 356
4. Conclusion and outlook — 393
Acknowledgments — 394
References — 394

7. Titanium catalyzed synthesis of amines and *N*-heterocycles — 405
Laurel L. Schafer, Manfred Manßen, Peter M. Edwards, Erica K.J. Lui, Samuel E. Griffin, and Christine R. Dunbar

1. Introduction — 405
2. Hydroamination — 407
3. Hydroaminoalkylation — 427
4. Titanium redox chemistry for catalytic amine synthesis — 441
References — 458

Contributors

Riccardo Carloni
Department of Pure and Applied Sciences, University of Urbino "Carlo Bo", Urbino, Italy

Rajarshi Dasgupta
Department of Chemistry, Indian Institute of Science Education and Research (IISER), Pune, India

F. Javier de la Mata
Department of Organic and Inorganic Chemistry, and Research Institute in Chemistry "Andrés M. del Río" (IQAR), University of Alcalá, Madrid; Networking Research Center on Bioengineering, Biomaterials and Nanomedicine (CIBER-BBN); Institute "Ramón y Cajal" for Health Research (IRYCIS), Spain

Christine R. Dunbar
Department of Chemistry, University of British Columbia, Vancouver, B.C, Canada

Peter M. Edwards
Department of Chemistry, University of British Columbia, Vancouver, B.C, Canada

Miguel A. Esteruelas
Departamento de Química Inorgánica, Instituto de Síntesis Química y Catálisis Homogénea (ISQCH), Centro de Innovación en Química Avanzada (ORFEO-CINQA), Universidad de Zaragoza-CSIC, Zaragoza, Spain

Sandra García-Gallego
Department of Organic and Inorganic Chemistry, and Research Institute in Chemistry "Andrés M. del Río" (IQAR), University of Alcalá, Madrid; Networking Research Center on Bioengineering, Biomaterials and Nanomedicine (CIBER-BBN); Institute "Ramón y Cajal" for Health Research (IRYCIS), Spain

Samuel E. Griffin
Department of Chemistry, University of British Columbia, Vancouver, B.C, Canada

Shabana Khan
Department of Chemistry, Indian Institute of Science Education and Research (IISER), Pune, India

Wenbo Li
Shanghai Key Laboratory of Green Chemistry and Chemical Processes, School of Chemistry and Molecular Engineering, East China Normal University, Shanghai, P.R. China

Erica K.J. Lui
Department of Chemistry, University of British Columbia, Vancouver, B.C, Canada

Manfred Manßen
Department of Chemistry, University of British Columbia, Vancouver, B.C, Canada

M. Trinidad Martín
Departamento de Química Inorgánica, Universidad de Sevilla, Sevilla, Spain

M. Carmen Nicasio
Departamento de Química Inorgánica, Universidad de Sevilla, Sevilla, Spain

Montserrat Oliván
Departamento de Química Inorgánica, Instituto de Síntesis Química y Catálisis Homogénea (ISQCH), Centro de Innovación en Química Avanzada (ORFEO-CINQA), Universidad de Zaragoza-CSIC, Zaragoza, Spain

Enrique Oñate
Departamento de Química Inorgánica, Instituto de Síntesis Química y Catálisis Homogénea (ISQCH), Centro de Innovación en Química Avanzada (ORFEO-CINQA), Universidad de Zaragoza-CSIC, Zaragoza, Spain

Paula Ortega
Department of Organic and Inorganic Chemistry, and Research Institute in Chemistry "Andrés M. del Río" (IQAR), University of Alcalá, Madrid; Networking Research Center on Bioengineering, Biomaterials and Nanomedicine (CIBER-BBN); Institute "Ramón y Cajal" for Health Research (IRYCIS), Spain

Riccardo Peloso
Instituto de Investigaciones Químicas (IIQ), Departamento de Química Inorgánica and Centro de Innovación en Química Avanzada (ORFEO-CINQA), Consejo Superior de Investigaciones Científicas (CSIC) and Universidad de Sevilla, Sevilla, Spain

Raquel J. Rama
Departamento de Química Inorgánica, Universidad de Sevilla, Sevilla, Spain

Natalia Sanz del Olmo
Department of Organic and Inorganic Chemistry, and Research Institute in Chemistry "Andrés M. del Río" (IQAR), University of Alcalá, Madrid; Networking Research Center on Bioengineering, Biomaterials and Nanomedicine (CIBER-BBN); Institute "Ramón y Cajal" for Health Research (IRYCIS), Spain

Laurel L. Schafer
Department of Chemistry, University of British Columbia, Vancouver, B.C, Canada

C. Vance Thompson
Department of Chemistry, University of Texas at San Antonio (UTSA), San Antonio, TX, United States

Zachary J. Tonzetich
Department of Chemistry, University of Texas at San Antonio (UTSA), San Antonio, TX, United States

Junliang Zhang
Department of Chemistry, Fudan University, Shanghai, P.R. China

Preface

Seven chapters of distinct areas of organometallic chemistry constitute this volume, ranging from synthetic and structural contributions to applications involving stoichiometric and catalytic transformations and biomedical uses as well.

De la Mata provides in Chapter 1 a current view of the uses of metallodendrimers with biomedical applications, showing the yet increasing number of potential uses of metal complexes. A more traditional but equally interesting topic is presented in Chapter 2, where Esteruelas discloses an update in sigma bond activation reactions promoted by osmium-hydride complexes. Group 14 members show up in this volume in Chapter 3, where Khan discusses the chemistry of *N*-heterocyclic germylenes and stannylenes, regarding their synthesis and reactivity, including catalysis. Base metals bearing pincer ligands, with one or more pyrrolyl moieties, constitute the area reviewed by Tonzetich in Chapter 4. This is followed by the chemistry of low-coordinate group 10 M(0) complexes with phosphine or *N*-heterocyclic carbene ligands is the topic presented in Chapter 5 by Nicasio. Next, Zhang has reviewed the use of palladium catalysts in asymmetric addition reactions in Chapter 6. To complete the volume, Schafer presents an account of the use of titanium for the synthesis of amines and *N*-containing heterocycles.

I wish to express my gratitude to all of them as the corresponding authors, as well as to the coauthors in the different chapters for their invaluable contribution. Finally, the work of the editorial team, headed by Shellie Bryant and Denny Mansingh, is also very much appreciated, particularly in this time where COVID-19 is affecting all kind of tasks.

PEDRO J. PÉREZ
Universidad de Huelva

CHAPTER ONE

Metallodendrimers as a promising tool in the biomedical field: An overview

Natalia Sanz del Olmo[a,b,c], Riccardo Carloni[d], Paula Ortega[a,b,c], Sandra García-Gallego[a,b,c,]*, F. Javier de la Mata[a,b,c,]*

[a]Department of Organic and Inorganic Chemistry, and Research Institute in Chemistry "Andrés M. del Río" (IQAR), University of Alcalá, Madrid, Spain
[b]Networking Research Center on Bioengineering, Biomaterials and Nanomedicine (CIBER-BBN), Spain
[c]Institute "Ramón y Cajal" for Health Research (IRYCIS), Spain
[d]Department of Pure and Applied Sciences, University of Urbino "Carlo Bo", Urbino, Italy
*Corresponding authors: e-mail address: sandra.garciagallego@uah.es; javier.delamata@uah.es

Contents

1. Introduction	1
2. Applications of metallodendrimers in biomedicine	5
2.1 Metallodendrimers in cancer treatment	5
2.2 Metallodendrimers as antimicrobial agents	28
3. Conclusions	44
Acknowledgments	45
References	45

1. Introduction

Since ancient times, inorganic compounds have been used to treat different diseases. The discovery of Arsfenamin to treat syphilis in 1910 by Paul Ehrlich launched the clinical use of metal-based drugs.[1] Although in some cases the treatment is carried out with inorganic salts of the corresponding metal, in most of them coordination complexes and organometallic derivatives are employed. The metal complexes present the ability to form strong interactions with the target biomolecule thanks to the combination of the coordination ability of metals, through covalent or ionic bond, with the unique stereoelectronic properties of the ligand. Moreover, the complexation of metals ions to organic ligands frequently reduces their toxicity

Table 1 Overview of common metals used in clinic, including examples of commercially available metallodrugs.

Metal	Therapeutic/diagnostic application	Metallodrug
Au	Rheumatoid arthritis, gout	Auranofin®
Ag	Antiseptic in burns and wounds	Silvadene®
Cu	Dermatitis, algaecide, fungicide, antioxidant	Copper aspirinate
Pt	Cancer	Cisplatin®
Gd	Contrast agent for Magnetic Resonance Imaging	Magnevist®
Zn	Peptic ulcer disease, dyspepsia	Polaprezinc®
Co	Synthesis of vitamin B_{12}	Cyanocobalamin
Li	Manic depression	Lithobid®
Bi	Stomach discomfort, diarrhea, skin injuries	Pepto-Bismol®
Hg	Antiseptic	Meralein sodium®

and improves both the bioavailability and the absorption of ions in tissues. The versatility of metal complexes, with a great variety of coordination numbers, geometries, structural diversity and kinetics of ligand exchange, places them as promising candidates in biomedicine (Table 1). Many other fields have also benefited from their catalytic properties, redox activities, Lewis acidity, magnetic and spectroscopic properties, radioactivity, etc. Nevertheless, the treatment with metallodrugs is often associated with a high toxicity and side effects,[2] thus highlighting the need to develop novel strategies with higher efficiency and selectivity. The potential accumulation of the metal ions in the human body could be harmful, leading to the development of exhaustive studies of their biodistribution and interaction with biological molecules.[3]

Nanotechnology has emerged as a promising alternative to traditional treatments. The nanometric size and the high surface area of these tools often lead to an increase in the therapeutic response and specificity. Since the first conceptualization of molecular machines and nanotechnology in 1959 by Richard Feynman,[4] the field has expanded exponentially and multiple strategies have been reported in the literature, including nanoparticles, liposomes, micelles and dendrimers.[5] Dendrimers and dendritic polymers emerged as a class of macromolecular structures, with an arborescent shape, highly branched and symmetrical, prepared through the repetition of a series of reaction steps that extend from a nucleus to the periphery. Such controlled synthesis, unique among the nanotechnological tools, produces monodisperse molecules with application-driven design. Together with their chemical stability, multivalence and possibility to on-demand attach therapeutic drugs, monitoring tags or other biomolecules, have boosted their use in the last decades not only in the biomedical field, but also in cosmetics,[6] catalysis[7] and photochemistry,[8] among others.

The field of dendritic polymers comprises different topologies—monodisperse dendrimers and dendrons, polydisperse dendritic polymers and hybrids—and different scaffolds—polyamidoamine (PAMAM), polypropyleneimine (PPI), carbosilane, polyesters, polyether, polyglycerol, and phosphorus dendrimers, among others.[9–11] Traditional synthetic approaches to dendrimers include the divergent, the convergent and the mixed synthesis methods. The divergent approach was described in 1978 by D. A. Tomalia and consists on a sequence of reactions that emerge from the core to the periphery.[12] Later, in 1990 C. Hawker and J. M. J. Fréchet developed a new method, in which the synthesis of the dendrimer was carried out in the opposite direction, called convergent synthesis.[13] The synthesis is mainly based on the attachment of dendrons to a multifunctional core, building the dendritic macromolecule from outside to inside. While the divergent synthesis might be suitable for large-scale production, the increasing steric hindrance in the growth to higher generations can compromise the purity and monodispersity of the final dendrimers and structural defects may appear. On the other hand, the convergent route reduces the possibility to get structural defects, but it is mainly limited to low-generation dendrimers, considering the steric hindrance in the attachment of big dendrons to the core. Finally, the mixed method emerged as an alternative to the previous two methods with the aim of reducing the disadvantages of both.[14] Recently, accelerated-approaches have remarkably simplified the synthetic routes to dendrimers and rely on the extraordinary properties of chemoselective and orthogonal reactions.[15,16] This growth concept makes the typical usage of activation steps obsolete, reducing the reaction time and increasing the total yield.

Considering their structural perfection, nanometric size and resemblance to proteins, dendrimers have been described as "artificial proteins." Dendrimers are in the size range of other biomolecules like DNA (2.4 nm), insulin (3 nm), hemoglobin (5.5 nm) or membrane lipids (around 5.5 nm).[17]

Furthermore, some dendrimers have the ability of mimicking actions of their biological analogues such as self-assembly, catalysis or light collection.[18] These similarities have led the scientist to conduct studies of these macromolecules in different biomedical applications. Dendrimers can act as therapeutic agents *per se* or as carriers of drugs or biomolecules through encapsulation, electrostatic interactions or covalent bonds. These macromolecules have been studied toward different diseases such as HIV infection,[19] malaria,[20] neurodegenerative diseases,[21] bacterial infections,[22] cancer and diagnosis,[23] among others. The reader is referred to excellent reviews in the literature.[17,24]

The combination of metal ions and dendritic scaffolds can lead to enhanced or even new properties, in the so-called metallodendrimers. The multivalency and monodisperse nature of dendrimers enable a controlled attachment of multiple metal entities in a single molecule, which has proved to be more effective than their mononuclear counterparts.[25] There are different positions in the dendritic framework susceptible of incorporating metals (Fig. 1). The synthetic route will be dictated by the location of the metal ion, being included during the growth of the dendritic scaffold or after, on a preformed dendrimer. Moreover, the location of the metal ion, together with the nature of the metal, will determine the final application. For example, metallodendrimers containing ruthenium, gold, copper, zinc, iron or technetium have been evaluated due to their therapeutic or diagnostic potential. This review summarizes the latest advances in the use of metallodendrimers in the field of biomedicine, focused on the treatment of cancer and infectious diseases as the most widespread uses. The reader is referred to an excellent review published in 2012 by G. Smith to previous state-of-the-art in the field of anticancer agents.[26]

The demand of multipurpose materials is increasingly changing the focus to heterofunctional dendritic molecules, bearing different moieties in a single scaffold. Therapeutic drugs, monitoring tags or targeting moieties can be attached in a single dendrimer and enhance the efficacy of the treatment. Several approaches have been reported in the literature, where the heterofunctionalities are introduced in the periphery or in the scaffold either randomly or in a controlled way (Fig. 1G–I). This review will also present different heterofunctional metallodendrimers, based on whether the heterofunctionalization has been performed only with metal complexes or if other molecules have been incorporated to provide additional properties such as water solubility or tracing tools. Surprisingly, despite the numerous possibilities that exist today in herofunctionalized dendrimers, only a few have been used in the field of metallodendrimers.

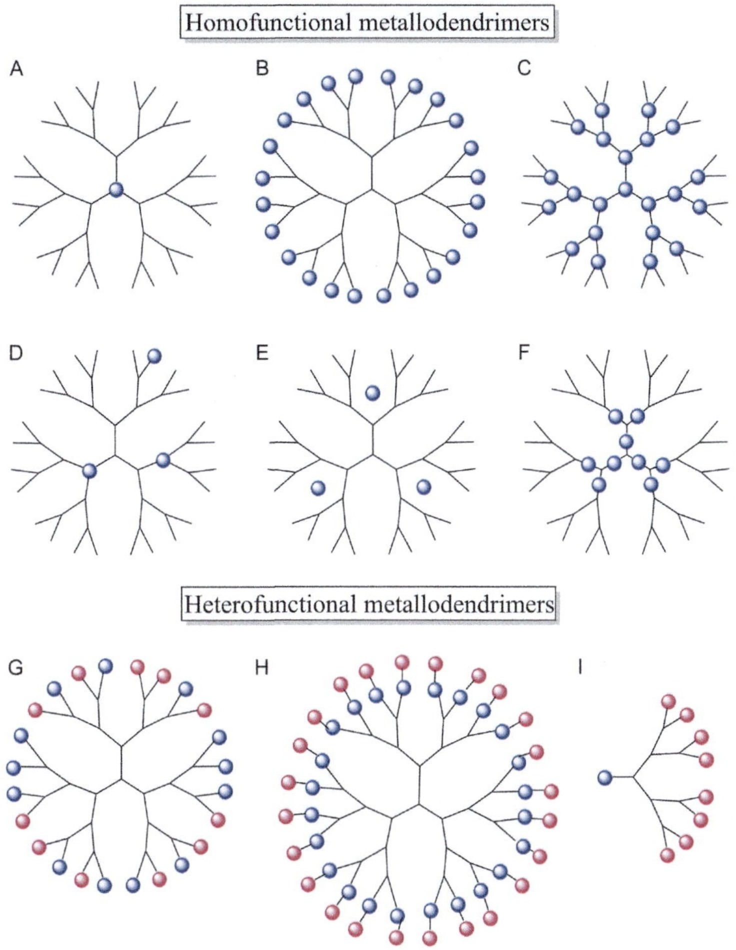

Fig. 1 Metallodendrimers classification attending to the metal position in the framework. (A) In the center of the skeleton; (B) in the periphery; (C) in the nodes of ramification; (D) in different points of the structure; (E) inside the cavities; (F) as building-blocks; (G) heterofunctional random; (H) heterofunctional layer-block; and (I) heterofunctional dendron.

2. Applications of metallodendrimers in biomedicine
2.1 Metallodendrimers in cancer treatment

Cancer is one of the main causes of death in the world. According to GLOBOCAN estimations for the year 2018, 18.1 million new cases would be diagnosed with cancer and this disease would be responsible of 9.6 million of deaths.[27] From a biological point of view, cancer arises from an uncontrolled cell proliferation that in some occasions acquires the ability to invade foreign tissues producing what we know today as metastasis, and often leads to the death of the patient. Due to the heterogeneous causes, it is difficult to find an effective and non-toxic treatment for cancer.

The use of metal complexes in cancer treatment started when Rosenberg discovered the potential of cisplatin in 1969.[28] Its mechanism of action resides in its ability of interacting with DNA, which avoids its replication, leading to cell death through apoptosis. Cisplatin and their second and third line analogues, carboplatin and oxaliplatin, are currently one of the most used strategies in clinic to treat patients with ovarian, lung, head, neck and bladder cancer. However, they also produce undesirable side effects and drug resistance related to their non-specificity.[29] In this sense, nanotechnological tools can greatly improve cancer treatment. Due to their nanometric size, these tools benefit from non-specific targeting to solid tumors, known as enhanced permeation and retention effect (Fig. 2).[30] The defective anatomy of the vasculature surrounding the tumor, highly disorganized with numerous dilations and pores, together with functional anomalies produce a high permeation and retention of these nanoparticles, thus increasing the therapeutic index.

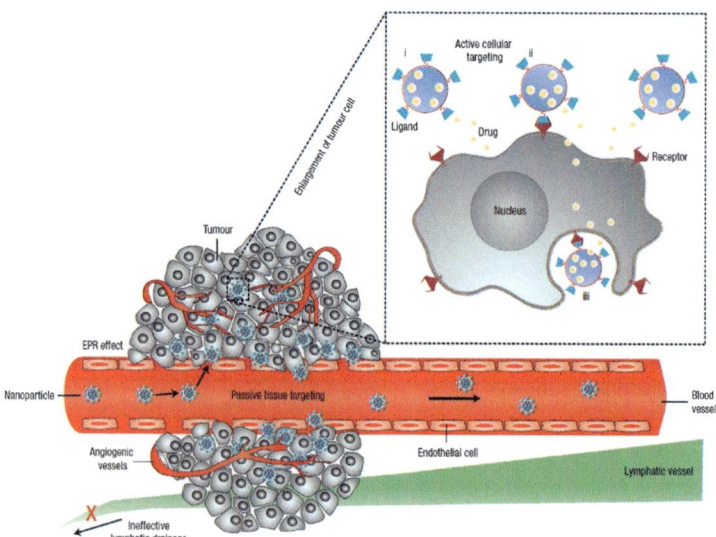

Fig. 2 Schematic representation of different mechanisms which nanocarriers use to deliver drugs to tumors, including passive tissue targeting (enhanced permeation and retention effect) and active cellular targeting. *Reprinted by permission from Springer Nature, Peer D, Karp JM, Hong S, Farokhzad OC, Margalit R, Langer R. Nanocarriers as an emerging platform for cancer therapy. Nat Nanotechnol. 2007;2 (12):751–760. Copyright 2007.*

In the search for new antitumor strategies, the combination of metal complexes and dendritic scaffolds seemed like a logic transition to explore. Different examples have been reported in the literature, where metallodendrimers act directly as drugs, or as carriers of drugs o genetic material in gene therapy. Furthermore, they have been studied as diagnostic tools or even as theranostic agents, combining both therapy and diagnosis. All these applications will be reviewed in the following sections.

2.1.1 Metallodendrimers as anticancer drugs

Considering the potential of platinum complexes as anticancer agents, a popular alternative has been the use of polymeric supports which enable the sustained release of the drug into the extracellular fluid.[31] The first metallodendrimers reported in the literature contained this drug in its structure, providing non-toxic and multivalent platforms. Fig. 3 presents the state-of-the-art timeline in the development of anticancer metallodendrimers, as highlighted by Smith in 2012.[26] In 1999, the first metallodendrimers with antitumor properties were presented to the scientific community, when Reedijk et al.[32] and Duncan et al.[33] designed platinum complexes conjugated with PPI and PAMAM dendrimers, respectively, showing high cytotoxic activities in diverse cancer cell lines. The carboxylate-decorated PAMAM dendrimer showed a high cytotoxicity in cisplatin-resistant cell lines, like an ovarian cancer cell line (SKOV3). However, the PPI metallodendrimer with four metal centers showed a low cytotoxicity in two mouse leukemia cells lines (L1210/0 and L1210/2) and seven human cancer cell lines, probably due to the high steric congestion hindering the cellular uptake. After this starting point, researchers introduced structural modifications with the goal of reducing some of the severe effects involved in the treatment with cisplatin. For example, the efforts of Gust et al.[34] on achieving a more selective system for breast cancer led to platinum G1-PPI metallodendrimers which improved the cellular uptake and DNA binding 20–700 times more than the free drug in breast cancer cells (MCF-7).

Besides platinum, other metal ions have been reported in the state-of-the-art of metallodendrimers, such as ruthenium(II), copper(II), gold(I) and palladium(II), as depicted in Fig. 3. It is worth highlighting that, in recent years, the research focus has shifted to such alternative metals, with few examples of new platinum metallodendrimers. A summary of the most promising ones and their cytotoxicity indices are included in Table 2 and will be discussed below.

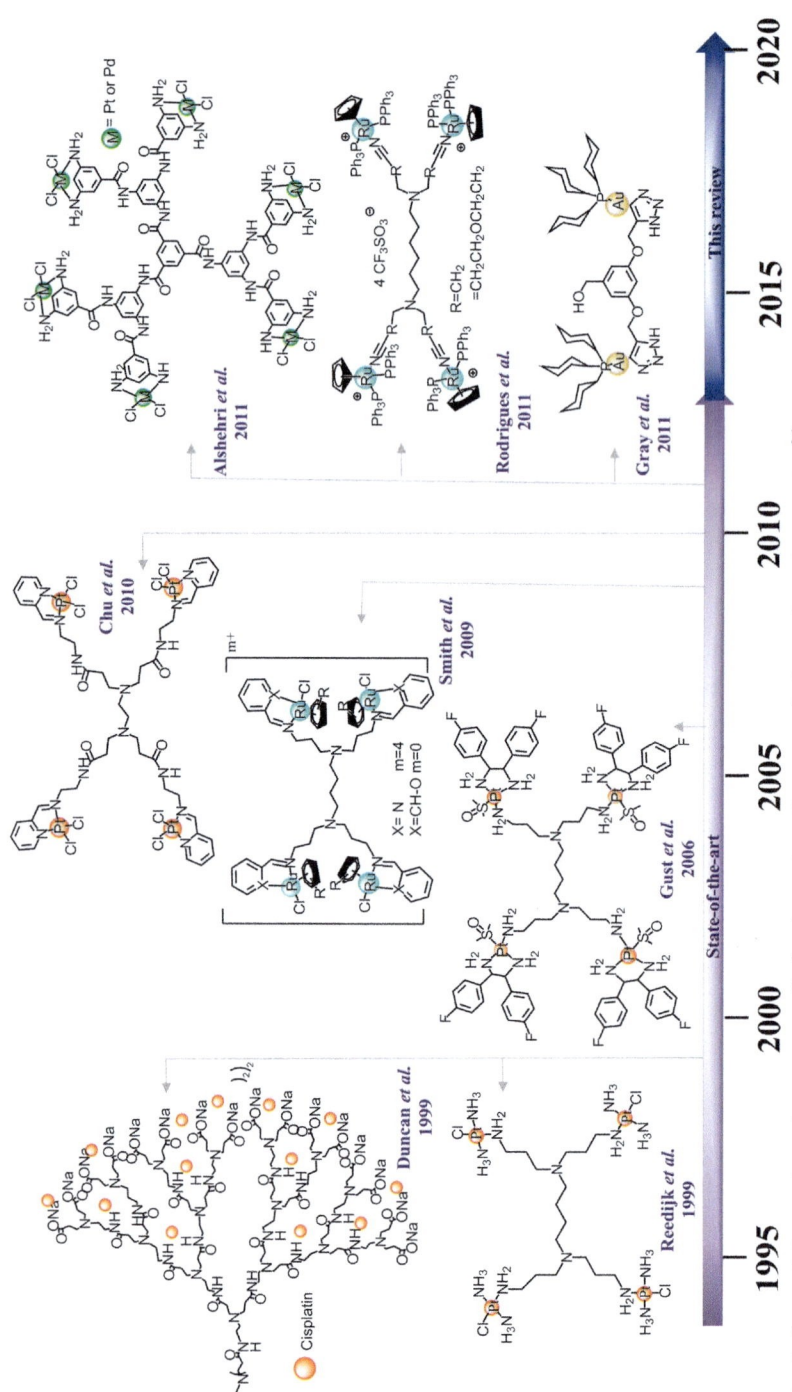

Fig. 3 State-of-the-art in anticancer metallodendrimers until 2012, including selected examples.[26]

Table 2 Overview of selected metallodendrimers including the cytotoxicity index IC$_{50}$ in different tumor and non-tumor cell lines, after 24 h treatment.

Entry	Metal	Dendritic scaffold	Metal complex	Number metals	Cell line (tumor, *non-tumor)	IC$_{50}$ ± SD (µM)	Ref.
1	Pt(IV)	G4-PAMAM		12	CH1/PA-1 A549 SW480	0.009 ± 0.002 0.15 ± 0.03 0.04 ± 0.02	35
2	Ru(II)	G4-PA		32	A2780 A2780cisR *HEK	0.8 ± 0.1 2.7 ± 0.1 2.6	36
3	Ru(II)	G1-CBS		4	HeLa MCF7 HT29 MDA-MB-231 PC-3 HL-60 * HEK-239T	8.1 ± 0.4 6.9 ± 0.5 5.6 ± 0.1 6.1 ± 1.0 7.8 ± 1.4 2.3 ± 0.4 10.7 ± 0.0	37,38
4	Ru(II)	G4-PA		32	A2780 A2780cisR *HEK293	5.0 ± 0.1[a] 17.3 ± 1.5[a] 12.0 ± 0.9[a]	39

Continued

Table 2 Overview of selected metallodendrimers including the cytotoxicity index IC$_{50}$ in different tumor and non-tumor cell lines, after 24 h treatment.—cont'd

Entry	Metal	Dendritic scaffold	Metal complex	Number metals	Cell line (tumor, *non-tumor)	IC$_{50}$±SD (μM)	Ref.
5	Ru(II)	G1-PA		4	Caco-2 CAL-72 MCF-7 A2780 A2780*isR* hMSC	3.4^a 0.6^a 2.5^a 0.1^a 0.3^a <0.05^a	[40]
6	Ru(II)	G1-CBS		4	HeLa MCF7 HT29 HCC1806 PC-3 *142BR	6.3±0.2 10.3±0.5 11.4±0.4 2.2±0.1 8.3±0.6 19.2±0.2	[41]
7	Cu(II)	G3-PD		48	KB HL60 HCT116 MCF7 OVCAR8 U87 *EPC *MRC5	0.46^a 0.58^a 0.30^a 0.41^a 0.65^a 0.86^a 1.36^a 0.80^a	[42]
8	Cu(II)	G1-CBS		4	HeLa PC-3 MCF-7 HCC1806 HT29 *142BR	1.7±0.5 3.4±0.2 2.1±0.2 1.9±0.1 9.3±0.5 7.7±0.5	[43]
9	Fe(II)	G1-PA		3	DLD-1 HT29	17.0±1.2 33.9±1.2	[44]

10	Re(I)	G1-PA	4		A431 DLD-1 A2780 *BJ	14.09 ± 2.23 10.18 ± 0.47 6.38 ± 1.18 17.69 ± 1.5	45
11	Au(III)	G3-PD	48		KB HL60 *EPC *MRC5	7.5 ± 7.5 3.3 ± 0.6 >1000 12.0 ± 5.0	46
12	Rh(III)	G2-PA	8		A2780 A2780cisR *HEK293	8.0 ± 0.5[a] 3.1 ± 0.3[a] 4.5 ± 0.5[a]	47
13	Ir(III)	G2-PA	8		A2780 A2780cisR *HEK	0.75 ± 0.01[a] 3.5 ± 0.3[a] 28.6 ± 1.3[a]	47
14	Os(II)	G2-PA	8		A2780 A2780cisR	16.4 ± 12.5[a] 43.5 ± 2.7[a]	48

[a]Values after 72 h treatment.
PLL, polylysine dendrimer; PA, polyamine dendrimer; CBS, carbosilane dendrimer; PD, phosphorous dendrimer.

2.1.1.1 Platinum(II, IV) metallodendrimers

Among the recent trends on the design of anticancer platinum metallodendrimers, two different approaches are prevalent. First, the use of dendritic scaffolds as carriers of cisplatin or analogues through electrostatic interactions or encapsulation. In 2013, Palakurthi et al. reported a new family of biotin-conjugated PAMAM metallodendrimers functionalized with amino and carboxylic groups in their periphery with the ability to act as potential carriers of cisplatin.[49] All dendrimers studied showed better anticancer activities than cisplatin in a wide variety of tumor cancer cell lines (OVCAR-3, SKOV, A2780, CP70), as well as higher expression of apoptotic genes. All these properties, besides their higher potential to form adducts with DNA, placed them as promising candidates in this field. A similar strategy was recently proposed by Wheate et al., using a cationic PAMAM dendrimer as a carrier of PHENSS, a potent platinum anticancer drug developed by Aldrich-Wright, previously encapsulated with *p*-sulfonatocalix[4]arene.[50] Unfortunately, the final dendritic nanocomplex did not potentiate the anticancer properties but placed it as a possible vehicle to increase the selectivity of the drug and decrease the side effects.

The second prevalent approach relies on the covalent attachment of platinum complexes to the dendritic scaffold, leading to new metallodendrimers with anticancer activity *per se*. In 2016, Xu et al. developed three different generations (G1, G2 and G3) of poly-L-lysine dendrimers with platinum coordinated to selenium in the core and amino groups in the periphery (Fig. 4A).[51] The coordination of platinum was carried out through the replacement of the chloro and amino ligands in the cisplatin complex leading to the formation of a selenium–platinum active site. These dendrimers showed better cellular uptake than cisplatin allowing the reduction of the dosage, as well as better anticancer activities *in vitro*. Importantly, the increase in generation led to a decrease in the anticancer activity due to the reduced availability of the metal center. *In vivo* experiments carried out in Balb/c mice with breast cancer showed that all metallodendrimers had similar activity than cisplatin but did not affect the mouse weight, unlike cisplatin (Fig. 4B). In the same year, Keppler et al. conjugated an oxaliplatin analogue to G2 and G4 PAMAM dendrimers decorated with amino groups, which showed improved cytotoxicity in some cell lines (CH1/PA-1, A549 and SW480) (Table 2, Entry 1).[35]

2.1.1.2 Ruthenium(II) metallodendrimers

In pursuit of new mechanisms of action that overcome cancer cell resistance to drugs, ruthenium has stood out during the last years. The main reason is

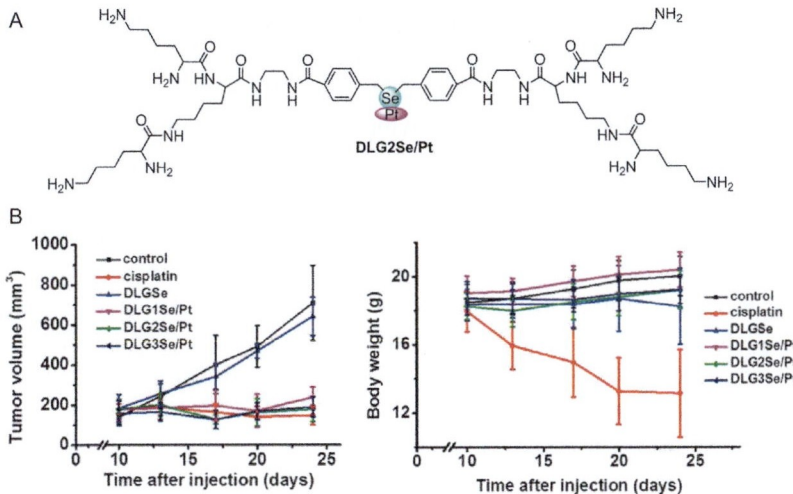

Fig. 4 Antitumor activity of Pt(II) poly-L-lysine metallodendrimers.[51] (A) Structure of the second-generation metallodendrimer. (B) Changes in tumor volume and body weight in breast (4T1) cancer mice models after the treatment with cisplatin, dendrimer and three different generations of metallodendrimers. *Reprinted with permission from Li T, Smet M, Dehaen W, Xu H. Selenium–platinum coordination dendrimers with controlled anti-cancer activity. ACS Appl Mater Interfaces. 2016;8(6):3609–3614. Copyright 2016 American Chemical Society.*

the wide variety of oxidation states (+2, +3, +4) and the ability to reduce systemic toxicity through the mimicking of iron behavior and binding to some biological molecules, including transferrin. Taking into account that some cancer cells have the transferrin receptor overexpressed a high level of ruthenium complexes will be delivered preferentially to these cells. Currently, there are two ruthenium complexes in clinical trials, namely NAMI-A H$_2$Im[trans-RuCl$_4$(DMSO)(Him)] (Him=imidazole) and KP1019 H$_2$Ind[trans-RuCl$_4$(Hind)$_2$] (Hind=indazole), both comprising Ru(III). Ru(III) complexes can behave as prodrugs, being reduced to Ru(II) in the solid tumor where a reducing environment is created due to the low oxygen content. On the contrary, only Ru(II) metallodendrimers have been reported in the literature, probably due to the protective nature of the dendrimer which enables a direct delivery of the active metal.

In the search of new metallodendritic structures, the imine-type ligands have become popular, since these can act as anchor points of various metals centers. In 2013, PPI metallodendrimers functionalized with RAPTA (ruthenium(II)-arene-1,3,5-triaza-7-phosphatricyclo) complexes were developed and evaluated as anticancer agents.[36] Smith et al. presented a library of dendrimers from first to fourth generation and functionalized with *N*-monodentate, specifically pyridine, and *N,O*-, *N,N*-chelating imine ligands and complexes with ruthenium(II)-arene-PTA where the arene derivate can be *p*-cymene or hexamethylbenzene. They found a correlation between the size of the metallodendrimer and the activity, being the complex with 32 phenol-imine ligands and *p*-cymene the most active, with IC$_{50}$ values in the low μM range in two ovarian cancer cell lines (A2780 and A2780*cis*R) (Table 2, Entry 2). The dendritic scaffolds improved the selectivity *in vitro* toward tumor cell lines, compared to their mononuclear analogues. In addition, they demonstrated that the presence of PTA ligands in the metal complexes improved the overall activity probably due to a better interaction with DNA.

Using the same chelating ligands but supported on a carbosilane scaffold, de la Mata et al. presented in 2016 different families of Ru(II) carbosilane metallodendrimers employing [Ru(η^6-*p*-cymene)Cl$_2$]$_2$ as precursor and evaluated their anticancer activity in a wide variety of tumor cell lines.[37] The influence of the lipophilic dendrimer was clearly observed, being the first-generation dendrimer with four functional groups the most efficient, activity that did not improve even if the number of metal centers doubled in the second generation. The reported studies of reactivity toward some biomolecules like Human Serum Albumin (HSA), Cathepsin-B and DNA confirmed a different mechanism of action, compared to cisplatin, and *in vitro* antimetastasic activity. In 2019, further *in vitro* studies— proliferation, cell cycle, cytotoxicity, cell adhesion—were performed on selected first-generation metallodendrimers. Overall, the assays concluded that the metallodendrimer with iminopyridine ligands G1-[[NCPh(*o*-N)Ru(η^6-*p*-cymene)Cl]Cl]$_4$ was an outstanding molecule with IC$_{50}$ values in the range 5.6–8.1 μM in different human cancer cell lines (Table 2, Entry 3). The authors evaluated the antitumor behavior of this metallodendrimer in an *ex vivo* mice model of human prostate cancer, inoculating untreated or treated PC-3 cells subcutaneously into two different groups of immunodepressed mice. A remarkable 82% smaller tumor size was observed in the treated mice, compared to non-treated group, at the end of the experiment (Fig. 5A). An *in vivo* mice model was later established, using athymic male nude mice nu/nu with subcutaneously injected PC-3 cells, in order to use a scenario as closer to clinical practice as possible. In this case, the subcutaneous treatment with this metallodendrimer at a daily dose of 5 mg/kg inhibited up to 36% of the tumor growth (Fig. 5A) and no appreciable negative effect in mice health was observed

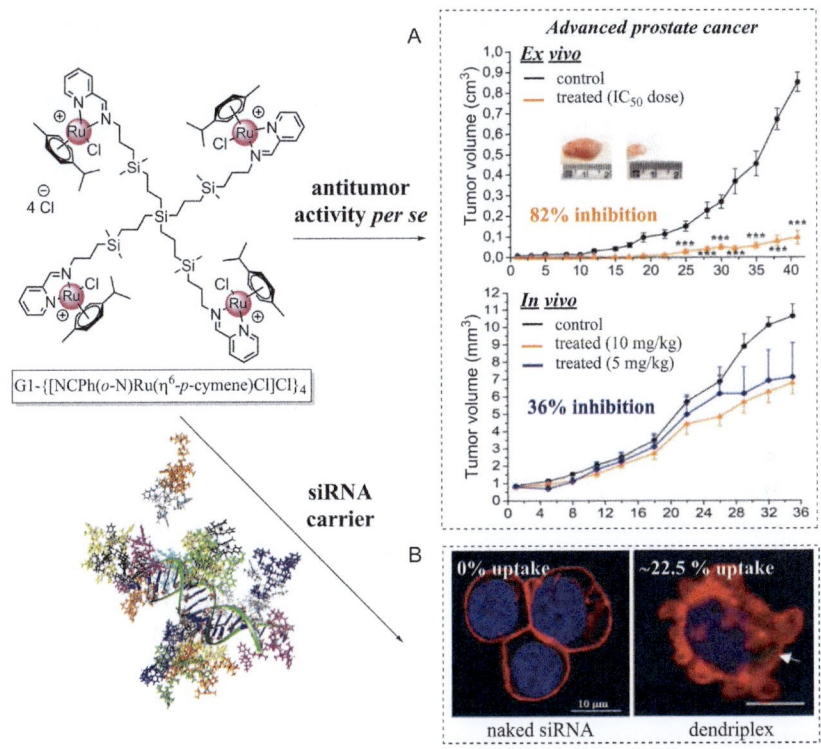

Fig. 5 Reported anticancer strategies employing the Ru(II) carbosilane metallodendrimer G1-[[NCPh(o-N)Ru(η⁶-p-cymene)Cl]Cl]₄. (A) Antitumor activity *per se* in advanced prostate cancer mice models. (B) Carrier of antitumor siRNA. *Panel (A): Figures adapted from Maroto-Díaz M, Sanz del Olmo N, Muñoz-Moreno L, et al. In vitro and in vivo evaluation of first-generation carbosilane arene Ru(II)-metallodendrimers in advanced prostate cancer. Eur Polym J. 2019;113:229–235. Copyright 2019, with permission from Elsevier; Panel (B) Figures adapted from Michlewska S, Ionov M, Maroto M, et al. Ruthenium dendrimers as carriers for anticancer siRNA. J Inorg Biochem. 2018;181:18–27. Copyright 2018, with permission from Elsevier.*

during the treatment.[52] Aiming to provide insight into the mechanism of action of these Ru(II) carbosilane metallodendrimers, Bryszewska et al. evaluated their antitumor effect in the human leukemia cell line HL-60.[38] Again, they observed a remarkable cytotoxicity for the first-generation metallodendrimer G1-[[NCPh(o-N)Ru(η⁶-p-cymene)Cl]Cl]₄ (Table 2, Entry 3). Circular dichroism studies using TMA-DPH and DPH probes showed the metallodendrimers interaction with the hydrophobic and hydrophilic regions of the cell membranes. Moreover, the first-generation metallodendrimer increased the production of Reactive Oxygen Species

(ROS) and decreased the mitochondria potential, confirming the activation of apoptosis pathway, as well as the DNA damage and the increase in caspase level. The treatment with this metallodendrimer produced visible changes in the cell structure, and they observed a chromatin condensation with changes in mitochondrial shape and appearance of multivesicular and lamellar bodies, characteristic of early apoptosis.

Other dendritic scaffolds used by Smith et al. in 2015 to anchor Ru(II)-ethylene-glycol complexes were G1–G4 pyridylimine-based poly(propylene)dendrimers.[39] The antiproliferative effect was evaluated *in vitro* in two ovarian cancer cell lines (the cisplatin sensitive A2780 and the cisplatin-resistant A2780cisR) and compared with monometallic complexes. The results showed that the monomer and low generation dendrimers were inactive, whereas the higher generation analogues exhibited antitumor activity (Table 2, Entry 4).

As an alternative cancer treatment, metallodendrimers have also been used as siRNA carriers to selectively silence aberrantly activated oncogenes. The lack of effective and cheap carriers of siRNA remains a major problem in gene silencing, where the use of dendrimers has shown promising results.[53] Cationic PAMAM, phosphorous and carbosilane dendrimers, among others, efficiently interact with negatively charged siRNA through electrostatic interactions.[54] Nonetheless, there are very few examples in which the nanocarrier contains metals in its structure. In 2018, Bryszewska and co-workers studied the interactions between ruthenium(II) carbosilane metallodendrimers and small siRNA,[55] aiming to observe an enhanced anticancer activity arising from the combination of the metallodendrimer and the siRNA. Two different families of ruthenium metallodendrimers were studied, comprising pyridine or iminopyridine ligands. The authors confirmed the *in vitro* interaction between the metallodendrimers and the siRNA, the protection from nucleases degradation as well as the transfection in human promyelocytic leukemia cell line (HL60), observing that the dendriplexes crossed the cell membrane and penetrated into the cell, unlike naked siRNA (Fig. 5B). The dendriplex uptake was more efficient for iminopyridine-bearing dendrimers, due to their cationic nature, and for second-generation counterparts, finding up to 30% loaded cells for the most efficient metallodendrimer G2-[[NCPh(o-N)Ru(η^6-p-cymene)Cl]Cl]$_8$, compared to the ca. 23% of the first-generation counterpart.

Besides the dendritic scaffold and the peripheral moieties, the ligands on the metal coordination sphere clearly influence the antitumor activity of the final metallodendrimer. Among the different arene ligands on the Ru(II) ion, cyclopentadienyl (Cp or η^5-C$_5$H$_5$) showed promising results. Continuing their work on poly(alkylidenimine) dendrimers functionalized with the metal complex [Ru(η^5-C$_5$H$_5$)(PPh$_3$)$_2$]$^+$,[56] in 2018 Rodrigues and co-workers evaluated the antitumor activity in a broad variety of tumor cells, including colorectal, osteosarcoma, breast and ovarian (Table 2, Entry 5).[40] The metallodendrimers exhibited high cytotoxicity, with IC$_{50}$ values in the range 0.1–3.4 μM after 72 h treatment, higher than the precursor dendrimers, the complex [Ru(η^5-C$_5$H$_5$)(PPh$_3$)$_2$Cl] and cisplatin. In addition, promising activity was observed in primary human mesenchymal stem cells (hMSCs), involved in tumor progression and drug resistance, with IC$_{50}$ < 0.05 μM.

The organometallic ligand cyclopentadienyl and the phosphine, 1,3,5-Triaza-7-phosphatricyclo[3.3.1.13.7]decane (PTA), were included in a new family of Ru(II) carbosilane metallodendrimers in order to improve the activity, solubility and stability.[41] Both families shared the carbosilane dendritic scaffold and peripheral N,N-chelating imine ligands, as well as similarly potent cytotoxic activity toward multiple cancer cell lines, including advanced prostate, breast, colorectal and cervix, higher than the activity observed toward non-tumor fibroblasts (Table 2, Entry 6). In order to corroborate the antitumor activity of this ruthenium dendrimer, an *in vivo* experiment was designed in athymic mice with advanced prostate cancer injecting a dose of 10 mg/kg intravenously once a week and achieving 25% of inhibition of the tumor growth compared to the non-treated mice. No negative symptoms, such as weight loss, were observed during the experiment (Fig. 6).

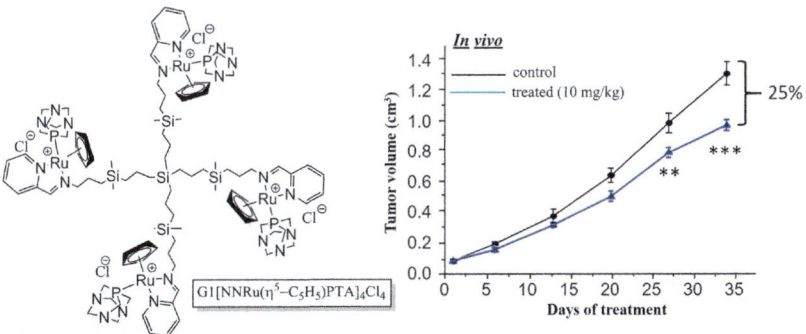

Fig. 6 Antitumor activity of Ru(II) carbosilane metallodendrimer G1{[NNRu(5-C$_5$H$_5$) PTA]Cl}$_4$ in an *in vivo* advanced prostate cancer mice model.[41]

2.1.1.3 Copper(II) metallodendrimers

Copper is a biocompatible metal, endogenously present in the human body. Copper(II) mononuclear complexes have been explored as anticancer agents showing promising results.[57] Its cytotoxicity is related to the high redox activity and the alteration of Reactive Oxygen Species (ROS) balance.[58] Copper is also involved in tumor angiogenesis at early stages. In 2013, Majoral and co-workers demonstrated the anticancer activity of new Cu(II)-conjugated phosphorus dendrimers.[42] Three different families were prepared bearing ligands with nitrogen donor atoms, i.e., N-(pyridin-2-ylmethylene) ethanamine, N-(di(pyridin-2-yl)methylene)ethanamine and 2-(2-(methylenehydrazinyl)pyridine, able to coordinate $CuCl_2$. The resultant first-, second- and third-generation metallodendrimers presented 12, 24 and 48 bound Cu(II) ions, respectively. Overall, the study concluded that both the nature and the number of the peripheral groups play a major role in the antitumor potency of the dendrimers and the metal coordination boosted this effect. The metallodendrimers produced an increasing inhibition of the proliferation of human promyelocytic leukemia cells (HL60) when the number of functional groups increased, being the third-generation dendrimer the most promising one. Interestingly, the third-generation metallodendrimer decorated with N-(pyridin-2-ylmethylene)ethanamine and 48 Cu(II) ions exhibited certain preference for cancer cells, such as HL60 and human cervical cancer derivative cell line (KB), compared to the non-cancer cells used as control, after 72 h treatment (Table 2, Entry 7).

The promising results obtained with Ru(II) carbosilane metallodendrimers decorated with N,N- and N,O-chelating ligands encouraged de la Mata and co-workers to generate the Cu(II) analogues and evaluate their antitumor activity. In 2017, the dendritic scaffolds were used to chelate $CuCl_2$.[43] Unlike the phosphorous Cu(II) metallodendrimers, the first-generation carbosilane metallodendrimers—with four metal ions—exhibited the highest antitumor activity among the different dendritic generations. This behavior is consistent with that observed for the Ru(II) carbosilane counterparts and confirms the dramatic influence of the lipophilic scaffold. In order to improve the water solubility of the Cu(II) metallodendrimers, in 2019 the dendritic scaffolds were used to chelate $Cu(NO_3)_2$.[59] The NO_3^- groups are more labile than the Cl^- ligands, and are susceptible to be released after dissolution in water creating an overall positive charge in the metallodendrimer. The study showed that the exchange from chloride to nitrate ligands increased not only the water solubility but also the general cytotoxic activity (Table 2, Entry 8), probably due to the positive charge of the new metallodendrimers, which interact and consequently destabilize the cell membrane, according to Electron Paramagnetic Resonance (EPR) studies. Further *ex vivo* studies in an advanced prostate cancer mice model (athymic male nude mice nu/nu) demonstrated that the first-generation carbosilane dendrimer functionalized with Cu(NO$_3$)$_2$ produced a significant inhibition of the tumor growth, reaching up to 37% smaller tumor size compared to non-treated mice. Copper metallodendrimers open new avenues in cancer treatment being more economic and more biocompatible than traditional metallodrugs.

2.1.1.4 Iron(II) metallodendrimers

Similar to copper, iron is also found in traces in the human body and takes part in numerous biological processes. This metal has shown interesting anticancer properties; in fact, ferrocene was the first organometallic compound reported for its antiproliferative effect.[60] Ferrocenyl derivatives might be oxidized in the cell via Fenton pathway generating free radicals that lead to DNA damage and final cell apoptosis as a possible mechanism of action.[61]

Despite the interest in including ferrocene moieties in a multivalent scaffold, surprisingly, only few examples are found in the literature. In recent years, Smith and co-workers presented tri- and tetrametallic systems comprising polypropylenimine tetramine or tris(2-aminoethyl)amine cores functionalized with a ferrocenyl aldehyde by Schiff condensation followed by reductive amination.[44] The Fe(II) metallodendrimers exhibited moderate IC$_{50}$ values in human colorectal cancer cell lines (Table 2, Entry 9). Both compounds were capable of penetrating inside the cell and producing early apoptosis, but only the trinuclear metallodendrimer increased the ROS level in tumor cells.

2.1.1.5 Rhenium(I) metallodendrimers

Recently, rhenium has attracted a widespread attention from the biomedical field due to the high variety of oxidation states and geometries, and the ability of coordinating different ligands.[62] In particular, the stable Re(I) tricarbonyl complexes have become quite popular,[63] inducing cell death in a different manner compared to cisplatin and many other drugs, suggesting a novel mode of action different from the traditional apoptosis, necrosis, paraptosis and autophagy. In 2018, Smith and co-workers presented two different first-generation Re(I) PPI metallodendrimers,[45] comprising 3 and 4 [Re(bpy)(CO)$_3$] complexes bound to the dendrimer through a pyridine moiety. The cytotoxicity of these metallodendrimers against epithelial, colon and ovarian carcinoma was moderate, with IC$_{50}$ values in the micromolar range and selectivity toward cancer cells in comparison to the mononuclear counterpart used as a control (Table 2, Entry 10). The tetranuclear metallodendrimer exhibited higher cytotoxicity than cisplatin in all cell lines tested. Mechanistic studies demonstrated that these rhenium metallodendrimers boost apoptotic cell death by modulating the expression of Bax-α, being the tetranuclear dendrimer the most efficient.

2.1.1.6 Gold(III) metallodendrimers

Gold has been used from ancient times, and its potential as anticancer agent has been proved in *in vitro* and *in vivo* experiments.[64] Despite sharing the same electronic configuration and similar properties, gold(III) and platinum(II) complexes frequently follow different mechanism of action. Gold complexes exhibit better selectivity and potency against cancer cells due to their weaker DNA-binding activity and higher affinity to protein targets through sulfhydryl, thiol and selenocysteine groups.[65]

Majoral and co-workers published in 2017 an original third-generation phosphorous metallodendrimer comprising 48 iminopyridine end groups complexing Au(III), namely 1G$_3$-[Au$_{48}$][AuCl$_4$]$_{48}$.[46] This metallodendrimer showed IC$_{50}$ values in the nanomolar range in different tumor cell lines, while a very weak effect was observed against non-cancer cells (Table 2, Entry 11). In order to corroborate that the activity arised from the metallodendritic complex, AuCl$_3$ was also studied, and it was observed that it retained a low intrinsic activity. In addition, the physical mixture of AuCl$_3$ and the precursor dendrimer did not potentiate the intrinsic antiproliferative effect of the dendrimer compared to that of the metallodendrimer 1G$_3$[Au$_{48}$][AuCl$_4$]$_{48}$. Moreover, the authors expanded the study performing different structural modifications, including the evaluation of non-complexed dendrimers with iminopyridine and PEG, gold metallodendrimers with PEG and dendrimers with free iminopyridine without metal, showing the best results with the metallodendrimer 1G$_3$-[Au$_{40}$-PEG$_8$][AuCl$_4$]$_{40}$ in HL60 (IC$_{50}$ = 1.7 ± 0.5 nM).

2.1.1.7 Osmium(II), Iridium(III), Rhodium(III) metallodendrimers

Recently, metals such as iridium, osmium and rhodium have caught the attention in the search for new metallodrugs due to their intrinsic anticancer activity.[66–68] In 2014 Smith et al. reported new families of rhodium(III) and iridium(III) metallodendrimers of first- and second-generation naphthaldimine-decorated PPIscaffolds and with Cp* (pentamethylcyclopentadienyl) as ligand of the metal complex. The anticancer activity was measured *in vitro* in two different ovarian cancer cell lines (A2780 and A2780cisR) where the second-generation complexes with eight functional groups stood out as the most promising ones, being Rh(III) complexes better than their Ir(III) counterparts (Table 2, Entries 12 and 13).[47] The activity found for the new rhodium and iridium derivatives is comparable with the antitumor activity of arene-ruthenium(II) complexes attached to similar dendritic scaffolds.[69]

Osmium analogues of potent anticancer drugs, such as Ruthenium-NAMI-A, present moderate antiproliferative activity *in vitro* in colon carcinoma (HT-29) and breast carcinoma (SK-BR-3).[70] With the objective to implement the activity of these derivatives, in 2015 Smith et al.[48] synthetized new families of neutral and cationic half-sandwich Os(II)-arene DAB-PPI metallodendrimers, namely [DAB-PPI-{(η^6-*p*-cym)Os((C_7H_5NO)-κ^2-N,O)Cl}$_n$], [DAB-PPI-{(η^6-*p*-cym)Os(($C_6H_5N_2$)-κ^2-N,N)Cl}$_n$][PF$_6$]$_n$ and [DAB-PPI-{(η^6-*p*-cym)Os((C_7H_5NO)-κ^2-N,O)PTA}$_n$]. *In vitro* studies against the cisplatin sensitive (A2780) and cisplatin-resistant (A2780cisR) ovarian cancer cell lines showed that neutral dendrimers were inactive while cationic metallodendrimers displayed moderate activity which improved when the counterion was PTA and not Cl (Table 2, Entry 14).

2.1.1.8 Heterofunctional metallodendrimers

An effective strategy in cancer treatment relies on combination therapy. It has been shown that the administration of several drugs with different mechanisms of action, used in their optimal dose, has a better effect than if they are administrated independently. In this sense, dendrimers are optimal platforms to provide a precise and controlled combination therapy in a single scaffold, in the so-called heterofunctional dendrimers.

Despite the multiple advantages of designing heterofunctional metallodendrimers with antitumor activity, very few examples can be found in the literature. In 2014, Smith et al. presented the synthesis of first- and second-generation PPI metallodendrimers bearing heterometallic ferrocenyl-derived *N*,*O*-*p*-cymene-Ru(II)-PTA-salicylaldimine and *N*,*N*-*p*-cymene-Ru(II)-2-pyridylimine (Fig. 7A).[71] Preliminary biological studies revealed that most of the compounds at equi-iron concentrations of 5 µM decreased the growth of the ovarian human cancer cell lines A2780 and A2780cisR by more than 50% and showed certain selectivity for these ovarian cancer cell lines,

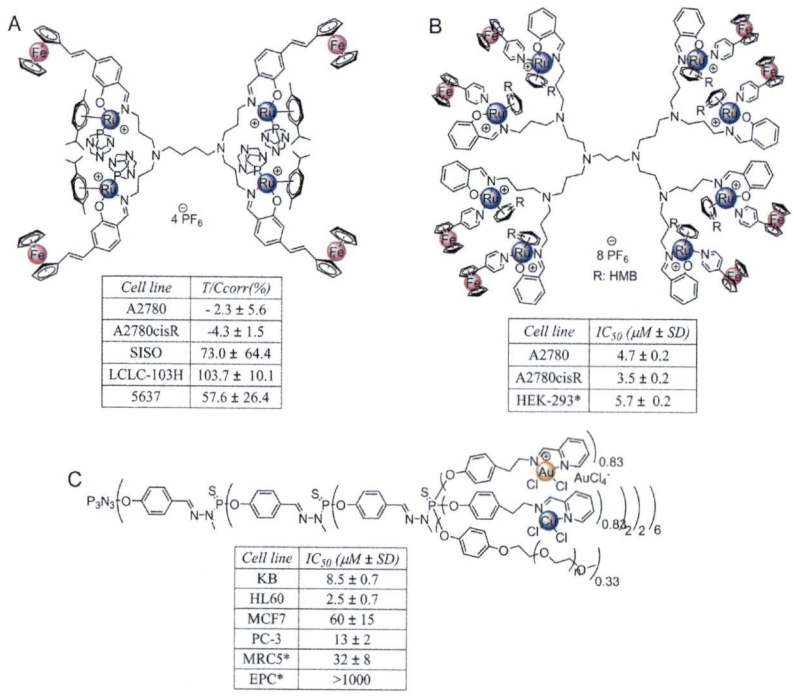

Fig. 7 Examples of heterometallic dendrimers and cytotoxicity indices in cancer and non-cancer cells, expressed as (A) T/Corr (%) at 96 h exposure to compound at 5 μM Fe relative to untreated control,[71] and (B and C) IC$_{50}$ after 72 h treatment.[46,72]

compared to other cancer cell lines. Higher cytotoxicity was observed with metallodendrimers with N,O-p-cymene ligands. Using a similar strategy but expanding the number of metal atoms, in 2016 new ferrocenyl PPI metallodendrimers were reported containing ruthenium(II)-p-cymene, ruthenium(II)–HMB (where HMB = hexamethylbenzene), rhodium(III)–Cp* or iridium(III)–Cp* moieties.[72] The cytotoxicity of these systems was studied against A2780 and A2780cisR cells and non-tumorigenic human embryonic kidney cells (HEK-293), observing an increase in the activity in the highest generations. Particularly the second-generation dendrimer with ruthenium(II)–HMB emerged as the most promising candidate, being less toxic in non-cancer cells (Fig. 7B). UV–Vis studies showed that these metallodendrimers produce a non-covalent interaction with DNA as a possible mechanism of action. In 2017, Majoral and co-workers also explored the heterometallic strategy in the cancer field. As an expansion of their work in Au(III) phosphorous metallodendrimers,[46] they reported a third-generation metallodendrimer bearing 20 gold(III) complexes and 20 copper(II) complexes, to evaluate a possible synergistic effect, and 8 PEG molecules to improve the water solubility and biocompatibility (Fig. 7C). The authors ruled out any additive effect, suggesting the existence of a threshold in the anticancer activity of Au(III) metallodendrimers.

The design of heterofunctional metallodendrimers goes beyond the heterometallic strategy, and the metal complexes can further be combined with other antitumor moieties or tracing moieties (Fig. 8). For example, Grabchev et al. have developed several families of Cu(II) and Zn(II) poly(propyleneamine) metallodendrimers decorated with different 1,8-naphtalimides,[73–77] where the metal coordination is performed via the tertiary amine groups at the dendrimer inner shell and the outer naphtalimide shell is available for additional therapeutic or fluorescent response. Even though they were mainly evaluated as antibacterial agents as it will be described in Section 2.2.1, some of these metallodendrimers also exhibited antitumor activity. For example, the second-generation metallodendrimers,[75] decorated with four 4-bromo-1,8-napthtalimide peripheral moieties and two metal ions in the internal shell, presented enhanced cytotoxicity toward human non-small cell lung cancer (A549), triple negative breast cancer (MDA-MB-231) and carcinoma of the uterine cervix (HeLa), compared to the precursor dendrimer. The metallodendrimers exhibited IC$_{50}$ values in the range 15–30 µM after 72 h treatment, with slightly lower values for the Zn(II) derivative (Fig. 8A).

The inclusion of tracing moieties in a heterofunctional dendrimer can simplify the challenge of unraveling its mechanism of action and provide accurate and complete information about their behavior in a biological environment. In 2019, de la Mata, Bryszewska and co-workers presented a first-generation ruthenium(II) carbosilane metallodendrimer with fluorescein statistically attached in one of the branches.[78] The attachment of a single fluorochrome enables the monitoring within the cell but at the same time maximizing the antitumor activity arising from the Ru(II) complexes. The uptake by two different cancer cell lines, leukemia HL60 [78] and human prostate PC-3,[52] was evaluated through confocal microscopy, confirming the ability of this compound to cross the cell membrane and remain in the cytoplasm even 24 h after treatment, unlike the well-known DNA-binding mechanism of platinum complexes (Fig. 8B1). A similar strategy was used to attach the spin-label 2,2,6,6-tetramethylpiperidine-1-oxyl (TEMPO) to carbosilane Ru(II) metallodendrimers and conduct a complete study by EPR to analyze the affinity of these metallodendrimers for cell membranes.[79] Two different types of model membranes were used:

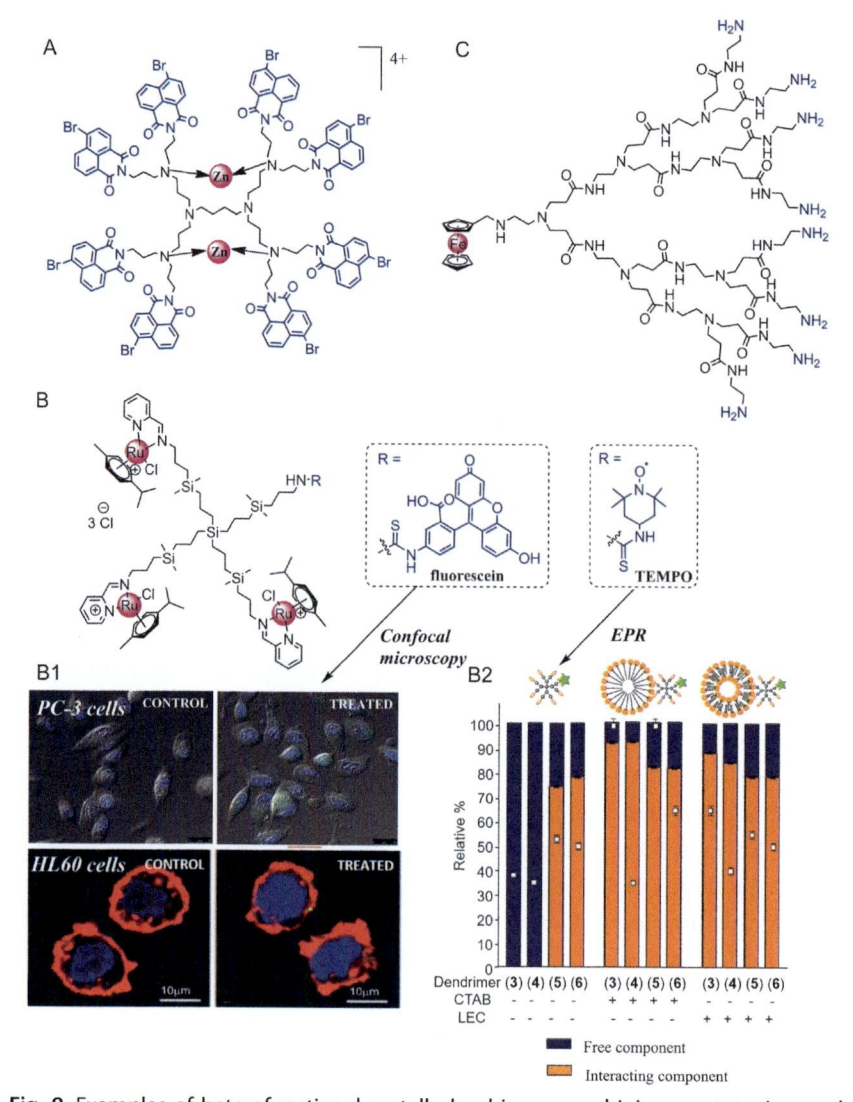

Fig. 8 Examples of heterofunctional metallodendrimers combining a cytotoxic metal complex and additional moieties for improvement of activity, traceability or solubility. (A) Layered-block Zn(II) metallodendrimer.[75] (B) Random Ru(II) metallodendrimer, including a fluorescein moiety for confocal microscopy (B1) or a TEMPO label for EPR evaluation (B2). (C) Ferrocene metallodendron.[80] Panel (B1): *Figures adapted from Maroto-Díaz M, Sanz del Olmo N, Muñoz-Moreno L, et al. In vitro and in vivo evaluation of first-generation carbosilane arene Ru(II)-metallodendrimers in advanced prostate cancer. Eur Polym J. 2019;113:229–235. Copyright 2019, with permission from Elsevier and Michlewska S, Kubczak M, Maroto-Díaz M, et al. Synthesis and characterization of FITC labelled ruthenium dendrimer as a prospective anticancer drug. Biomolecules 2019;9(9): E411. Copyright 2019, with permission from MDPI.* Panel (B2): *Figures adapted from Carloni R, Sanz del Olmo N, Ortega P, et al. Exploring the interactions of ruthenium (II) carbosilane metallodendrimers and precursors with model cell membranes through a dual spin-label spin-probe technique using EPR. Biomolecules. 2019;9(10):540. Copyright 2019, with permission from MDPI*

cetyltrimethylammonium bromide micelles (CTAB) and lecithin liposomes (LEC). The study provided complementary information from (1) the metallodendrimer point of view, containing the TEMPO label; and (2) the model membrane point of view, using 4-(N,N-dimethyl-N-dodecyl)ammonium-2,2,6,6-tetramethylpiperidine-1-oxyl bromide (CAT12), probe that can be inserted in the membrane. The authors confirmed the partial insertion of the surface groups attached to the dendrimer skeleton into the model membranes, mainly CTAB micelles, and the presence of both types of interactions (hydrophilic and hydrophobic) whereas, in the case of LEC, it is possible to observe a prevalence of polar interactions (Fig. 8B2). Any change in the dendritic structure can modify their ability of interacting with membranes, and, therefore, their anticancer activity.

The inherent heterofunctional nature of dendrons has also been employed to design metallodendrons with improved antitumor activity. In 2013, Şenel and co-workers presented a family of PAMAM dendrons with ferrocene in the focal point and amine groups in the periphery (Fig. 8C).[80] The dendrons inhibited the proliferation of stomach adenocarcinoma cells (AGS) via apoptosis and necrosis, increasing with the dose and the generation of the dendron. A similar strategy was recently reported by de la Mata and co-workers, where carbosilane dendrons comprised a [Ru(η^6-p-cymene)] fragment at the focal point coordinated through a N-monodentate, a N,N-chelate or an N,O-chelate ligand, and multiple $-$NMe$_2$ groups in the periphery.[81] These dendrons could potentially be evaluated as antitumor agents in future assays.

2.1.2 Metallodendrimers in the diagnosis and treatment monitoring

Imaging techniques are the mainstay in the diagnosis of numerous diseases. The exclusive properties provided by modern imaging methods in the study of the composition of the human body placed them among the most relevant advances of the century. There are a large number of imaging techniques,[82] and some of them provide remarkable information in the field of cancer. For example, the differences between tumor and normal tissues can be established by Magnetic Resonance Imaging (MRI). In order to improve the MRI signal, contrast agents are often used such as gadolinium chelates. Other techniques, like Single-Photon Emission Computed Tomography (SPECT) and Positron Emission Tomography (PET), involve the use of radiometals. Different metallodendrimers have been developed as diagnostic tools for MRI, optical and nuclear medical imaging, as described below.

In 2014, Shen et al. reported a tumor-targeting biodegradable polyester dendrimer conjugated with gadolinium and functionalized with PEG and folic acid as a MRI contrast agent, namely FA-PEG-G2-DTPA-Gd.[83] The PEG increased the blood circulation time whereas the folic acid increased the tumor selectivity, due to the overexpressed folate receptors in most tumors. This metallodendrimer exhibited a much higher contrast enhancement for longer times than Magnevist®, which is currently used in clinic as a gadolinium contrast agent. Whereas Magnevist® contrast signal starts decreasing 5 min post injection, the dendrimer prolongs it up to 15 min before its biodegradation and excretion in urine. In addition, 60 min post injection a weak contrast was observed between tumor and other tissues in mice using Magnevist® or the non-targeting metallodendrimer PEG-G2-DTPA-Gd; however, the contrast in mice treated with FA-PEG-G2-DTPA-Gd was strong, while in normal tissues the signal deceased with the excretion of the contrast agent (Fig. 9A).

Aiming for improved SPECT imaging, Adronov and co-workers reported in 2015 a fifth-generation polyester dendrimer functionalized with mPEG chains of different molecular weights at the periphery and dipicolylamine technetium(I) at the core.[84] *In vivo* experiments in healthy rats showed that this metallodendrimer circulated in blood for up 24 h. Moreover, the inoculation of mice with Human Lung Squamous Cell Carcinoma (H520) showed that the dendrimer could accumulate in the tumor for 6 h post injection, probably due to the enhanced permeation and retention effect that characterize these systems (Fig. 9B1). In parallel, Felder-Flesch and co-workers reported the synthesis of ^{111}In labeled PAMAM dendrons derived from DOTA and functionalized with PEG and with a melanin-targeting ligand, and studied the *in vitro* and *in vivo* targeting efficacy in murine melanoma models.[85] They found a correlation between the multivalence of the dendrimers and the tumor uptake, the highest being 12.7 ± 1.6% of the total injected dose per gram of tissue (ID/g) at 4 h post intravenous injection of the second-generation probe, compared to the 1.5 ± 0.5 ID/g for a non-dendritic counterpart (Fig. 9B2). The authors highlight the importance of multivalent small nanoprobes for imaging sensitivity enhancement and increased delivery of radiation doses to tumors.

2.1.3 Metallodendrimers as theranostic agents
Current trends in the biomedical field focus on the design of new theranostic agents, which combine diagnostic and therapeutic agents in a single

Metallodendrimers as a promising tool in the biomedical field 27

A *MRI contrast agent*

B *Radio imaging*

B1 *Technetium*

H: heart; L: lungs;
B: bladder; T: tumor

B2 *Indium*

targeting agent

imaging agent

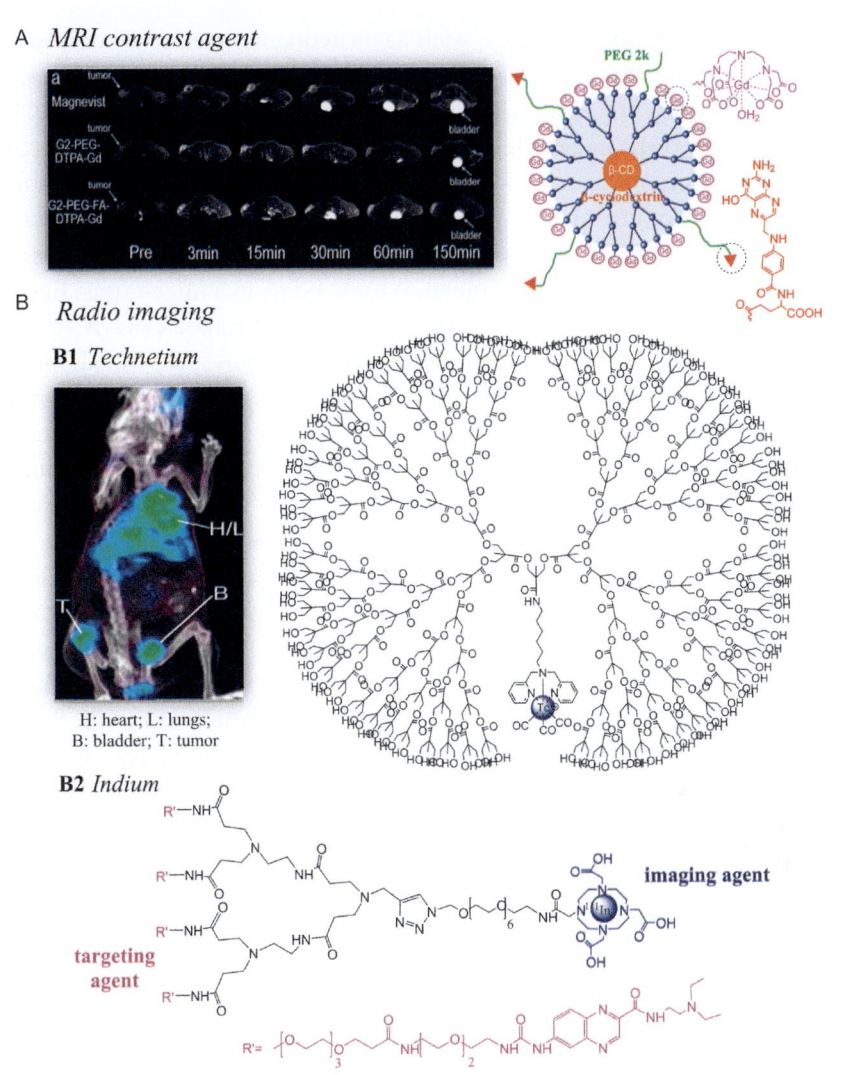

Fig. 9 Recent examples of metallodendrimers as imaging agents. (A) Bis-MPA polyester Gd metallodendrimer as contrast agent for MRI, including 2D axial images contrast to noise ratio (CNR) and tumors in mice injected with FA-PEG-G2-DTPA-Gd or Magnevist at 0.1 mmol/kg.[83] (B1) Bis-MPA polyester Tc metallodendrimer for Scintigraphic-CT image of H520 tumor model 6 h post injection of [99mTcDPA-G5-(PEG750)]+ (9.2 MBq).[84] (B2)[111] In labeled PAMAM dendron bearing melanin-targeting ligands.[85] *Panel (A): Figures adapted from Ye M, Qian Y, Tang J, Hu H, Sui M, Shen Y. Targeted biodegradable dendritic MRI contrast agent for enhanced tumor imaging. J Control Release. 2013;169(3):239–245. Copyright 2013, with permission from Elsevier, McNelles SA, Knight SD, Janzen N, Valliant JF, Adronov A. Synthesis, radiolabeling, and in vivo imaging of PEGylated high-generation polyester dendrimers. Biomacromolecules 2015;16(9):3033–3041. Copyright 2015, with permission from ACS.*

platform. In these smart systems, each agent works synergistically and efficiently. Dendrimers are ideal nanoplatforms for this purpose, enabling a controlled and precise conjugation of the different moieties. In 2015, Chen et al. used cationic Ru(II) metallodendrimers to coat selenium nanoparticles decorated with anionic L- and D-arginine, which provided a chiral nature to the nanoparticle (Fig. 10A).[86] The multifunctional nanoparticle was evaluated as carrier of MDR-siRNA (Ru@L-SeNPs-siRNA) into tumor cells, observing that the intravenous injection each 3 days in a mice model bearing adenocarcinoma human alveolar basal epithelial tumor cells (A549R) produced 79.4% inhibition of tumor growth. Furthermore, the intrinsic fluorescence of the multivalent Ru(II) complexes enabled a real-time monitoring in a xenograft mice model, observing a higher intensity in the tumor compared to the rest of the organs 4 h after the last injection. Another recent example of a powerful theranostic nanosystem has been reported by Shi, Majoral et al. and consists on a fifth-generation Cu(II) PAMAM metallodendrimer (G5.NHAc-Pyr/Cu(II)) (Fig. 10B).[87] In addition to the promising anticancer properties of Cu(II), the presence of unpaired electrons in its outermost orbital has proposed Cu(II) as an alternative T1-weighted MR contrast agent. The authors confirmed the anticancer properties of this metallodendrimer, which increased the ROS levels, then produced a cell cycle S-phase arrest, apoptosis and finally induced cell death *in vitro*. Furthermore, two *in vivo* experiments—a xenografted subcutaneous tumor model and lung metastatic nodules in a bloodstream metastasis model-confirmed the compound accumulation in the tumor through enhanced permeation and retention effect and the suitability of this metallodendrimer as MRI agent. Nowadays, it is well known that radiotherapy changes the tumor vasculature and favors the accumulation of nanoparticles in the tumor. This fact has been also observed with G5.NHAc-Pyr/Cu(II), where the MRI performance in tumor-bearing mice can be significantly improved after the radiation of tumors observing an increment in the accumulation of this compound in the tumor environment (Fig. 10B).

2.2 Metallodendrimers as antimicrobial agents

Ever since Alexander Fleming discovered penicillin in 1928, the antibiotic consumption has grown worldwide at neck-breaking speed. Undoubtedly, the discovery of this seeming panacea represented a milestone of medicine for the whole humanity but has ultimately led to a serious problem of multidrug-resistant bacteria. As an example, 14% of *S. aureus* hospital strains

Metallodendrimers as a promising tool in the biomedical field　　29

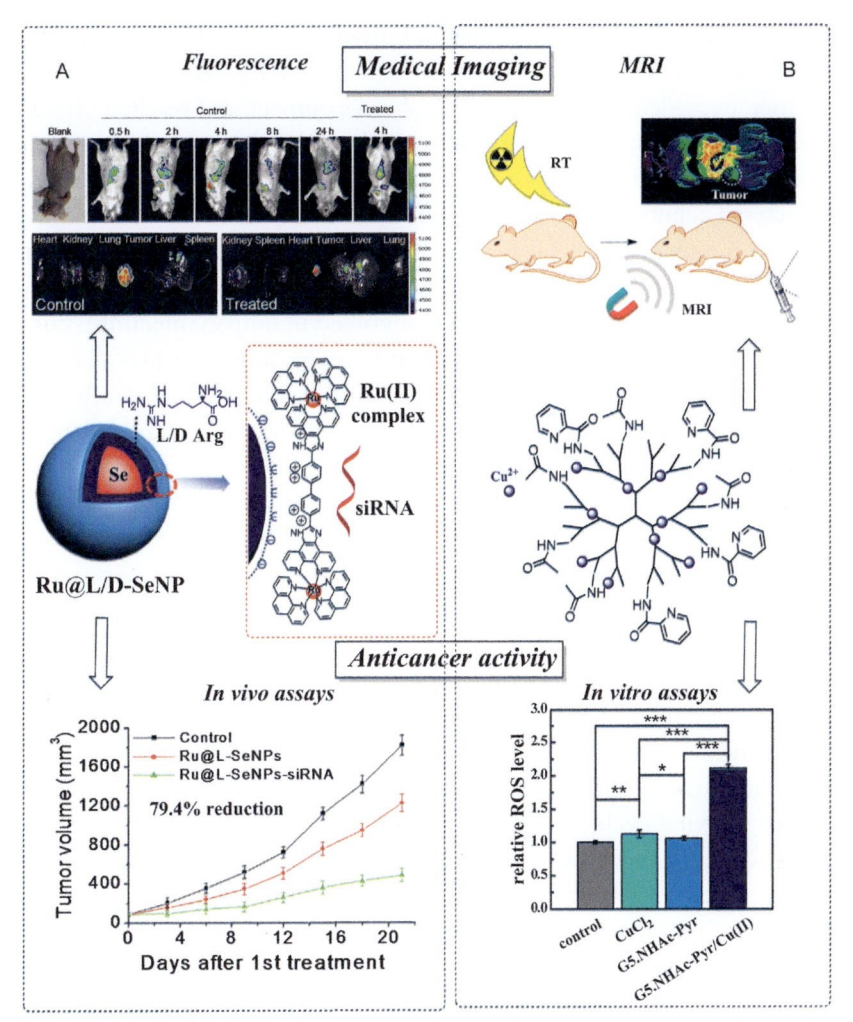

Fig. 10 Dual activity of two metallodendritic-based theranostic agents. (A) Selenium NPs (Ru@L-SeNPs-siRNA).[86] (B) Copper(II) metallodendrimers G5.NHAc Pyr/Cu(II).[87] Panel (A): Figure adapted from Chen Q, Qianqian Y, Liu Y, et al. Multifunctional selenium nanoparticles: chiral selectivity of delivering MDR-siRNA for reversal of multidrug resistance and real-time biofluorescence imaging. Nanomedicine (N Y, NY, US). 2015;11(7):1773–1784. Copyright 2015, Elsevier. *Panel (B): Figure adapted from Fan Y, Zhang J, Shi M, et al. Poly(amidoamine) dendrimer-coordinated copper(II) complexes as a theranostic nanoplatform for the radiotherapy-enhanced magnetic resonance imaging and chemotherapy of tumors and tumor metastasis. Nano Lett 2019;19(2):1216–1226. Copyright 2019, ACS.*

became resistant 4 years after penicillin started to be used in clinic, and the percentage increased to 59% just 4 years later. Between the 1980s and 90s, resistance exceeded 80% in non-clinical environment and reached 95% in the majority of hospitals.[88] It is estimated that by 2050, 10 million people per year will die due to untreatable infections. The development of new antibacterial agents is therefore one of the most important and challenging tasks that the pharmaceutical field is currently facing. Biofilm formation is one of the main hurdles that new antimicrobial drugs need to overtake.[89] This bacterial mode of growth, highly involved in human infections, consists of a matrix made of extracellular polymeric substance produced by microbes, exhibiting an altered phenotype. The biofilm acts as a barrier hindering the penetration of antimicrobial agents and offering a favorable environment for microbial growth.

The interest in the antimicrobial activity of dendrimers has boosted in the last 20 years.[90] Their versatility and multivalency enable an accurate and application-oriented design to reach maximum efficiency. In most cases, antibacterial dendrimers rely on an adequate cationic charge distribution and amphiphilicity to disrupt the negatively charged microbe membrane in both Gram-positive and Gram-negative bacteria.[91–93] A major issue arises on the high toxicity displayed also toward mammalian cells. On the contrary, antiviral dendrimers mainly exhibit anionic, sugar or peptide moieties which can interfere in the virus-host interaction and avoid the infection at early stages of the viral cycle.[94,95]

To overcome dendrimer limitations and further improve the therapeutic response, researchers have shifted the focus to the combination of metals and dendrimers. The attachment of metal atoms into biologically active molecules has been widely reported as a useful strategy to overcome microbial resistance and improve activity.[96] In this field, the first metal complex dates back to the last century, when Paul Ehrlich and his co-workers developed an organoarsenic compound to treat syphilis. The following sections will provide a complete overview of the different metallodendrimers reported as antimicrobial agents in the literature, toward bacteria, yeast, virus and parasite infections.

2.2.1 Metallodendrimers as antibacterial agents

Even though the inclusion of antibacterial metals in dendritic scaffolds dates back to 2000s, few examples have been described during this 20-year period. An overview of selected metallodendrimers is presented in Table 3, including indices of the antimicrobial activity such as the Minimum Inhibitory

Table 3 Overview of selected metallodendrimers and their antibacterial activity indicated as Minimum Inhibitory Concentration (MIC) and Minimum Bactericidal Concentration (MBC) values on different bacterial strains.

Entry	Metal	Dendritic scaffold	Metal complex	Number metals	Bacteria	MIC/MBC (mg/L)	Ref.
1	Ag(I)	G5-PAMAM		[a]	S. aureus P. aeruginosa E. coli	12.0[a,b] 10.5[a,b] 8.7[a,b]	97
2	Ag(I)	G1-PETIM		6	S. aureus MRSA	41.7/– 26.0/–	98
3	Cu(II)	G1-CBS		4	E. coli S. aureus S. aureus (BF)[c]	4/4 4/8 8/4[d]	99
4	Ru(II)	G1-CBS		4	E. coli S. aureus S. aureus (BF)[c]	16/16 4/4 32/32[d]	99
5	Pd(II)	G2-Polyamide		6	E. coli S. typhi B. subtilis S. aureus	70/– 102/– 84/– 82/–	100

[a]Not reported.
[b]Sensitivity indicated as inhibition area (in mm).
[c]BF stands for biofilm mode of growth.
[d]For biofilm-forming bacteria, the Minimum Biofilm Inhibitory Concentration (MBIC) is reported.

Concentration (MIC), the Minimum Bactericidal Concentration (MBC) and the Minimum Biofilm Inhibitory Concentration (MBIC). These indices represent the minimal concentration that inhibits the growth of the microorganisms (MIC), kills the microorganism (MBC) or inhibits the formation of biofilm but not the growth (MBIC).

2.2.1.1 Silver(I) metallodendrimers

Silver is probably the most ancient antibacterial agent ever used by humans. Nowadays, different antibacterial drugs to treat external wounds or antibiotic coatings in medical devices rely on silver-based metallodrugs. It has been described that Ag(I) ions interfere with the nutrient transport chain in microorganisms, as well as bind to bacteria's genetic material.[101] Despite this excellent and specific mechanism of action, microorganisms have developed resistance to silver drugs[102] and demand new approaches to treat infections.

The first antibacterial metallodendrimers, described in 2000 by Balogh et al. comprised polyamidoamine scaffolds which complexed Ag(I) ions, forming stable and well characterized bonds (Table 3, Entry 1).[97] Generation 4 and 5 PAMAM dendrimers decorated with 192 hydroxyl and 256 carboxylate groups respectively were complexed with silver acetate. The dendrimers enabled an extremely high local concentration of silver cations within a relatively thin shell at the periphery while the internal tertiary nitrogen atoms could form stable silver complexes. Nevertheless, the authors indicate that the metallodendrimers slowly photolyze into Ag(0) containing dendrimer-silver nanocomposites (Fig. 11A). Using a standard agar overlay method, the diffusible antimicrobial activity was evaluated against planktonic *P. aeruginosa*, *E. coli* and *S. aureus*. Both metallodendrimers and nanocomposites exhibited comparable or higher antibacterial activity than the $AgNO_3$ used as control. Interestingly, the antimicrobial activity was superior in carboxylate-functional dendrimers, due to the higher surface concentration of silver ions—256 carboxylate groups around a 54 Å sphere—eager to interact with the bacterial membranes. The study also revealed the need for accessibility to silver, considering that those dendrimers with mostly internally conjugated silver ions produced a lower activity.

A similar strategy was recently used by Govender et al. to generate the silver salts of propyletherimine (PETIM) dendrimers (Table 3, Entry 2).[98] The low-generation dendron and dendrimers comprised carboxylate moieties at the surface with Ag(I) counterions (Fig. 11B). The PETIM silver salts exhibited lower toxicity toward healthy cells than a comparable concentration of $AgNO_3$, probably due to the reduced net surface charge. The authors

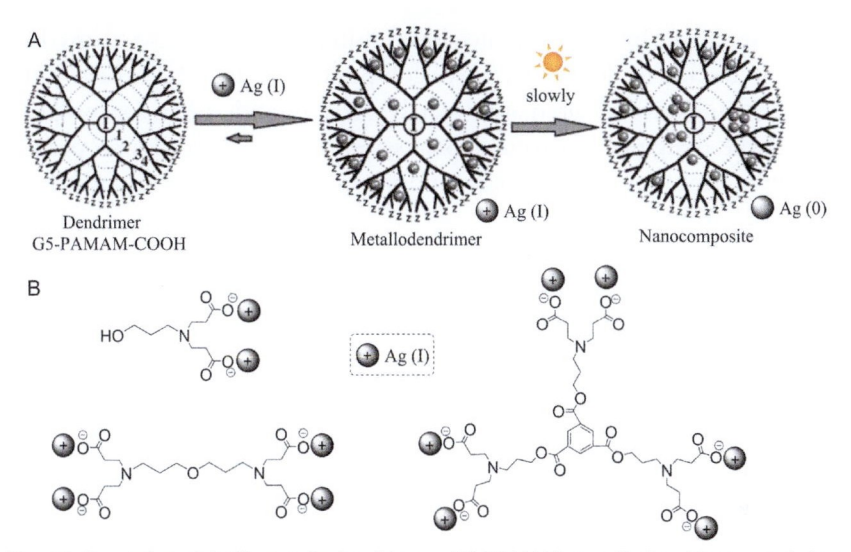

Fig. 11 Examples of Ag(I) metallodendrimers. (A) PAMAM metallodendrimers and slow conversion to nanocomposites.[97] (B) PETIM metallodendrimers.[98] *Panel (A): With permission from American Chemical Society. Balogh L, Swanson DR, Tomalia DA, Hagnauer GL, McManus AT. Dendrimer–silver complexes and nanocomposites as antimicrobial agents. Nano Lett. 2001;1(1):18–21. Copyright 2001.*

demonstrated a direct relation between the number of Ag(I) ions and the antibacterial activity; the more potent response was observed with the aromatic cored metallodendrimer—featuring six Ag(I) ions bound on its surface—which decreased for the oxygen cored metallodendrimer—with only four ions—and even more for the metallodendron with only two ions. In order to analyze the combined effects of the dendritic compounds and silver nitrate, the Fractional Inhibitory Concentration index (ΣFIC) was calculated. The index showed no antagonist effects and, remarkably, synergist effect toward *S. aureus* for the aromatic cored metallodendrimer and toward *MRSA* for the oxygen cored counterpart. Overall, they presented a promising approach considering that the low dendritic generation simplifies the synthetic route, thus reducing time, costs and potential toxicity due to the low number of metal ions.

The antibacterial properties of silver, enhanced by the fine control of dendrimer scaffold, seemed indeed like the best route to get new drugs to overcome the problems of the traditional class of antimicrobial agents. However, like many similar compounds, these fell into the "dendrimer paradox" of very high medical expectation which fades away due to the hurdles arising during the development of new drugs.[103] Indeed, since this first example of antibacterial metallodendrimers in 2000, the focus has shifted to the use of dendrimer/silver composites due to the synthetic simplicity, the inclusion of alternative metals which provide additional properties and the design of heterofunctional systems that expand the properties and applications of metal-based dendrimers, as it will be described in Section 2.2.1.5.

2.2.1.2 Copper(II) metallodendrimers

Copper is a very promising metal, biocompatible, versatile and cheap, featuring potent antimicrobial activity against viruses, yeasts and bacteria.[104,105] The use of copper in medicine expanded in the last two centuries, with a variety of inorganic copper compounds used to treat eczemas, tubercular infections, syphilis and antibacterial infections.[106] Furthermore, the intrinsic paramagnetic activity of its +2 oxidation state enables the study and monitoring of Cu(II) complexes through Electron Paramagnetic Resonance (EPR), resulting in an accurate chemico-physical characterization.

The attractive properties of copper—cost-effective, antimicrobial and easy to monitor—encouraged de la Mata et al. to evaluate in 2019 the antibacterial properties of Cu(II) carbosilane metallodendrimers bearing iminopyridine ligands (Table 3, Entry 3), previously reported as antitumoral agents.[43,59] The study not only showed the promising antibacterial activity toward Gram-positive and Gram-negative bacteria—including biofilm mode of growth—but most importantly confirmed the positive influence of the following parameters on the antibacterial activity[99]: (1) The *metal complexation* to the ligand, unlike other metallodrugs which exhibit antibacterial activity only after the metal release; (2) the *lipophilic scaffold*, being the first-generation metallodendrimers the most potent members in the family unlike most other dendritic scaffolds which require higher generations; (3) structural parameters such as the *metal ion and counterion*, exhibiting different activity between Ru(II) vs Cu(II) complexes and chloride vs nitrate counterion, confirming additional mechanisms besides the one related to the cationic charge at the dendrimer surface (Fig. 12). Furthermore, the authors proved that carbosilane metallodendrimers did not produce hemolysis at the MIC concentrations and can be safely used as antibacterial agents.

2.2.1.3 Ruthenium(II) metallodendrimers

Ruthenium has long proved to be one of the most promising metals to be incorporated in antimicrobial agents, due to its interaction with DNA.[107] Ruthenium complexes exhibit different mechanisms of action,[108] including

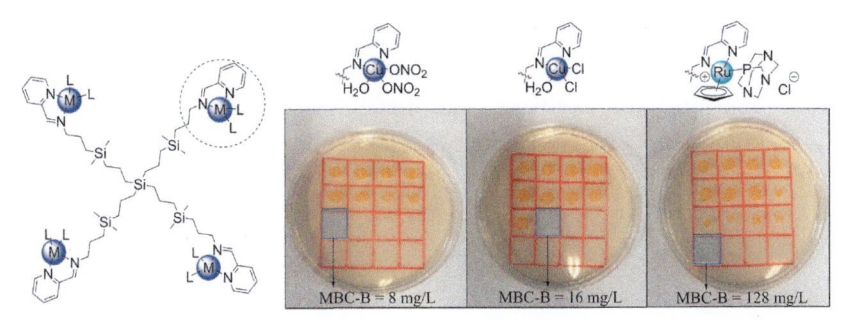

Fig. 12 Effect of different first-generation carbosilane metallodendrimers in preventing the formation of *S. aureus* biofilms, calculating the Minimum Bactericidal Concentration for Biofilms (MBC-B). *Figure adapted from Lamazares C, Sanz Del Olmo N, Ortega P, Gomez R, Soliveri J, de la Mata FJ, García-Gallego S and Copa-Patiño JL. Antibacterial effect of carbosilane metallodendrimers in planktonic cells of gram-positive and gram-negative bacteria and Staphylococcus aureus biofilm. Biomolecules 2019;9(9):E405. Copyright 2019, MDPI.*

the interaction to DNA and the inhibition of DNA processing enzymes. In general, Ru(II) complexes exhibit more potent antibacterial activity toward Gram-positive than Gram-negative bacteria, being highly influenced by the charge and lipophilicity of the complex.

The first antibacterial Ru(II) metallodendrimers reported were part of a broader study carried out by de la Mata et al.[99], aiming to evaluate the influence of different structural parameters on the antibacterial activity (Fig. 12). Similar to their copper counterparts described in the previous section, the Ru(II) complexes were chelated through the iminopyridine ligands at the periphery of the carbosilane dendrimers.[41] The first-generation Ru(II) metallodendrimer showed potent antibacterial activity against planktonic *S. aureus* and *E. coli* (Table 3, Entry 4), even higher than the analogous Cu(II) metallodendrimers. Furthermore, it inhibited the formation of *S. aureus* biofilm at low concentrations (MBIC 32 mg/L) but even lower values were accomplished with the Cu(II) metallodendrimers (MBIC 8 mg/L).

2.2.1.4 Platinum(II) and Palladium(II) metallodendrimers

Platinum and palladium complexes have been widely described as efficient antitumor agents.[109,110] However, fewer reports appear on the antimicrobial activity of these metallodrugs.

In 2012, Alshehri et al. reported the synthesis of platinum(II) and palladium(II) aromatic polyamide metallodendrimers.[100] Each metal ion was coordinated through two amino groups at the dendrimer periphery, a total of three metal ions in first-generation dendrimers and six metal ions in second-generation counterparts. The metallodendrimers exhibited potent antibacterial activity against different Gram-positive and Gram-negative strains, higher than the non-complexed dendrimer, and especially high for Pd(II) derivatives. Indeed, the second-generation Pd(II) metallodendrimer exhibited the most potent activity, comparable to the commercially available antibiotic streptomycin (Table 3, Entry 5).

2.2.1.5 Heterofunctional metallodendrimers and metallodendrons

The field of antibacterial treatments has also benefited from attaching different moieties in a single molecule. Enhancing the therapeutic response, reducing bacteria resistance or increasing the solubility are among the different aims of these multipurpose materials. Several approaches have been reported so far and are described in this section, with relevant examples depicted in Fig. 13.

Since 2009, Grabchev et al. have developed several families of Cu(II) and Zn(II) poly(propyleneamine) metallodendrimers decorated with different 1,8-naphtalimides (unsubstituted,[73] 4-bromo-,[74,75] 4-amino-[76]; 4-(N,N-dimethylaminoethoxy)[77]; and acridine,[111] all biologically active moieties with promising antibacterial activity and fluorescent properties. Similarly to copper, the antimicrobial properties of zinc have been known for long time. For example, ZnO fine particles have deodorizing and antibacterial activity and are therefore exploited in the production of cotton fabrics and oral products.[114] These properties come in handy in the treatment of skin conditions, like sunburn or rush, due to Zn antiseptic activity.

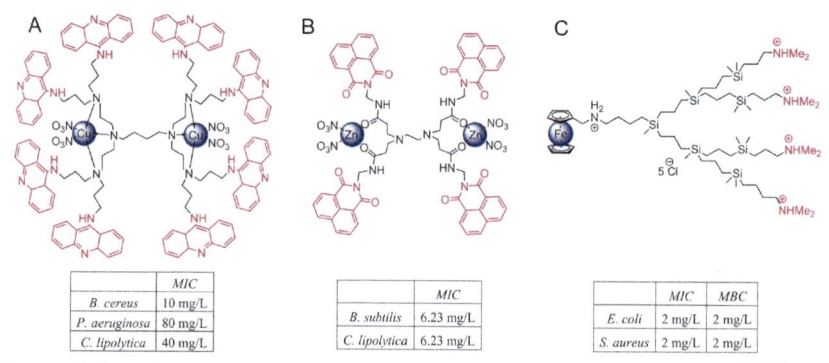

Fig. 13 Selected examples of heterofunctional metallodendrimers (A and B)[111,112] and metallodendron (C),[113] including MIC and MBC values in different bacterial strains.

The poly(propyleneamine) complexes are clear examples of heterofunctional metallodendrimers, considering that the metal coordination is performed via the tertiary amine groups at the dendrimer inner shell, comprising 1, 2 or 4 metal ions in G1, G2 and G3 dendrimers, respectively, being the outer shell available for further therapeutic response. The authors evaluated the antibacterial effect toward different Gram-positive bacteria (*B. subtilis*, *B. cereus*, *S. lutea* and *M. luteus*), Gram-negative bacteria (*P. aeruginosa*, *E. coli*, *A. johnsonii* and *Xanthomonas oryzae*) and yeast strains (*C. lipolytica* and *S. cerevisiae*). In general, Zn(II) complexes exhibited slightly higher antibacterial effect compared to the Cu(II) counterparts and superior activity to antimicrobial drugs such as tetracycline and nystatin. The authors also demonstrated the influence of the substituent at the 1,8-naphthalimide ring: the –Br group in the second-generation metallodendrimer [$Zn_2(D)(NO_3)_4$] reveals MIC values in the range 500–2000 μg/mL, while the –NH_2 in the first-generation metallodendrimer [$Zn(D)(NO_3)_2$] produces values in the range 50–200 μg/mL. The third-generation counterpart [$Zn_4(D)(NO_3)_8$] decreases MIC values even further and displayed the highest inhibition toward *B. subtilis*, *E. coli* and *P. aeruginosa* with MIC ~50 μg/mL. Similarly to 1,8-naphtalimides, acridine moieties have also been attached to a second-generation poly(propyleneamine) dendrimer and used as platforms for antimicrobial Cu(II) metallodendrimers. Compared to the dendrimer alone, the Cu(II) complex exhibited an enhanced antimicrobial activity against *B. cereus*, *P. aeruginosa* and *C. lipolytica* (Fig. 13A). Deposition of dendrimers on the surface of cotton fabric increased the hydrophobicity of the textile and prevented the formation of bacterial biofilm—inhibiting about 90% of the growth of *B. cereus* and more than 50% of the growth of *P. aeruginosa*.[111] The authors explain the enhanced antimicrobial activity of these metallodendrimers through the Overtone's concept of cell permeability and the Tweedy's chelation theory.[76] The chelation reduces the polarity of the metal ion, increases the delocalization of electrons over the chelate ring and thus increases the lipophilicity of the metallodendrimers, favoring the penetration into the lipid cell membrane.

The influence of the dendritic scaffold was also evaluated, decorating PAMAM scaffolds with 1,8-naphtalimides and generating the corresponding Cu(II) and Zn(II) metallodendrimers (Fig. 13B).[112] In this case, the metal coordination is produced through the either the carbonyl groups of the internal amide bonds, in the case of Zn(II), or the ethylenediamine core of two different molecules, in the case of Cu(II), as evidenced through EPR analysis. SEM analysis indicated that the metal coordination

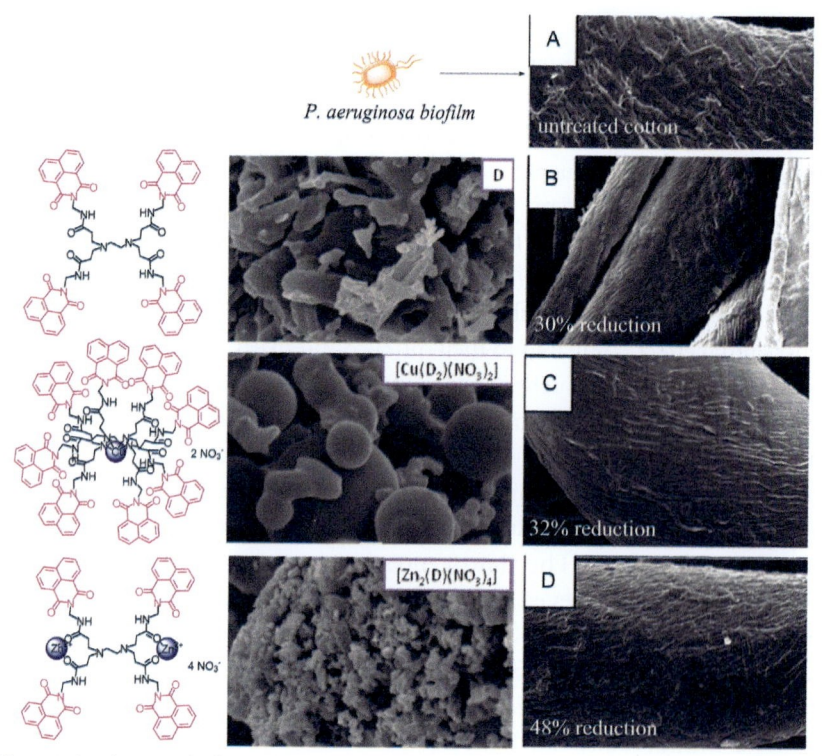

Fig. 14 Antibacterial effect of PAMAM metallodendrimers on cotton fabrics. On the left, SEM micrographs of the dendrimer and the corresponding metallodendrimers. On the right, SEM images of cotton fabrics tested against P. aeruginosa biofilm.[112] *Used with permission of The Royal Society of Chemistry, from* Impact of Cu(II) and Zn(II) ions on the functional properties of new PAMAM metallodendrimers, *Copyright 2018.*[112]

to the dendrimer significantly changed the surface morphology (Fig. 14). The smooth and thick laminar structure of the dendrimer is exchanged for a microspherical structure in the copper complex with some agglomerates and for a highly porous rod structure in the zinc derivative, which is later translated into an increased antibacterial activity due to an easier penetration into the cell walls. Again, the Zn(II) complex exhibited slightly higher antibacterial effect compared to the Cu(II) counterpart, with MIC values at the low mg/L range for Gram-positive bacteria. The negligible inhibition toward Gram-negative bacteria can be explained considering the additional outer membrane and slightly different cell wall structure of these bacteria,[115] as well as the different nature of the little holes in intact phospholipidic bilayer where non-charged dendrimers are adsorbed.[116]

The authors also reported the antimicrobial efficacy of cotton fabrics treated with the dendrimers, arising from both slow diffusion of the dendrimers from the cotton fabrics to the medium, a direct contact with the microbes as well as an increase in hydrophobicity that prevents the formation and proliferation of bacterial biofilms. As example, the Zn(II) metallodendrimer reduced the growth of *C. lipolytica* (100%), *B. cereus* (60%) and *P. aeruginosa* (48%) (Fig. 14).

A different strategy relies on the use of dendrons, as the most simple heterofunctional dendritic entity. The first antibacterial metallodendron, reported in 2015, comprised a poly(propyl ether imine) scaffold bearing a hydroxyl group in the focal point and two Ag(I) carboxylate moieties. As previously described in Section 2.2.1.1, this metallodendron displayed moderate antibacterial activity against *S. aureus* and methicillin-resistant *S. aureus*, only observing an additive effect between the dendron and AgNO$_3$ in the first case. The antibacterial effect was clearly improved using the metallodendrimer counterparts bearing 4 or 6 Ag(I) complexes.

In parallel, Gómez et al. designed a carbosilane dendron comprising a ferrocene moiety in the focal point and multiple ammonium groups in the periphery (Fig. 13C).[113] Ferrocene is a Fe(II) "sandwich" complex, where the metal ion is coordinated between two cyclopentadienyl anions (Fe(η^5-C$_5$H$_5$)$_2$), with antimicrobial activity. Even though the mechanism of action is not clearly defined, it might be related to the triggering of Fenton's reaction and the production of ROS from hydrogen peroxide in biological environments.[117] Unlike the previous silver dendron, the metal is now located in the focal point and anchored through a covalent bond in a real organometallic complex. Furthermore, the antibacterial activity of the ferrocene is reinforced by the multiple cationic groups in the periphery. The Fe(II) metallodendrons exhibited a potent antibacterial activity toward Gram-negative (*E. coli*) and Gram-positive (*S. aureus*) bacteria, especially remarkable for the first-generation complex, which exhibited one of the lowest MIC (2 mg/L) for the examined strains of the whole metallodendrimer field. Surprisingly, the activity decreased when increasing the generation, and NMR DOSY experiments suggested that the large hydrodynamic volume of the G3 metallodendron reduced the interactions with the bacteria membrane. The high activity in both types of bacteria confirm the potential broad spectrum antibacterial activity of these metallodendrons, unlike many other organometallic compounds featuring ferrocene,[118] highlighting once again the importance of the dendritic scaffold.

2.2.2 Metallodendrimers as antiviral agents

Antiviral metallodendrimers aim to combine the activity of the peripheral anionic, sugar or peptide moieties, with the promising properties of the metals, in order to interfere in the different steps of the virus cycle.[94,119] Despite the few examples reported in the literature in the field of antiviral metallodendrimers, the rationale behind the design of these metallodrugs provides useful insight into most important parameters influencing the antiviral behavior.

2.2.2.1 Copper(II) metallodendrimers

Polyanionic microbicides are known to interfere in the first steps of the HIV replicative cycle, inhibiting the adsorption and fusion of the virus to the cell. Dendrimers decorated with anionic moieties have been widely evaluated as antiviral agents against HIV, such as carbosilane[120–122] and the polylysine-based SPL7013 (VivaGel®).[123] Considering that several metal complexes are known to interfere in this process as well as in subsequent steps of the cycle, such as the viral DNA integration or viral protein processing by HIV protease, de la Mata et al. combined both strategies developing Cu(II) carbosilane metallodendrimers with carboxylate and sulfonate peripheral groups (Fig. 15A).[124] In order to evaluate a possible synergic inhibitory effect between the metal atoms and the anionic groups, the dendrimers were modified with both stoichiometric and sub-stoichiometric amounts of copper, the latter leading to statistically functionalized metallodendrimers with an overall negative charge. The UV–Vis and EPR study revealed that carboxylate decorated dendrimers chelate the Cu(II) ion in a square planar $CuNO_3$ geometry, while the sulfonate decorated counterparts probably form dimers at the internal–external dendrimer layer and at the higher copper concentrations the ions are coordinated by the sulfonate groups and water at the external layer. Second-generation dendrimers showed preventive and therapeutic activity against HIV infection, as revealed the study performed in HEC-1A cells—as a model for the first cellular barrier in vaginal epithelia—and PBMC—as main target of HIV and second barrier model. The authors concluded that the carboxylate systems are more efficient and the metal coordination can further enhance the inhibitory effect of the anionic dendrimer, and this effect is more important as the metal amount is increased.

The exchange of the dendritic scaffold from carbosilane to poly(propyleneimine) with ethylenediamino core revealed a remarkable strategy to precisely control the heterofunctionalization of the anionic dendrimers.[125] In this case, an accurate specific pattern in the coordination of

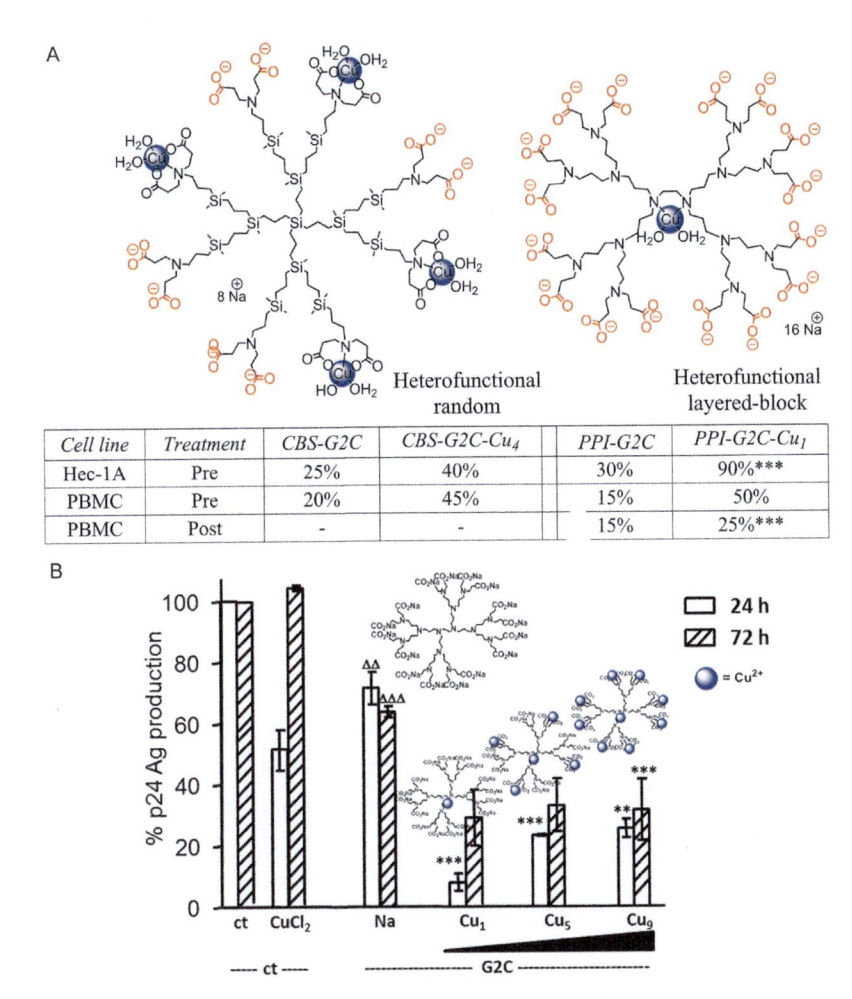

Metallodendrimers as a promising tool in the biomedical field   41

Cell line	Treatment	CBS-G2C	CBS-G2C-Cu₄	PPI-G2C	PPI-G2C-Cu₁
Hec-1A	Pre	25%	40%	30%	90%***
PBMC	Pre	20%	45%	15%	50%
PBMC	Post	-	-	15%	25%***

Fig. 15 (A) Selected examples of heterofunctional Cu(II) metallodendrimers with antiviral activity against HIV-1. The inserted table shows the percentage of inhibition of HIV-1 infection after 1 h pre/post-treatment with the compounds at 10 µM. (B) Decrease in inhibitory action when increasing the metal content, at different dendrimer:metal ratio 1:0 (Na), 1:1 (Cu₁), 1:5 (Cu₅) and 1:9 (Cu₉). *Figure (B) reprinted with permission from García-Gallego S, Díaz L, Jiménez JL, Gómez R, de la Mata FJ, Muñoz-Fernández MÁ. HIV-1 antiviral behavior of anionic PPI metallo-dendrimers with EDA core. Eur J Med Chem 2015;98:139–148. Copyright 2015, with permission from Elsevier.*

metal atoms was discovered using Cu(II) as a probe: at 1:1 M ratio between Cu(II) and the dendrimer, the metal exhibits a CuN₂O₂ pattern at the core of the dendrimer, while the increase in Cu(II) concentration indicated a peripheral CuNO₃ coordination in a square planar coordination which ultimately forms CuO₄ complexes at the highest concentrations tested. The authors demonstrated a different interaction for sulfonate dendrimers, with a weaker interaction toward nitrogen sites and stronger interaction with the oxygen in the sulfonate group. The antiviral effect toward HIV-1 was dependent upon the number, type and location of the metal within the dendritic scaffold (Fig. 15B).[126] Importantly, the highest activity was obtained with the Cu(II) metallodendrimer comprising a single metal ion in the core, considering that all anionic groups were available for the interaction instead of being blocked by chelating the metal atoms. A remarkable increase from 30% to 90% inhibition of HIV-1 infection in Hec-1A was produced by complexing a single Cu(II) ion to the second-generation dendrimer bearing 16 carboxylate groups.

2.2.2.2 Nickel(II), Cobalt(II) and Zinc(II) metallodendrimers

As part of the study regarding anionic PPI metallodendrimers with ethylenediamine core,[126] the influence of the nature of the metal ion on the antiviral activity against HIV was evaluated. Accordingly, the carboxylate and sulfonate dendrimers were reacted with Ni(II), Co(II), Cu(II) and Zn(II) in different dendrimer:metal ratios and the resultant metallodendrimers were studied as antiviral agents. Hec-1A and VK-2 cell lines were used as model of the first barrier against HIV-1 infection and PBMC as model of the second barrier, and both preventive and therapeutic behaviors were studied. Important conclusions were drawn from this study, stating the positive influence of the following parameters on the antiviral activity: (a) the *metal coordination* to the dendritic scaffold, significantly increasing the biocompatibility of the free metal ions as well as the antiviral effect of the precursor dendrimer; (b) the *low metal:dendrimer ratio* (1:1), enabling a cooperative effect between the metal and the free anionic groups and increasing the inhibition from the anionic dendrimer (e.g., 50% for sulfonate and 30% for carboxylate in Hec-1A cells at 24 h) to the metallodendrimer (40–60% for sulfonate and 80–90% for carboxylate in Hec-1A cells). (c) The *nature of the metal* ion, determining the inhibitory action (specially high for Co(II), Cu(II) and Zn(II), negative for Ni(II)) and even certain HIV-1 strain and co-receptor specificity. Overall, the PPI metallodendrimers showed a high preventive inhibitory action in the first barrier model, avoiding virus internalization inside cells and inhibiting different viral strains, as well as a dual preventive-therapeutic behavior in the second barrier model. A rational design of such metallodendrimers opens new avenues for the production of versatile and efficient treatments against HIV-1 infection.

2.2.3 Metallodendrimers as antiparasitic agents

The multivalent nature of metallodendrimers has also been employed to treat parasite infections, enabling an increased interaction to a multi-receptor target. Malaria is a highly prevalent infectious disease produced by the parasite *Plasmodium falciparum*, which has evolved to generate resistance to traditional antimalarial drugs such as chloroquine. Aiming to alternative treatments with novel modes of action, different metallodrugs have emerged, such as ferroquine, a ferrocene-containing analogue of chloroquine.[127] This metallodrug combines the inhibitory action of the hemozoin formation from the chloroquine moiety and the ferrocenyl production of toxic OH radicals via a Fenton-like reaction.

Smith and co-workers have broad experience in the development of antiparasitic poly(propyleneimine) metallodendrimers. In 2011, they reported the synthesis of first-generation PPI dendrimers decorated with ferrocenyl-thiosemicarbazones.[128] The conjugation of this motif to the PPI scaffold is presumed to enhance the accumulation of the thiosemicarbazone drug into parasite-infected erythrocytes, due to the recognition of the polyamine scaffold by specific polyamine transporters.[129] The ferrocenyl metallodendrimers showed high antiplasmodial activity against the chloroquine-resistant W2 strain of *P. falciparum*, in the range 2–7 µM, compared to the 20 µM of the non-conjugated ligands. Furthermore, at 10 µM concentration, they inhibited up to 62% the growth of the parasite *Trichomonas vaginalis*, a protozoan responsible for the sexually transmitted disease trichomoniasis. The presence of the lipophilic ferrocene moiety may contribute to the enhanced activity but still the activity was not comparable to the drug metronidazole, producing 100% inhibition.[130] In 2016, a new family of ferrocenyl metallodendrimers was reported, bearing an aromatic extension of the thiosemicarbazone moiety (Fig. 16A).[131] First- and second-generation metallodendrimers were screened against the H37Rv strain of *Mycobacterium tuberculosis* and the chloroquine-sensitive NF54 of *P. falciparum*, displaying moderate activity compared to commercial drugs. The second-generation complex performed slightly better, with MIC_{90} values of 41.7 µM toward *M. tuberculosis* and IC_{50} 32.8 µM toward *P. falciparum*. Recently, Smith et al. developed a PPI dendrimer comprising a single ferroquine moiety in one of the branches, which exhibited IC_{50} of 0.634 ± 0.043 µM and 0.545 ± 0.030 µM toward the NF54 and K1 strains of *P. falciparum*, respectively. This is an appreciable activity, though lower than chloroquine, and maintains the activity even in the

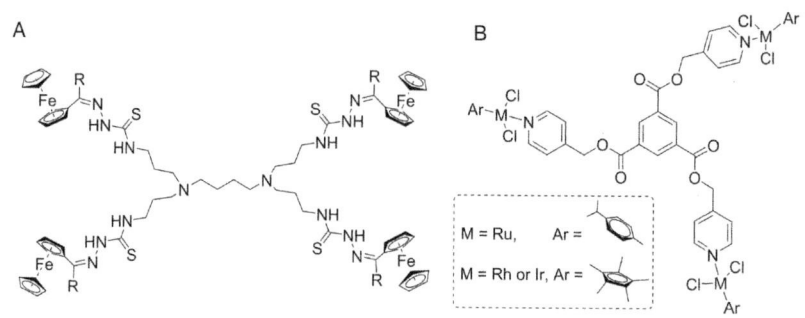

Fig. 16 Examples of antiparasite metallodendrimers. (A) Polynuclear ferrocenylthiosemicarbazone complexes[131] and (B) polynuclear organometallic Ru(II), Rh(III) and Ir(III) pyridyl ester complexes.[133]

chloroquine-resistant strain, suggesting the influence of the polyamine or ferrocenyl moiety on the improvement of the resistance indices.[132]

Smith's group has also explored alternative organometallic complexes of Ru(II), Rh(II) and Ir(II) using first-generation aromatic polyester dendrimers bearing pyridine ligands (Fig. 16B). The obtained trinuclear metallodendrimers exhibited moderate to high antiplasmodial activities toward the chloroquine-sensitive strain NF54 of *P. falciparum* (IC$_{50}$ 5–10 μM) and G3 strain of *Trichomonas vaginalis* (34–67% inhibition at 25 μM, 45–96% inhibition at 50 μM).[133]

3. Conclusions

Metallodendrimers have proven to be an original and powerful tool to fight cancer or infectious diseases. The inclusion of several metals centers in one single molecule enhances the effect produced by their mononuclear analogues. Although platinum was a trend in the cancer treatment, recently, other alternatives have emerged which involve the use of other metals such as ruthenium, copper, gold, among others, that have shown very promising activities together with less toxicity. The strategies that have stood out recently in the cancer field have been the use of metallodendrimers as anticancer drugs, as carriers of genetic material or as diagnostic agents. There are different parameters that have a strong influence in the final activity of these metallodendrimers, not only the nature of the metal center, its number and its oxidation state, but also the type of ligands surrounding it, the skeleton of the dendritic systems and the topology of these compounds. In general, the multivalency in metallodendrimers help to improve the activity of these compounds compared to the metallic mononuclear complexes.

In the fight against infectious diseases, dendritic scaffolds decorated with silver, copper, ruthenium, palladium or platinum complexes have shown their possibilities to act as antibacterial agents. In many cases the antibacterial activity is determined not only by the metal but also by the dendritic structure, where its nature and topology play an important role. In these cases it seems that the hydrophilic/hydrophobic balance in the different parts of these metallodendrimers is crucial for their activity. An important recent achievement is that the use of combination therapy, that is, combination of dendritic structures with traditional antibiotic drugs or other molecules with antimicrobial activity, could be a new therapeutic approach that might help to reduce resistance to conventional antibiotics.

The multivalence present in dendrimers has been exploited to design innovative heterofunctional systems bearing different types of metal complexes or other interesting molecules. Such are the advances that have taken place in recent years that theranostic systems have been designed with application in both areas diagnostic and treatment, showing very promising results that open up a very interesting line of research.

Acknowledgments

Authors thanks funding by grants from CTQ2017-86224-P (MINECO), consortiums IMMUNOTHERCAN-CM B2017/BMD-3733 and NANODENDMED II-CM ref. B2017/BMD-3703, the Comunidad de Madrid Research Talent Attraction Program 2017-T2/IND-5243, project SBPLY/17/180501/000358 Junta de Comunidades de Castilla-la Mancha (JCCM). CIBER-BBN is an initiative funded by the VI National R&D&i Plan 2008-2011, Iniciativa Ingenio 2010, the Consolider Program, and CIBER Actions and financed by the Instituto de Salud Carlos III with assistance from the European Regional Development Fund. This work has been supported partially by a EUROPARTNER: Strengthening and spreading international partnership activities of the Faculty of Biology and Environmental Protection for interdisciplinary research and innovation of the University of Lodz Programme: NAWA International Academic Partnership Programme. This publication is based upon work from COST Action CA 17140 "Cancer Nanomedicine from the Bench to the Bedside" supported by COST (European Cooperation in Science and Technology).

References

1. Sakurai H. Overview and frontier for the development of metallopharmaceutics. *J Health Sci*. 2010;56(2):129–143.
2. Slobodan N, Isidora S, Katarina R, Maja S, Jovana J. Toxic effects of metallopharmaceuticals. *Serb J Exp Clin Res*. 2017;18(3):191–194.

3. Huang Y, Havert M, Gavin D, et al. Biodistribution studies: understanding international expectations. *Mol Ther Methods Clin Dev.* 2016;3:16022.
4. Feynman R. There is plenty ol room at the bottom. In: Gilbert HD, ed. *Miniaturization.* Reinhold; 1961:282–296.
5. Sharma A, Kakkar A. Designing dendrimer and miktoarm polymer based multi-tasking nanocarriers for efficient medical therapy. *Molecules.* 2015;20(9):16987–17015.
6. Raj S, Jose S, Sumod US, Sabitha M. Nanotechnology in cosmetics: opportunities and challenges. *J Pharm Bioallied Sci.* 2012;4(3):186–193.
7. Caminade A-M. Inorganic dendrimers: recent advances for catalysis, nanomaterials, and nanomedicine. *Chem Soc Rev.* 2016;45(19):5174–5186.
8. Sakurai H, Maruyama T, Arai T. Photochemistry and aggregation behavior of triethylene glycol (TEG) terminated stilbene dendrimers. *Photochem Photobiol Sci.* 2017;16(10):1490–1494.
9. Dufès C, Uchegbu IF, Schätzlein AG. Dendrimers in gene delivery. *Adv Drug Deliv Rev.* 2005;57(15):2177–2202.
10. Gillies E, Fréchet JMJ. Dendrimers and dendritic polymers in drug delivery. *Drug Discov Today.* 2005;10(1):35–43.
11. Scherrenberg R, Coussens B, van Vliet P, et al. The molecular characteristics of poly(propyleneimine) dendrimers as studied with small-angle neutron scattering, viscosimetry, and molecular dynamics. *Macromolecules (Washington DC, US).* 1998;31(2):456–461.
12. Tomalia D, Baker H, Dewald JR, et al. A new class of polymers: Starburst-dendritic macromolecules. *Polym J (Tokyo, Jpn).* 1985;17:117–132.
13. Hawker CJ, Fréchet JMJ. Preparation of polymers with controlled molecular architecture. A new convergent approach to dendritic macromolecules. *J Am Chem Soc.* 1990;112(21):7638–7647.
14. Inoue K. Functional dendrimers, hyperbranched and star polymers. *Prog Polym Sci.* 2000;25(4):453–571.
15. García-Gallego S, Andrén OCJ, Malkoch M. Accelerated chemoselective reactions to sequence-controlled heterolayered dendrimers. *J Am Chem Soc.* 2020;142(3):1501–1509.
16. Walter MV, Malkoch M. Simplifying the synthesis of dendrimers: accelerated approaches. *Chem Soc Rev.* 2012;41(13):4593–4609.
17. Svenson S, Tomalia D. Dendrimers in biomedical applications—reflections on the field. *Adv Drug Deliv Rev.* 2006;57(15):2106–2129.
18. Yang X, Shang H, Ding C, Li J. Recent developments and applications of bioinspired dendritic polymers. *Polym Chem.* 2015;6(5):668–680.
19. Relaño-Rodríguez I, Juárez-Sánchez R, Pavicic C, Muñoz E, Muñoz-Fernández MA. Polyanionic carbosilane dendrimers as a new adjuvant in combination with latency reversal agents for HIV treatment. *J Nanobiotechnol.* 2019;17(1):69.
20. Movellan J, Urbán P, Moles E, et al. Amphiphilic dendritic derivatives as nanocarriers for the targeted delivery of antimalarial drugs. *Biomaterials.* 2014;35(27):7940–7950.
21. Florendo M, Figacz A, Srinageshwar B, et al. Use of polyamidoamine dendrimers in brain diseases. *Molecules.* 2018;23(9):E2238.
22. Meyers SR, Juhn FS, Griset AP, Luman NR, Grinstaff MW. Anionic amphiphilic dendrimers as antibacterial agents. *J Am Chem Soc.* 2008;130(44):14444–14445.
23. Sampathkumar SG, Yarema K. Dendrimers in cancer treatment and diagnosis. In: Kumar CSSR, ed. *Nanotechnologies for the Life Sciences.* Wiley-VCH; 2007:vol 7.
24. Klajnert B, Peng L, Cena V, eds. *Dendrimers in Biomedical Applications.* RSC Publishing; 2013.

25. Shingu T, Chumbalkar V, Gwak H-S, et al. The polynuclear platinum BBR3610 induces G2/M arrest and autophagy early and apoptosis late in glioma cells. *Neuro-Oncology (Cary, NC, US)*. 2010;12:1269–1277.
26. Govender P, Therrien B, Smith G. Bio-metallodendrimers—emerging strategies in metal-based drug design. *Eur J Inorg Chem*. 2012;2012(17):2853–2862.
27. Bray F, Ferlay J, Soerjomataram I, Siegel R, Torre L, Jemal A. Global cancer statistics 2018: GLOBOCAN estimates of incidence and mortality worldwide for 36 cancers in 185 countries. *Ca-Cancer J Clin*. 2018;68(6):394–424.
28. Rosenberg B, Vancamp L, Trosko JE, Mansour VH. Platinum dompounds: a new class of potent antitumour agents. *Nature (London, UK)*. 1969;222(5191):385–386.
29. Oun R, Moussa Y, Wheate N. The side effects of platinum-based chemotherapy drugs: a review for chemists. *Dalton Trans*. 2018;47(19):6645–6653.
30. Maeda H, Wu J, Sawa T, Matsumura Y, Hori K. Tumor vascular permeability and the EPR effect in macromolecular therapeutics: a review. *J Control Release*. 2000;65:271–284.
31. Howell B, Fan D, Rakesh L. Nanoscale dendrimer-platinum conjugates as multivalent antitumor drugs. In: Abd-El-Aziz AS, Carraher CE, Pittman CU, Zeldin M, eds. *Inorganic and Organometallic Macromolecules*. New York, NY: Springer; 2008:269–294.
32. Jansen BAJ, van der Zwan J, Reedijk J, den Dulk H, Brouwer J. A tetranuclear platinum compound designed to overcome cisplatin resistance. *Eur J Inorg Chem*. 1999;1999(9):1429–1433.
33. Malik N, Evagorou EG, Duncan R. Dendrimer-platinate. *Anticancer Drugs*. 1999;10(8):767–776.
34. Kapp T, Dullin A, Gust R. Platinum(II)-dendrimer conjugates: synthesis and investigations on cytotoxicity, cellular distribution, platinum release, DNA, and protein binding. *Bioconjug Chem*. 2010;21:328–337.
35. Sommerfeld NS, Hejl M, Klose MHM, et al. Low-generation polyamidoamine dendrimers as drug carriers for platinum(IV) complexes. *Eur J Inorg Chem*. 2017;2017(12):1713–1720.
36. Govender P, Sudding LC, Clavel CM, Dyson PJ, Therrien B, Smith GS. The influence of RAPTA moieties on the antiproliferative activity of peripheral-functionalised poly(salicylaldiminato) metallodendrimers. *Dalton Trans*. 2013;42(4):1267–1277.
37. Maroto-Díaz M, Elie BT, Gómez-Sal P, et al. Synthesis and anticancer activity of carbosilane metallodendrimers based on arene ruthenium(II) complexes. *Dalton Trans*. 2016;45(16):7049–7066.
38. Michlewska S, Ionov M, Maroto-Díaz M, et al. Ruthenium dendrimers against acute promyelocytic leukemia: *in vitro* studies on HL-60 cells. *Future Med Chem*. 2019;11(14):1741–1756.
39. Govender P, Riedel T, Dyson PJ, Smith GS. Higher generation cationic *N,N*-ruthenium(II)-ethylene-glycol-derived metallodendrimers: synthesis, characterization and cytotoxicity. *J Organomet Chem*. 2015;799–800:38–44.
40. Gouveia M, Figueira J, Jardim M, et al. Poly(alkylidenimine) dendrimers functionalized with the organometallic moiety $[Ru(\eta^5\text{-}C_5H_5)(PPh_3)_2]^+$ as promising drugs against cisplatin-resistant cancer cells and human mesenchymal stem cells. *Molecules*. 2018;23(6):E1471.
41. de la Mata FJ, Gómez R, Ortega P, et al. Inventors; PCT/ES2019/070416, 14 June, assignee. *Metalodendrímeros de naturaleza carbosilano conteniendo rutenio y cobre coordinados a ligandos base de Schiff, su preparación y sus usos*. 2019.
42. El Brahmi N, El Kazzouli S, Mignani SM, et al. Original multivalent copper(II)-conjugated phosphorus dendrimers and corresponding mononuclear copper(II) complexes with antitumoral activities. *Mol Pharm*. 2013;10(4):1459–1464.

43. Sanz del Olmo N, Maroto-Díaz M, Gómez R, et al. Carbosilane metallodendrimers based on copper (II) complexes: synthesis, EPR characterization and anticancer activity. *J Inorg Biochem*. 2017;177:211–218.
44. Baartzes N, Szabo C, Cenariu M, et al. In vitro antitumour activity of two ferrocenyl metallodendrimers in a colon cancer cell line. *Inorg Chem Commun*. 2018;98:75–79.
45. Giffard D, Fischer-Fodor E, Vlad C, Achimas-Cadariu P, Smith G. Synthesis and antitumour evaluation of mono- and multinuclear [2 + 1] tricarbonylrhenium(I) complexes. *Eur J Med Chem*. 2018;157:773–781.
46. Mignani S, El Brahmi N, El Kazzouli S, et al. Original multivalent gold (III) and dual gold(III)-copper(II) conjugated phosphorus dendrimers as potent antitumoral and antimicrobial agents. *Mol Pharm*. 2017;14(11):4087–4097.
47. Sudding LC, Payne R, Govender P, et al. Evaluation of the in vitro anticancer activity of cyclometalated half-sandwich rhodium and iridium complexes coordinated to naphthaldimine-based poly(propyleneimine) dendritic scaffolds. *J Organomet Chem*. 2014;774:79–85.
48. Govender P, Edafe F, Makhubela BCE, Dyson PJ, Therrien B, Smith GS. Neutral and cationic osmium(II)-arene metallodendrimers: synthesis, characterisation and anticancer activity. *Inorg Chim Acta*. 2014;409:112–120.
49. Yellepeddi VK, Vangara KK, Palakurthi S. Poly(amido)amine (PAMAM) dendrimer–cisplatin complexes for chemotherapy of cisplatin-resistant ovarian cancer cells. *J Nanopart Res*. 2013;15(9):1897.
50. Pang CT, Ammit AJ, Ong YQE, Wheate NJ. Para-Sulfonatocalix[4]arene and polyamidoamine dendrimer nanocomplexes as delivery vehicles for a novel platinum anticancer agent. *J Inorg Biochem*. 2017;176:1–7.
51. Li T, Smet M, Dehaen W, Xu H. Selenium–platinum coordination dendrimers with controlled anti-cancer activity. *ACS Appl Mater Interfaces*. 2016;8(6):3609–3614.
52. Maroto-Díaz M, Sanz del Olmo N, Muñoz-Moreno L, et al. In vitro and in vivo evaluation of first-generation carbosilane arene Ru(II)-metallodendrimers in advanced prostate cancer. *Eur Polym J*. 2019;113:229–235.
53. Wu J, Huang W, He Z. Dendrimers as carriers for siRNA delivery and gene silencing: a review. *Sci World J*. 2013;2013:630654.
54. Ionov M, Lazniewska J, Dzmitruk V, et al. Anticancer siRNA cocktails as a novel tool to treat cancer cells. Part (A). Mechanisms of interaction. *Int J Pharm (Amsterdam, Neth)*. 2015;485(1–2):261–269.
55. Michlewska S, Ionov M, Maroto M, et al. Ruthenium dendrimers as carriers for anticancer siRNA. *J Inorg Biochem*. 2018;181:18–27.
56. Rodrigues J, Jardim MG, Figueira J, Gouveia M, Tomás H, Rissanen K. Poly(alkylidenamines) dendrimers as scaffolds for the preparation of low-generation ruthenium based metallodendrimers. *New J Chem*. 2011;35(10):1938–1943.
57. Santini C, Pellei M, Gandin V, Porchia M, Tisato F, Marzano C. Advances in copper complexes as anticancer agents. *Chem Rev*. 2014;114(1):815–862.
58. Hussain A, AlAjmi MF, Rehman MT, et al. Copper(II) complexes as potential anticancer and Nonsteroidal anti-inflammatory agents: in vitro and in vivo studies. *Sci Rep*. 2019;9(1):5237.
59. Sanz del Olmo N, Carloni R, Bajo A, et al. Insight into the antitumor activity of carbosilane Cu(II)-metallodendrimers through their interaction with biological membrane models. *Nanoscale*. 2019;11(28):13330–13342.
60. Köpf-Maier P, Köpf H, Neuse EW. Ferrocenium salts—the first antineoplastic iron compounds. *Angew Chem Int Ed Engl*. 1984;23(6):456–457.
61. Neuse E. Macromolecular ferrocene compounds as cancer drug models. *J Inorg Organomet Polym Mater*. 2005;15:3–31.

62. Konkankit CC, Marker SC, Knopf KM, Wilson JJ. Anticancer activity of complexes of the third row transition metals, rhenium, osmium, and iridium. *Dalton Trans.* 2018;47(30):9934–9974.
63. Lee LC-C, Leung K-K, Lo KK-W. Recent development of luminescent rhenium(I) tricarbonyl polypyridine complexes as cellular imaging reagents, anticancer drugs, and antibacterial agents. *Dalton Trans.* 2017;46(47):16357–16380.
64. Nardon C, Boscutti G, Fregona D. Beyond platinums: gold complexes as anticancer agents. *Anticancer Res.* 2014;34:487–492.
65. Yeo CI, Ooi KK, Tiekink ERT. Gold-based medicine: a paradigm shift in anti-cancer therapy? *Molecules (Basel, Switzerland).* 2018;23(6):1410.
66. Büchel GE, Stepanenko IN, Hejl M, Jakupec MA, Keppler BK, Arion VB. En route to osmium analogues of KP1019: synthesis, structure, spectroscopic properties and antiproliferative activity of trans-[Os(IV)Cl$_4$(Hazole)$_2$]. *Inorg Chem.* 2011;50(16):7690–7697.
67. Dömötör O, Aicher S, Schmidlehner M, et al. Antitumor pentamethylcyclopentadienyl rhodium complexes of maltol and allomaltol: synthesis, solution speciation and bioactivity. *J Inorg Biochem.* 2014;134:57–65.
68. Liu Z, Sadler PJ. Organoiridium complexes: anticancer agents and catalysts. *Acc Chem Res.* 2014;47(4):1174–1185.
69. Govender P, Renfrew AK, Clavel CM, Dyson PJ, Therrien B, Smith GS. Antiproliferative activity of chelating *N,O*- and *N,N*-ruthenium(II) arene functionalised poly(propyleneimine) dendrimer scaffolds. *Dalton Trans.* 2011;40(5):1158–1167.
70. Cebrián-Losantos B, Krokhin AA, Stepanenko IN, et al. Osmium NAMI-A analogues: synthesis, structural and spectroscopic characterization, and antiproliferative properties. *Inorg Chem.* 2007;46(12):5023–5033.
71. Govender P, Lemmerhirt H, Hutton AT, Therrien B, Bednarski PJ, Smith GS. First- and second-generation heterometallic dendrimers containing ferrocenyl–ruthenium(II)–arene motifs: synthesis, structure, electrochemistry, and preliminary cell proliferation studies. *Organometallics.* 2014;33(19):5535–5545.
72. Govender P, Riedel T, Dyson PJ, Smith GS. Regulating the anticancer properties of organometallic dendrimers using pyridylferrocene entities: synthesis, cytotoxicity and DNA binding studies. *Dalton Trans.* 2016;45(23):9529–9539.
73. Refat MS, El-Deen IM, Grabchev I, Anwer ZM, El-Ghol S. Spectroscopic characterizations and biological studies on newly synthesized Cu^{2+} and Zn^{2+} complexes of first and second generation dendrimers. *Spectrochim Acta A.* 2009;72(4):772–782.
74. Staneva D, Vasileva-Tonkova E, Makki MSI, et al. Synthesis and spectral characterization of a new PPA dendrimer modified with 4-bromo-1,8-naphthalimide and *in vitro* antimicrobial activity of its Cu(II) and Zn(II) metal complexes. *Tetrahedron.* 2015;71(7):1080–1087.
75. Grabchev I, Staneva D, Vasileva-Tonkova E, et al. Antimicrobial and anticancer activity of new poly(propyleneamine) metallodendrimers. *J Polym Res.* 2017;24:210.
76. Grabchev I, Yordanova S, Vasileva-Tonkova E, Bosch P, Stoyanov S. Poly(propylenamine) dendrimers modified with 4-amino-1,8-naphthalimide: synthesis, characterization and in vitro microbiological tests of their Cu(II) and Zn(II) complexes. *Inorg Chim Acta.* 2015;438:179–188.
77. Staneva D, Grabchev I, Bosch P, Vasileva-Tonkova E, Kukeva R, Stoyanova R. Synthesis, characterisaion and antimicrobial activity of polypropylenamine metallodendrimers modified with 1,8-naphthalimides. *J Mol Struct.* 2018;1164:363–369.
78. Michlewska S, Kubczak M, Maroto-Díaz M, et al. Synthesis and characterization of FITC labelled ruthenium dendrimer as a prospective anticancer drug. *Biomolecules.* 2019;9(9):E411.

79. Carloni R, Sanz del Olmo N, Ortega P, et al. Exploring the interactions of ruthenium (II) carbosilane metallodendrimers and precursors with model cell membranes through a dual spin-label spin-probe technique using EPR. *Biomolecules.* 2019;9(10):E540.
80. Sağır T, Işık S, Şenel M. Ferrocene incorporated PAMAM dendrons: synthesis, characterization, and anti-cancer activity against AGS cell line. *Med Chem Res.* 2013; 22(10):4867–4876.
81. Maroto-Díaz M, Sanz del Olmo N, García-Gallego S, Gómez R, Ortega P, de la Mata FJ. Synthesis and structural characterization of carbosilane ruthenium(II) metallodendrons containing cymene units. *J Organomet Chem.* 2019;901:120942.
82. Kherlopian AR, Song T, Duan Q, et al. A review of imaging techniques for systems biology. *BMC Syst Biol.* 2008;2:74.
83. Ye M, Qian Y, Tang J, Hu H, Sui M, Shen Y. Targeted biodegradable dendritic MRI contrast agent for enhanced tumor imaging. *J Control Release.* 2013;169(3):239–245.
84. McNelles SA, Knight SD, Janzen N, Valliant JF, Adronov A. Synthesis, radiolabeling, and *in vivo* imaging of PEGylated high-generation polyester dendrimers. *Biomacromolecules.* 2015;16(9):3033–3041.
85. Parat A, Kryza D, Degoul F, et al. Radiolabeled dendritic probes as tools for high *in vivo* tumor targeting: application to melanoma. *J Mater Chem B.* 2015;3(12):2560–2571.
86. Chen Q, Qianqian Y, Liu Y, et al. Multifunctional selenium nanoparticles: chiral selectivity of delivering MDR-siRNA for reversal of multidrug resistance and real-time biofluorescence imaging. *Nanomedicine (NY, US).* 2015;11(7):1773–1784.
87. Fan Y, Zhang J, Shi M, et al. Poly(amidoamine) dendrimer-coordinated copper(II) complexes as a theranostic nanoplatform for the radiotherapy-enhanced magnetic resonance imaging and chemotherapy of tumors and tumor metastasis. *Nano Lett.* 2019;19(2):1216–1226.
88. Heymann DL, Prentice T, Reinders LT. *The World Health Report 2007: A Safer Future: Global Public Health Security in the 21st Century.* Geneva, Switzerland: WHO; 2007.
89. Denyer SP, Gorman SP, Sussman M. *Microbial Biofilms: Formation and Control.* London, UK: Wiley-Blackwell Scientific Publications LTD; 1993.
90. García-Gallego S, Franci G, Falanga A, et al. Function oriented molecular design: dendrimers as novel antimicrobials. *Molecules.* 2017;22(10):E1581.
91. Castonguay A, Ladd E, van de Ven TGM, Kakkar A. Dendrimers as bactericides. *New J Chem.* 2012;36(2):199–204.
92. Tülü M, Ertürk AS. Bobbarala V, ed. *Dendrimers as Antibacterial Agents.* IntechOpen; 2012.
93. Lind TK, Polcyn P, Zielinska P, Cárdenas M, Urbanczyk-Lipkowska Z. On the antimicrobial activity of various peptide-based dendrimers of similar architecture. *Molecules.* 2015;20(1):738–753.
94. Rosa Borges A, Schengrund CL. Dendrimers and antivirals: a review. *Curr Drug Targets Infect Disord.* 2005;5(3):247–254.
95. Jiménez JL, Pion M, de la Mata FJ, et al. Dendrimers as topical microbicides with activity against HIV. *New J Chem.* 2012;36(2):299–309.
96. Scott P. Antimicrobial metallodrugs. In: Simpson DH, Lo KK-W, eds. *Inorganic and Organometallic Transition Metal Complexes With Biological Molecules and Living Cells.* Academic Press; 2017:205–243.
97. Balogh L, Swanson DR, Tomalia DA, Hagnauer GL, McManus AT. Dendrimer–silver complexes and nanocomposites as antimicrobial agents. *Nano Lett.* 2001;1(1):18–21.
98. Suleman N, Kalhapure RS, Mocktar C, Rambharose S, Singh M, Govender T. Silver salts of carboxylic acid terminated generation 1 poly (propyl ether imine) (PETIM) dendron and dendrimers as antimicrobial agents against *S. aureus* and MRSA. *RSC Adv.* 2015;5(44):34967–34978.

99. Llamazares C, Sanz Del Olmo N, Ortega P, Gomez R, Soliveri J, de la Mata FJ, García-Gallego S, Copa-Patiño JL. Antibacterial effect of carbosilane metallodendrimers in planktonic cells of Gram-positive and Gram-negative bacteria and Staphylococcus aureus biofilm. *Biomolecules.* 2019;9(9):E405.
100. Ahamad T, Mapolie SF, Alshehri SM. Synthesis and characterization of polyamide metallodendrimers and their anti-bacterial and anti-tumor activities. *Med Chem Res.* 2012;21(8):2023–2031.
101. Maillard JY, Hartemann P. Silver as an antimicrobial: facts and gaps in knowledge. *Crit Rev Microbiol.* 2013;39(4):373–383.
102. Panáček A, Kvítek L, Smékalová M, et al. Bacterial resistance to silver nanoparticles and how to overcome it. *Nat Nanotechnol.* 2018;13(1):65–71.
103. Svenson S. The dendrimer paradox–high medical expectations but poor clinical translation. *Chem Soc Rev.* 2015;44(12):4131–4144.
104. Vincent M, Hartemann P, Engels-Deutsch M. Antimicrobial applications of copper. *Int J Hyg Environ Health.* 2016;219(7, pt A):585–591.
105. Grass G, Rensing C, Solioz M. Metallic copper as an antimicrobial surface. *Appl Environ Microbiol.* 2011;77(5):1541–1547.
106. Dalecki AG, Crawford CL, Wolschendorf F. Copper and antibiotics: discovery, modes of action, and opportunities for medicinal applications. In: Poole RK, ed. *Advances in Microbial Physiology.* Academic Press; 2017:193–260. vol 70.
107. Li F, Collins JG, Keene FR. Ruthenium complexes as antimicrobial agents. *Chem Soc Rev.* 2015;44(8):2529–2542.
108. Brabec V, Kasparkova J. Ruthenium coordination compounds of biological and biomedical significance. DNA binding agents. *Coord. Chem Rev.* 2018;376:75–94.
109. Katarzyna M, Anna S, Zielinska-Blizniewska H, Ireneusz M. An evaluation of the antioxidant and anticancer properties of complex compounds of copper (II), platinum (II), palladium (II) and ruthenium (III) for use in cancer therapy. *Mini-Rev Med Chem.* 2018;18(16):1373–1381.
110. Abu-Surrah AS, Kettunen M. Platinum group antitumor chemistry: design and development of new anticancer drugs complementary to cisplatin. *Curr Med Chem.* 2006;13(11):1337–1357.
111. Bosch P, Staneva D, Vasileva-Tonkova E, et al. New poly(propylene imine) dendrimer modified with acridine and its Cu(II) complex: synthesis, characterization and antimicrobial activity. *Materials.* 2019;12(18):3020.
112. Grabchev I, Vasileva-Tonkova E, Staneva D, Bosch P, Kukeva R, Stoyanova R. Impact of Cu(II) and Zn(II) ions on the functional properties of new PAMAM metallodendrimers. *New J Chem.* 2018;42(10):7853–7862.
113. Lozano-Cruz T, Ortega P, Batanero B, et al. Synthesis, characterization and antibacterial behavior of water-soluble carbosilane dendrons containing ferrocene at the focal point. *Dalton Trans.* 2015;44(44):19294–19304.
114. Lallo da Silva B, Abucafy MP. Relationship between structure and antimicrobial activity of zinc oxide nanoparticles: an overview. *Int J Nanomedicine.* 2019;2019(14):9395–9410.
115. Denyer SP, Maillard J-Y. Cellular impermeability and uptake of biocides and antibiotics in Gram-negative bacteria. *J Appl Microbiol.* 2002;92(s1):35S–45S.
116. Mecke A, Uppuluri S, Sassanella TM, et al. Direct observation of lipid bilayer disruption by poly(amidoamine) dendrimers. *Chem Phys Lipids.* 2004;132(1):3–14.
117. Acevedo-Morantes CY, Meléndez E, Singh SP, Ramírez-Vick JE. Cytotoxicity and reactive oxygen species generated by ferrocenium and ferrocene on MCF7 and MCF10A cell lines. *J Cancer Sci Ther.* 2012;4:271–275.

118. Scutaru D, Mazilu I, Vâță M, et al. Heterodisubstituted derivatives of ferrocene. Ferrocene-containing penicillins and cephalosporins. *J Organomet Chem*. 1991; 401(1):87–90.
119. Sepúlveda-Crespo D, Gómez R, de la Mata FJ, Jiménez JL, Muñoz-Fernández MA. Polyanionic carbosilane dendrimer-conjugated antiviral drugs as efficient microbicides: recent trends and developments in HIV treatment/therapy. *Nanomedicine (NY, US)*. 2015;11(6):1481–1498.
120. Rasines B, Sánchez-Nieves J, Maiolo M, et al. Synthesis, structure and molecular modelling of anionic carbosilane dendrimers. *Dalton Trans*. 2012;41(41):12733–12748.
121. Arnáiz E, Vacas-Córdoba E, Galán M, et al. Synthesis of anionic carbosilane dendrimers via "click chemistry" and their antiviral properties against HIV. *J Polym Sci A Polym Chem*. 2014;52(8):1099–1112.
122. Galán M, Sánchez Rodríguez J, Jiménez JL, et al. Synthesis of new anionic carbosilane dendrimers via thiol–ene chemistry and their antiviral behaviour. *Org Biomol Chem*. 2014;12(20):3222–3237.
123. Price CF, Tyssen D, Sonza S, et al. SPL7013 Gel (VivaGel(R)) retains potent HIV-1 and HSV-2 inhibitory activity following vaginal administration in humans. *PLoS One*. 2011;6(9):e24095.
124. Galán M, Sánchez-Rodríguez J, Cangiotti M, et al. Antiviral properties against HIV of water soluble copper carbosilane dendrimers and their EPR characterization. *Curr Med Chem*. 2012;19(29):4984–4994.
125. García-Gallego S, Cangiotti M, Fiorani L, et al. Anionic sulfonated and carboxylated PPI dendrimers with the EDA core: synthesis and characterization of selective metal complexing agents. *Dalton Trans*. 2013;42(16):5874–5889.
126. García-Gallego S, Díaz L, Jiménez JL, Gómez R, de la Mata FJ, Muñoz-Fernández MÁ. HIV-1 antiviral behavior of anionic PPI metallo-dendrimers with EDA core. *Eur J Med Chem*. 2015;98:139–148.
127. Biot C, Glorian G, Maciejewski LA, Brocard JS. Synthesis and antimalarial activity *in vitro* and *in vivo* of a new ferrocene-chloroquine analogue. *J Med Chem*. 1997;40(23): 3715–3718.
128. Khanye SD, Gut J, Rosenthal PJ, Chibale K, Smith GS. Ferrocenylthiosemicarbazones conjugated to a poly(propyleneimine) dendrimer scaffold: synthesis and *in vitro* antimalarial activity. *J Organomet Chem*. 2011;696(21):3296–3300.
129. Müller IB, Das Gupta R, Lüersen K, Wrenger C, Walter RD. Assessing the polyamine metabolism of *Plasmodium falciparum* as chemotherapeutic target. *Mol Biochem Parasitol*. 2008;160(1):1–7.
130. Stringer T, Taylor D, de Kock C, et al. Synthesis, characterization, antiparasitic and cytotoxic evaluation of thioureas conjugated to polyamine scaffolds. *Eur J Med Chem*. 2013;69:90–98.
131. Baartzes N, Stringer T, Okombo J, et al. Mono- and polynuclear ferrocenylthiosemicarbazones: synthesis, characterisation and antimicrobial evaluation. *J Organomet Chem*. 2016;819:166–172.
132. Stringer T, Wiesner L, Smith GS. Ferroquine-derived polyamines that target resistant *Plasmodium falciparum*. *Eur J Med Chem*. 2019;179:78–83.
133. Chellan P, Land KM, Shokar A, et al. Synthesis and evaluation of new polynuclear organometallic Ru(II), Rh(III) and Ir(III) pyridyl ester complexes as *in vitro* antiparasitic and antitumor agents. *Dalton Trans*. 2014;43(2):513–526.

CHAPTER TWO

Sigma-bond activation reactions induced by unsaturated Os(IV)-hydride complexes

Miguel A. Esteruelas*, Montserrat Oliván, Enrique Oñate
Departamento de Química Inorgánica, Instituto de Síntesis Química y Catálisis Homogénea (ISQCH), Centro de Innovación en Química Avanzada (ORFEO-CINQA), Universidad de Zaragoza-CSIC, Zaragoza, Spain
*Corresponding author: e-mail address: maester@unizar.es

Contents

1. Introduction	54
2. H—H bond activation	56
3. B—H bond activation	58
4. C—H bond activation	61
4.1 C(sp³)-H bond activation of saturated heterocycles	61
4.2 Cycloolefin-promoted C(sp³)-H bond activation of phosphines	62
4.3 C(sp²)-H bond activation of olefins, enynes, acyclic dienes, allenes, and allenedienes	64
4.4 C(sp²)-H bond activation and subsequent C—C bond formation	69
4.5 Formation of N—H wingtip NHC ligands by tautomerization of heterocycles	71
5. C—C bond activation	73
6. C—O bond activation	76
7. C-halogen bond activation	77
8. Si—H bond activation	78
9. Sn—H bond activation	79
10. N—H bond activation	83
11. O—H bond activation	87
12. O—N bond activation: Azavinylidene compounds	89
13. Rupture of the oxygen molecule	97
14. Cl—H bond activation	98
15. Conclusions	98
Acknowledgments	100
References	100

Advances in Organometallic Chemistry, Volume 74
ISSN 0065-3055
https://doi.org/10.1016/bs.adomc.2020.04.002

1. Introduction

The σ-bond activation reactions promoted by transition metal complexes can be grouped in three classes: oxidative addition, also denoted as homolytic cleavage, heterolytic cleavage, and σ-bond metathesis. The first step for all them is the coordination of that bond to the transition metal, to form a $L_nM(\eta^2$-X-Y$)$ σ-complex (Scheme 1). Because the interaction between the σ-bond and the metal center is usually weak, the starting complex must be unsaturated. Saturated species prevent the σ-bond activation processes. The coordination implies σ-donation from the σ(XY) orbital to empty orbitals of the metal and back-bonding from the metal to the σ* (XY) orbital.[1] Basic metal centers increase the back-donation, favoring the oxidative addition of the X-Y bond (**a**), while acidic metal centers enhance the σ-donation to the metal for promoting the heterolytic cleavage of the X-Y bond (**b**). The cation acceptor can be an external base or a group with free electron pairs in the sphere of the metal center.[2] This latter case is denoted as activation via *"metal-ligand cooperation"* by some authors, which have proposed it as a fundamental step in relevant catalytic reactions,[3–8] although recent results from DFT calculations are challenging the proposal.[9–16] The third form of cleavage, the σ-bond metathesis, is described as a concerted process via a four-center, four-electron transition state (**c**).[17,18] In contrast to **a**, activations **b** and **c** avoid the formal 2-e oxidation of the metal.

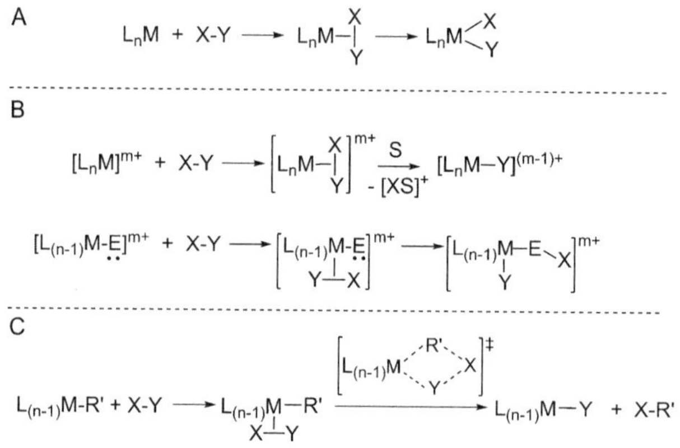

Scheme 1 Modes of σ-bond activation reactions.

Unsaturated osmium(IV)-hydride complexes are six-coordinate d^4 species, which bear at least a π-donor ligand and are generally stabilized by bulky phosphines. They undergo distortion from the octahedral geometry to destabilize one orbital from the t_{2g} set and simultaneously to stabilize some occupied orbital. Thus, they prefer to be diamagnetic.[19,20] The distortions mainly give rise to two geometries type (Chart 1). The structures of complexes $OsH_2Cl_2(P^iPr_3)_2$ and $OsH_3Cl(P^iPr_3)_2$ are representatives of them. In the beginning, the structure of $OsH_2Cl_2(P^iPr_3)_2$ was viewed as a distorted square antiprism of ideal D_{4d} symmetry with two vacant coordination sites.[21] Later, it was described as a bicapped tetrahedron or as a trigonal prism.[22] The structure of $OsH_3Cl(P^iPr_3)_2$ essentially displays C_{2v} symmetry.[23] This distortion partially cancels the electron deficiency at the metal center, which receives electron density through the σ-bonds with the hydride ligands and additionally from a lone pair of the halide via a π bond.[24]

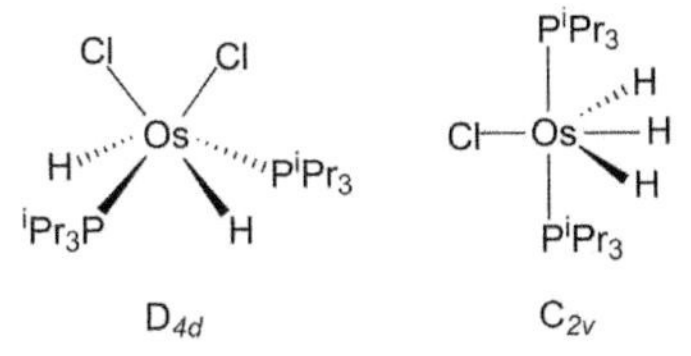

Chart 1 Geometries of unsaturated osmium(IV)-hydride complexes.

These osmium(IV) compounds induce the activation of H—H, H—B, C—H, Si—H, N—H, or C—X (X=Cl, Br, I) among other σ-bonds and some multiple bonds such as that of molecular oxygen. The induction occurs because they promote the cleavage of the bond by themselves or because lose H_2 or HCl to generate 14-valence electrons transitory osmium(II) intermediates ($OsCl_2(PR_3)_2$ or $OsHX(PR_3)_2$), which are in many cases the real active species. In this context, the choice of the solvent of the reaction is crucial in the planning of the activation.[25] The reaction solvent determines the stability of the precursor and decides the formation of the osmium(II) intermediate, which governs the activation.

The relevance of the σ-bond cleavage in the modern chemistry along with the electronic (fairly acidic metal center) and structural (unusual geometries among the platinum group metal derivatives) peculiarities of the unsaturated osmium(IV)-hydride complexes convert the reactions of σ-bond activation, induced by these compounds, in a fascinating subject in organometallic chemistry. This review contextualizes the reactions of this class performed between 1991 and 2019.

2. H—H bond activation

Complexes OsH$_2$X$_2$(P^{i}Pr$_3$)$_2$ (X=Cl (**1**), Br (**2**), I (**3**)) coordinate a hydrogen molecule in dichloromethane, to give the dihydride-elongated dihydrogen derivatives OsH$_2$X$_2$(η^2-H$_2$)(P^{i}Pr$_3$)$_2$ (X=Cl (**4**), Br (**5**), I (**6**)).[26] Crystallographic and spectroscopic evidences suggest that the donor atoms around the metal center form a dodecahedron. One of the perpendicular trapezoidal planes contains the phosphorous atoms of the phosphines and the hydrogen molecule, whereas the halides and the hydride ligands lie in the other one.[27] In agreement with the acidity of the metal center in these compounds, the coordinated hydrogen molecule undergoes heterolytic cleavage releasing HX, to give the related polyhydrides OsH$_3$X(η^2-H$_2$)(P^{i}Pr$_3$)$_2$ (X=Cl (**7**), Br (**8**), I (**9**)), under hydrogen atmosphere.[23,26] The replacement of a halide by hydride reduces the back-donation power of the metal center to the σ* orbital of the coordinated hydrogen molecule. As a consequence, the latter forms a Kubas-type dihydrogen ligand in **7–9**. Its dissociation leads to the unsaturated trihydrides OsH$_3$X(P^{i}Pr$_3$)$_2$ (X=Cl (**10**), Br (**11**), I (**12**)), according to Scheme 2. The hydride ligands of **10–12** display quantum exchange coupling in addition to thermal exchange.[23]

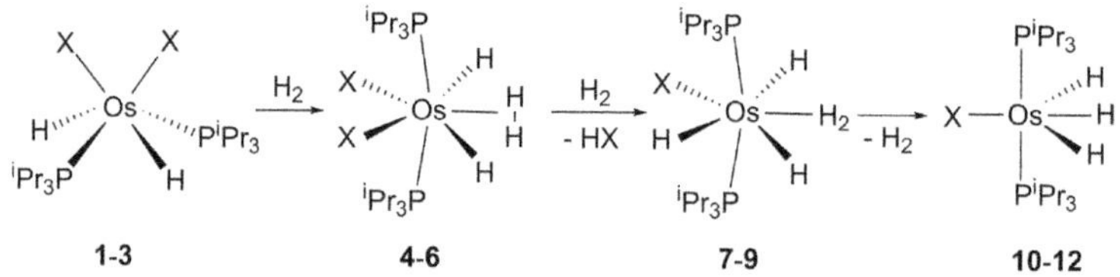

X = Cl (**1, 4, 7, 10**), Br (**2, 5, 8, 11**), I (**3, 6, 9, 12**)

Scheme 2 Reactions of **1–3** with molecular dihydrogen.

Pincer ligands also stabilize unsaturated osmium(IV)-hydride complexes, which activate molecular hydrogen. Reaction of [Et$_4$N]$_2$OsCl$_6$ with the diphosphine 2,6-(CH$_2$P^{t}Bu$_2$)$_2$C$_6$H$_4$, in 2-propanol, at 100 °C, under hydrogen atmosphere leads to the dihydride derivative OsH$_2$Cl{κ^3-P,C,P-[2,6-(CH$_2$P^{t}Bu$_2$)$_2$C$_6$H$_3$]} (**13**) in 76% yield. This compound, which displays a similar structure to that of the trihydride **10**, reacts with hydrogen in methylcyclohexane-d$_{14}$ to afford two complexes (Scheme 3). The kinetic bis(elongated dihydrogen) product OsCl(η^2-H$_2$)$_2${κ^3-P,C,P-[2,6-(CH$_2$P^{t}Bu$_2$)$_2$C$_6$H$_3$]} (**14**) is formed below −70 °C. On warming to −30 °C, one of the coordinated hydrogen molecules undergoes homolytic cleavage to give the *cis*-dihydride-elongated dihydrogen derivative OsH$_2$Cl(η^2-H$_2$){κ^3-P,C,P-[2,6-(CH$_2$P^{t}Bu$_2$)$_2$C$_6$H$_3$]} (**15**). Although for this compound two different pentagonal bipyramidal structures can be considered, H$_2$ *trans* to Cl and H$_2$ *trans* to phenyl, it is more likely that the hydrogen molecule is situated *trans* to the ligand possessing the strongest *trans*-influence, while the dihydride unit is more likely to be *trans* to the ligand of weak *trans*-influence and good π-donor chloride.[28]

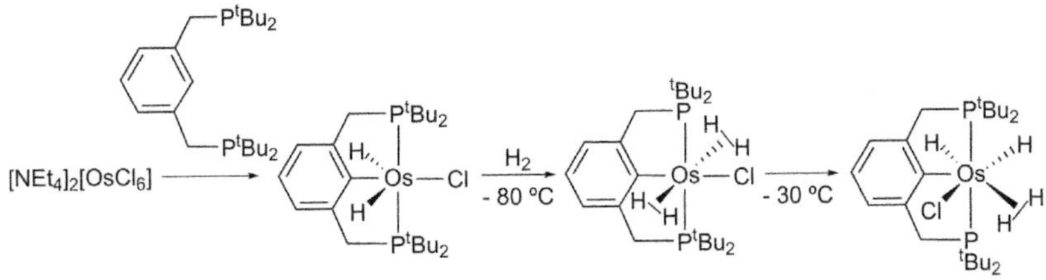

Scheme 3 Reaction of the P,C,P-pincer complex **14** with molecular hydrogen.

[Et$_4$N]$_2$OsCl$_6$ also reacts with the aminediphosphine HN(C$_2$H$_4$P^{i}Pr$_2$)$_2$ (Scheme 4). The reaction affords the trihydride derivative OsH$_3$Cl{κ^3-P,N,P-[HN(C$_2$H$_4$P^{i}Pr$_2$)$_2$]} (**16**), in 2-propanol or 2-pentanol. Upon treatment with K^{t}BuO or NaN(SiMe$_3$)$_2$, complex **16** undergoes dehydrochlorination to afford the P,N,P-pincer counterpart of **10** OsH$_3${κ^3-P,N,P-[N(C$_2$H$_4$P^{i}Pr$_2$)$_2$]} (**17**), which heterolytically activates molecular hydrogen. The activation leads to the tetrahydride OsH$_4${κ^3-P,N,P-[HN(C$_2$H$_4$P^{i}Pr$_2$)$_2$]} (**18**). The latter is an efficient catalyst for the hydrogen transfer from alcohols to ketones, the acceptorless dehydrogenative coupling of primary alcohols to esters and the monoalkylation of primary amines with primary alcohols.[29]

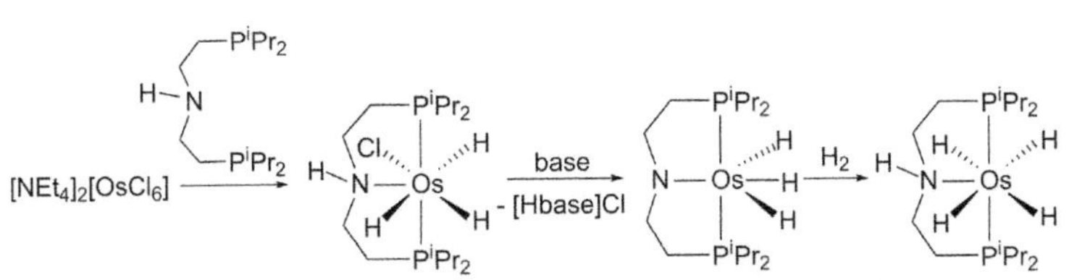

Scheme 4 Preparation of the P,N,P-pincer complex **17** and its reaction with molecular hydrogen.

3. B—H bond activation

Trihydride **10** shows a marked tendency to release H_2 to afford the 14-valence electrons species $OsHCl(P^iPr_3)_2$ (**A**), which is also formed from the dihydride-dichloride $OsH_2Cl_2(P^iPr_3)_2$ (**1**) by reductive loss of HCl. This unsaturated species is responsible for activating the B—H bond of boranes and amine-boranes using **10** or **1** (Scheme 5).

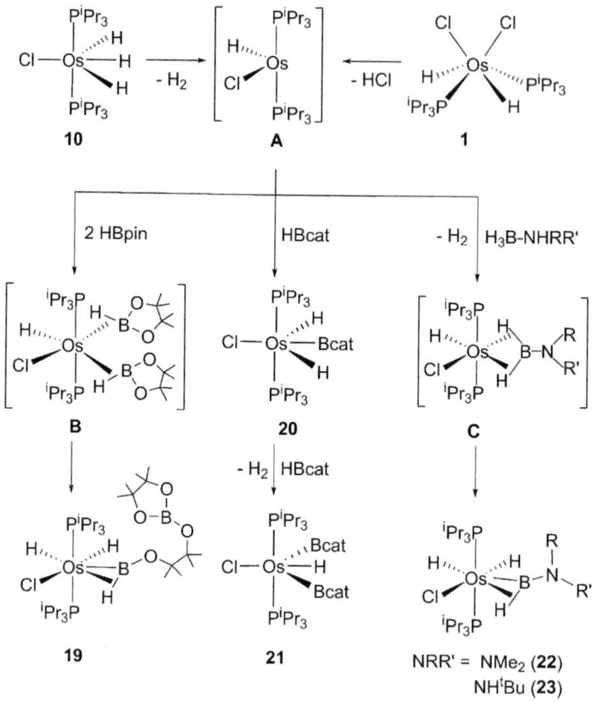

Scheme 5 B—H bond activation reactions promoted by **1** and **10**.

The reaction of **A** with pinacolborane (HBpin) yields the borinium derivative $OsH_2Cl(\eta^2\text{-H-BOCMe}_2CMe_2OBpin)(P^iPr_3)_2$ (**19**)[30]; its formation seems to take place via intermediate $OsHCl(HBpin)_2(P^iPr_3)_2$ (**B**), which can be generated by coordination of the B—H bonds of two molecules of HBpin to the osmium atom of **A**. Once coordinated the boranes, the heterolytic B—H bond activation of one of them gives **19**, using an oxygen atom of the other one as an external base. In contrast to HBpin, catecholborane (HBcat) affords boryl derivatives. The dihydride compound

OsH$_2$Cl(Bcat)(P^{i}Pr$_3$)$_2$ (**20**) is initially formed, as a result of the homolytic addition of the B—H bond to the osmium atom of **A**. Complex **20** reacts with a second molecule of HBcat to give the hydride-bis(boryl) derivative OsHCl(Bcat)$_2$(P^{i}Pr$_3$)$_2$ (**21**) and molecular hydrogen.[31] Amine-boranes undergo dehydrogenation in the presence of **A** to form aminoboranes. The hydride transfer from the aminoboranes to its metal center leads to the aminoborinium compounds OsH$_2$Cl(η^2-H-BNRR′)(P^{i}Pr$_3$)$_2$ (NRR′ = NMe$_2$ (**22**), NH^{t}Bu (**23**)),[30] most probably via the bis(σ-B-H) intermediates OsHCl(η^2,η^2-H$_2$BNRR′)(P^{i}Pr$_3$)$_2$ (**C**).

DFT calculations suggest that the major contribution to the interaction between the metal fragment and the borinium ligands is electrostatic (57%). This reveals a high degree of polarization for the osmium-borinium bond, which agrees well with the high electronegativity of the osmium atom and suggests a significant positive partial charge on the ligand in a consistent manner with the borinium denomination and the calculated hybridization at the boron atom (sp$^{1.3-1.6}$). The orbital term of the interaction is the result of three contributions: two σ-interactions and a π-interaction. The most important σ-interaction, accounting for the 61% of the total ΔE_{orb} value, involves a charge flow from the metal center to the boron atom; the other one is a donor-acceptor interaction from the doubly occupied σ(B—H) orbital to the metal fragment (about 10% of ΔE_{orb}). The π-interaction is also significant: its associated energy for the deformation density corresponds to about 21% of the total orbital attractions.[32]

The activation of HBpin and HBcat promoted by an unsaturated cationic osmium(IV)-dihydride generated in situ has been also reported (Scheme 6). Trihydride OsH$_3$Cl{κ^3-P,O,P-[xant(P^{i}Pr$_2$)$_2$]} (**24**; xant(P^{i}Pr$_2$)$_2$ = 9,9-dimethyl-4,5-bis(diisopropylphosphino)xanthene) adds the proton of HBF$_4$ to afford the compressed dihydride-dihydrogen complex [OsH$_2$(η^2-H$_2$)Cl{κ^3-P,O,P-[xant(P^{i}Pr$_2$)$_2$]}]BF$_4$ (**25**), which is stable under hydrogen atmosphere. Under argon, it dissociates the coordinated hydrogen molecule to afford the 16-valence electrons osmium(IV)-dihydride [OsH$_2$Cl{κ^3-P,O,P-[xant(P^{i}Pr$_2$)$_2$]}]BF$_4$ (**D**), which rapidly reaches an equilibrium with the dimer [(OsH$_2${κ^3-P,O,P-[xant(P^{i}Pr$_2$)$_2$]})$_2$(μ-Cl)$_2$](BF$_4$)$_2$ (**26**). The unsaturated dihydride **D** heterolytically activates the B—H bond of HBpin and HBcat using the [BF$_4$]$^-$ anion as an external base. The release of molecular hydrogen generates a square-planar pyramidal metal fragment OsHCl{κ^3-P,O,P-[xant(P^{i}Pr$_2$)$_2$]} (**E**), which coordinates the B—H bond of a second molecule of borane to yield OsHCl(η^2-H-BR$_2$){κ^3-P,O,P-[xant(P^{i}Pr$_2$)$_2$]} (BR$_2$ = Bpin (**27**), Bcat (**28**)).[33]

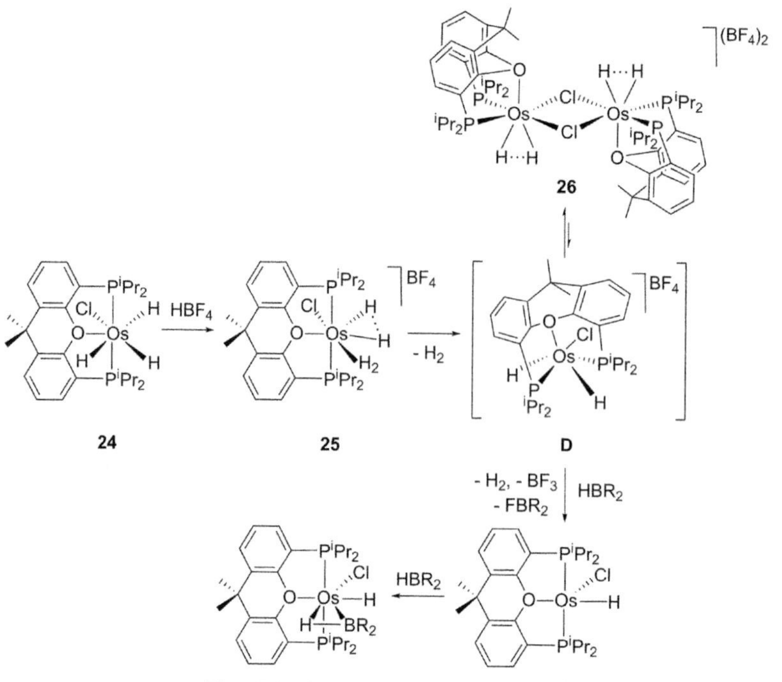

Scheme 6 B—H bond activation of boranes promoted by a P,O,P-pincer system.

The interaction between the metal center and the coordinated B—H bond in **27** and **28** has been analyzed, from X-ray diffraction and theoretical points of view, and compared with that in the dihydrides OsH$_2$(η^2-H-BR$_2$){κ^3-P,O,P-[xant(P^{i}Pr$_2$)$_2$]} (BR$_2$ = Bpin (**29**), Bcat (**30**)). X-ray diffraction analysis and DFT-optimized structures of **27** and **28** reveal distances between the coordinated B and H atoms of the borane in the range of 1.6–1.7 Å, which support an elongated σ-borane character for these compounds. Atoms in molecules (AIM) analysis displays a triangular topology for the OsHB unit involving Os—B, Os—H, and B—H bond critical points and a ring critical point.[34] In contrast to **27** and **28**, the structures of the dihydrides **29** and **30** show B—H distances for the coordinated B—H bond in the range of 1.4–1.5 Å, which indicate character of σ-borane derivatives. In agreement with this, the AIM analysis for the OsHB unit only displays Os—B and B—H bond critical points; i.e., they lack a similar topology.[33,34] The electron withdrawing ability of the chloride ligand increases the donation from the σ(B—H) orbital to the osmium atom, whereas its π-donor power favors the d(Os)-to-σ*(B—H) back-donation.

4. C—H bond activation
4.1 C(sp³)-H bond activation of saturated heterocycles

Trihydride complex **10** doubly dehydrogenates a carbon α to the oxygen atom of tetrahydrofuran, in the presence of *tert*-butylethylene, to afford the carbene derivative OsHCl[=CO(CH$_2$)$_3$](P^{i}Pr$_3$)$_2$ (**31**) and the alkane. In benzene, under 1 atm of hydrogen, complex **31** gives the *trans*-hydride-dihydrogen OsHCl(η^2-H$_2$)[=CO(CH$_2$)$_3$](P^{i}Pr$_3$)$_2$ (**32**) in equilibrium with the trihydride OsH$_3$Cl[=CO(CH$_2$)$_3$](P^{i}Pr$_3$)$_2$ (**33**). The relative population of these species (95:5) at 20 °C indicates a thermodynamic preference for the *trans*-hydride-dihydrogen redox isomer (Scheme 7).[35]

Scheme 7 C(sp³)-H bond activation of tetrahydrofuran promoted by **10**.

Trihydride **10** also reacts with pyrrolidine (Scheme 8). However, there are significant differences between this activation and that shown in Scheme 7. In contrast to tetrahydrofuran, pyrrolidine initially coordinates to the metal center as amine to yield the saturated trihydride OsH$_3$Cl(NHC$_4$H$_8$)(P^{i}Pr$_3$)$_2$ (**34**).

Scheme 8 C(sp³)-H bond activation of pyrrolidine promoted by **10** and **1**.

The latter slowly loses molecular hydrogen to give the 16-valence electrons intermediate OsHCl(NHC$_4$H$_8$)(P^{i}Pr$_3$)$_2$ (35), which can be also formed by dehydrochlorination of 1 with 2 equiv. of the heterocycle. The metal center of 35 promotes a double C$_\alpha$-H bond activation of the coordinated pyrrolidine, which in a sequential manner generates complexes OsH$_2$Cl[CH(NH)C$_3$H$_6$](P^{i}Pr$_3$)$_2$ (36) and OsH$_3$Cl[=CNH(CH$_2$)$_3$](P^{i}Pr$_3$)$_2$ (37).

4.2 Cycloolefin-promoted C(sp^3)-H bond activation of phosphines

1,5-Cyclooctadiene (COD), 2,5-norbornadiene (NBD), and tetrafluorobenzobarrelene (TFB) promote the dehydrogenation of a phosphine ligand at 1. The C(sp^3)-H activations are accompanied by C—C bond activation and C—C and C—P bond formation reactions, depending upon the experimental conditions and the diolefin.[36]

Complex 1 reacts with 2 equiv. of COD, in refluxing toluene, to give the isopropenylphosphine derivative OsCl$_2$(η^4-COD){η^3-P[C(CH$_3$)=CH$_2$]^{i}Pr$_2$} (38), as a consequence of the hydrogen transfer from an isopropyl substituent of the phosphine to a molecule of the diolefin, which yields cyclooctene. In addition, the coordination of the other diene molecule and the release of the other triisopropylphosphine ligand and molecular hydrogen take place (Scheme 9).

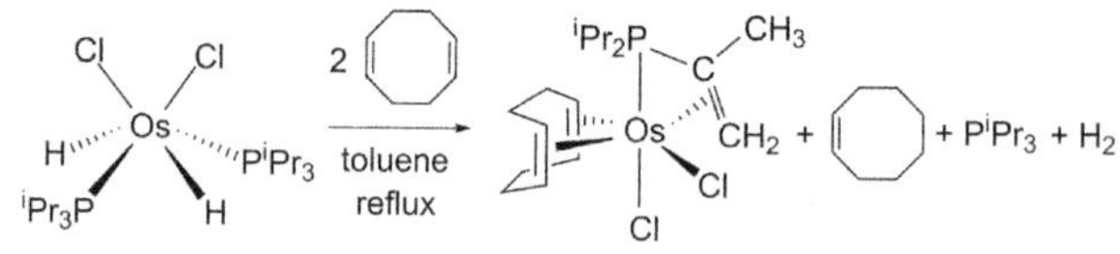

Scheme 9 C(sp^3)-H bond activation of triisopropylphosphine promoted by 1,5-cyclooctadiene.

The C(sp^3)-H bond activation of the diene is favored with regard to that of the phosphine at lower temperature. Thus, at 85 °C, treatment of toluene solutions of 1 with COD leads to a mixture of 38 and a new compound, OsHCl$_2$(η^4-C$_8$H$_{11}$P^{i}Pr$_3$)(P^{i}Pr$_3$) (39), in a 30:65 ratio. The formation of 39 has been rationalized according to Scheme 10. The initial η^2-coordination of the diene to the osmium atom of 1 affords the dihydrogen intermediate OsCl$_2$(η^2-H$_2$)(η^2-COD)(P^{i}Pr$_3$)$_2$ (F), which dissociates the coordinated hydrogen molecule to give OsCl$_2$(η^4-COD)(P^{i}Pr$_3$)$_2$ (G). The subsequent dissociation of a phosphine from G leads to the 16-valence electrons species OsCl$_2$(η^4-COD)(P^{i}Pr$_3$) (H), which generates the hydride complex

OsHCl$_2$(η^3-C$_8$H$_{11}$)(P^{i}Pr$_3$) (**I**) by a C(sp^3)-H bond activation reaction on a CH$_2$ group of the coordinated diolefin. The addition of the phosphine, released in the formation of **H**, to one of the terminal carbon atoms of the allyl unit of **I** leads to the reaction product.

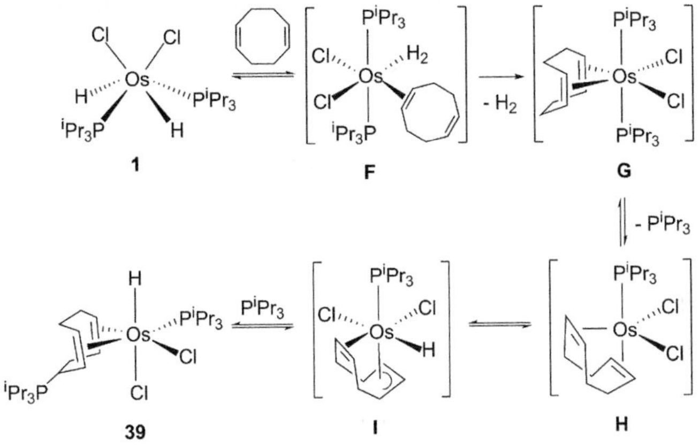

Scheme 10 Mechanism for the formation of complex **39**.

2,5-Norbornadiene is not activated at 85 °C, in toluene. However, it promotes the dehydrogenation of the phosphine to afford the NBD-counterpart of **38**, OsCl$_2$(η^4-NBD){η^3-P[C(CH$_3$)=CH$_2$]^{i}Pr$_2$} (**40**). Tetrafluorobenzobarrelene is not C—H activated either. In this case, the dehydrogenation of the phosphine leads to the mixture of the propenylphosphine isomers OsCl$_2$(η^4-TFB){η^3-P[C(CH$_3$)=CH$_2$]^{i}Pr$_2$} and OsCl$_2$(η^4-TFB){η^3-P(CH$_2$CH=CH$_2$)^{i}Pr$_2$} (**41** and **42**, respectively, in Scheme 11).

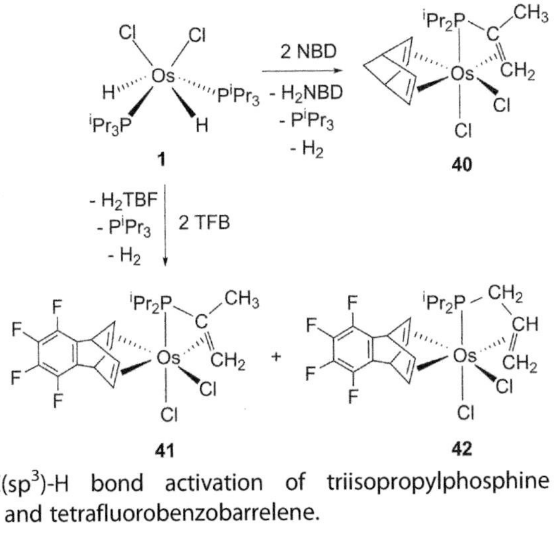

Scheme 11 C(sp^3)-H bond activation of triisopropylphosphine promoted by norbornadiene and tetrafluorobenzobarrelene.

The formation of the isopropenylphosphine complexes **38**, **40**, and **41** and the allylphosphine derivative **42** can be understood as two competitive β-elimination processes on the metalated group of a metalated triisopropylphosphine ligand (Scheme 12). The β-elimination of hydrogen leads to **38**, **40**, and **41** (route a), while the β-elimination of the methyl group (route b) gives rise to **42**.

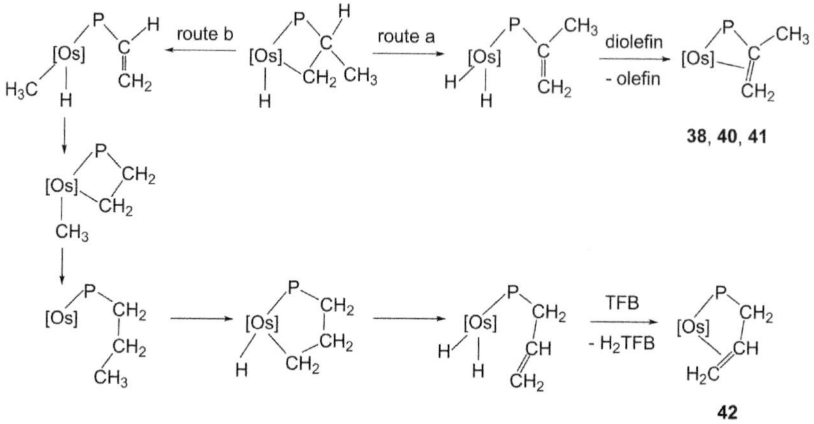

Scheme 12 Rationalization of the formation of **38** and **40–42**.

4.3 C(sp²)-H bond activation of olefins, enynes, acyclic dienes, allenes, and allenedienes

Dihydride-dichloride **1** reacts with propylene and styrene to give equimolecular amounts of the alkylidyne derivatives OsHCl$_2$(≡CCH$_2$R) (P^{i}Pr$_3$)$_2$ (R=Me (**43**), Ph (**44**)) and the hydrogenated olefin.[37] The reactions can be rationalized according to Scheme 13. Complex **1** is initially

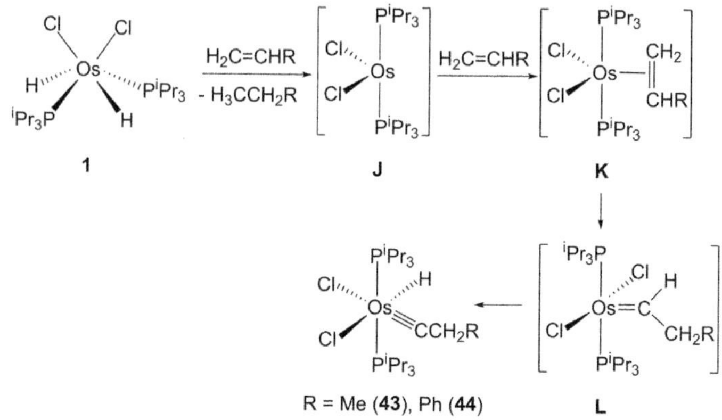

R = Me (**43**), Ph (**44**)

Scheme 13 Reactions of **1** with olefins.

dehydrogenated by the olefin, which yields the alkane. The resulting 14-valence electrons species $OsCl_2(P^iPr_3)_2$ (**J**) coordinates a second equiv. of olefin to afford the π-olefin intermediates $OsCl_2(\eta^2\text{-}CH_2\text{=}CHR)(P^iPr_3)_2$ (**K**), which undergo a M(olefin)-to-M(alkylidene) rearrangement. The latter leads to compounds $OsCl_2(\text{=}CHCH_2R)(P^iPr_3)_2$ (**L**). The 1,2-hydrogen shift from the C_α atom of the alkylidene to the metal center gives the reaction products **43** and **44**.

Complex **1** also reacts with 2-methyl-1-hexen-3-yne. In this case, the reaction leads to the hydride-alkenylalkylidyne $OsHCl_2\{\equiv CC(Me)\text{=}CH^nPr\}(P^iPr_3)_2$ (**45**). Its formation has been rationalized according to Scheme 14, on the basis of DFT calculations. The reduction of the C—C triple bond of the enyne affords the η^2-diolefin intermediate $OsCl_2\{\eta^2\text{-}CH_2\text{=}C(Me)CH\text{=}CHEt\}(P^iPr_3)_2$ (**M**), which undergoes the activation of both terminal $C(sp^2)$-H bonds. One of the hydrogen atoms migrates to the CHEt-carbon atom, through the metal, to give the alkenylalkylidene species $OsCl_2\{\text{=}CHC(Me)\text{=}CH^nPr\}(P^iPr_3)_2$ (**N**). The other one goes to the metal center via a C_α-to-Os 1,2-hydrogen shift on **N**. In agreement with this, the treatment of **1** with 2 equiv. of 2,4-dimethyl-1,3-pentadiene gives rise to the hydrogenation of 1 equiv. of diene and the formation of $OsHCl_2\{\equiv CC(Me)\text{=}CH^iPr\}(P^iPr_3)_2$ (**46**).[38]

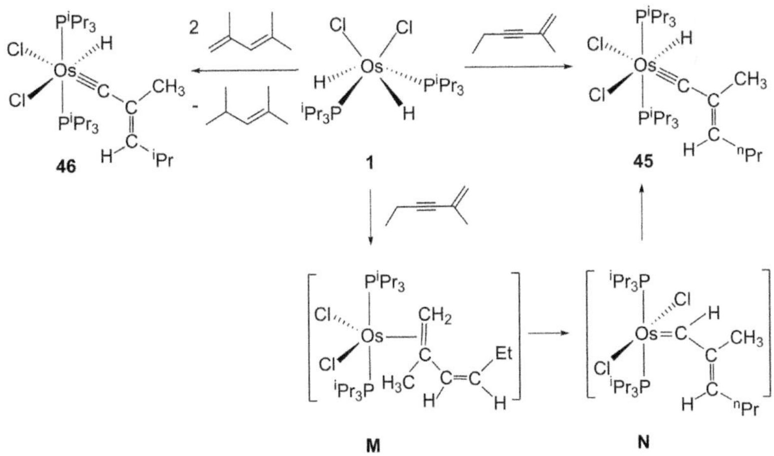

Scheme 14 Reactions of **1** with enynes and acyclic dienes.

Allenes react in a similar manner to olefins, enynes, and acyclic dienes. Reactions of **1** with 1,1-dimethylallene and 1-methyl-1-(trimethylsilyl)allene initially give 1 equiv. of olefin and the π-allene derivatives $OsCl_2\{\eta^2\text{-}CH_2\text{=}C\text{=}C(Me)R\}(P^iPr_3)_2$ (R = Me (**47**), Me_3Si (**48**)), which evolve into the respective hydride-alkenylalkylidynes OsHCl$_2${≡CCH=C(Me)R}(P^{i}Pr$_3$)$_2$ (R=Me (**49**), Me$_3$Si (**50**)). The transformation involves two different 1,2-hydrogen shifts, as a consequence of two C(sp^2)-H bond activations (Scheme 15). The first 1,2-hydrogen shift has an activation energy higher than the second one and takes place through the hydride-osmacyclopropene intermediates **O**, which evolve to the alkenylalkylidenes OsCl$_2${=CHCH=C(Me)R}(P^{i}Pr$_3$)$_2$ (**P**). Intermediates **P**, like **N**, undergo an α-hydrogen elimination to yield the hydride-alkenylalkylidyne reaction products.[39]

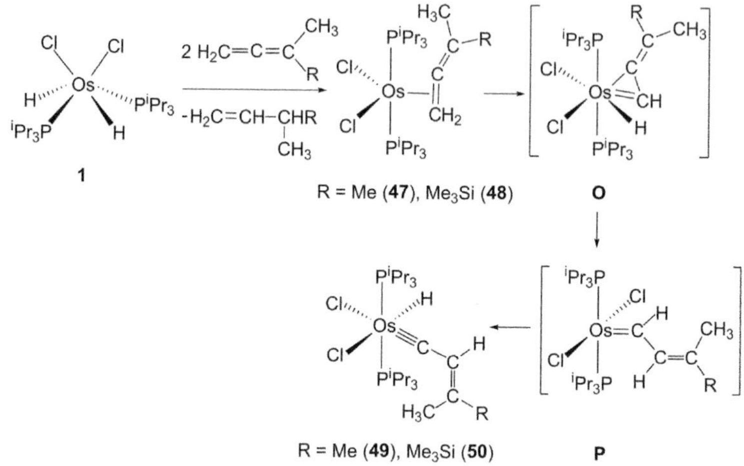

Scheme 15 Reactions of **1** with allenes.

Kinetic studies[40] and DFT calculations[41] have demonstrated that these transformations from alkylidene to hydride-alkylidyne follow the same electronic pattern as the oxidative additions from d^8 to d^6; i.e., π-donor halides destabilize the monosubstituted alkylidene and favor the hydride-alkylidyne form. For example, the sequential replacement of the chloride ligands of complexes OsHCl$_2${≡CCH=CR$_2$}(P^{i}Pr$_3$)$_2$ by acetonitrile molecules produces a sequential decrease of the activation energy for the hydride migration from the metal center to the alkylidyne carbon atom, as a consequence of the gradual decrease of the electron richness of the metal center. The control of the position of this redox equilibrium has allowed performing sequential C—C coupling reactions on the coordination sphere of the metal center, to build a variety of interesting organic fragments.[42]

The reactions of the allenedienes collected in Chart 2 with **1** have been also studied. The results uncovered an interesting interplay between

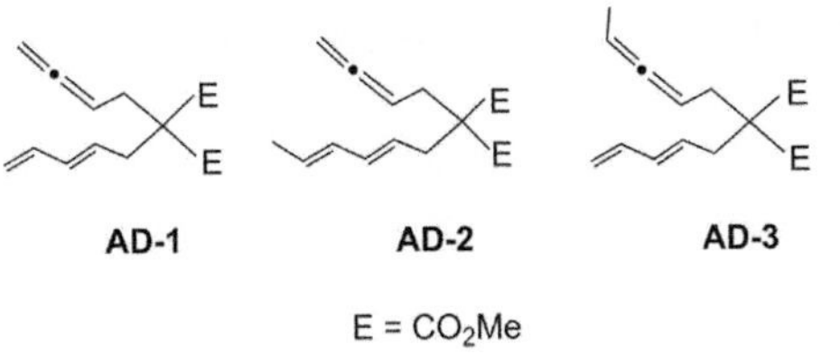

Chart 2 Studied allenedienes.

coordination and reactivity. Thus, different types of products are observed, depending on the substitution of the allene and of the terminal double bond of the diene.[43]

The reaction of **1** with 2 equiv. of **AD-1** at room temperature leads to a 0.4:0.4:0.2 mixture of the organic products **AD-4**, **AD-5**, and **AD-6** and the π-allene complex **51** (Scheme 16). The trienes **AD-4** and **AD-5** are the result of the hydrides transfer from **1** to the internal and terminal double bonds, respectively, of the allene moiety of the allenediene, whereas the tetrahydroindene derivative **AD-6** results from an intramolecular (4+2) cycloaddition reaction of **AD-1** promoted by **J**, arising from hydride transfer processes. Complex **51** is generated from the coordination of the terminal double bond of the allene unit of the second equivalent of **AD-1** to the

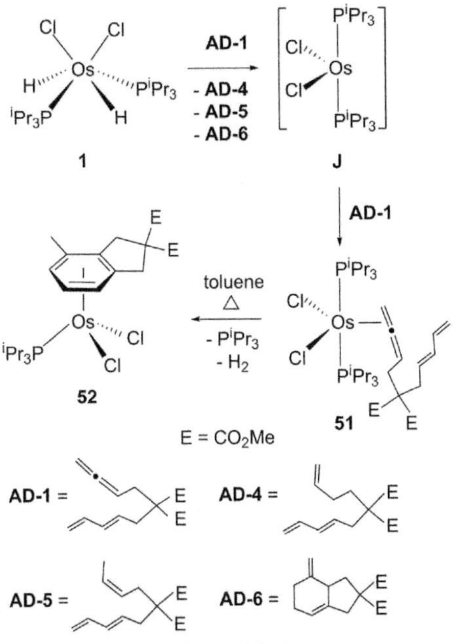

Scheme 16 Reaction of **1** with allenediene **AD-1**.

osmium atom of **J**. Complex **51** is moderately stable. In toluene solution evolves to the arene derivative **52** and free phosphine. The arene ligand results from the intramolecular (4 + 2) cycloaddition of the coordinated allenediene and the subsequent aromatization of the resulting tetrahydroindene, with the loss of a hydrogen molecule.

The reaction of **1** with 2 equiv. of **AD-2** at room temperature gives a 0.5:0.5 mixture of the trienes **AD-7** and **AD-8**, related to **AD-4** and **AD-5**, and the π-allene derivative **53** (Scheme 17). Complex **53** is also moderately stable in solution, like **51**. However, in contrast to the latter, in toluene, it does not evolve into an arene species similar to **52**. The presence of a methyl substituent at the terminal double bond of the diene moiety prevents the coordination of this double bond and therefore the (4 + 2) cycloaddition. It yields a complex mixture of compounds, being the main component (about 50%) the hydride-alkenylalkylidyne **54**. Its formation is consistent with Scheme 15.

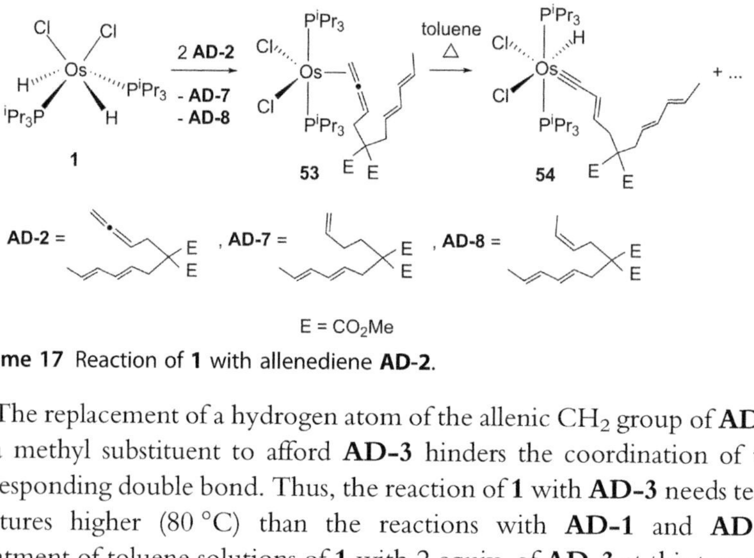

Scheme 17 Reaction of **1** with allenediene **AD-2**.

The replacement of a hydrogen atom of the allenic CH_2 group of **AD-1** by a methyl substituent to afford **AD-3** hinders the coordination of the corresponding double bond. Thus, the reaction of **1** with **AD-3** needs temperatures higher (80 °C) than the reactions with **AD-1** and **AD-2**. Treatment of toluene solutions of **1** with 2 equiv. of **AD-3** at this temperature leads to the bicycle **AD-9** together with the arene complex **55** and hydrogenated products, in the ratio shown in Scheme 18. Bicycle **AD-9** is a result of the intramolecular (2 + 2) cycloaddition of **AD-3**, whereas the arene ligand of **55** results from an intramolecular (4 + 2) cycloaddition of **AD-3** and subsequent aromatization of the resulting tetrahydroindene, in a process which resembles that shown in Scheme 16. The difference in behavior between **AD-3** and **AD-1** can be understood in terms of a competitive coordination of either allenic double bond to the osmium atom.

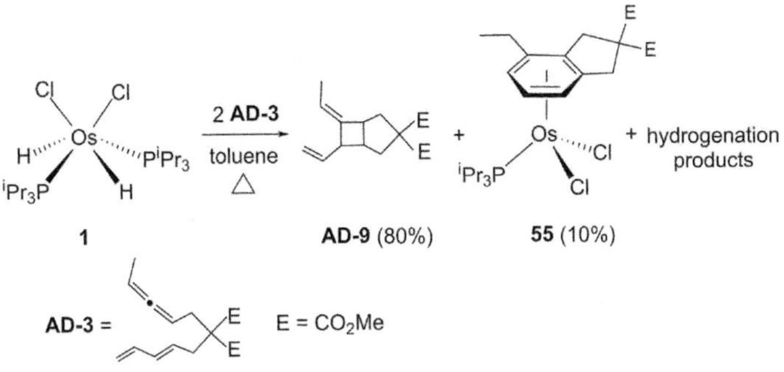

Scheme 18 Reaction of **1** with allenediene **AD-3**.

The (2+2) cycloaddition can be accomplished in a catalytic manner. Heating of benzene solutions of **AD-3** with 10% of **1** produces about 50% conversion to the bicycle **AD-9**. Unfortunately, the simultaneous formation of **55** sequesters the active catalyst and suppresses further turnovers.

4.4 C(sp^2)-H bond activation and subsequent C—C bond formation

Complex **1** reacts with 2 equiv. of 2-vinylpyridine to give molecular hydrogen, [HP^{i}Pr$_3$]Cl, and compound **56**. The latter is isolated as a 6:4 mixture of the isomers **56a** and **56b** shown in Scheme 19.[44] Complex **56** bears two molecules of the starting heterocycle. One of them is metalated, as a result of the C(sp^2)-H bond activation of the CH$_2$ group of vinyl substituent, while the other one is coordinated to the metal center by the nitrogen atom and the vinylic C—C double bond. Complex **56** reacts with Tl(OTf) (OTf=trifluoromethanesulfonate). The reaction produces the replacement of chloride by trifluoromethanesulfonate to selectively afford **57**. Under atmospheric pressure of acetylene, complex **57** is converted into compound **58**. Its formation is a one-pot synthesis of multiple complex reactions. In addition to a 1,3-hydrogen shift, three selective C—C coupling processes are assembled to give this species: (i) coupling of the substituents of the pyridines to afford an Os{κ^3-N,N,C(5)-[py-C(8)H=C(7)H-C(6)H$_2$-C(5)H-py]}moiety, (ii) coupling on the osmium coordination sphere of two acetylene molecules to give an Os{=C(4)H-C(3)H=C(2)H-C(1)H=}-osmacyclopentatriene unit, and (iii) C(4)-C(5) bond formation by migratory insertion of the Os=C(4) double bond of the osmacyclopentatriene unit into the Os-C(5) single bond of the Os{κ^3-N,N,C(5)-[py-C(8)H=C(7)H-C(6)H$_2$-C(5)H-py]} moiety. The 1,3-hydrogen shift takes place between C(6) and C(4).

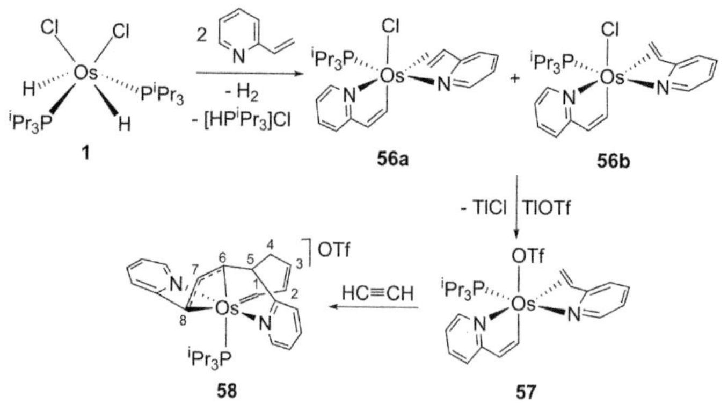

Scheme 19 2-Vinylpyridine-acetylene multiple coupling induced by **1**.

The transformation from **57** to **58** reveals that the insertion of the η²-coordinated –CH=CH₂ group into the bond between the metal center and the metalated substituent is favored with regard to the insertion of acetylene. The latter occurs in absence of coordinated vinylpyridine (Scheme 20). Under an atmosphere of carbon monoxide the OTf⁻ anion and the η²-coordinated vinyl group of **57** are displaced by CO to give the *cis*-dicarbonyl salt **59**. In dichloromethane, the cation of **59** slowly dissociates 2-vinylpyridine and coordinates the anion to afford **60**. Under atmospheric pressure of acetylene, complex **60** reacts with the alkyne to yield **61**, which contains a metalated 2-butadienylpyridine ligand. The butadienyl moiety is the result of the insertion of the C—C triple bond of acetylene into the osmium-vinyl bond.

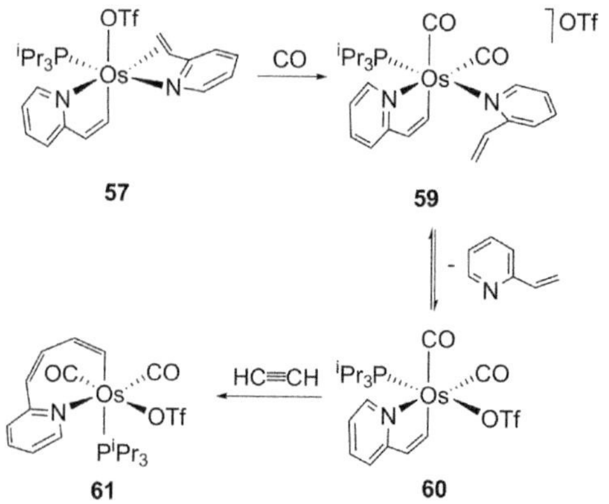

Scheme 20 2-Vinylpyridine-acetylene coupling induced by **57** and carbon monoxide.

4.5 Formation of N—H wingtip NHC ligands by tautomerization of heterocycles

Pyridine and quinoline have NHC-tautomers, which lie about 40 and 44 kcal mol^{-1}, respectively, above the usual ones (Scheme 21).

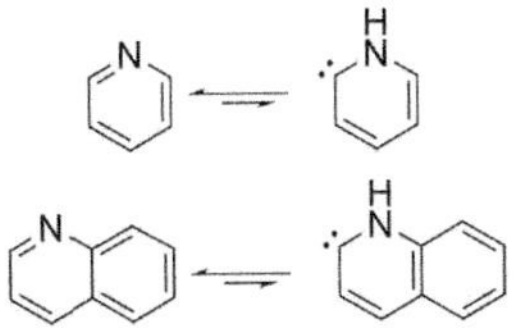

Scheme 21 NHC-tautomers of pyridine and quinoline.

Dihydride-dichloride complex **1** promotes the tautomeric transformation of this class of heterocycles and stabilizes the NHC-tautomer. Thus, its reactions with 2-methylpyridine,[45] 8-methylquinoline, quinoline,[46] and benzo[*h*]quinoline[47] lead to the respective NHC-derivatives **62–65** (Scheme 22), which undergo an additional stabilization as a consequence of an intramolecular Cl···H–N hydrogen bond. A dihydride-to-elongated dihydrogen transformation also takes place during the tautomerizations.

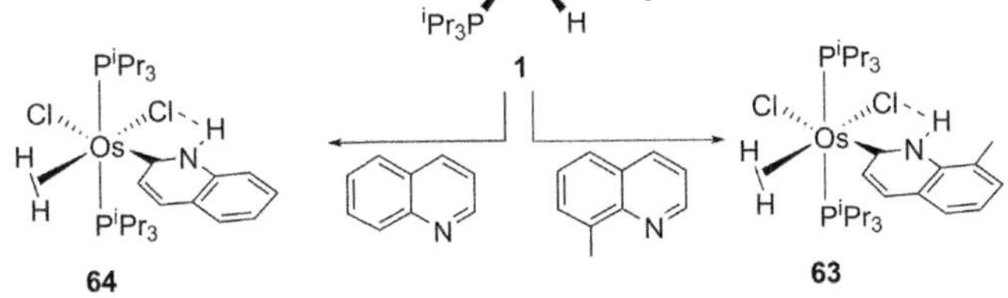

Scheme 22 Reactions of **1** with N-heterocycles: Stabilization of NHC-tautomers of pyridines and quinolines by metal coordination.

DFT calculations[45] on the tautomerization of 2-methylpyridine promoted by the model compound OsH$_2$Cl$_2$(PMe$_3$)$_2$ (B3PW91) suggest that the formation of **62–65** is a process of three stages: (i) intermolecular osmium

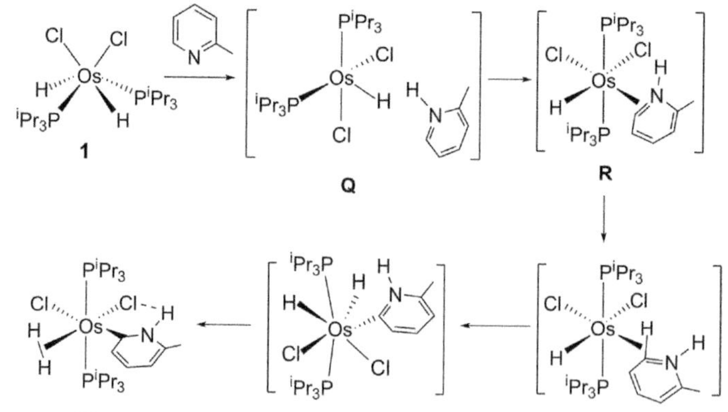

Scheme 23 Mechanism of the tautomerization of 2-methylpyridine.

to nitrogen hydrogen migration, (ii) C_α-H_α bond activation of the resulting protonated heterocycle to afford a dihydride species, and (iii) dihydride-dihydrogen intramolecular reduction (Scheme 23).

The intermolecular osmium to nitrogen hydrogen migration leads to [HN-Heterocycle]$^+$[OsHCl$_2$(P^{i}Pr$_3$)$_2$]$^-$ (**Q**) cation-anion pairs. The C_α-H_α bond activation of the cation by the anion subsequently occurs. This process is initiated by the coordination of a C—N bond of the protonated heterocycles to the metal center, resulting η^2-C,N-intermediates **R**. Then, the slippage of the metal center from the C_α-N bond to the C_α-H_α one gives **S**, which undergoes the activation of the coordinated C—H bond. Finally, the resulting dihydride **T** tautomerizes to give the elongated dihydrogen products.

The steric hindrance experienced between the heterocycle and the phosphines of **1** is determinant for the tautomerization. Thus, the presence of the methyl substituent in 2-methylpyridine is required for the transformation of the heterocycle. In its absence the 1,2-hydrogen shift does not take place. In contrast to 2-methylpyridine, pyridine reacts with **1** to give OsCl$_2$(py)$_3$(P^{i}Pr$_3$). In agreement with this, it has been observed that benzophenone,[48] benzylideneacetophenone, and benzylideneacetone[49] promote the reduction and subsequent tautomerization of the 2-vinylpyridine ligand of the hydride elongated dihydrogen **66** to afford **67–69**, while the smaller methyl vinyl ketone induces the reduction of the coordinated double bond and the release of 2-ethylpyridine to form **70** (Scheme 24).

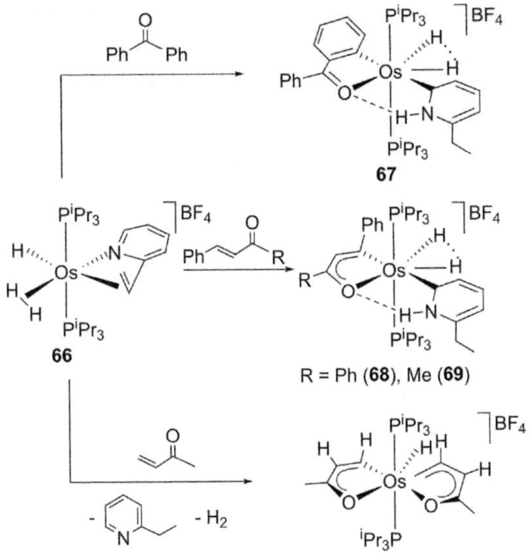

Scheme 24 Reactions of **66** with ketones: Reduction and NHC-tautomers of 2-vinylpyridine.

5. C—C bond activation

Hydride-alkylidyne and -alkenylalkylidyne type compounds shown in Schemes 13–15 are also formed by reaction of the dihydride-dichloride complex **1** with terminal alkynes and alkynols, respectively.[50] They are the osmium-counterparts of the well-known Grubbs's catalysts for olefin metathesis.[51] The ruthenium-alkylidene complexes and the osmium-hydride-alkylidyne derivatives are both parts of the same redox equilibrium (Scheme 25). Ruthenium, more oxidizing than osmium, favors the reduced form. Consequently, osmium is more reducing than ruthenium and, preferring saturated species, stabilizes the saturated oxidized form.[42]

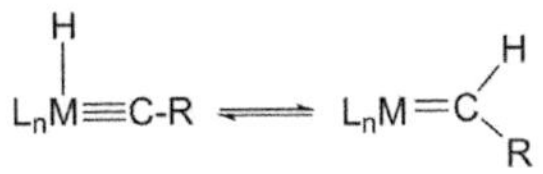

Scheme 25 Redox equilibrium M-alkylidene/M-hydride-alkylidyne.

A distal amide function at an appropriate position of an alkylic chain of a terminal alkyne changes the reactivity of the latter with regard to **1**. In contrast to other terminal alkynes, *N*-phenylhex-5-ynamide and *N*-phenylhept-6-ynamide react with **1** to give the alkylidene derivatives

OsCl$_2${=C(CH$_3$)(CH$_2$)$_n$(CO)NHPh}(P^{i}Pr$_3$)$_2$ (n=3 (**71**), 4 (**72**)), as a result of the addition of both hydride ligands of **1** to the terminal carbon atom of the alkynes (Scheme 26). The transformation also occurs in substrates in which the alkyne and the amide are connected through the nitrogen instead of the carbonyl group. In fact, the reaction of **1** with N-(pent-4-yn-1-yl)benzamide leads to OsCl$_2${=C(CH$_3$)(CH$_2$)$_3$NH(CO)Ph}(P^{i}Pr$_3$)$_2$ (**73**).[52]

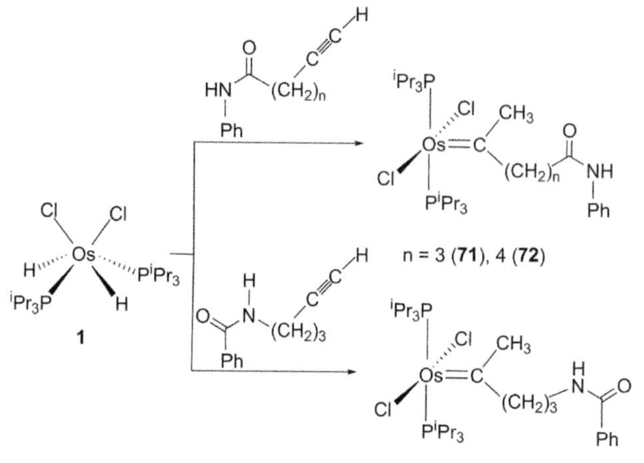

Scheme 26 Reactions of **1** with N-phenylhex-5-ynamide and N-phenylhept-6-ynamide.

These alkylidene compounds are intermediate species in the osmium-mediated rupture of the C—C triple bond of the initial alkynes. Thus, they are unstable and undergo the rupture of the C$_\alpha$-CH$_3$ bond of the alkylidene to evolve into six-coordinated hydride-alkylidyne derivatives, in agreement with the strongly reducing character of osmium and its marked preference to form saturated species.

Complex **71** experiences two different transformations, which are competitive in dichloromethane at 60 °C (Scheme 27). The migration of the methyl group from the C$_\alpha$ atom of the alkylidene to the metal center, to afford intermediate **U**, and the subsequent methylidene extrusion yield the hydride-alkylidyne derivative OsHCl$_2${≡C(CH$_2$)$_3$C(O)NHPh}(P^{i}Pr$_3$)$_2$ (**74**), containing a linear spacer between the alkylidyne C$_\alpha$ atom and the amide (pathway **a**). On the other hand, the activation of one of the C—H bonds of the C$_\beta$H$_2$ group and a concerted 1,2-methyl shift in the resulting osmacyclopropene **V** give OsHCl$_2${≡CCH(CH$_3$)(CH$_2$)$_2$C

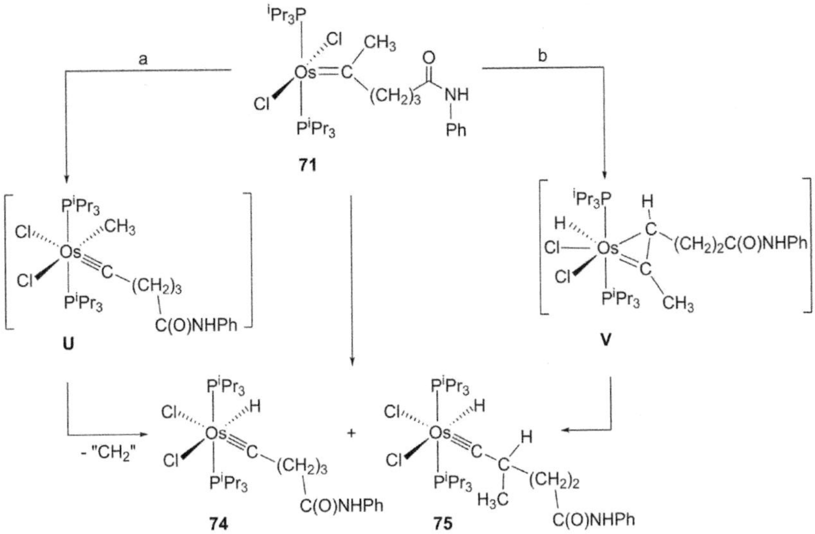

Scheme 27 Transformations of complex **71** into hydride-alkylidyne derivatives.

(O)NHPh}(P^{i}Pr$_3$)$_2$ (**75**), with a branched spacer between the alkylidyne C$_\alpha$ atom and the amide (pathway b). As a result of both processes, a 10:7 mixture of both hydride-alkylidynes is formed.

The relative position of the carbonyl and the NH groups in the amide influences the behavior of the alkylidene. In contrast to **71**, complex **73** selectively evolves into OsHCl$_2${≡C(CH$_2$)$_3$NHC(O)Ph}(P^{i}Pr$_3$)$_2$ (**76**), the benzamide counterpart of **74**, containing a linear spacer (Scheme 28).

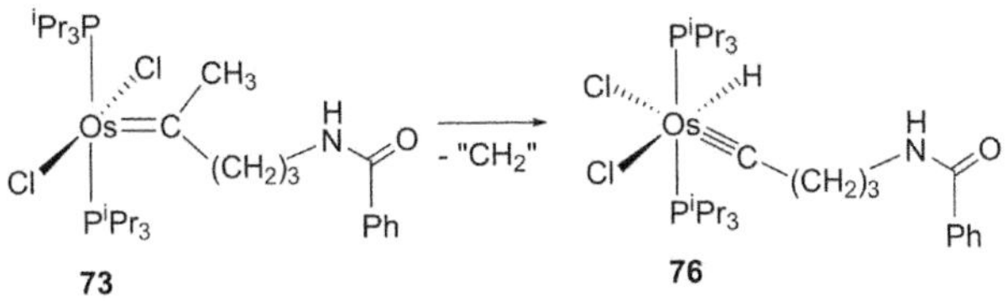

Scheme 28 Transformations of complex **73** into a hydride-alkylidyne derivative.

These elusive five-coordinate osmium-alkylidene compounds undergo metathesis with allenes, being in this way a useful entry to interesting dicarbon-disubstituted vinylidene derivatives (Scheme 29). Treatment of complex **71** with 3-methyl-1,2-butadiene and cyclohexylallene yields OsCl$_2$(=C=CR$_2$)(P^{i}Pr$_3$)$_2$ (CR$_2$=CMe$_2$ (**77**), CHCy (**78**)).

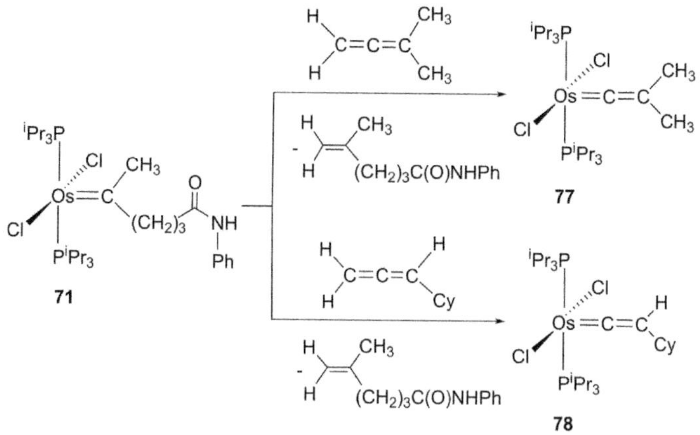

Scheme 29 Reactions of **71** with allenes.

6. C—O bond activation

Trihydride complex **10** adds vinyl ethers to form η^2-olefin adducts, which evolve to carbene derivatives in the presence of additional vinyl ether. Subsequent R-dependent reactions involve C(sp^2)-OR bond cleavage, to make either alkylidyne or vinylidene complexes (Scheme 30).[53]

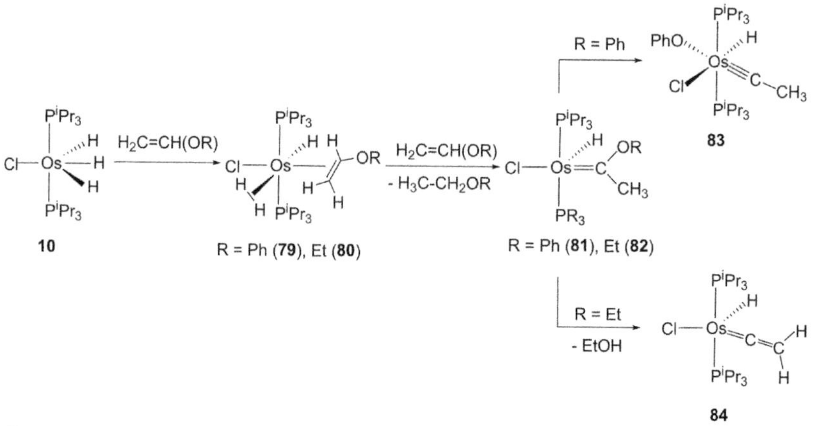

Scheme 30 Reactions of **10** with vinyl ethers.

The addition of phenyl vinyl ether and ethyl vinyl ether, to toluene solutions of **10**, at −80 °C leads to the elongated dihydrogen-π-olefin intermediates OsHCl(η^2-H$_2$){η^2-CH$_2$=C(OR)H}(P^{i}Pr$_3$)$_2$ (R=Ph (**79**),

Et (**80**)). At about −10 °C, in the presence of additional vinyl ether, these compounds generate ethyl phenyl ether, or diethyl ether, and form the unsaturated Fischer-type carbene derivatives OsHCl{=C(OR)Me}(P^{i}Pr$_3$)$_2$ (R=Ph (**81**), Et (**82**)), which are unstable at room temperature. Again, in agreement with the strongly reducing character of osmium and its marked preference to form saturated species, complex **81** undergoes a C$_\alpha$-to-Os migration of the phenoxy group to give the alkylidyne complex OsHCl(OPh){≡CMe}(P^{i}Pr$_3$)$_2$ (**83**). In contrast to the latter, compound **82** eliminates ROH to afford OsHCl(=C=CH$_2$)(P^{i}Pr$_3$)$_2$ (**84**), which is one of the few five-coordinate hydride-vinylidenes known.[54,55]

There is a ring constraint against intramolecular migration to make an alkylidyne when the vinyl ether is cyclic. Thus, the reaction of **10** with 2,3-dihydrofuran gives tetrahydrofuran and the hydride-cyclic carbene derivative **31** (Scheme 31), which is stable in contrast to **81** and **82**.[53]

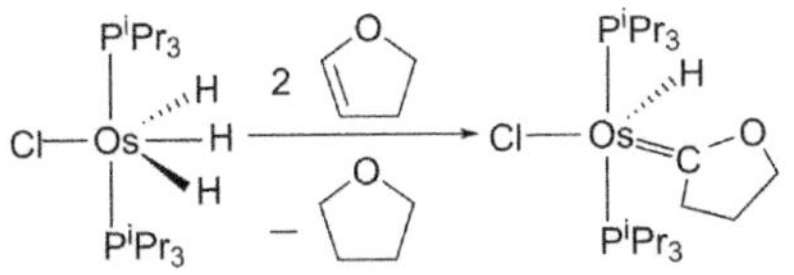

Scheme 31 Reaction of **10** with 2,3-dihydrofuran.

7. C-halogen bond activation

Dihydride-dichloride complex **1** undergoes halide exchange at room temperature, by simple dissolving in ethyl bromide or methyl iodide. The exchanges lead to the dibromide and diiodide derivatives **2** and **3**, respectively, via the heteroleptic dihalides **85** and **86** (Scheme 32). The reaction rate depends upon the solvent and is faster for methyl iodide than for ethyl bromide, in agreement with a C-halide bond dissociation energy (BDE) smaller for methyl iodide than for ethyl bromide. In both cases, the rate of the halide exchange is drastically increased by using trimethylsilyl halide reagents. Typically, BDE(C—Cl) is about 28 kcal mol^{-1} greater than BDE(C—I). The difference is even greater for the silicon-halide bonds: BDE(Si—Cl) − BDE(Si—I) ≈ 35 kcal mol^{-1}. This increased driving force appears to accelerate the reaction. Thus, halide exchange occurs very rapidly with only a slight excess of Me$_3$Si-halide in dichloromethane.[56]

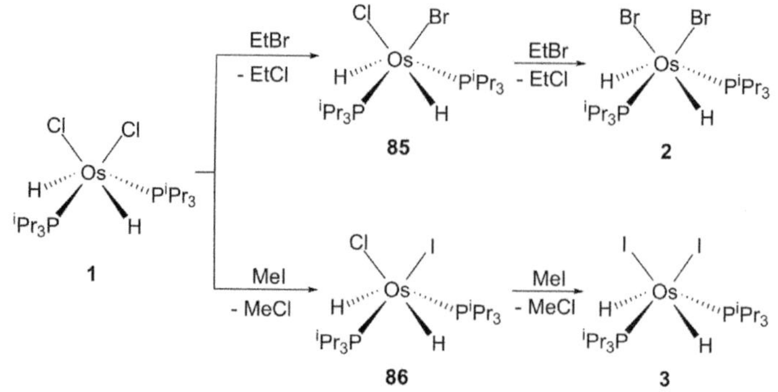

Scheme 32 C-halogen bond activation reactions promoted by **1**.

The mechanism of the C-halide activation appears to involve halocarbon coordination to osmium as a pre-equilibrium step. By means of a four-centered transition state, the resulting species undergo combined Os-to-C chloride and C-to-Os halide migrations.

8. Si—H bond activation

Treatment of [Et$_4$N]$_2$[OsCl$_6$] with 1,5-bis(di-*tert*-butylphosphino)pentane, in the presence of triethylamine, in pentan-2-ol, at 140 °C, under hydrogen atmosphere leads to the cyclometalated compound OsH$_2$Cl{κ^3-P,C,P-[CH(C$_2$H$_4$P^{t}Bu$_2$)$_2$]} (**87**), which undergoes thermal reductive dehydrogenation, to afford the osmium(II)-carbene derivative OsHCl{κ^3-P,C,P-[=C(C$_2$H$_4$P^{t}Bu$_2$)$_2$]} (**88**), according to Scheme 33.[57]

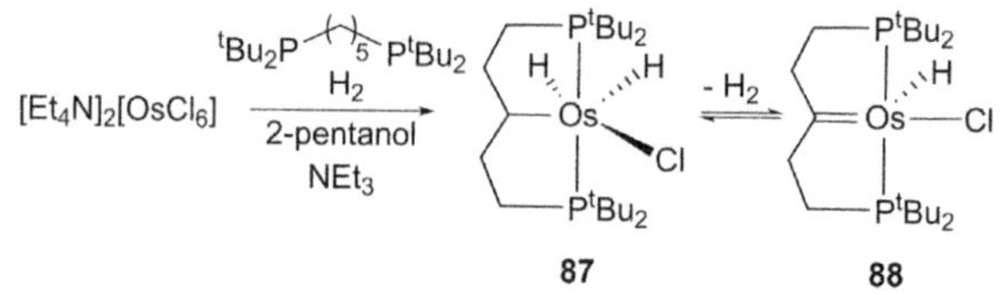

Scheme 33 Formation of **87** and **88**.

Complex **87** activates PhSiH$_3$ (Scheme 34).[58] Addition of 2 equiv. of the silane to toluene solutions of the dihydride gives the base-stabilized silylene OsH$_5$(SiPh$_2$Cl){κ^2-Si,P-[SiH$_2$P^{t}Bu$_2$(CH$_2$)$_5$P^{t}Bu$_2$]} (**89**). During the process, complex **87** and two molecules of PhSiH$_3$ undergo a series of redistribution reactions culminating in the net hydrogenation of the Os-C(sp^3) bond and the generation of a silyl, one phosphine-stabilized silylene, and five

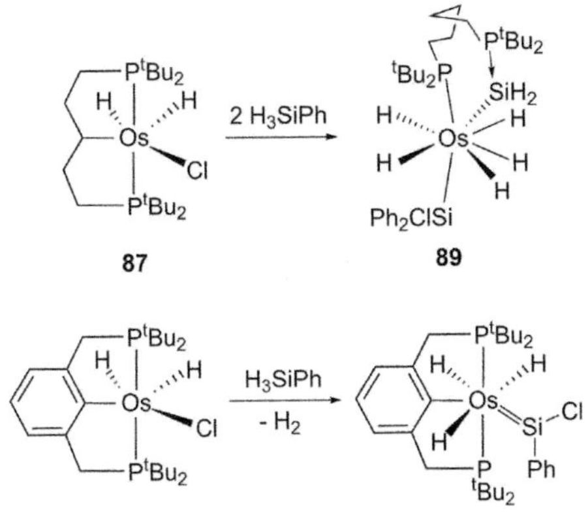

Scheme 34 Si—H bond activation reactions promoted by Os-PCP complexes.

hydride ligands. The donor atoms around the metal center define a dodecahedron consisting of two orthogonal trapezoidal planes. One of them contains the Si-silyl atom and the osmium-coordinated phosphorous atom at B sites and two hydride ligands, whereas the other plane contains three hydrides and the Si atom of the base-stabilized silylene. The unexpected behavior of **87** seems to be a consequence of the flexibility of the linker between the P^{t}Bu$_2$ groups. In contrast to the latter, complex OsH$_2$Cl{κ^3-P, C,P-[2,6-(CH$_2$P^{t}Bu$_2$)$_2$C$_6$H$_3$]} (**13**) bearing a pincer with a more rigid backbone reacts with PhSiH$_3$ to form the trihydride-free silylene derivative OsH$_3$(=SiClPh){κ^3-P,C,P-[2,6-(CH$_2$P^{t}Bu$_2$)$_2$C$_6$H$_3$]} (**90**) and molecular hydrogen.

9. Sn—H bond activation

Dihydride-dichloride compound **1** also activates the Sn—H bond of Ph$_3$SnH. This activation has allowed to develop a rich chemistry of stannyl- and bis(stannyl)-polyhydride derivatives, including pentahydrides, tetrahydrides, trihydrides and compressed dihydride species.

Complex **1** reacts with 2 equiv. of Ph$_3$SnH to afford the tetrahydride-stannyl compound OsH$_4$Cl(SnPh$_3$)(P^{i}Pr$_3$)$_2$ (**91**) and Ph$_3$SnCl. It has been proposed that the reaction involves the initial oxidative addition of the Ph$_3$Sn-H bond to **1** to give the osmium(VI) intermediate OsH$_3$Cl$_2$(SnPh$_3$)(P^{i}Pr$_3$)$_2$ (**W**), which evolves by reductive elimination of Ph$_3$SnCl into trihydride **10**. Thus, the subsequent oxidative addition of a second molecule of Ph$_3$SnH to the latter could yield **91** (Scheme 35).

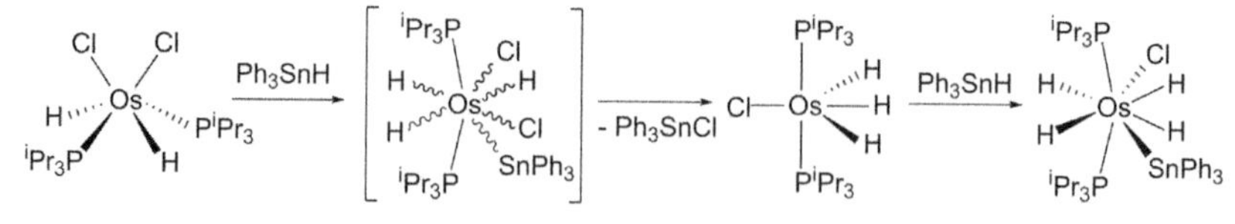

Scheme 35 Reaction of **1** with Ph$_3$SnH.

The structure of this compound is the expected dodecahedron with the bulky ligands at B sites of the orthogonal trapezoidal planes.[59]

Complex **91** reacts with two new molecules of Ph$_3$SnH to give Ph$_3$SnCl, H$_2$, and the tetrahydride-bis(stannyl) derivative OsH$_4$(SnPh$_3$)$_2$(P^{i}Pr$_3$)$_2$ (**92**). This species is a rare example of a bis(stannyl) compound with the transition metal in high oxidation state. A distinguishing feature of its structure is the P-Os-Sn angle in both orthogonal trapezoidal planes of 122.56(3)°, which is significantly smaller than the related angle in other eight-coordinate osmium-polyhydride complexes (145–156°). This seems to be the result of the steric hindrance experienced by the phosphine and stannyl ligands of different planes.[60] In the presence of diphenylacetylene, compound **91** affords the trihydride-stannyl derivative OsH$_3$(SnClPh$_2$){κ^3-C,C,P-[CH$_2$=C(CH$_3$)P^{i}Pr$_2$]}(P^{i}Pr$_3$) (**93**), cis-stilbene, and benzene. Its structure in the solid state can be rationalized as a highly distorted pentagonal bipyramid, with the phosphorous atom of triisopropylphosphine and the midpoint of the C—C double bond of the isopropenyl group of the dehydrogenated phosphine occupying axial positions. During the reaction, four different processes are assembled: (i) dehydrogenation of one isopropyl group of one triisopropylphosphine, (ii) reduction of diphenylacetylene to give cis-stilbene, (iii) hydrogenolysis of a phenyl group of the triphenylstannyl ligand, and (iv) chloride migration from the transition metal to the tin atom. The reaction appears to be radical-promoted, since the formation of **93** is inhibited in the presence of hydroquinone. Under hydrogen atmosphere, complex **93** yields the pentahydride-stannyl compound OsH$_5$(SnClPh$_2$)(P^{i}Pr$_3$)$_2$ (**94**), as a result of the reduction of the coordinated olefin and a d^4-d^2 oxidative addition of H$_2$ (Scheme 36). The donor atoms

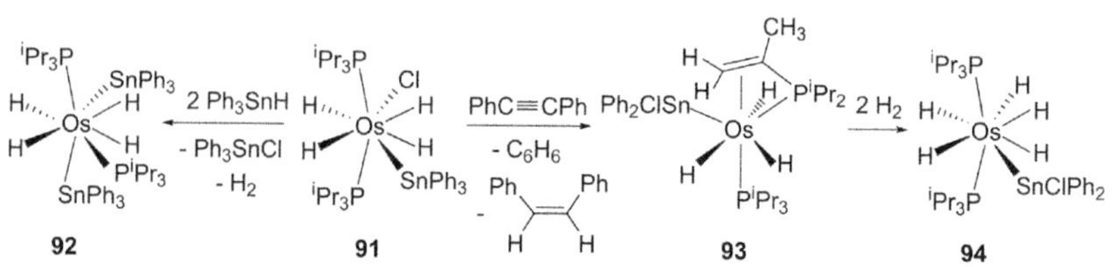

Scheme 36 Reactions of **91** with Ph$_3$SnH and diphenylacetylene.

around the metal center of **94** adopt the expected dodecahedral disposition with the bulky ligands at the B sites of the trapezoidal planes.[59]

The coordinated olefin of **93** also undergoes reduction in the presence of benzoic acid (Scheme 37).[61] The addition of the latter to **93** initially leads to the compressed dihydride-stannyl derivative OsH$_2$(SnClPh$_2$)(κ^2-O$_2$CPh)(P^{i}Pr$_3$)$_2$ (**95**). In toluene, the tin atom exchanges a chloride with the transition metal which in turns gives back one of the oxygen atoms of the carboxylate group, to afford OsH$_2$Cl{κ^2-O,Sn-[OC(Ph)OSnPh$_2$]}(P^{i}Pr$_3$)$_2$ (**96**).

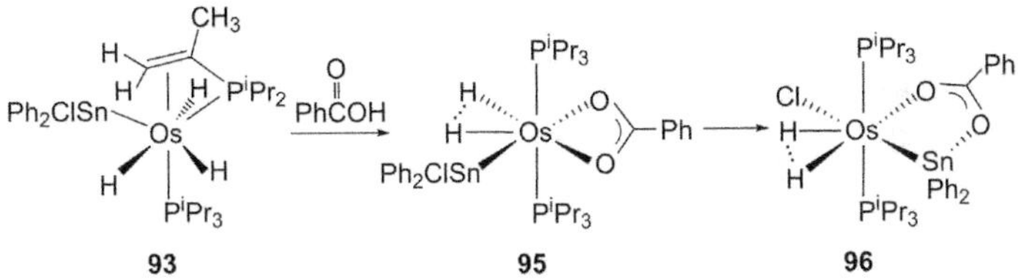

Scheme 37 Reaction of **93** with benzoic acid.

The formation of **94** and **95** is consistent with the transformation of **93** into the 14-valence electrons monohydride OsH(SnClPh$_2$)(P^{i}Pr$_3$)$_2$ (**X**), a stannyl-counterpart of **A**, which adds two hydrogen molecules or one molecule of benzoic acid. This functionally equivalent promotes the chelate-assisted C(sp^2)-H bond activation of aromatic-imines, -ketones, and -aldehydes; α,β-unsaturated-ketones, and -aldehydes; 2-vinylpyridine; and (E)-N-(phenylmethylene)-2-pyridinamine (Scheme 38). The reaction of **93** with benzophenone imine leads to OsH$_2$(SnClPh$_2$){κ^2-N,C-[NHC(Ph)C$_6$H$_4$]}(P^{i}Pr$_3$)$_2$ (**97**), whereas acetophenone and benzophenone afford OsH$_2$(SnClPh$_2$){κ^2-O,C-[OC(R)C$_6$H$_4$]}(P^{i}Pr$_3$)$_2$ (R=Me (**98**), Ph (**99**)). The *ortho*-CH bond activation is favored with regard to the *ortho*-CF bond activation. Thus, the reactions with 2,3,4,5,6-pentafluorobenzophenone and 2-fluoroacetophenone give OsH$_2$(SnClPh$_2$){κ^2-O,C-[OC(C$_6$F$_5$)C$_6$H$_4$]}(P^{i}Pr$_3$)$_2$ (**100**) and OsH$_2$(SnClPh$_2$){κ^2-O,C-[OC(Me)C$_6$FH$_3$]}(P^{i}Pr$_3$)$_2$ (**101**), respectively.[62] The *ortho*-CH bond activation is also favored over the OC-H bond activation in benzaldehydes. Treatment of **93** with these substrates yields the corresponding *ortho*-metalated compounds OsH$_2$(SnClPh$_2$){κ^2-O,C-[OC(H)C$_6$RH$_3$]}(P^{i}Pr$_3$)$_2$ (R=H (**102**), *p*-OCH$_3$ (**103**), *p*-CF$_3$ (**104**), *m*-CF$_3$ (**105**)).[63] α,β-Unsaturated-ketones and -aldehydes generate osmafuran derivatives; methyl vinyl ketone and benzylideneacetone afford OsH$_2$(SnClPh$_2$){κ^2-O,C-[OC(CH$_3$)CHCR]}(P^{i}Pr$_3$)$_2$ (R=H (**106**),

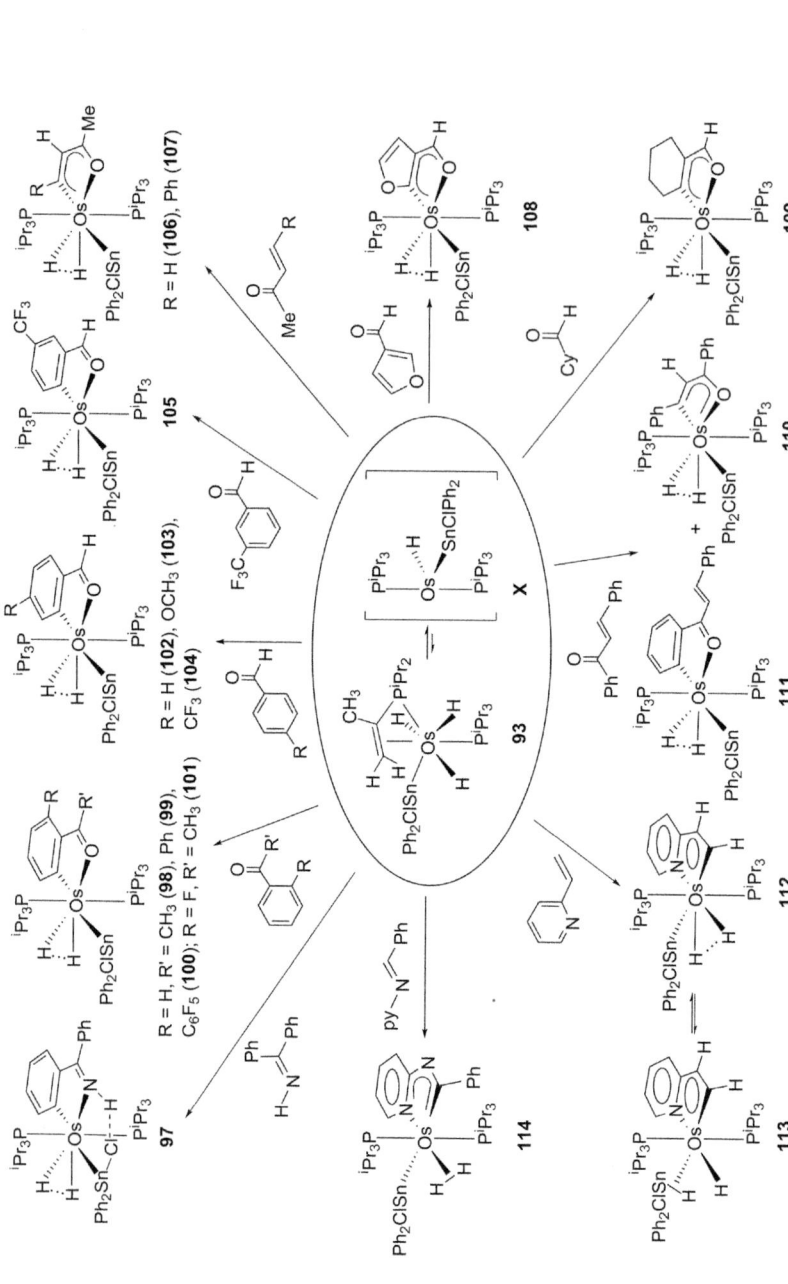

Scheme 38 C—H bond activation reactions promoted by 93.

Ph (**107**)),[64] whereas 3-furaldehyde and 1-cyclohexene-1-carboxaldehyde form OsH$_2$(SnClPh$_2$){κ^2-O,C-[OCHC$_4$(O)H$_2$]}(P^{i}Pr$_3$)$_2$ (**108**) and OsH$_2$(SnClPh$_2$){κ^2-O,C-[OCHC$_6$H$_8$]}(P^{i}Pr$_3$)$_2$ (**109**), respectively.[63] As expected, monohydride **X** activates both β-olefinic- and *ortho*-CH bonds of benzylideneacetophenone to give the osmafuran OsH$_2$(SnClPh$_2$){κ^2-O,C-[OC(Ph)CHCPh]}(P^{i}Pr$_3$)$_2$ (**110**) and the osmaisobenzofuran OsH$_2$(SnClPh$_2$){κ^2-O,C-[OC(CH=CHPh)C$_6$H$_4$]}(P^{i}Pr$_3$)$_2$ (**111**). The activation of the olefinic moiety is kinetically preferred. However, the osmaisobenzofuran complex **111** is the product of thermodynamic control. The activation of 2-vinylpyridine gives OsH$_2$(SnClPh$_2$){κ^2-N,C-(pyCHCH)}(P^{i}Pr$_3$)$_2$ (**112**) in equilibrium with OsH{κ^2-N,C-(pyCHCH)}(η^2-H-SnClPh$_2$)(P^{i}Pr$_3$)$_2$ (**113**), where the stannane is bonded to the transition metal by a Os-H-Sn three-centered bond. The activation of (*E*)-*N*-(phenylmethylene)-2-pyridinamine leads to Os(SnClPh$_2$){κ^2-N,C-(pyNCPh)}(η^2-H$_2$)(P^{i}Pr$_3$)$_2$ (**114**). In this case, the OsH$_2$ unit appears form an elongated dihydrogen.[64]

10. N—H bond activation

Dihydride-dichloride complex **1** activates one N—H bond of each NH$_2$ group of 1,2-phenylenediamine, in the presence of triethylamine, to give the six-coordinate d^4-dihydride derivative OsH$_2${κ^2-N,N-(*o*-NH-C$_6$H$_4$-NH)}(P^{i}Pr$_3$)$_2$ (**115**), containing an osmabenzimidazolium core. The planarity and length equalization of the bicycle along with negative nucleus-independent chemical shifts (NICS) values calculated for both rings and the aromatic MO delocalization indicate that, as the organic counterpart benzimidazolium cation, it is aromatic (Scheme 39).[65] A detailed analysis of the core interactions reveals that from the five frontier orbitals of the [OsH$_2$(P^{i}Pr$_3$)$_2$]$^+$ fragment (LUMO+2, LUMO+1, LUMO, HOMO, and HOMO-1), LUMO+2 and LUMO are involved in σ-interactions, whereas LUMO+1 is engaged in the delocalization of the π-electron system of the metalabicycle; i.e., the metal center does not have

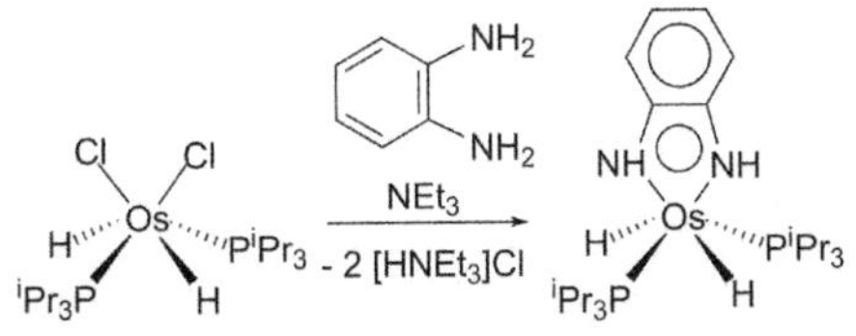

Scheme 39 N—H bond activation of 1,2-phenylenediamine promoted by **1**.

an empty orbital to interact with Lewis bases and therefore, in contrast to other six-coordinate d^4-dihydride-osmium compounds,[25,66,67] complex **115** is inert in the presence of these ligands.[65]

Complex **115** reduces the cation [FeCp$_2$]$^+$. The addition of 1 equiv. of the PF$_6$-salt of this cation to acetonitrile solutions of **115** gives rise to a 1:1 mixture of the monohydride [OsH{κ^2-N,N-(o-NH-C$_6$H$_4$-NH)}(CH$_3$CN)(P^{i}Pr$_3$)$_2$]PF$_6$ (**116**) and the trihydride [OsH$_3${κ^2-N,N-(o-NH-C$_6$H$_4$-NH)}(P^{i}Pr$_3$)$_2$]PF$_6$ (**117**), in addition to Fe(η^5-C$_5$H$_5$)$_2$ (Scheme 40). The first of them is the result of the substitution of one of the hydride ligands by an acetonitrile molecule, while the second one is a consequence of the protonation of the metal center. In this context, it should be mentioned that HOMO-1 of the [OsH$_2$(P^{i}Pr$_3$)$_2$]$^+$ fragment remains nonbonding in **115**. Bicycles of **116** and **117** also appear to be aromatic.

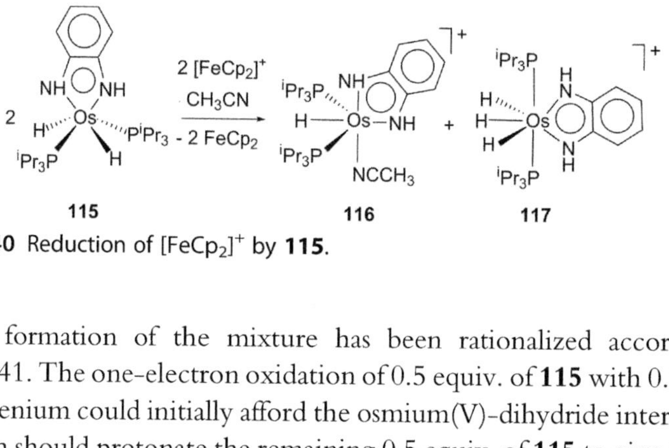

Scheme 40 Reduction of [FeCp$_2$]$^+$ by **115**.

The formation of the mixture has been rationalized according to Scheme 41. The one-electron oxidation of 0.5 equiv. of **115** with 0.5 equiv. of ferrocenium could initially afford the osmium(V)-dihydride intermediate **Y**, which should protonate the remaining 0.5 equiv. of **115** to give **117** and the 15-valence electrons osmium(III)-hydride **Z**. Then, the coordination of the solvent to the metal center of the latter could afford **Aa**, which should reduce the remaining 0.5 equiv. of ferrocenium to yield **116**.

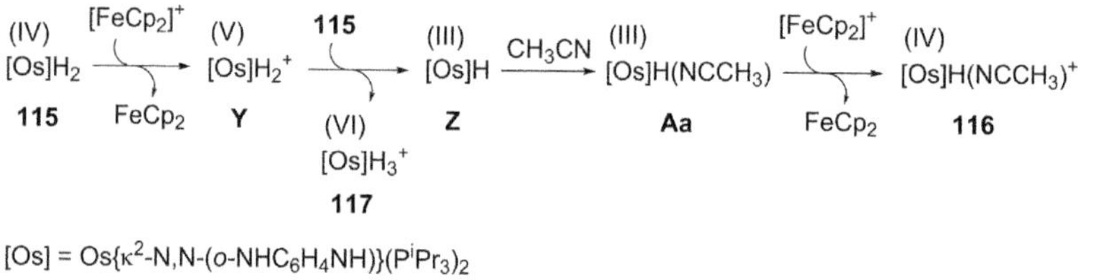

[Os] = Os{κ^2-N,N-(o-NHC$_6$H$_4$NH)}(P^{i}Pr$_3$)$_2$

Scheme 41 Rationalization of the formation of **116** and **117**.

Trihydride **10** activates 1,2-phenylenediamine and *N*-methyl-1,2-phenylenediamine. The reaction with the first of them gives **115**, whereas the reaction with the second one affords OsH$_2$\{κ^2-N,N-(*o*-NH-C$_6$H$_4$-NMe)\}(P^{i}Pr$_3$)$_2$ (**118**). Catechol replaces the chelated diamine from the osmium atom of these compounds. Thus, its addition in excess to toluene solutions of **115** and **118** leads to OsH$_2$\{κ^2-O,O-(*o*-O-C$_6$H$_4$-O)\}(P^{i}Pr$_3$)$_2$ (**119**), according to Scheme 42.[68]

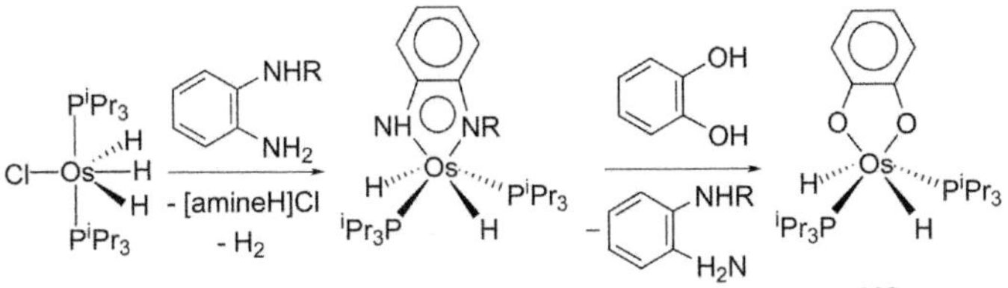

Scheme 42 N—H bond activation reactions of 1,2-phenylenediamine and *N*-methyl-1,2-phenylenediamine promoted by **10**.

Dihydride-dichloride **1** also activates the N—H bond of 1,3-bis(6'-methyl-2'-pyridylimino)isoindoline (HBMePI), which is a sterically demanding version of 1,3-bis(2-pyridylimino)isoindoline (HBPI).[69] The activation generates the isoindolinate anion BMePI and HCl. The latter is removed from the reaction medium with K^{t}BuO. In contrast to BPI, which generally acts as κ^3-(N$_{py}$,N$_{iso}$,N$_{py}$)$_{mer}$,[70] anion BMePI coordinates κ^2-N$_{py}$, N$_{imine}$ to the osmium atom. The reason for this novel preference is thermodynamic, since it gives rise to the most stable structure from all the possible options.[69] Accordingly, treatment of 2-propanol solutions of **1** with 0.5 equiv. of HBMePI, in the presence of 0.5 equiv. of K^{t}BuO, at room temperature gives the salt [\{OsH$_3$(P^{i}Pr$_3$)$_2$\}$_2$\{μ-(κ^2-N$_{py}$,N$_{imine}$)$_2$-BMePI\}]Cl (**120**), containing a cation that can be formally described as the result of the coordination of an [OsH$_3$(P^{i}Pr$_3$)$_2$]$^+$ fragment to the free chelate-N$_{py}$,N$_{imine}$ moiety of complex OsH$_3$(κ^2-N$_{py}$,N$_{imine}$-BMePI)(P^{i}Pr$_3$)$_2$ (Scheme 43). The latter is an efficient catalyst precursor for the acceptorless and base-free dehydrogenation of secondary and primary alcohols and cyclic and lineal amines. The primary alcohols afford aldehydes. The amount of H$_2$ released per gram of heterocycle depends upon the presence of a methyl group adjacent to the nitrogen atom, the position of the nitrogen atom in the heterocycle, and the size of the heterocycle.

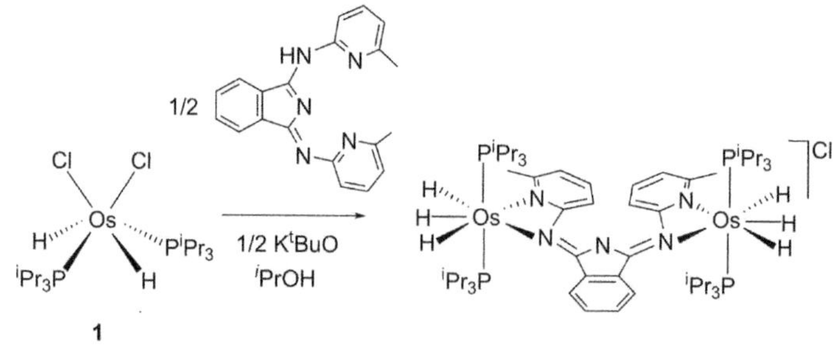

Scheme 43 N—H bond activation of 1,3-bis(6′-methyl-2′-pyridylimino)isoindoline promoted by **1**.

A more efficient catalyst for the base-free and acceptorless dehydrogenation of alcohols has been prepared by N—H bond activation of 3-(2-pyridyl)pyrazol (3-py-pzH) and subsequent coordination of the Ir(η^4-COD) fragment to the resulting N,N,N-osmaligand (Scheme 44).[71] Complex **1** reacts with 3-py-pzH in toluene under reflux, to give the osmium(II) derivative Os{κ^2-N,N-(3-py-pz)}$_2$(3-py-pzH)(P^{i}Pr$_3$) (**121**). This compound can be deprotonated by the bridging methoxy groups of the dimer [Ir(μ-OMe)(η^4-COD)]$_2$. The deprotonation produces the coordination of two free nitrogen atoms of the metalloligand to iridium to afford the heterobimetallic derivative **122**.

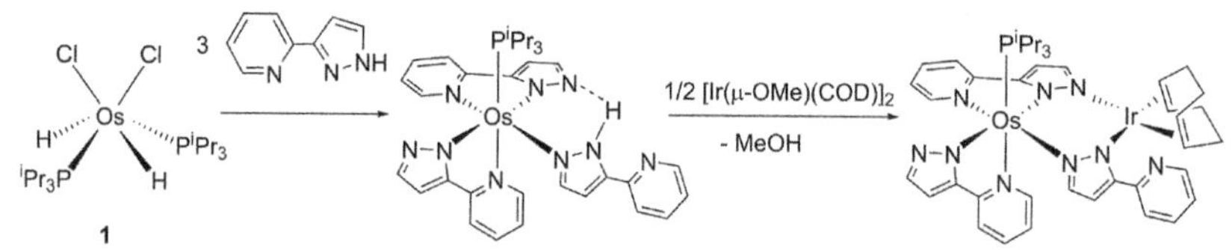

Scheme 44 N—H activation of 3-(2-pyridyl)pyrazol promoted by **1**.

The iridium center of **122** catalyzes the dehydrogenation of secondary alcohols to ketones, primary alcohols to aldehydes or esters, and diols to lactones. Cyclooctatriene was detected during the catalysis, suggesting that the active species is an iridium(III)-dihydride compound with the anionic osmaligand κ^3-(N,N,N)$_{mer}$-coordinated. The presence of a phenyl group in the substrates favors the dehydrogenations. The products of the reactions of benzyl alcohols bearing a substituent at the phenyl group shows a marked dependence upon the Hammett σ_p value of the substituent, which is consistent with the transitory formation of hemiacetals in the dehydrogenative homocoupling to generate esters.

11. O—H bond activation

Complex **1** activates the O—H bond of acetone oxime and cyclohexanone oxime, at room temperature, in the presence of triethylamine (Scheme 45). The reactions give the oximate derivatives OsH$_2$Cl{κ^2-O, N-(ON=CR$_2$)}(P^{i}Pr$_3$)$_2$ (CR$_2$=CMe$_2$ (**123**), CC$_5$H$_{10}$ (**124**)). The structure of these compounds, determined by X-ray diffraction analysis, has been rationalized as a distorted pentagonal bipyramid with the phosphines occupying two relative *trans* positions. The perpendicular plane is formed by the hydrides, the chloride and the oximate group. The latter acts as bidentate ligand with a bite angle of 36.6(1)°, which strongly deviates from the ideal value of 72°.[72]

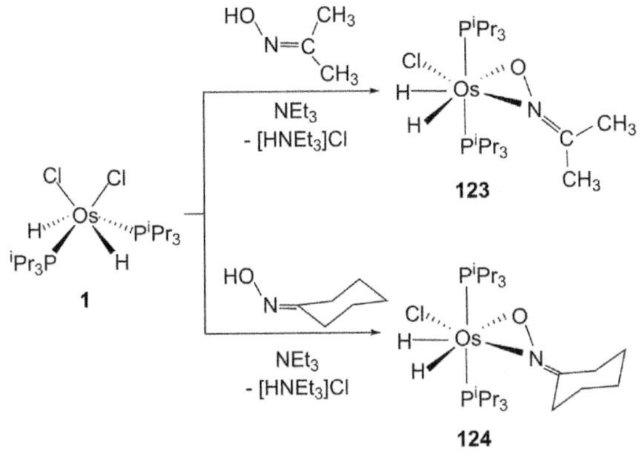

Scheme 45 O—H activation of oximes promoted by complex **1**.

Complex **1** also activates O—H bonds of metal-nucleosides, in the presence of triethylamine. The reactions lead to dinuclear species formed by two different metal fragments (Scheme 46). Treatment of **1** with complexes methylurinate **125** and urinate **126**, containing nucleosides derived from ribose, in the presence of the base leads to the dinuclear species OsH$_3$(P^{i}Pr$_3$)$_2$(nucleobase)-(ribose)OsH$_2$(P^{i}Pr$_3$)$_2$ **127** and **128**, as a result of the replacement of the proton of the OH functional groups of the ribose five-membered ring by the [OsH$_2$(P^{i}Pr$_3$)$_2$]$^{2+}$ metal fragment.[73] Similarly, the reactions with the complexes rhodium- and iridium-purine **129** and **130** afford the heterobimetallic compounds **131** and **132**, respectively, whereas the iridium-pyrimidine **133** gives **134**.[74]

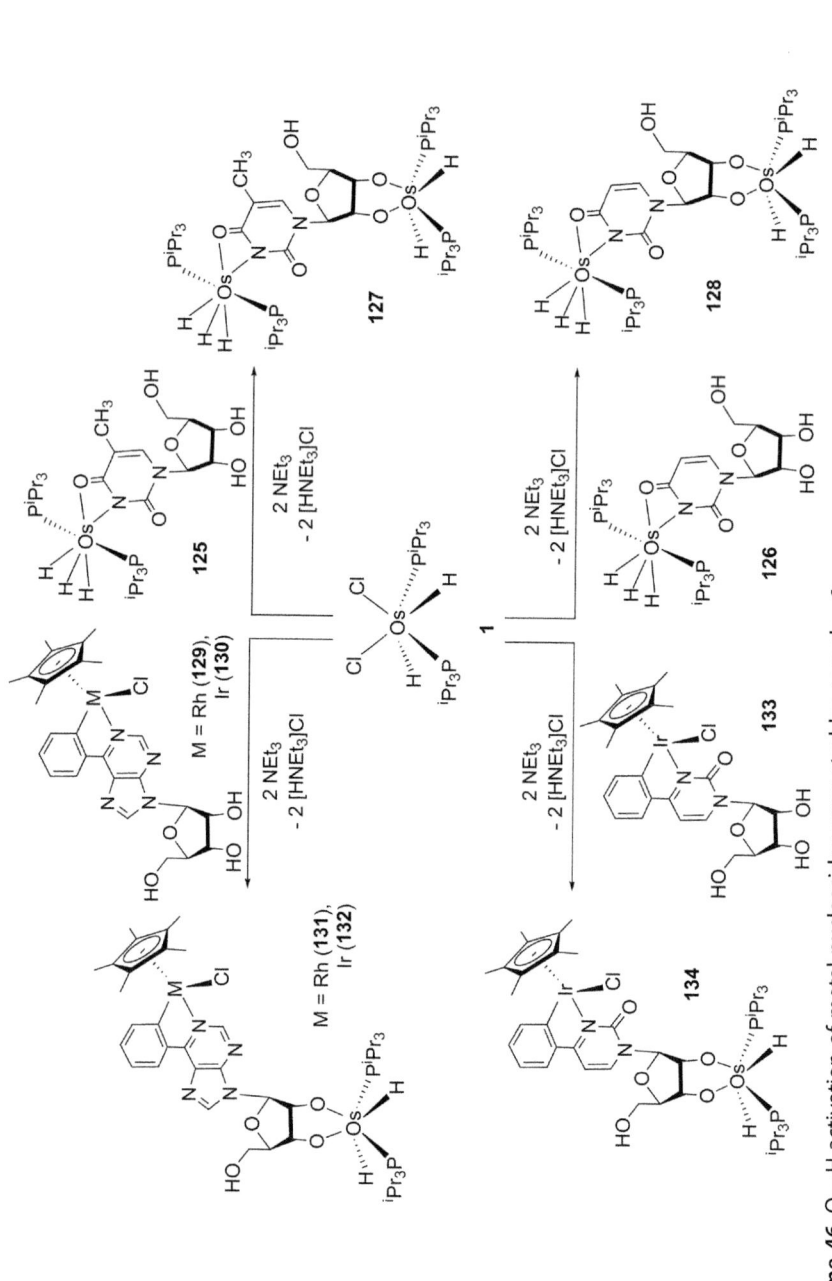

Scheme 46 O—H activation of metal-nucleosides promoted by complex 1.

12. O—N bond activation: Azavinylidene compounds

Treatment of toluene solutions of **1** with cyclohexanone oxime, in the absence of triethylamine, under reflux leads to the hydride azavinylidene derivative OsHCl$_2$(=N=CC$_5$H$_{10}$)}(P^{i}Pr$_3$)$_2$ (**135**) and water (Scheme 47). The reaction takes place via the oximate complex **124**, which is initially formed along with HCl according to Scheme 45. In agreement with this, it has been also observed that the addition of HCl to **123** affords OsHCl$_2$(=N=CMe$_2$)}(P^{i}Pr$_3$)$_2$ (**136**).[75] The azavinylidene ligands of **135** and **136** are stable toward hydrolysis. Both compounds react with AgOTf in the presence of water to afford the salts [OsHCl(=N=CR$_2$)(H$_2$O)(P^{i}Pr$_3$)$_2$]OTf (CR$_2$=CC$_5$H$_{10}$ (**137**), CMe$_2$ (**138**)).[76]

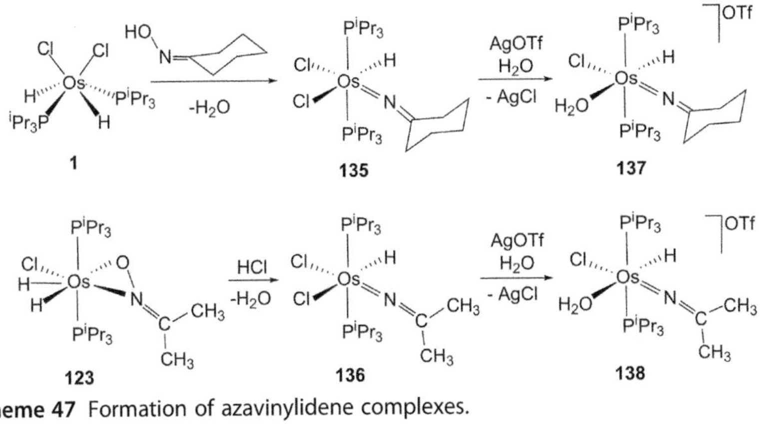

Scheme 47 Formation of azavinylidene complexes.

The structures of **135** and **138** have been determined by X-ray diffraction analysis. Although they are six-coordinate diamagnetic species of a formally d^4-ion, the geometries around the metal center are octahedral with the phosphines occupying *trans* positions. The hydride and azavinylidene ligands lie at the perpendicular plane disposed mutually *cis*. The chloride ligands appear to play a main role in this ligand distribution. In contrast to **135** and **138**, the trihydride derivative OsH$_3$(=N=CHPh)}(P^{i}Pr$_3$)$_2$ (**139**), which does not bear any chloride ligand, displays a structure of *Cs* symmetry which resembles those of **10–12** with the azavinylidene group at the position of the halide. To gain insight into the reason of this difference and the influence having in the reactivity, the osmium-azavinylidene bonding situations in **139** and a model hydride-dichloride-counterpart

OsHCl$_2$(=N=CHPh)(P^{i}Pr$_3$)$_2$ (**140**) were analyzed, by means of the energy decomposition analysis—Natural orbitals for chemical valence (EDA-NOCV) method, using two different bonding schemes: the donor-acceptor bonding in their electronic singlet state, with charged fragments ([OsH$_3$(P^{i}Pr$_3$)$_2$]$^+$, [OsHCl$_2$(P^{i}Pr$_3$)$_2$]$^+$, and [N=CHPh]$^-$), and electron sharing mixed with donor-acceptor bonding where the fragments ([OsH$_3$(P^{i}Pr$_3$)$_2$]·, [OsHCl$_2$(P^{i}Pr$_3$)$_2$]·, and [N=CHPh]·) were calculated in their doublet state. According to the results of the analysis (Table 1) the donor-acceptor description of the Os—N bond dominates over mixed electron-sharing/dative bonding in **139**. However, this situation is markedly different in **140**; for it, the mixed electron-sharing/donor-acceptor bonding better describes the Os—N interaction.[77] In this context, it should be mentioned that quantum chemical calculations using DFT at the BP86-D3(BJ)/

Table 1 Results of EDA-NOCV method computed at the ZORA-BP86-D3/TZ2P//BP86-D3/def2-SPV level.

	Donor-acceptor	Electron-sharing/ Donor-acceptor	Donor-acceptor	Electron-sharing/ Donor-acceptor
Fragments	[Os]$^+$ [N=CHPh]$^-$	[Os]· (d) [N=CHPh]· (d)	[Os]$^+$ [N=CHPh]$^-$	[Os]· (d) [N=CHPh]· (d)
ΔE_{int}	−202.2	−112.0	−258.5	−117.4
ΔE_{Pauli}	244.2	210.0	347.1	293.6
ΔE_{elstat}[a]	−277.4 (62.2)	−145.9 (45.3)	−313.0 (52.4)	−189.5 (46.1)
ΔE_{orb}[a]	−154.7 (34.6)	−161.9 (50.3)	−268.9 (45.0)	−205.8 (50.1)
ΔE_{disp}[a]	−14.3 (3.2)	−14.3 (4.4)	−15.7 (2.6)	−15.7 (3.8)
$\Delta E_{orb}(\rho^1)$[b]	−50.9 (32.9)	−81.9 (50.6)	−127.0 (47.2)	−92.5 (44.9)
$\Delta E_{orb}(\rho^2)$[b]	−27.7 (17.9)	−28.5 (17.6)	−35.0 (13.5)	−37.6 (18.3)
$\Delta E_{orb}(\rho^3)$[b]	−47.0 (30.4)	−36.9 (22.8)	−69.5 (7.9)	−57.3 (27.9)
$\Delta E_{orb}(rest)$[b]	−29.1 (18.8)	−14.6 (9.0)	−37.4 (13.9)	−18.4 (8.9)

[a]The values in parentheses indicate the percentage to the total attractive interaction energy: $\Delta E_{elstat} + \Delta E_{orb} + \Delta E_{disp}$.
[b]The values in parentheses give the percentage contribution to the total orbital interactions ΔE_{orb}. The smallest ΔE_{orb} value indicates the most faithful description of the type of the binding.

def2-TZVPP level of theory have revealed that electrophilic alkylidynes, like those of **43** and **44** (Scheme 13), also engage in a mixture of dative bonding (σ-donation and π-backdonation) and one electron-sharing π-bond. The EDA-NOCV calculations of nucleophilic alkylidynes using open-shell species in their quartet electronic state gave ΔE_{orb} values similar to those of neutral fragments in their electronic doublet state.[78] For **139** and **140**, a pure electron-sharing bonding with the fragments in the quartet state afforded the highest ΔE_{orb} and was therefore discarded.[77]

Azavinylidenes are usually viewed as α-nitrogen counterparts of vinylidenes. However, they compare better with alkylidynes according to that previously mentioned. Vinylidenes are two-electron donor ligands with a standardized reactivity, which is dominated by the addition of nucleophiles at the α-carbon, whereas the electrophiles attack at the β-carbon atom.[79] In contrast both azavinylidenes and alkylidynes are three-electron donors and display two different electronic situations, depending upon the metal fragment, which give rise to different chemical behaviors, although the difference between nitrogen and carbon generates significant changes in reactivity.

The differences in the osmium-azavinylidene bonding situation are also reflected in the charges on the atoms of the C=N bond and in the hydride ligands (Chart 3). The N atom of the azavinylidene ligand of **139** supports a negative charge that is approximately twice that on the N atom of **140**. On the other hand, while the C atom of the azavinylidene of **139** is slightly positive, that of **140** is slightly negative. Furthermore, interestingly, one of the hydride ligands of the trihydride is slightly positive, whereas the other two and that of **140** are slightly negative.

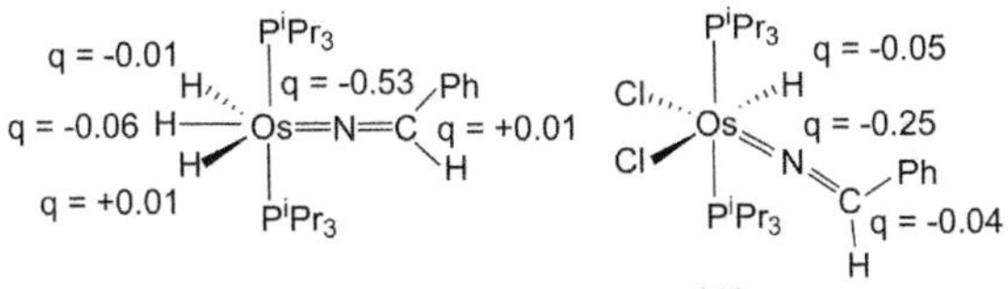

Chart 3 Computed (BP86-D3/def2-SVP level) natural bond orbital (NBO) partial charges for **139** and **140**.

The electronic differences between **139** and **140** translate into the chemical behaviors. In contrast to **135–138**, complex **139** is not stable in the presence of water. The addition of 1 equiv. of water to the toluene or tetrahydrofuran solutions of the latter rapidly produces its isomerization to the orthometalated phenylmethanime derivative OsH$_3${κ^2-N,C-(NH=CHC$_6$H$_4$)}(P^{i}Pr$_3$)$_2$ (**141**). The participation of water in the reaction was confirmed by means of the addition of D$_2$O instead of H$_2$O, which afforded selectively and quantitatively OsH$_3${κ^2-N,C-(ND=CHC$_6$H$_4$)}(P^{i}Pr$_3$)$_2$ (**141-d$_1$**). The formation of **141** involves the migration of the electrophilic hydride from the metal center to the nucleophilic N atom of the azavinylidene group to afford an unsaturated dihydride-osmium-aldimine intermediate **Ab**, which evolves by oxidative addition of an *ortho*-CH bond of the phenyl substituent of the aldimine (Scheme 48). According to the generation of **141-d$_1$**, the hydride migration is promoted by water, which acts as a proton shuttle. This was confirmed by DFT calculations (B3LYP-D3/SDD/6-31G**). The direct migration of the hydride to the N atom takes place with activation energy of 31.4 kcal mol^{-1}, whereas the proton shuttle formed by three water molecules, consecutively associated by means of hydrogen bonds, reduces the barrier for the hydride migration to 19.6 kcal mol^{-1}.

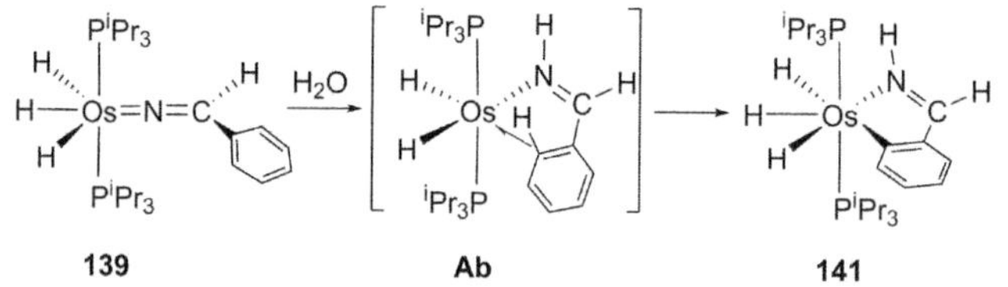

Scheme 48 Isomerization of **139–141**.

The pair formed by the electrophilic hydride and the nitrogen atom of the azavinylidene of **139** promotes the heterolytic rupture of the B—H bond of HBpin (Scheme 49). In toluene, at 80 °C, the reaction of the trihydride with the borane affords H$_2$ and a dihydride-osmium-boryl(aldimine) species **Ac**, which similarly to **Ab** undergoes the oxidative addition of an *ortho*-CH bond of the phenyl substituent of the aldimine to give OsH$_3${κ^2-N,C-[N(Bpin)=CHC$_6$H$_4$]}(P^{i}Pr$_3$)$_2$ (**142**). In agreement with the formation of **142**, complex **139** also isomerizes into **141** under 1 atm of H$_2$.[77]

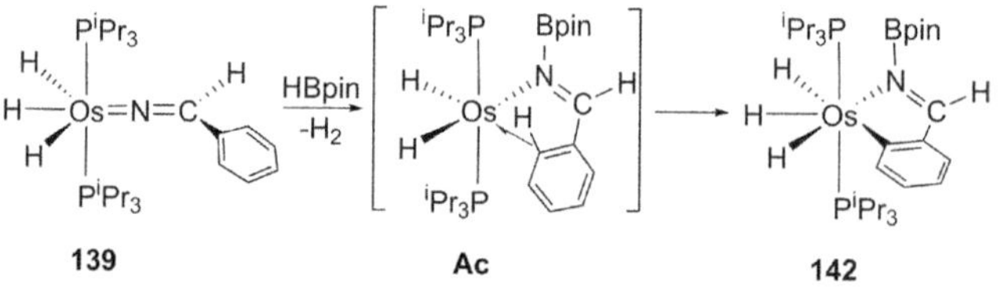

Scheme 49 B—H bond activation of HBpin promoted by **139**.

Azavinylidene ligands of octahedral complexes **135** and **136** promote an interesting intramolecular C(sp^2)-H bond activation (Scheme 50). Treatment of their dichloromethane solutions with AgOTf, at room temperature, and the subsequent addition of phenylacetylene to the resulting solution at −25 °C leads to the alkenyl-azavinylidene derivatives [Os{(E)-CH=CHPh}Cl(=N=CR$_2$)(P^{i}Pr$_3$)$_2$]OTf (CR$_2$=CC$_5$H$_{10}$ (**143**), CMe$_2$ (**144**)), as a result of the insertion of the C—C triple bond of the alkyne into the Os—H bond of the starting azavinylidene complexes. The metal center of these species is saturated through an agostic bond with the alkenyl H$_\beta$ atom. Addition of NaCl to the tetrahydrofuran solutions of **143** and **144**, at −30 °C, produces the split of the agostic interaction and the formation of the neutral six-coordinate compounds Os{(E)-CH=CHPh}Cl$_2$(=N=CR$_2$)(P^{i}Pr$_3$)$_2$ (CR$_2$=CC$_5$H$_{10}$ (**145**), CMe$_2$ (**146**)), which evolve into the imine-vinylidene derivatives OsCl$_2$(=C=CHPh)(NH=CR$_2$)(P^{i}Pr$_3$)$_2$ (CR$_2$=CC$_5$H$_{10}$ (**147**), CMe$_2$ (**148**)) as a consequence of a C$_\alpha$-to-N hydrogen migration.[80]

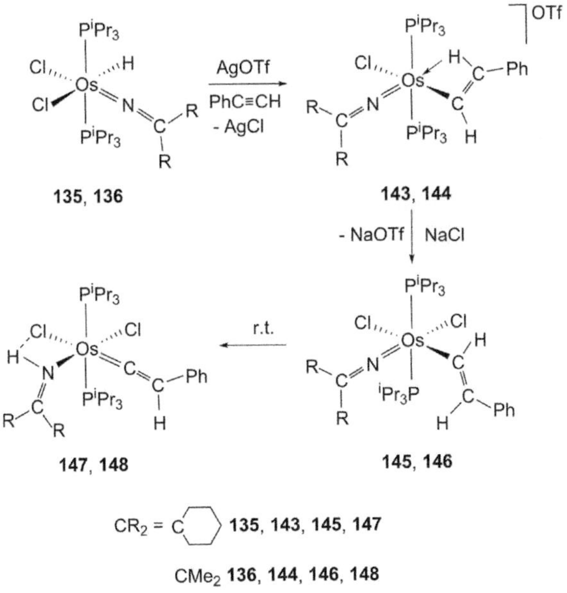

Scheme 50 Transformation of **135** and **136** in **147** and **148**: formation of imine-vinylidene derivatives.

The phenyl group stabilizes the vinylidene ligand of **147** and **148**. Thus, in contrast to phenyl, cyclohexyl promotes an additional N-to-C$_\beta$ hydrogen migration in the imine-vinylidene derivative, which affords an

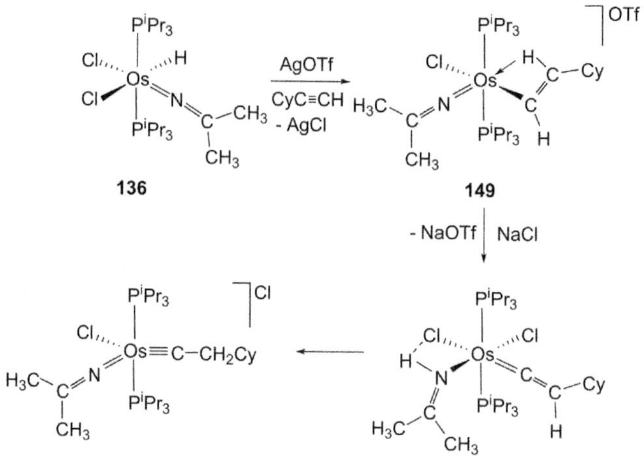

Scheme 51 Transformation of **136** in **151**: formation of azavinylidene-alkylidyne derivatives.

azavinylidene-alkylidyne compound (Scheme 51). The addition of cyclohexylacetylene, at −25 °C, to the dichloromethane solutions resulting from the treatment of **136** with AgOTf affords [Os{(E)-CH=CHCy}Cl(=N=CMe$_2$)(P^{i}Pr$_3$)$_2$]OTf (**149**), related to **143** and **144**. Similarly to them, it evolves into the neutral imine-vinylidene OsCl$_2$(=C=CHCy)(NH=CMe$_2$)(P^{i}Pr$_3$)$_2$ (**150**) in the presence of NaCl, at −30 °C, in tetrahydrofuran. In contrast to **147** and **148**, complex **150** is not stable in dichloromethane at room temperature and isomerizes to the azavinylidene-alkylidyne salt [OsCl(=N=CMe$_2$)(≡CCH$_2$Cy)(P^{i}Pr$_3$)$_2$]Cl (**151**).[81]

Addition of acetylene, 1-pentyne, 2-methyl-1-buten-3-yne, 1-phenyl-2-propyn-1-ol, 2-phenyl-3-butyn-2-ol, and 1,1-diphenyl-2-propyn-1-ol, at room temperature, to dichloromethane solutions resulting from the treatment of **136** with AgOTf also afford azavinylidene-alkylidyne derivatives, which give azavinylidene-vinylidenes by deprotonation (Scheme 52). Acetylene and 1-pentyne generate the alkylalkylidyne complexes [OsCl(=N=CMe$_2$)(≡CCH$_2$R)(P^{i}Pr$_3$)$_2$]OTf (R=H (**152**), ^{n}Pr (**153**)), whereas the enyne and the alkynols give the respective alkenylalkylidyne compounds [OsCl(=N=CMe$_2$)(≡CCH=CMe$_2$)(P^{i}Pr$_3$)$_2$]OTf (**154**) and [OsCl(=N=CMe$_2$)(≡CCH=CPhR)(P^{i}Pr$_3$)$_2$]OTf (R=H (**155**), Me (**156**), Ph (**157**)). The deprotonation of alkylidyne ligands of **152** and **153** with MeLi leads to the vinylidene complexes OsCl(=N=CMe$_2$)(=C=CHR)(P^{i}Pr$_3$)$_2$ (R=H (**158**), ^{n}Pr (**159**)), whereas the alkenylalkylidyne

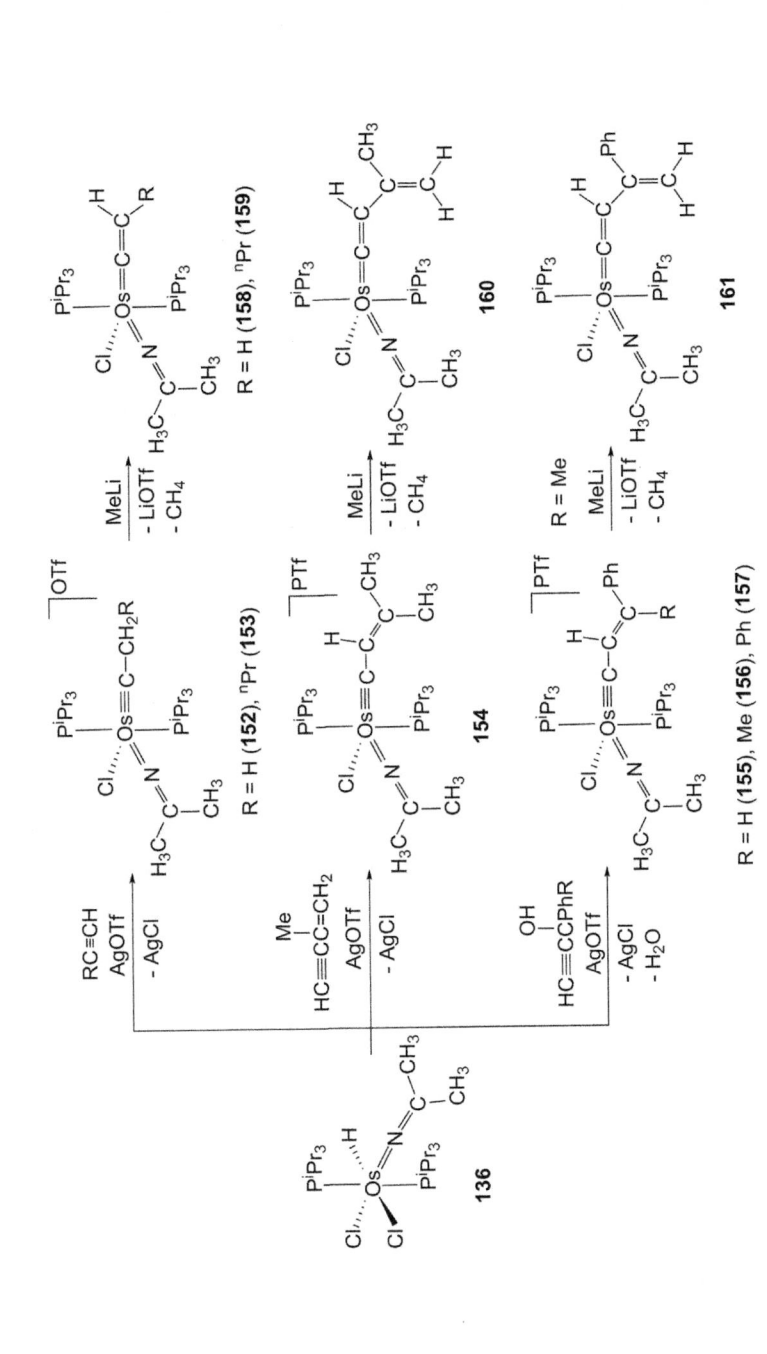

Scheme 52 Reactions of **136** with alkynes, enynes, and alkynoles.

ligands of **154** and **156** give the alkenylvinylidenes of derivatives OsCl(=N=CMe$_2$){=C=CHC(R)=CH$_2$}(P^{i}Pr$_3$)$_2$ (R=Me (**160**), Ph (**161**)) under the same conditions. Treatment of the imine-vinylidene **147** and **148** with ^{n}BuLi yields the azavinylidene-vinylidene OsCl(=N=CR$_2$)(=C=CHPh)(P^{i}Pr$_3$)$_2$ (CR$_2$=CC$_5$H$_{10}$ (**162**), CMe$_2$ (**163**)), related to **158** and **159**.[82]

The coordination of chloride to the metal center of **143** and **144** is certainly responsible of the C$_\alpha$-to-N hydrogen migration, which leads to the imine-vinylidene derivatives **147** and **148**. Thus, in contrast to chloride, carbon monoxide promotes the coupling between the azavinylidene and styryl ligands (Scheme 53).[83] Under 1 atm of this gas, complexes **143** and **144** evolve to the respective Δ^2-1,2 azaosmetine compounds **164** and **165**. Their formation has been rationalized as intramolecular [2+2] cycloaddition reactions between the Os—N and C—C double bonds. The π-acceptor character of the initially coordinated carbonyl ligand appears to excite the π-donor nature of the azavinylidene group, which increases the double character of the Os—N bond and therefore favors the cyclization. The deprotonation of the C(sp^3) atom of the four-membered heterometalaring with MeLi produces a Δ^2-to-Δ^3 transformation of the azaosmetine to give **166** and **167**, which react with molecular hydrogen to afford the well-known dihydride-dihydrogen derivative OsH$_2$(η^2-H$_2$)(CO)(P^{i}Pr$_3$)$_2$[84] and the corresponding 2-aza-1,3-butadienes.[83]

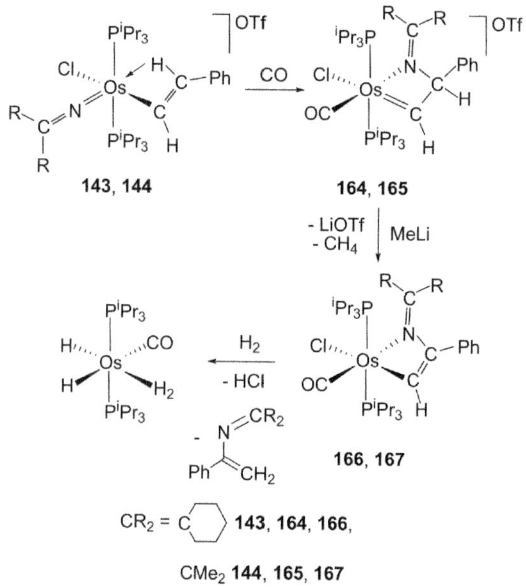

Scheme 53 Carbon monoxide promoted azavinylide-styryl coupling.

13. Rupture of the oxygen molecule

Oxo complexes are extremely important due to their applications for the *cis*-hydroxylation of alkenes to *cis*-diols. The reduction of the oxo compound by the organic substrate can directly occur. However, in general, the resultant reduced metallic species cannot be reoxidized by molecule oxygen but require strong oxidants.[85] Given the advantages of air or dioxygen as final oxidant, the metal-promoted activation of the O$_2$ molecule is a reaction of great interest.

Dihydride-dichloride complex **1** is capable of activating the O—O double bond of molecular oxygen from the air or pure oxygen, to give the osmium(VI)-dioxo derivative *trans*-OsO$_2$Cl$_2$(P^{i}Pr$_3$)$_2$ (**168**), in toluene, at room temperature. Subsequent treatment of toluene solutions of **168** with ^{n}BuLi affords *trans*-OsO$_2$(P^{i}Pr$_3$)$_2$ (**169**) and *n*-octane (Scheme 54).[86]

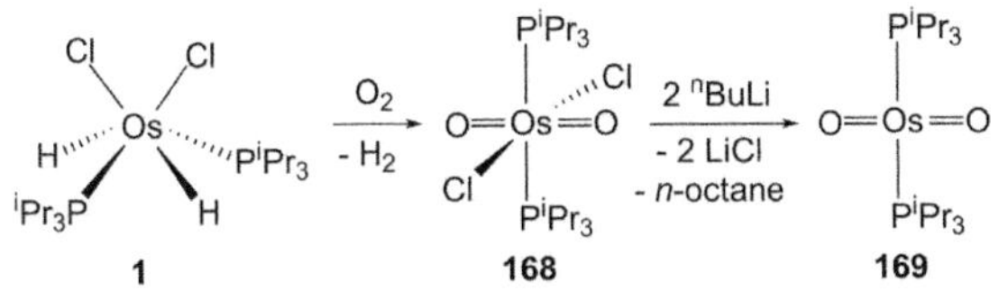

Scheme 54 Reaction of **1** with oxygen.

The geometry of **168** is octahedral with O-Os-O, P-Os-P, and Cl-Os-Cl angles of 180°, whereas the geometry of **169** is square-planar and can be viewed as derived from that of **168** by loss of the chloride ligands and shortening of the Os—O and Os—P bonds. Thus, the O-Os-O and P-Os-P angles are also 180°, while the Os—O (1.743(2) Å) and Os—P (2.4178(9) Å) bond lengths are about 0.03 and 0.1 Å, respectively, shorter than the Os—O (1.767(6) Å) and Os—P (2.519(2) Å) distances found in **168**. DFT Calculations on **169** revealed that the electron filling pattern is consistent with a formal d^4 electron count for the metal ion. The occupied HOMO and HOMO-1 have mainly d character of nonbonding nature. The three next empty orbitals (LUMO, LUMO+1, and LUMO+2) are of metal-ligand antibonding character. The LUMO and LUMO+1 result from π-antibonding interactions between π-donor *p* orbitals of the oxygen atoms and *d* orbitals of the metal, whereas LUMO+2 is σ-phosphorous-osmium antibonding. The high energy of the empty orbitals leads to a sizable HOMO-LUMO gap, which is in agreement with the stability of this unusual four-coordinate molecule.

14. Cl—H bond activation

This bond activation has been invoked to rationalize the formation of the trichloride-azavinylidene complex OsCl$_3$(=N=CC$_5$H$_{10}$)}(P^{i}Pr$_3$)$_2$ (**170**), as a side product of the reaction shown in Scheme 47, which yields **135** as main product. The HCl molecule generated in the formation of the initial oximate intermediate, complex **124**, is activated by the starting complex **1**, to give H$_2$ and the hydride-trichloride OsHCl$_3$(P^{i}Pr$_3$)$_2$ (**171**). The subsequent reaction of the latter with cyclohexanone oxime should afford **170** and water, by a process similar to the formation of **135** (Scheme 55).[75] The hydride-trichloride complex **171** is usually prepared as a blue-green solid, by treatment of **1** with 1 equiv. of N-chlorosuccinimide in chloroform at room temperature.[27]

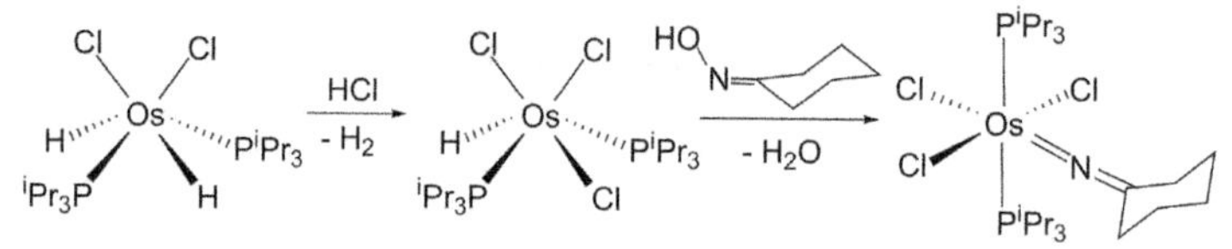

Scheme 55 Formation of the trichloride-azavinylidene **170**.

The osmium–azavinylidene bonding situation in **170** is similar to that of the hydride-dichloride derivatives **135** and **136**; i.e., the Os—N interaction is well described by means of a mixed electron-sharing/donor-acceptor bonding.[77]

15. Conclusions

The number of unsaturated osmium(IV)-hydride complexes is certainly limited. However, their chemistry is very rich. They display a notable ability to activate symmetrical and asymmetrical σ-bonds of a wide range of organic and inorganic molecules. A characteristic of the activation processes is that they do not result in an increase of the formal oxidation state of the metal center. This is a consequence of their dihydride or trihydride nature, which undergo an intramolecular reduction to afford saturated Kubas-type or elongated dihydrogen species, when coordinate the σ-bond. Before the activation, these intermediates dissociate the Kubas-type dihydrogen or eliminate a HX molecule as consequence of the heterolytic cleavage of the elongated dihydrogen. In this way, even the homolytic cleavage of the coordinated σ-bond does not increase the formal metal oxidation state over that of the starting compound.

Dihydride-dichloride complex **1** is the compound of this family showing the widest variety of reactions. Thus, it has become one of the cornerstones in the development of the modern osmium organometallic chemistry. Its ability to activate $C(sp^2)$-H and $C(sp^2)$-C bonds of unsaturated organic substrates and to generate Os—C multiple bonds is noticeable. The variety of stannyl-polyhydrides that its reactions with Ph_3SnH have provided and the ability of some of them to activate $C(sp^2)$-H bonds should be also pointed out. The formation of azavinylidenes with an interesting reactivity, which promotes C-to-N and N-to-C hydrogen migrations to convert single Os—C bonds into double and subsequently into triple bonds, should be also mentioned, as well as its rupture capacity of molecular oxygen.

Unsaturated osmium(IV)-hydride complexes have been usually stabilized with bulky phosphines. This has been certainly one of the reasons that have limited the number of available compounds of this class. New ligands are necessary to reach further development in the chemistry of these type of complexes. In this context, the use of pincers is promising, according to the results obtained with pincer-diphosphines in Si—H bond activation reactions.[58] N-heterocyclic carbenes (NHCs) are other ligands of interest, which could stabilize these species improving their properties. Some mixed phosphine-osmium-NHC complexes have been prepared in recent years and their chemical behavior is currently in study.[87]

The revised reactions suggest two lines of future application. On one hand, the possibility of stabilizing compounds with both electrophilic and nucleophilic centers, as the trihydride azavinylidene **139**, opens new reactivity pathways out the metal coordination sphere, which can result of interest to design novel types of catalysis. On the other, the incorporation of two different metal fragments by means of the O—H bond activation of metal-nucleosides could be used as a way for the orthogonal functionalization of oligonucleotides, taking advantage of the particular reactivity and properties of each metal fragment to promote specific metal transformations.

In conclusion, unsaturated osmium(IV)-hydride complexes are a class of compounds now stabilized by bulky phosphines, which promote a wide range of σ-bond activation reactions and could have interesting new future applications if the number and variety of ligands for their stabilization is increased.

Acknowledgments

Financial support from the MINECO of Spain (Projects CTQ2017-82935-P and RED2018-102387-T (AEI/FEDER, UE)), Gobierno de Aragón (Group E06_20R and project LMP148_18), FEDER, and the European Social Fund is acknowledged.

References

1. Kubas GJ. *Metal Dihydrogen and σ-Bond Complexes*. New York: Kluwer Academic/Plenum Publishers; 2001.
2. Esteruelas MA, López AM, Oliván M. Polyhydrides of platinum group metals: nonclassical interactions and σ-bond activation reactions. *Chem Rev*. 2016;116: 8770–8847.
3. Clapham SE, Hadzovic A, Morris RH. Mechanisms of the H_2-hydrogenation and transfer hydrogenation of polar bonds catalyzed by ruthenium hydride complexes. *Coord Chem Rev*. 2004;248:2201–2237.
4. Grützmacher H. Cooperating ligands in catalysis. *Angew Chem Int Ed*. 2008;47: 1814–1818.
5. Khusnutdinova JR, Milstein D. Metal-ligand cooperation. *Angew Chem Int Ed*. 2015;54: 12236–12273.
6. Morris RH. Exploiting metal-ligand bifunctional reactions in the design of iron asymmetric hydrogenation catalysts. *Acc Chem Res*. 2015;48:1494–1502.
7. Alig L, Fritz M, Schneider S. First-row transition metal (De)hydrogenation catalysis based on functional pincer ligands. *Chem Rev*. 2019;119:2681–2751.
8. Higashi T, Kusumoto S, Nozaki K. Cleavage of Si-H, B-H, and C-H bonds by metal-ligand cooperation. *Chem Rev*. 2019;119:10393–10402.
9. Bertoli M, Choualeb A, Lough AJ, Moore B, Spasyuk D, Gusev DG. Osmium and ruthenium catalysts for dehydrogenation of alcohols. *Organometallics*. 2011;30: 3479–3482.
10. Spasyuk D, Gusev DG. Acceptorless dehydrogenative coupling of ethanol and hydrogenation of esters and imines. *Organometallics*. 2012;31:5239–5242.
11. Dub PA, Henson NJ, Martin RL, Gordon JC. Unravelling the mechanism of the asymmetric hydrogenation of acetophenone by [RuX_2(diphosphine)(1,2-diamine)] catalysts. *J Am Chem Soc*. 2014;136:3505–3521.
12. Spasyuk D, Vicent C, Gusev DG. Chemoselective hydrogenation of carbonyl compounds and acceptorless dehydrogenative coupling of alcohols. *J Am Chem Soc*. 2015;137:3743–3746.
13. Dub PA, Gordon JC. The mechanism of enantioselective ketone reduction with Noyori and Noyori-Ikariya bifunctional catalysts. *Dalton Trans*. 2016;45:6756–6781.
14. Gusev DG. Dehydrogenative coupling of ethanol and ester hydrogenation catalyzed by pincer-type YNP complexes. *ACS Catal*. 2016;6:6967–6981.
15. Gusev DG. Rethinking the dehydrogenative amide synthesis. *ACS Catal*. 2017;7: 6656–6662.
16. Dub PA, Gordon JC. Metal-ligand bifunctional catalysis: the "accepted" mechanism, the issue of concertedness, and the function of the ligand in catalytic cycles involving hydrogen atoms. *ACS Catal*. 2017;7:6635–6655.
17. Perutz RN, Sabo-Etienne S. The σ-CAM mechanism: σ complexes as the basis of σ-bond metathesis at late-transition-metal centers. *Angew Chem Int Ed*. 2007;46: 2578–2592.
18. Waterman R. σ-Bond metathesis: a 30-year retrospective. *Organometallics*. 2013;32: 7249–7263.

19. Templeton JL, Ward BC. Molecular structure of Mo(CO)$_2$[S$_2$CN-i-Pr$_2$]$_2$. A Trigonal-prismatic electron-deficient molybdenum(II) carbonyl derivative. *J Am Chem Soc.* 1980;102:6568–6569.
20. Kubáček P, Hoffmann R. Deformations from octahedral geometry in d^4 transition-metal complexes. *J Am Chem Soc.* 1981;103:4320–4332.
21. Aracama M, Esteruelas MA, Lahoz FJ, et al. Synthesis, reactivity, molecular structure, and catalytic activity of the novel dichlorodihydridoosmium(IV) complexes OsH$_2$Cl$_2$(PR$_3$)$_2$ (PR$_3$ = P-i-Pr$_3$, PMe-t-Bu$_2$). *Inorg Chem.* 1991;30:288–293.
22. Maseras F, Eisenstein O. Opposing steric and electronic contributions in OsCl$_2$H$_2$(PPr$^i_3$)$_2$. A theoretical study of an unusual structure. *New J Chem.* 1998;22:5–9.
23. Gusev DG, Kuhlman R, Sini G, Eisenstein O, Caulton KG. Distinct structures for ruthenium and osmium hydrido halides: Os(H)$_3$X(P^{i}Pr$_3$)$_2$ (X = Cl, Br, I) are non-octahedral classical trihydrides with exchange coupling. *J Am Chem Soc.* 1994;116:2685–2686.
24. Kuhlman R, Clot E, Leforestier C, Streib WE, Eisenstein O, Caulton KG. Quantum exchange coupling: a hypersensitive indicator of weak interactions. *J Am Chem Soc.* 1997;119:10153–10169.
25. Esteruelas MA, Fuertes S, Oliván M, Oñate E. Behavior of OsH$_2$Cl$_2$(P^{i}Pr$_3$)$_2$ in aceto-nitrile: the importance of the small details. *Organometallics.* 2009;28:1582–1585.
26. Gusev DG, Kuznetsov VF, Eremenko IL, Berke H. An unusual example of H$_2$ coordination by a d^4 metal center: reactions between OsH$_2$Cl$_2$(P-i-Pr$_3$)$_2$ and H$_2$. *J Am Chem Soc.* 1993;115:5831–5832.
27. Kuhlman R, Gusev DG, Eremenko IL, Berke H, Huffman JC, Caulton KG. Dihydrogen addition to (P^{i}Pr$_3$)$_2$OsX$_n$H$_{4-n}$. *J Organomet Chem.* 1997;536–537:139–147.
28. Gusev DG, Dolgushin FM, Antipin MY. Cyclometalated osmium complexes containing a tridentate PCP ligand. *Organometallics.* 2001;20:1001–1007.
29. Bertoli M, Choualeb A, Gusev DG, Lough AJ, Major Q, Moore B. PNP pincer osmium polyhydrides for catalytic dehydrogenation of primary alcohols. *Dalton Trans.* 2011;40:8941–8949.
30. Esteruelas MA, Fernández-Alvarez FJ, López AM, Mora M, Oñate E. Borinium cations as σ-B-H ligands in osmium complexes. *J Am Chem Soc.* 2010;132:5600–5601.
31. Esteruelas MA, Fernández I, López AM, Mora M, Oñate E. Preparation, structure, bonding, and preliminary reactivity of a six-coordinate d^4 osmium – boryl complex. *Organometallics.* 2012;31:4646–4649.
32. Buil ML, Cardo JJF, Esteruelas MA, Fernández I, Oñate E. Unprecedented addition of tetrahydroborate to an osmium-carbon triple bond. *Organometallics.* 2014;33:2689–2692.
33. Esteruelas MA, Fernández I, García-Yebra C, Martín J, Oñate E. Elongated σ-borane versus σ-borane in pincer-POP-osmium complexes. *Organometallics.* 2017;36:2298–2307.
34. Babón JC, Esteruelas MA, Fernández I, López AM, Oñate E. Evidence for a bis(elongated σ)-dihydrideborate coordinated to osmium. *Inorg Chem.* 2018;57:4482–4491.
35. Ferrando-Miguel G, Coalter III JN, Gérard H, Huffman JC, Eisenstein O, Caulton KG. Geminal dehydrogenation of ether and amine C(sp^3)H$_2$ groups by electron-rich Ru(II) and Os. *New J Chem.* 2002;26:687–700.
36. Edwards AJ, Esteruelas MA, Lahoz FJ, et al. Reactivity of OsH$_2$Cl$_2$(P^{i}Pr$_3$)$_2$ toward diolefins: new reactions involving C-H and C-C activation and C-C and C-P bond formation processes. *Organometallics.* 1997;16:1316–1325.
37. Spivak GJ, Coalter JN, Oliván M, Eisenstein O, Caulton KG. Osmium converts terminal olefins to carbynes: α-hydrogen migration redox isomers with reversed stability for ruthenium and for osmium. *Organometallics.* 1998;17:999–1001.

38. Collado A, Esteruelas MA, Oñate E. Hydride alkenylcarbyne osmium complexes versus cyclopentadienyl type half-sandwich ruthenium derivatives. *Organometallics*. 2011;30: 1930–1941.
39. Collado A, Esteruelas MA, López F, Mascareñas JL, Oñate E, Trillo B. C-H bond activation of terminal allenes: formation of hydride-alkenylcarbyne-osmium and disubstituted vinylidene-ruthenium derivatives. *Organometallics*. 2010;29:4966–4974.
40. Buil ML, Cardo JJF, Esteruelas MA, Oñate E. Square-planar alkylidyne-osmium and five-coordinate alkylidene-osmium complexes: controlling the transformation from hydride-alkylidyne to alkylidene. *J Am Chem Soc*. 2016;138:9720–9728.
41. Bolaño T, Castarlenas R, Esteruelas MA, Modrego FJ, Oñate E. Hydride-alkenylcarbyne to alkenylcarbene transformation in bisphosphine-osmium complexes. *J Am Chem Soc*. 2005;127:11184–11195.
42. Bolaño T, Esteruelas MA, Oñate E. Osmium-carbon multiple bonds: reduction and C-C coupling reactions. *J Organomet Chem*. 2011;696:3911–3923.
43. Collado A, Esteruelas MA, Gulías M, Mascareñas JL, Oñate E. Reactions of an osmium(IV) complex with allenedienes: coordination and intramolecular cycloadditions. *Organometallics*. 2012;31:4450–4458.
44. Esteruelas MA, Fernández-Alvarez FJ, Oliván M, Oñate E. C-H bond activation and subsequent C-C bond formation promoted by osmium: 2-vinylpyridine-acetylene couplings. *J Am Chem Soc*. 2006;128:4596–4597.
45. Esteruelas MA, Fernández-Alvarez FJ, Oñate E. NH-tautomerization of 2-substituted pyridines and quinolines on osmium and ruthenium: determining factors and mechanism. *Organometallics*. 2008;27:6236–6244.
46. Esteruelas MA, Fernández-Alvarez FJ, Oñate E. Stabilization of NH tautomers of quinolines by osmium and ruthenium. *J Am Chem Soc*. 2006;128:13044–13045.
47. Esteruelas MA, Fernández-Alvarez FJ, Oñate E. Osmium and ruthenium complexes containing an N-heterocyclic carbene ligand derived from Benzo[h]quinoline. *Organometallics*. 2007;26:5239–5245.
48. Buil ML, Esteruelas MA, Garcés K, Oliván M, Oñate E. Understanding the formation of N-H tautomers from α-substituted pyridines: tautomerization of 2-ethylpyridine promoted by osmium. *J Am Chem Soc*. 2007;129:10998–10999.
49. Buil ML, Esteruelas MA, Garcés K, Oliván M, Oñate E. $C_\beta(sp^2)$-H bond activation of α,β-unsaturated ketones promoted by a hydride-elongated dihydrogen complex: formation of osmafuran derivatives with carbene, carbyne, and NH-tautomerized α-substituted pyridine ligands. *Organometallics*. 2008;27:4680–4690.
50. Espuelas J, Esteruelas MA, Lahoz FJ, Oro LA, Ruiz N. Synthesis of new hydride-carbyne and hydride-vinylcarbyne complexes of osmium(II) by reaction of $OsH_2Cl_2(P-i-Pr_3)_2$ with terminal alkynes. *J Am Chem Soc*. 1993;115:4683–4689.
51. Herbert MB, Grubbs RH. Z-selective cross metathesis with ruthenium catalysts: synthetic applications and mechanistic implications. *Angew Chem Int Ed*. 2015;54: 5018–5024.
52. Casanova N, Esteruelas MA, Gulías M, Larramona C, Mascareñas JL, Oñate E. Amide-directed formation of five-coordinate osmium alkylidenes from alkynes. *Organometallics*. 2016;35:91–99.
53. Ferrando G, Gérard H, Spivak GJ, et al. Facile $C(sp^2)$/OR bond cleavage by Ru or Os. *Inorg Chem*. 2001;40:6610–6621.
54. Bourgault M, Castillo A, Esteruelas MA, Oñate E, Ruiz N. Synthesis, spectroscopic characterization, and reactivity of the unusual five-coordinate hydrido-vinylidene complex $OsHCl(C=CHPh)(PiPr_3)_2$: precursor for dioxygen activation. *Organometallics*. 1997;16:636–645.
55. Oliván M, Clot E, Eisenstein O, Caulton KG. Hydride is not a spectator ligand in the formation of hydrido vinylidene from terminal alkyne and ruthenium and osmium hydrides: mechanistic differences. *Organometallics*. 1998;17:3091–3100.

56. Gusev DG, Kuhlman R, Rambo JR, Berke H, Eisenstein O, Caulton KG. Structural and dynamic properties of OsH$_2$X$_2$L$_2$ (X = Cl, Br, I; L = P^{i}Pr$_3$) complexes: interconversion between remarkable non-octahedral isomers. *J Am Chem Soc*. 1995;117: 281–292.
57. Gusev DG, Lough AJ. Double C-H activation on osmium and ruthenium centers: carbene vs olefin products. *Organometallics*. 2002;21:2601–2603.
58. Gusev DG, Fontaine F-G, Lough AJ, Zargarian D. Polyhydrido(silylene) osmium and silyl(dinitrogen)ruthenium products through redistribution of phenylsilane with osmium and ruthenium pincer complexes. *Angew Chem Int Ed*. 2003;42: 216–219.
59. Esteruelas MA, Lledós A, Maseras F, et al. Preparation and characterization of osmium-stannyl polyhydrides: d^4-d^2 oxidative addition of neutral molecules in a late transition metal. *Organometallics*. 2003;22:2087–2096.
60. Esteruelas MA, Lledós A, Maresca O, Oliván M, Oñate E, Tajada MA. Preparation and full characterization of a tetrahydride-bis(stannyl)-osmium(VI) derivative. *Organometallics*. 2004;23:1453–1456.
61. Eguillor B, Esteruelas MA, Oliván M. Preparation, X-ray structures, and NMR spectra of elongated dihydrogen complexes with four- and five-coordinate tin centers. *Organometallics*. 2006;25:4691–4694.
62. Esteruelas MA, Lledós A, Oliván M, Oñate E, Tajada MA, Ujaque G. Ortho-CH activation of aromatic ketones, partially fluorinated aromatic ketones, and aromatic imines by a trihydride- stannyl-osmium(IV) complex. *Organometallics*. 2003;22:3753–3765.
63. Eguillor B, Esteruelas MA, Oliván M, Oñate E. C$_\beta$-H activation of aldehydes promoted by an osmium complex. *Organometallics*. 2004;23:6015–6024.
64. Eguillor B, Esteruelas MA, Oliván M, Oñate E. C(sp^2)–H activation of RCH═E-py (E = CH, N) and RCH═CHC(O)R' substrates promoted by a highly unsaturated osmium-monohydride complex. *Organometallics*. 2005;24:1428–1438.
65. Baya M, Esteruelas MA, Oñate E. Analysis of the aromaticity of osmabicycles analogous to the benzimidazolium cation. *Organometallics*. 2011;30:4404–4408.
66. Barea G, Esteruelas MA, Lledós A, López AM, Tolosa JI. Synthesis and spectroscopic and theoretical characterization of the elongated dihydrogen complex OsCl$_2$(η^2-H$_2$) (NH = CPh$_2$)(P^{i}Pr$_3$)$_2$. *Inorg Chem*. 1998;37:5033–5035.
67. Esteruelas MA, Lahoz FJ, Oro LA, Oñate E. Ruiz N syntheses, spectroscopic characterizations, and X-ray structures of new Os(η^2-H$_2$) compounds containing azole ligands. *Inorg Chem*. 1994;33:787–792.
68. Ferrando-Miguel G, Wu P, Huffman JC, Caulton KG. New d^4 dihydrides of Ru(IV) and Os(IV) with π-donor ligands: M(H)$_2$(chelate)(P^{i}Pr$_3$)$_2$ with chelate = ortho-XYC$_6$H$_4$ with X, Y = O, NR; R = H or CH$_3$. *New J Chem*. 2005;29:193–204.
69. Buil ML, Esteruelas MA, Gay MP, et al. Osmium catalysts for acceptorless and base-free dehydrogenation of alcohols and amines: unusual coordination modes of a BPI anion. *Organometallics*. 2018;37:603–617.
70. Csonka R, Speier G, Kaizer J. Isoindoline-derived ligands and applications. *RSC Adv*. 2015;5:18401–18419.
71. Alabau RG, Esteruelas MA, Martínez A, Oliván M, Oñate E. Base-free and acceptorless dehydrogenation of alcohols catalyzed by an iridium complex stabilized by a N,N, N-osmaligand. *Organometallics*. 2018;37:2732–2740.
72. Castarlenas R, Esteruelas MA, Gutiérrez-Puebla E, et al. Synthesis and characterization of OsH$_2$Cl[κN,κO-(ON═CR$_2$)](P^{i}Pr$_3$)$_2$ (CR$_2$ = C(CH$_2$)$_4$CH$_2$, R = CH$_3$): influence of the L$_2$ ligand on the nature of the H$_2$ unit in OsH$_2$ClL$_2$(PiPr$_3$)$_2$ (L$_2$ = ON═CR$_2$, NH═C(Ph)C$_6$H$_4$) complexes. *Organometallics*. 1999;18:4296–4303.
73. Esteruelas MA, García-Raboso J, Oliván M, Oñate E. N-H and N-C bond activation of pyrimidinic nucleobases and nucleosides promoted by an osmium polyhydride. *Inorg Chem*. 2012;51:5975–5984.

74. Valencia M, Merinero AD, Lorenzo-Aparicio C, et al. Osmium-promoted σ–bond activation reactions on nucleosides. *Organometallics*. 2020;39:312–323.
75. Castarlenas R, Esteruelas MA, Gutiérrez-Puebla E, et al. Synthesis, characterization, and theoretical study of stable hydride-azavinylidene osmium(IV) complexes. *Organometallics*. 2000;19:3100–3108.
76. Castarlenas R, Esteruelas MA, Jean Y, Lledós A, Oñate E, Tomàs J. Formation and stereochemistry of octahedral cationic hydride-azavinylidene osmium(IV) complexes. *Eur J Inorg Chem*. 2001;2001:2871–2883.
77. Babón JC, Esteruelas MA, Fernández I, López AM, Oñate E. Reduction of benzonitriles via osmium-azavinylidene intermediates bearing nucleophilic and electrophilic centers. *Inorg Chem*. 2019;58:8673–8684.
78. Jerabek P, Schwerdtfeger P, Frenking G. Dative and electron-sharing bonding in transition metal compounds. *J Comput Chem*. 2019;40:247–264.
79. Esteruelas MA, López AM, Oliván M. Osmium-carbon double bonds: formation and reactions. *Coord Chem Rev*. 2007;251:795–840.
80. Castarlenas R, Esteruelas MA, Oñate E. Formation of imine-vinylidene-osmium(II) derivatives by hydrogen transfer from alkenyl ligands to azavinylidene groups in alkenyl-azavinylidene-osmium(IV) complexes. *Organometallics*. 2000;19:5454–5463.
81. Castarlenas R, Esteruelas MA, Oñate E. One-pot synthesis for osmium(II) azavinylidene-carbyne and azavinylidene-alkenylcarbyne complexes starting from an osmium(II) hydride-azavinylidene compound. *Organometallics*. 2001;20:3283–3292.
82. Castarlenas R, Esteruelas MA, Gutiérrez-Puebla E, Oñate E. Reactivity of the imine-vinylidene complexes OsCl$_2$(=C=CHPh)(NH=CR$_2$)(P^{i}Pr$_3$)$_2$ [CR$_2$=CMe$_2$, C(CH$_2$)$_4$CH$_2$]. *Organometallics*. 2001;20:1545–1554.
83. Castarlenas R, Esteruelas MA, Oñate E. Δ^2- and Δ^3-Azaosmetine complexes as intermediates in the stoichiometric imination of phenylacetylene with oximes. *Organometallics*. 2001;20:2294–2302.
84. Esteruelas MA, López AM, Mora M, Oñate E. Ammonia-borane dehydrogenation promoted by an osmium dihydride complex: kinetics and mechanism. *ACS Catal*. 2015;5:187–191.
85. Wang C. Vicinal *anti*-dioxygenation of alkenes. *Asian J Org Chem*. 2018;7:509–521.
86. Esteruelas MA, Modrego FJ, Oñate E, Royo E. Dioxygen activation by an osmium-dihydride: preparation and characterization of a d^4 square-planar complex. *J Am Chem Soc*. 2003;125:13344–13345.
87. Buil ML, Cardo JJF, Esteruelas MA, Fernández I, Oñate E. An entry to stable mixed phosphine-osmium-NHC polyhydrides. *Inorg Chem*. 2016;55:5062–5070.

CHAPTER THREE

N-heterocyclic germylenes and stannylenes: Synthesis, reactivity and catalytic application in a nutshell

Rajarshi Dasgupta, Shabana Khan*
Department of Chemistry, Indian Institute of Science Education and Research (IISER), Pune, India
*Corresponding author: e-mail address: shabana@iiserpune.ac.in

Contents

1. Introduction	106
2. Synthesis of heavier tetrelylenes [germylenes and stannylenes]	107
2.1 Synthesis of carbon substituted germylenes and stannylenes	107
2.2 Synthesis of N-heterocyclic heavier tetrelylenes [NHGe: and NHSn:]	110
2.3 Synthesis of germanium(II) and tin(II) hydrides	114
2.4 Synthesis, structural and reaction chemistry of N-boryl substituent supported N-heterocyclic germylene and stannylene	115
2.5 A unique homo-conjugation stabilized germylene	118
2.6 Fluorenyl tethered N-heterocyclic stannylene: Synthesis and reactivity	120
2.7 N-heterocyclic stannylene supported by planar chiral [2.2] *para*-cyclophane	122
2.8 PGeP and PSnP pincer compounds as flexible dynamic and versatile ligands	123
2.9 Synthesis and reactivity of a cyclic(alkyl)(amino)germylene [cAAGe]	126
3. Reactivity of heavier tetrelylenes	127
3.1 Activation of H_2, NH_3, H_2O and alkynes	128
3.2 Activation of CO_2 using germanium(II) and tin(II) hydrides	133
3.3 Activation of white phosphorus by germylene	134
3.4 C—F bond activation by heavier tetrelylenes	135
4. N-heterocyclic germylenes and stannylenes as organocatalyst for basic organic transformations	137
4.1 Hydroboration reaction catalyzed by heavier N-heterocyclic tetrelylenes	137
4.2 Cyanosilylation reaction catalyzed by heavier N-heterocyclic tetrelylenes	143
5. Summary at a glance	145
References	145

Advances in Organometallic Chemistry, Volume 74
ISSN 0065-3055
https://doi.org/10.1016/bs.adomc.2020.04.001

1. Introduction

Recent decades have witnessed carbenes as important synthetic intermediates for a variety of organic transformations, which led to a large number of discoveries for stable carbene species. The chemistry of carbene has been widely explored and well understood.[1] However much of attention is now being paid to the heavier analogues of carbene namely silylenes, germylenes, stannylenes and plumbylenes. These heavier tetrelylenes having various similarities and differences with carbenes might have considerable applications in fundamental and applied chemistry. In this book chapter, we will be mainly discussing over N-heterocyclic germylenes and stannylenes for their synthesis, reactivities and role as catalysts for various fundamental organic transformations. The valency of the central atom for the heavier tetrelylenes R_2M (M = Ge, Sn) is two.[2] Regarding the M(II) species as the principal quantum number increases down the group, the stability of the tetrelylene increases (Pb > Sn > Ge > Si). The heavier tetrelylenes, in comparison to carbenes, have a lesser ability to form hybrid orbitals and prefer ns^2np^2 as a stable electronic configuration. The ground state of the heavier tetrelylenes is singlet in nature in comparison to the triplet ground state of carbenes.[3] Furthermore, down the group, the electronic stability of the monomeric species (MR_2) increases compared to the respective dimeric species ($R_2M=MR_2$). Thus with proper steric and electronic stabilization, it has been possible to synthesize monomeric heavier tetrelylenes successfully. However, in the absence of such stabilization, some N-heterocyclic germylenes, stannylenes and plumbylenes suffer from disproportionation reaction leading to the deposition of elemental germanium, tin and lead, respectively.[4] Owing to the singlet ground state of heavier tetrelylenes, they have a vacant p-orbital and a lone pair of electrons. The lone pair present in their valence orbital is relatively inert due to its pronounced s-character, while the presence of vacant p-orbital makes them highly reactive. Synthesis of heavier N-heterocyclic tetrelylenes are synthetically extremely challenging and can be realized only using appropriate steric and electronic factors. Previously these heavier tetrelylenes were evidenced using gas phase kinetics or spectroscopic detection in matrices at low temperature.[5,6] But recent developments lead to the successful solid state isolation of these compounds at ambient temperature under inert condition. Now we will be elaborating on the synthesis of germylenes and stannylenes. Apart from N-heterocyclic tetrelylenes there are several examples of carbon-based germylenes and stannylenes. We have included them for the sake of completion.

2. Synthesis of heavier tetrelylenes [germylenes and stannylenes]

2.1 Synthesis of carbon substituted germylenes and stannylenes

The synthesis of heavier tetrelylenes having the central atom in +II oxidation state can be synthesized in various ways namely (a) reduction of $M^{(IV)}$ species as the precursors, (b) substitution of $M^{(II)}$ species a dihalo precursor of $M^{(II)}$ species with organo-metallic ligands such as ArLi/RLi as the nucleophile, (c) photo-chemical reductive elimination of a dimeric compounds $[R_2M=MR_2]$, (d) reduction of the corresponding dihalo precursors using reducing agents, e.g., lithium naphthalenide or KC_8, etc.[7] Carbon substituents unlike other heteroatom substituents N, O, P leads to less electronic perturbations. An optimum electronic and steric stabilization is required for the synthesis of these monomeric heavier tetrelylenes as greater steric stabilization would lead to dissociation. On the contrary, a lesser extent of steric and electronic stabilization would result in oligomerization leading to dimeric products.[8] Till to date, many stable carbon supported germylenes are reported. Initial attempts to prepare germylenes were started with the reduction of dibromogermanes (**1**) using reducing agents like lithium naphthalenide which leads to the formation of **4**. **4** can also be prepared by the photolysis of **2** or **3**. Important to note that **4** exists as a monomer in solution but dimer (**5**) in the solid state. The substitution pathway reaction of $GeCl_2$.dioxane, GeI_2 or $[(Me_3Si)_2N]Ge$ with RLi or RMgBr (R=Dip, Dis, etc.) led to the formation of **4** with simultaneous elimination of insoluble salts (Scheme 1). The first isolable monomeric stable dialkyl germylene (**6**) was synthesized by Jutzi's group. **6** is monomeric in both solid and solution state.[9] **7** and **8** are a few other examples of carbon substituted stable germylenes that exist as monomers in both solution and solid state.[10,11] These bulky carbon substituents supported germylenes are stable for some time and slowly decomposes at room temperature.

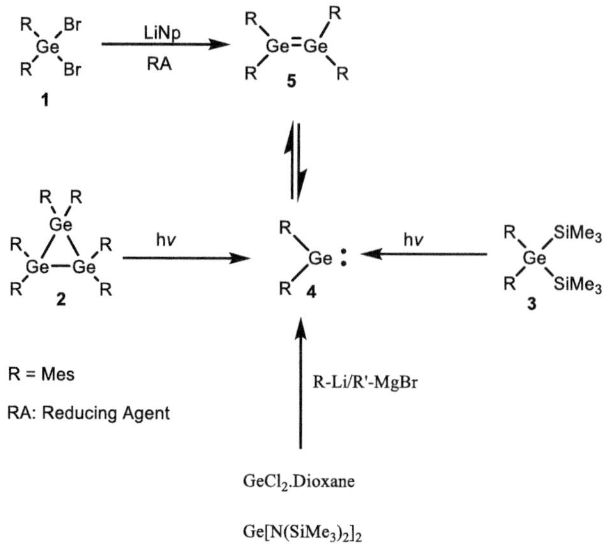

Scheme 1 Various methodologies for the synthesis of **4**.

This concludes that bulky substituents like Mes* groups can provide steric protection to overcome the lone pair-lone pair interactions thus preventing dimerization (Scheme 2).

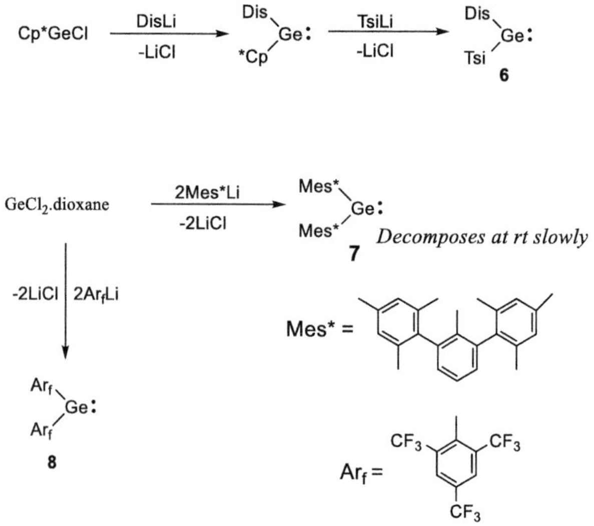

Scheme 2 Room temperature stable carbon substituted monomeric germylenes.

Focusing on carbon substituted stannylene, the +II species of Sn are intrinsically stable. Hence stable divalent $SnCl_2$ or $Sn[N(SiMe_3)_2]$ are starting materials for the synthesis of monomeric stable N-heterocyclic stannylene via substitution method. Alternative methods to prepare monomeric stable stannylenes involve the reduction of dihalo metallanes (R$_2$MX$_2$) [X = Cl, Br] and thermolysis or photolysis of cyclotristannane.[12] Some other examples of carbon substituted stable monomeric stannylenes were reported by Eaborn et al. (**9**) and Weidenbruch et al. **10** (Scheme 3).[13,14]

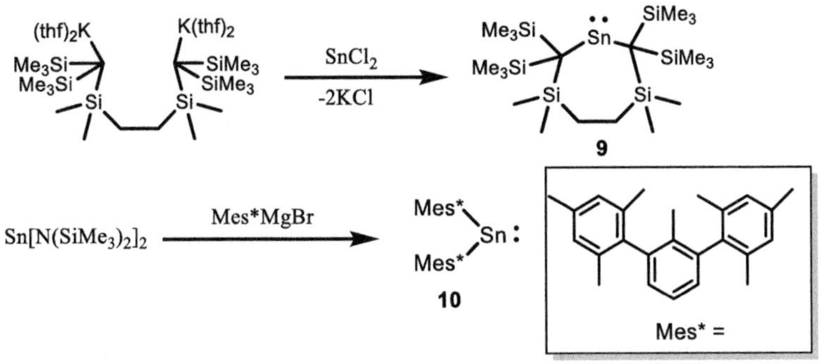

Scheme 3 Synthesis of carbon substituted stable monomeric stannylenes.

Tokitoh et al. devised a strategy to synthesize **11** and **12** by utilizing bulky Tbt and Tpp (Ar) groups [Tpp = 2,4,6-tris(1-ethypropyl)-phenyl]. The synthetic strategy for **12** involved the reaction of tetra-thia metallolanes with an excess of trialkyl phosphine (Ph$_3$P), [15] while **11** was prepared via salt metathesis method. However, these species were confirmed by trapping experiments. The first stable and well-defined monomeric dialkyl germylene (**16**) and stannylene (**17**) were prepared by using 1,1,4,4-tetrakis(trimethylsilyl) butane-1,4-diyl (**13**) as the ligand backbone.[16] Although **17** was prepared by the reaction of SnCl$_2$ with the dilithiated compound **13**, similar reactions with GeCl$_2$.dioxane, GeI$_2$ or Ge[N(SiMe$_3$)$_2$] failed to yield **16**. However successful synthesis reports came up by the reduction of the respective dihalogenated compounds (**14, 15**) with KC$_8$ at −65 °C (Scheme 4).[16]

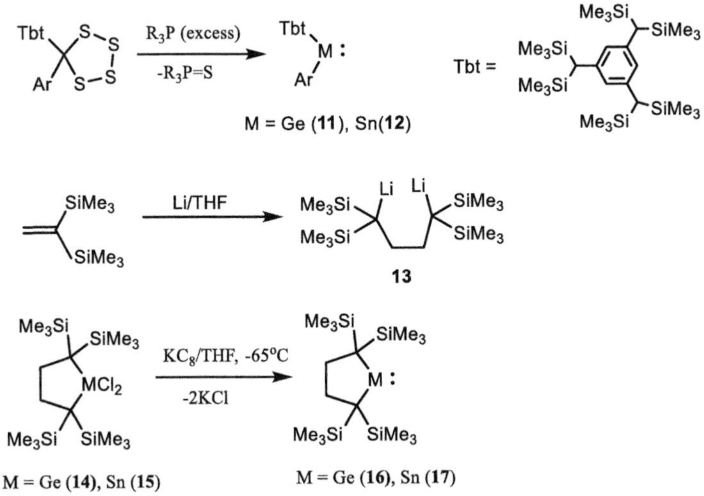

Scheme 4 Synthesis of dialkyl substituted germylenes (**11, 16**) and stannylenes (**12, 17**).

2.2 Synthesis of N-heterocyclic heavier tetrelylenes [NHGe: and NHSn:]

The research on heavier N-heterocyclic tetrelylenes has been triggered since the discovery and exploration of N-heterocyclic carbene by Arduengo III et al. in 1991.[17] Owing to their fascinating electronic structure and chemical properties they can act as spectator ligands for various catalytic cycles.[18] The introduction of heteroatom (N) in the ligand backbone helps to stabilize the reactive divalent state. Apart from steric stabilization by the introduction of the sterically hindered group, the heteroatom causes reduction of the σ-electron density at the tetrel center by the −I effect and by donating the π-electron donation of the unshared electron pairs from the nitrogen atom to the vacant p-orbital of the tetrel center. Heinicke et al. compared the stabilities of several N-heterocyclic germylenes and stannylenes substituted by benzo, naphtho and pyrido annulations.[19] It was referenced that benzo and naphtha provide sufficient thermodynamic stabilities whereas for pyrido, due to the formation of asymmetric 10π electronic system, it leads to kinetic destabilization.[20] Based on their electronic structures the heavier tetrelylenes can form complexes with Lewis acids, Lewis bases and transition metals, etc. This hypothesis suggests that heavier N-heterocyclic metallylenes [M=Ge, Sn] are very important for various complexation reactions and catalytic cycles. Veith et al. in 1982 synthesized the first N-heterocyclic germylene with a four-membered cyclic framework **18**.[21] The saturated analogue of N-heterocyclic germylene (**19**) was investigated for its chemical properties by Meller et al. Benz annulated germylenes are also reported in the literature (**20**).[22] Hermann et al. synthesized **21** based on Arduego's ligand system with Mes groups on N-atoms.[23a] Some pyridyl[b] (**22**) and benzo (**23**) substituted germylenes and their respective metal complexes were synthesized by Heinicke et al. Derivative of **21** was also prepared using 1,8 naphthalene as an ancillary ligand.[19] Cationic germylenes **24** and **25** were also isolated.[23b,c] Various mixed heteroatom substituted germylenes are also stabilized and it is observed that mixed heteroatom germylenes like **26** and **27** are more stable at room temperature.[24a] Recently, in 2016 an acyclic α-phosphinoamido germylene **28** was reported (Fig. 1) with PPh$_2$ substituents on N-atoms. This is the only available example of α-phosphinoamido-germylene. The reason could be attributed to the instability of such moieties.[24b]

Literature reports also prevail on polydentate N-substituted bis- or polygermylenes.[25] Oligomerization of Ge=N–R species helps to produce

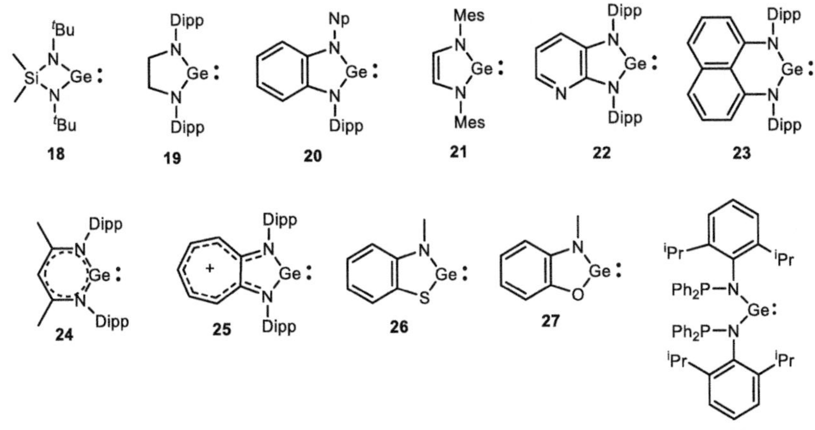

Fig. 1 Various N-heterocyclic germylenes and acyclic phosphinoamido-germylene.

cubane-type tetrakis(germylene) **29** along with some cage compounds (Fig. 2).[21,26] Bis(germylene) **30** containing two spiro-silicon atoms was the first bis(germylene) with two individual NHGe moieties. Planar Ge_2N_2 or Ge_3N_3 cores are present for cyclic tris(germylene) **31** and the germanazene **32** in the solid state.[27] The spacer separated acyclic bis(germylenes) (**33** and **34**) were prepared independently by Kobayashi et al.[28] and Braunschweig et al.,[29] respectively. The substitution reaction of R_2NGeCl with an organolithium compound was adopted to synthesize them. **33** has two benzimidazolin-2-germylene moieties and are linked by different bridging groups like $-CH_2$ spacers through the nitrogen atoms present in the ring.[28,29] Among the polygermylenes (**29–34**), only **34** has a flexible backbone allowing it to act in a chelating fashion.

The monomeric stannylene **35** is the first representative for N-heterocyclic stannylenes (NHSn) (Fig. 3), which was first prepared in 1975.[30] The base-stabilized cyclic stannylenes (type **36**) displayed intramolecular Sn⋯S/Sn⋯N interactions. These interactions were found significant for these molecules to exist as monomers in the solid state as well as in solution state. The first NHSn compound **37** with the tin atom being part of a conjugated π-system was reported in 1995 by Braunschweig et al.[30,31] The tin analogues of Arduengo's carbene (**39**) have recently been isolated (Fig. 3).[32] **38** does not exhibit planarity due to deviation of Sn atom from the plane of the 1,8-naphthalenediyl moiety.[33] Saturated NHSn compounds of type **40** having six- or seven-membered heterocycles show a

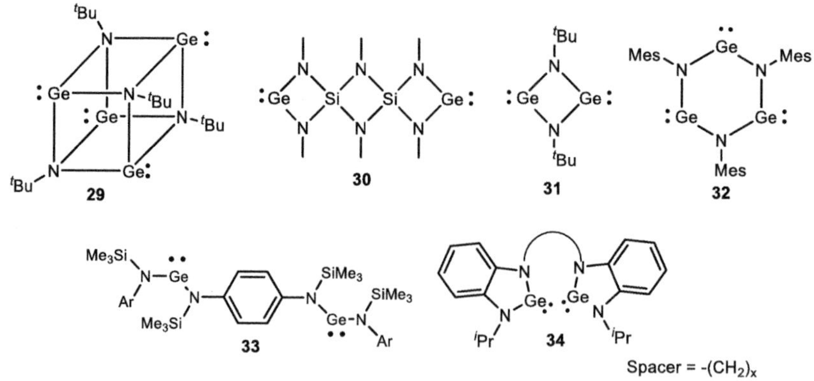

Fig. 2 Examples of bis- and polygermylenes.

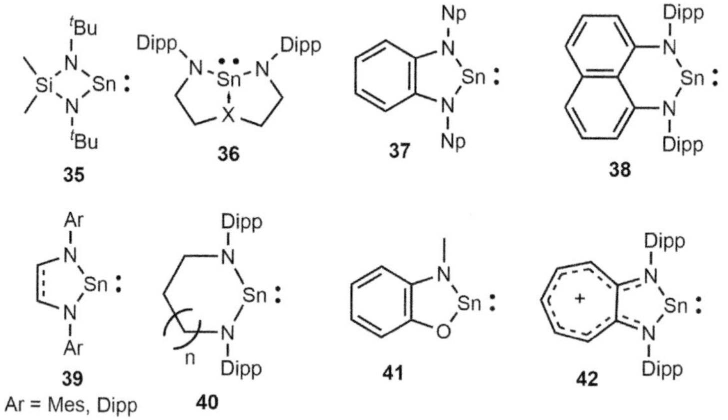

Fig. 3 Various N-heterocyclic stannylenes.

monomer/dimer equilibrium in the solution state. Moreover, **40** produces α,ω-bis(carbodiimides) via heterocumulene metathesis reaction with tert-butyl isocyanates.[34] The heteroatom stabilized cyclic stannylene **41** has recently been prepared where two heteroatoms (N, O) are present. **42** displays a cationic framework containing 10π-electron system along with a hetero-bicyclic planar C_7N_2Sn ring.[35]

Some selected polydentate N-substituted and N-heterocyclic stannylenes are depicted in Fig. 4. **43** was prepared by the reaction of an in situ generated monomeric reactive stannylene to yield the tin derivative

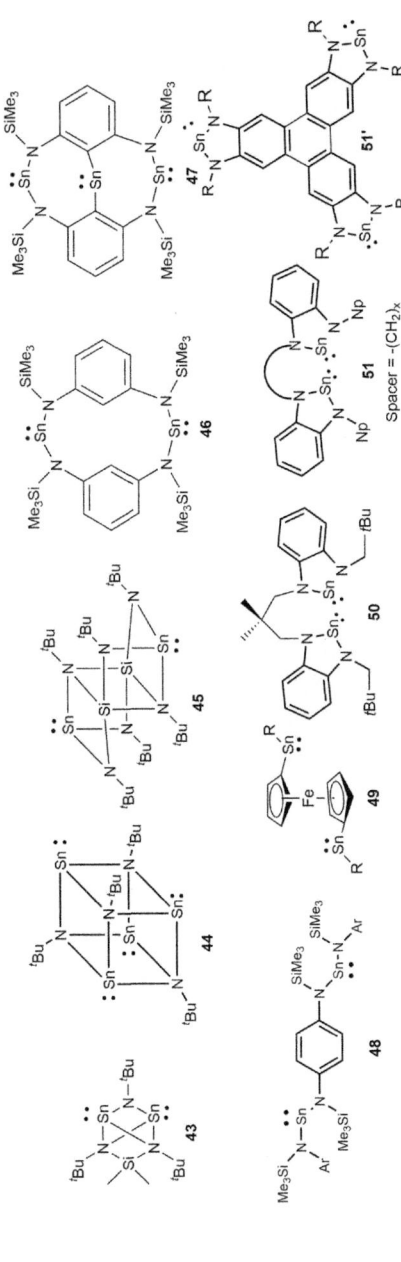

Fig. 4 Examples of bis- and poly-stannylenes.

Sn=N—R. **43** was heated to yield the cubane-type tetrakis(stannylene) **44**.[21] Each tin atom in **44** possesses a chemically active unshared electron pair, behaving as a Lewis base toward Lewis acids, e.g., AlCl$_3$.

45 displayed two additional intramolecular Sn⋯N interactions, leading to a pseudo-bis(cubane) skeleton. **46** was prepared by the reaction of 1,3-(RLiN)$_2$C$_6$H$_4$ with [Sn(μ-Cl)(SiMe$_3$)]$_2$. Sn[N(SiMe$_3$)$_2$]$_2$ helps in C—H and N—H bond activation in 1,3-(RHN)$_2$C$_6$H$_4$ giving **47** with two diamino- and one (dialkylstannyl)stannylene. Bis(stannylenes) of type **48** can be prepared in a similar synthetic fashion as the one used for the preparation of **33**. The stannylene-substituted ferrocenes **48** can react with transition metals in a chelating fashion.[36a] N-heterocyclic bis(stannylenes) (**50**) were isolated recently having various bridging units (—NH, CH$_2$). They were prepared from the reaction of suitable tetraamines and Sn[N(SiMe$_3$)$_2$]$_2$. The spacer separated bisstannylene **51**, with a –CH$_2$C(CH$_3$)$_2$CH$_2$– linking unit between two stannylene moieties, displays both intramolecular Sn⋯NMe$_2$ contacts [Sn⋯N distances 2.561 and 2.530 Å] and intermolecular interaction of tin(II) centers with a π-electronic cloud of benzene present in the adjacent molecule. These interactions ultimately lead to a polymeric arrangement of bis(stannylene) molecules in the solid state. Similarly, triphenylene-based planar tris-N-heterocyclic stannylenes (**51'**) were prepared from the corresponding 2,3,6,7,10,11-hexaamino-triphenylene and Sn[N(TMS)$_2$]$_2$.[36b] From the above mentioned discussion on the synthesis and bonding characteristics of heavier N-heterocyclic tetrelylenes, it can be concluded that these species can serve as ambiphilic ligands.

2.3 Synthesis of germanium(II) and tin(II) hydrides

The more fundamental low-valent E—H compounds (E = Ge, Sn) have shown diverse reactivities and are quite diverse as compared to the congeners in +IV oxidation state. These hydrides have been widely used for hydro functionalization catalysis. Focusing on Ge(II) hydrides, the first isolable Ge(II)-H **52** was reported by Roesky et al. **52** showed labile coordination to BH$_3$ which can be removed by utilizing PMe$_3$ base to afford free germanium hydride **53**.[37] The molecular structure of **53** was established by the same group in 2006.[37b] The further derivative of **53** was later reported by Jones et al. via salt metathesis reaction.[38] Roesky et al. further extended the ligand system and stabilized Ge(II)

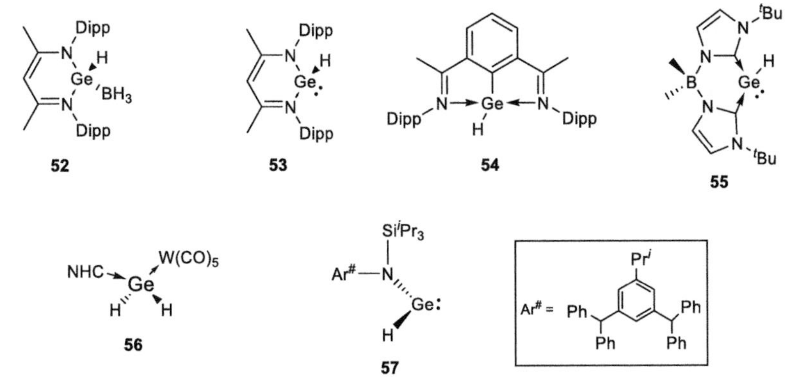

N-heterocyclic germylenes and stannylenes

Fig. 5 Selected examples of the stable germanium(II) hydrides.

hydride **54** using pincer based framework.[39] Driess et al. further reported Ge—H complex **55** supported by biscarbene borate ligands.[40] Rivard et al. also successfully synthesized transition metal supported germanium dihydride complex **56**.[41] Later, Jones et al. prepared an acyclic germanium hydride **57** stabilized by extremely bulky amido groups (Fig. 5).[42] Likewise, tin(II) hydrides are also garnering considerable attention. Monomeric base-stabilized tin(II) hydride **58** was reported by Wesemann et al.[43]. Roesky et al. also reported the tin analogues (**59, 60**) of the Ge(II) hydrides supported by a heteroatom/pincer framework, respectively.[37,39] Likewise a similar push-pull stabilization helped Rivard et al. to isolate transition metal stabilized tin(II) hydride **61**.[41] Jones et al. used bulky amido substituents to prepare the acyclic tin(II) hydride **62** which was demonstrated to give excellent catalytic activities (Fig. 6).[44]

2.4 Synthesis, structural and reaction chemistry of N-boryl substituent supported N-heterocyclic germylene and stannylene

Lappert and coworkers extensively worked to establish the chemistry of heavier tetrelylenes of the type EX_2 in 1970s.[45] The studies revealed that the lone pair of tetrelylenes center is nucleophilic while the presence of vacant pπ orbital exerts electrophilic behavior.[46] This ambiphillic nature helps in the activation of small molecules like H_2 and NH_3.[47] The main key for such reactivities is decreasing the HOMO-LUMO gap by suitable choice of pendant ligands. Recently, Aldridge et al. used

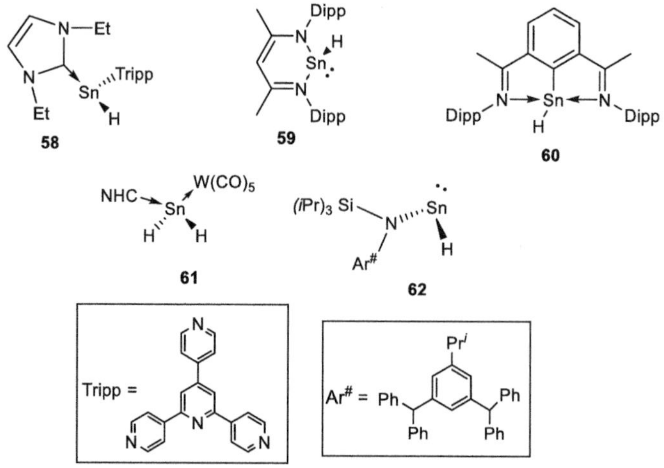

Fig. 6 A few reported stable tin(II) hydrides.

boryl group (—BR$_2$) to synthesize saturated N-heterocyclic germylenes (**63**, **65**) and stannylenes (**64**, **66**).[48] Compared to the general amido substituents, the presence of the pendant boryl function elevates the HOMO of the tetrelylene (due to their strong σ-donor capabilities). It also relatively lowers the energy of the LUMO by decrementing π-donation toward the heavier tetrel center. The presence of two additional bulky boryl N-substituents generates a very high peripheral steric loading in the vicinity of the germanium/tin center, which offers sterically protected "pocket" for further reaction chemistry.[49] The borylated ligands were further investigated for their subsequent conversion to the respective tetrelylenes. The two methodologies were employed: (i) A unimolecular reaction between bis(trimethylsilyl)amido derivatives, E{N(SiMe$_3$)$_2$}$_2$ (E=Ge, Sn); and (ii) deprotonation via reaction with an organolithium or -potassium base followed by metathesis with the corresponding electrophile, i.e., germanium or tin dihalide. Even at elevated temperature (ca. 110 °C) the former pathway showed no product formation. On the other hand, the use of an excess of n-BuLi in the presence of one equivalent of tetramethyl-ethylenediamine (TMEDA) in toluene resulted in the corresponding dilithioamide (Scheme 5). **63–66** were finally synthesized from the corresponding (dilithio)- or (dipotassio) amides via the addition of GeCl$_2$·dioxane or SnCl$_2$, respectively (Scheme 5).

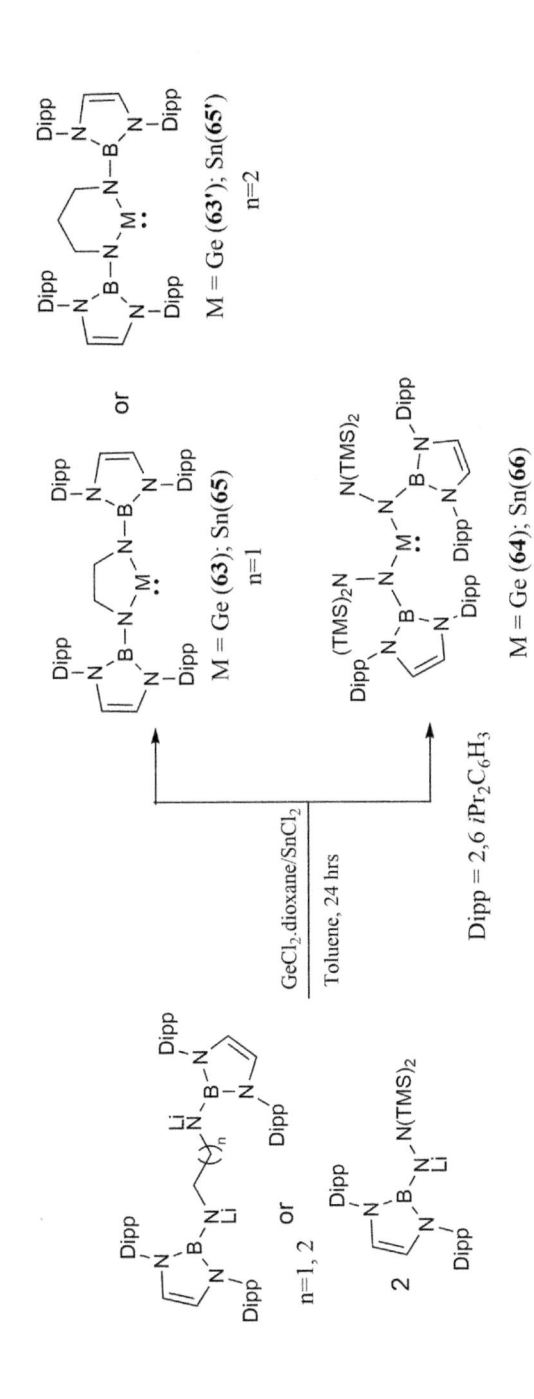

Scheme 5 Synthesis of N-boryl substituted germylenes (**63**, **65**) and stannylenes (**64**, **66**).

The reactivity of **63–66** was studied over a wide range of small molecules. Unlike ketones, the heavier analogues, e.g., silanone, germanone and stannone, etc., tend to oligomerize due to less effective pπ-overlap and greater differences in electronegativity. Till to date, only a few examples of stable unsupported silanones, germanones and stannones have been reported (i.e., ones not having additional stabilization by an electron-donating Lewis base or a transition metal fragment) and they typically feature steric stabilization using bulky ligand sets.[50] Most of the oxygen transfer reagents upon reaction with **63–66** did not produce any product. However, the reaction of **63** with two equivalent of Me$_3$NO produced an oxidative product displaying +IV oxidation state of germanium. The high reactivity of **63** was also witnessed upon reaction with pyridine N-oxide where a similar reactivity pattern is witnessed. As a concluding remark, heavier N-heterocyclic tetrelylenes supported by diazaboryl group were synthesized to generate a strong electron donating and sterically imposing coordination environment at the respective tetrel center.

2.5 A unique homo-conjugation stabilized germylene

The synthetic strategies for the isolation of germylenes are mainly based on stabilization through donation or synergistic effect between donor and acceptor. This stabilization helps to increase the coordination number of the germanium atom to three (donor-stabilized) or four (donor/acceptor stabilized). The research groups of Baceiredo/Kato and Aldridge have utilized N-heterocyclic carbene and phosphido-substituent to isolate germylenes, **67** and **68**, respectively (Fig. 7).[50,51]

67 and **68** were found very reactive and further used to activate small molecules. The highly reactive nature of **67** and **68** can be attributed to their unusual bonding. Here a brief discussion will be put forward on the synthesis of germylene **70**, which is stabilized by an interaction with a remote C=C double bond through homo-conjugation reported by Muller et al. in 2016.[52]

This stabilization pattern brings out intriguing structural consequences for unusual reactivity. **70** was prepared in good yield via the addition of

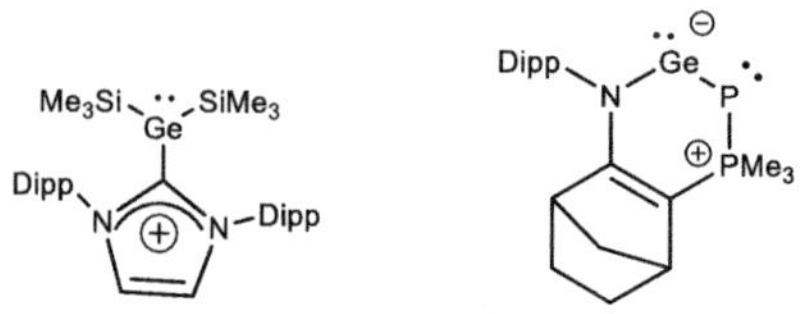

Fig. 7 Germylenes stabilized by substituent effects.

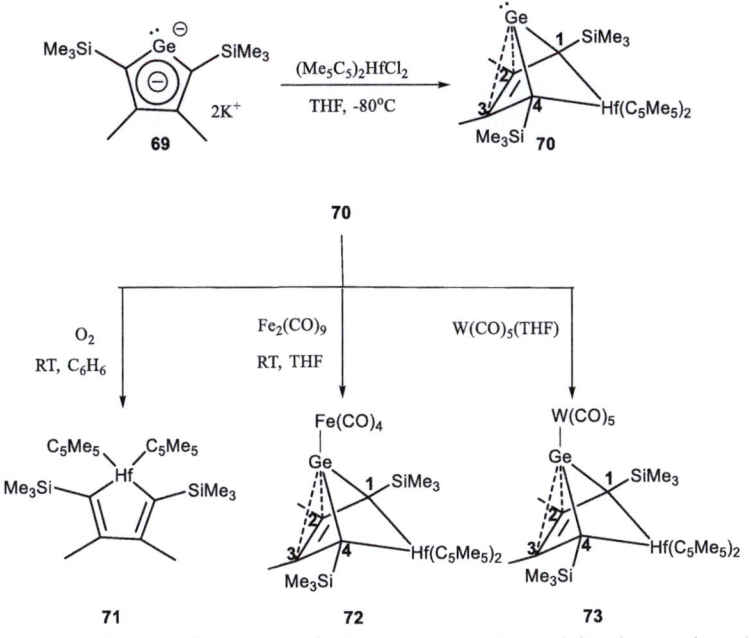

Scheme 6 Synthesis and reactivity of a homo-conjugation stabilized germylene (**70**).

hafnocene dichloride in THF to a solution of dipotassium germacyclopentadienediide K$_2$ (**69**) in THF at −80 °C (Scheme 6).[52] The X-ray analysis of **70** revealed that three independent molecules of **70** are present in one asymmetric unit. Each molecule showed similar unusual bicyclo [2.1.1]hexene-like structure with the Cp$_2$Hf fragment and the germanium atom occupying the bridging positions (Fig. 8).

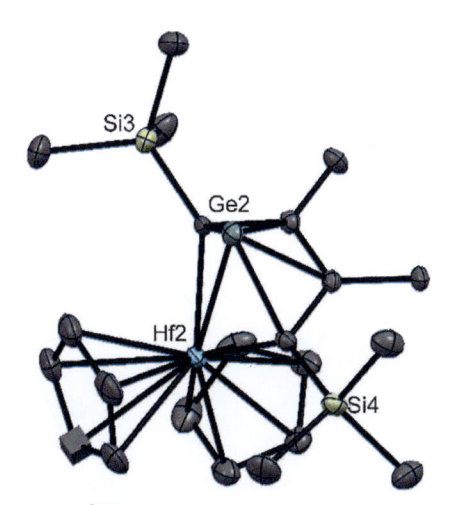

Fig. 8 Molecular structure of **70**.

70 reacts with oxygen to yield hafnocyclopentadiene (**71**) (Scheme 6). No reactivity takes place with relatively unactivated molecules like phenyl acetylene, e.g., 2,3-dimethyl butadiene and silanes, e.g., Ph$_3$SiH, Et$_3$SiH, and Ph$_2$SiH$_2$, etc. This indicates the limited electrophilic reactivity of **70**.[52] However, **70** leads to the formation of bimetallic germylene complexes, **72** and **73**, upon reaction with metal carbonyl complexes, e.g., Fe$_2$(CO)$_9$ and W(CO)$_5$(THF), respectively (Scheme 6). Structural parameters of **72** and **73** highlight the σ-donor properties of **70** with negligible π-back bonding (Fig. 9).[53]

2.6 Fluorenyl tethered N-heterocyclic stannylene: Synthesis and reactivity

The chemistry of tethered ligands is quite interesting for designing a new generation of homogeneous catalysts as combination of such ligands helps in the development of chelation chemistry.[54] A tethered ligand mainly consists of anionic moieties like cyclopentadienyl anion or a fused system such as indenyl/fluorenyl groups separated by hydrocarbon spacers. These can be anionic/neutral or even hard/soft according to Pearson's HSAB classification.[55] Anchoring groups such as alkoxide or amide, have been mostly used as tethered ligands for N-heterocyclic carbenes (NHCs).[3] Tethered ligands highly promote formation of bimetallic complex or 1,2 reaction across metal-ligand bond.[56] A recent report has demonstrated that low-valent Sn center of a phosphine stabilized stannylene can be ambiphilic in nature

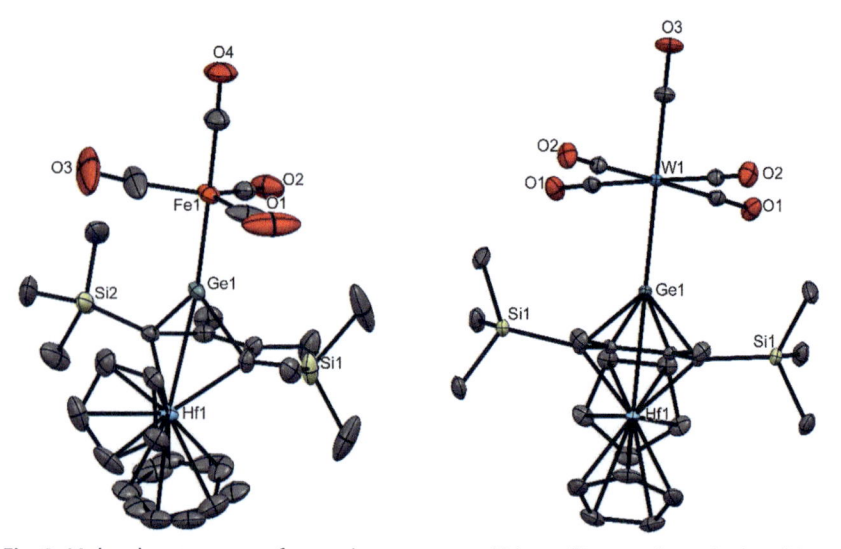

Fig. 9 Molecular structures of germylene supported bimetallic complexes (**72**) and (**73**).

while coordinating with palladium metal.[57] In the context of the anionic tethered ligands described below, recent work on protic NHSns demonstrated deprotonation of the N—H with NaH to give an anionic complex. Aiming for exploration of new bimetallic complexes for C—H activation, and recognizing that tin has a unique role in hetero-bimetallic catalysis, Mansell et al. started the development of tethered NHSn ligands to explore the new cooperative reactivity.[58] The synthesis of fluorenyl tethered stannylenes was accomplished using sterically encumbered diisopropylphenyl (Dipp) substituents (**76**). The diamine **74** was subjected to lithiation using *n*-BuLi which afforded trilithiated ligand precursor **75**. However, further reaction with SnCl$_2$ was not selective via transamination using [Sn{N(SiMe$_3$)$_2$}$_2$] (SnN″$_2$) (Scheme 7). **75** was reacted with SnCl$_2$ in diethyl ether to form a fluorenide substituted NHSn, but multinuclear NMR spectroscopy revealed that a non-selective reaction took place. Nonetheless, the reaction of SnN″$_2$ with **74** in a non-coordinating solvent afforded white insoluble precipitate of **76**. ^{119}Sn NMR spectrum of **76** showed a sharp resonance at +67 ppm which is at a much lower frequency as compared to previously characterized monomeric, 2-coordinate NHSns (e.g., +366 ppm for [Sn{N(Dipp)CH$_2$}$_2$]).[59] This region is very specific for 3-coordinate NHSns which are dimeric through dative N—Sn bonds.

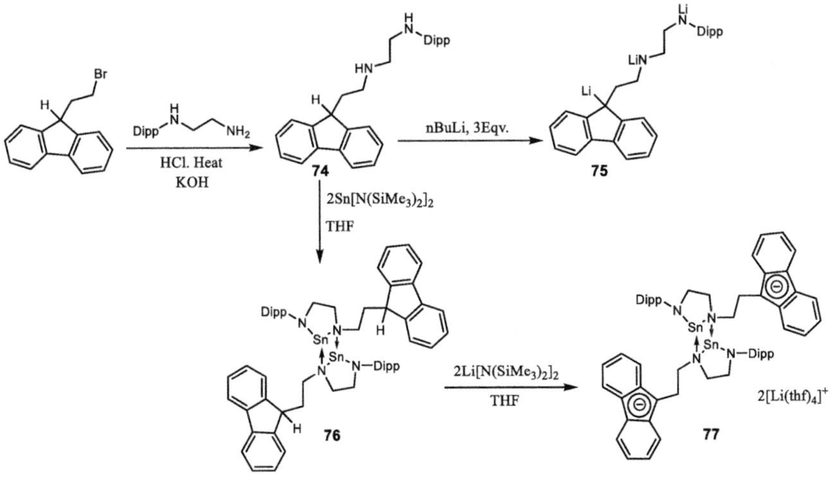

Scheme 7 Synthesis of fluorenyl tethered N-heterocyclic stannylene.

Deprotonation of **76** was readily achieved with Li[N(SiMe$_3$)$_2$]$_2$ in THF leading to the formation of the dianionic species **77**. **77** is more soluble than **76** in coordinating solvents. Reactivity of **77** was studied with rhodium metal to form bimetallic Sn-Rh compounds (**78–79**) (Scheme 8).

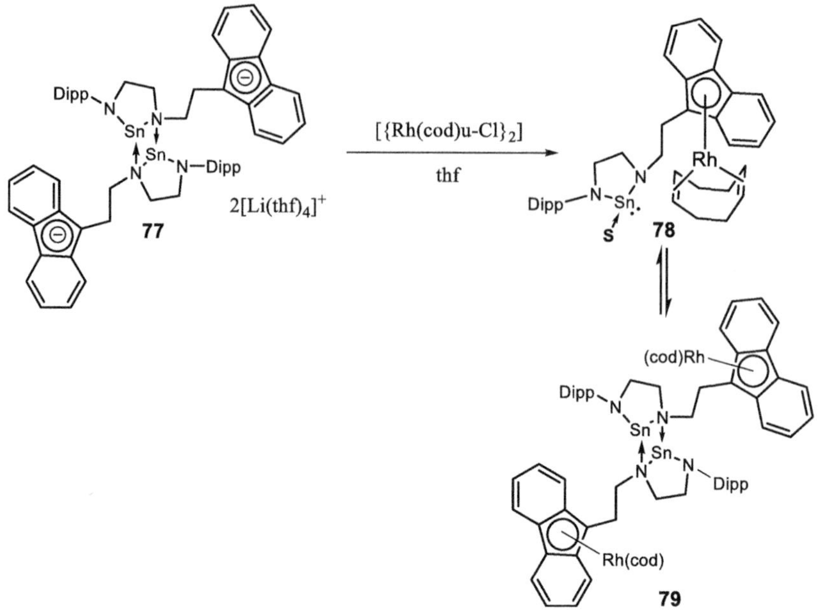

Scheme 8 Reactivity of **77** with rhodium salt.

76 was found unsuitable for coordination with rhodium as no reaction took place at room temperature even after many days and decomposition occurred upon heating. Therefore, the alternative route was taken and reaction of dianionic NHSn (**77**) with [{Rh(cod)(μ-Cl)}$_2$] was performed. NMR analysis along with X-ray diffraction studies confirmed simultaneous formation of **78** and **79**. **79** displayed a dimeric structure through dative Sn—N bonds with Rh(cod) moieties along with η^5 mode of bonding to each of the fluorenyl five-membered rings. Thus the entire chemistry demonstrates the synthesis of a fluorenyl tethered N-heterocyclic stannylene which was further used to stabilize transient Rh complexes.

2.7 N-heterocyclic stannylene supported by planar chiral [2.2] *para*-cyclophane

Since 1960 *para*-cyclophane has been known as a ligand for stabilizing heterocyclic molecules.[60] The [2.2] *para*-cyclophane is of highlighting importance as it provides topological chirality and overcomes the limitation of the central chirality.[61] NHCs bearing chiral substituents and even planar chiral NHCs are known, among which cyclophane stabilized NHC's show promising activities in organocatalysis.[62] The key to success seems to be the restriction in conformational flexibility, as Fürstner et al. demonstrated a carbene center being incorporated in the stereogenic unit.[63] In comparison to the N-heterocyclic carbenes, the heavier group 14 analogues still lack that dynamic progress. Nevertheless, there is a report on a monomeric planar chiral N-heterocyclic stannylene **83**, featuring a unique [2.2]paracyclophane backbone.[64] Glorius et al. successfully synthesized **83** where two atoms of the *para*-cyclophane backbone are NH-functionalized and these two nitrogen atoms are bound by the same tin atom[64] (Scheme 9).

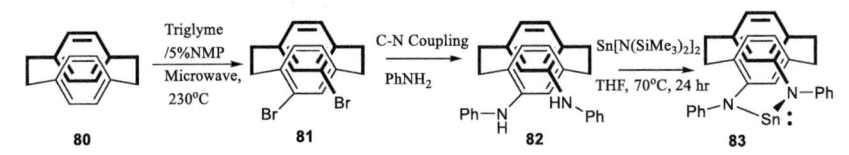

Scheme 9 N-heterocyclic stannylene (**83**) supported by [2.2]-*para*-cyclophane.

Glorius et al. followed a simple synthetic methodology as given in Scheme 9. As expected for a racemic mixture, (±) **83** crystallizes in the centrosymmetric space group $P2_1/n$ with both enantiomers simultaneously present in the same crystal lattice. The N—Sn—N angle expands with the size of the heterocycle and N—Sn—N angle (96.69(5)°) in (±)-**83** is slightly smaller than the N—Sn—N angle found in acyclic diaminostannylenes which ranges from 100° to 110°.[65] The molecular structure of **83** also reveals that the maximum steric hindrance is provided by the ancillary ligands.

2.8 PGeP and PSnP pincer compounds as flexible dynamic and versatile ligands

Pincer type ligands have been widely used in homogeneous catalysis since it provides extra stability via chelation effects.[66] A large number of reports prevail for the pincer based N-heterocyclic carbenes but pincer based heavier tetrelylenes are far less explored although the heavier N-heterocyclic tetrelylenes came into the scene far before NHC's.[67] First pincer based heavier tetrelylenes (PGeP/PSnP) were reported by Goicoechea and coworkers.[68] These pincer based heavier tetrelylenes are of great importance as they can be used to stabilize transition metal species as well as in homogeneous catalysis and can also display excellent luminescent characteristics.[69] A PNHNHP ligand devised by Thomas et al. was used to synthesize the phosphine stabilized heavier tetrelylenes **85**, **86**.[70] Treatment of **84** with two equivalents of KHMDS and the corresponding divalent metal salts, GeCl$_2$(dioxane) and SnCl$_2$, leads to the formation of **85/86**, respectively (Scheme 10). The pincer based germylene (**85**) and stannylene (**86**) were characterized by single crystal X-ray analysis. These pincer ligands were

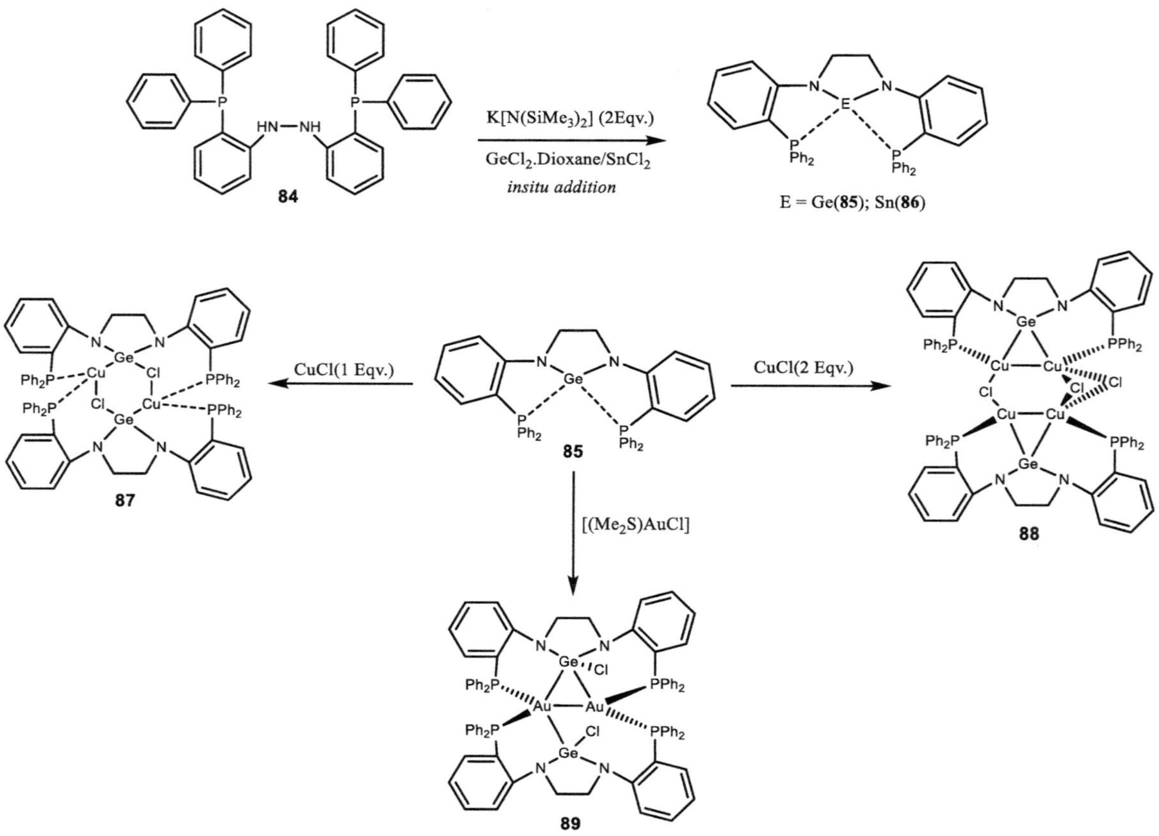

Scheme 10 Formation of PGeP (**85**) and PSnP (**86**) pincer compounds and further reaction with coinage metals.

N-heterocyclic germylenes and stannylenes 125

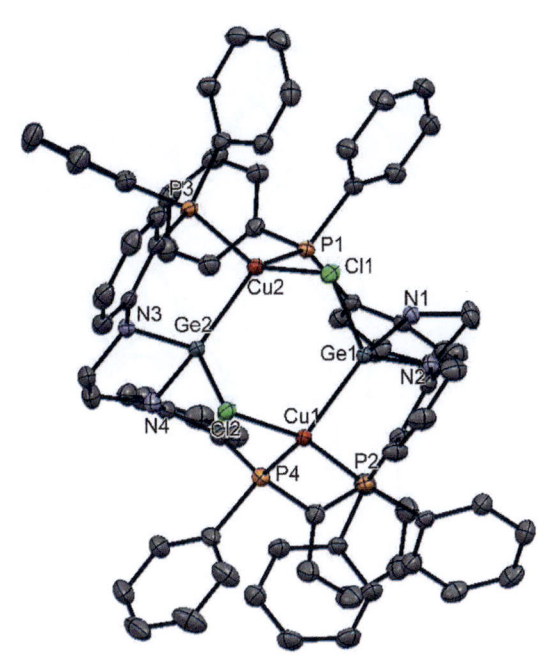

Fig. 10 Molecular structure of PGeP supported copper complex (**87**).

further subjected to study the coordination chemistry with transition metals mainly coinage metals (Cu, Ag, Au) (Scheme 10).[68] Reaction of **85** with CuCl in toluene resulted in a dimeric species [(PNNP)GeCu(μ-Cl)]$_2$ (**87**).

87 crystallized in $P2_1/c$ space group and the molecular structure revealed a six-membered inorganic ring containing [GeCuCl]$_2$ fragment in a boat conformation (Fig. 10).

An agglomerate cluster **88** is also formed when **85** reacts with two equivalent of CuCl. The reaction of **85** with [(Me$_2$S)AuCl] in benzene afforded a dimeric complex **89**.

The molecular structure of **89** disclosed a short Au1–Au2 distance of 2.8293^2 Å (Fig. 11), which indicates a significant aurophilic interaction. The geometry and coordination around Au(I) exhibit a distorted seesaw or trigonal bipyramidal geometry, which is unusual for Au(I) ions. Furthermore, Au$_2$ features a slightly distorted trigonal planar coordination sphere with almost equal Au—P distances. The coordination ability of **86** was also examined. A monomeric [{(PNNP)Sn}Pt(PPh$_3$)] (**90**) complex was formed upon a reaction of **86** with [Pt(C$_2$H$_4$)(PPh$_3$)$_2$] (Scheme 11).[68] Molecular structure of **90** reveals a distorted tetrahedral geometry at Pt center with two coordination sites occupied by —PPh$_2$ ligands.

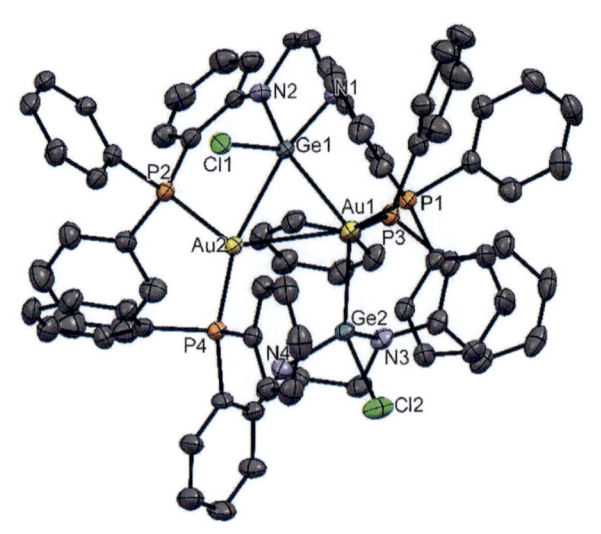

Fig. 11 Molecular structure of PGeP ligand supported Au complex (**89**).

Scheme 11 Reactivity of **86** with platinum complexes.

2.9 Synthesis and reactivity of a cyclic(alkyl)(amino)germylene [cAAGe]

After the pioneering works on stable germylenes[45] by Lappert et al., a vast amount of reports has come forward on the synthesis and reactivity of germylenes supported by a wide variety of ligands. Thermally stable N-heterocyclic germylenes[71] to cyclic dialkyl germylenes (**16**),[8] all these compounds find wide application.[72] Recently, Kinjo et al. isolated the first cyclic(alkyl)(amino)germylene (**93**).[73] The synthetic strategy involved the isolation of a α-β-unsaturated imine (**91**) via the condensation pathway first. Further, the treatment of **91** with GeCl$_2$.dioxane lead to the quantitative isolation of **92**. Finally, the reduction of **92** with potassium graphite afforded the desired cyclic(alkyl)(amino)germylene (**93**) (Scheme 12). A UV-visible spectrum of **93** in hexane showed a broad peak at 368 nm, which is attributed to the HOMO-LUMO transition responsible for the red color of **93**. The synthesis of **93** highlighted the fact that the substitution of a π-donating and σ-withdrawing amino group with a σ-donating trimethyl silyl group

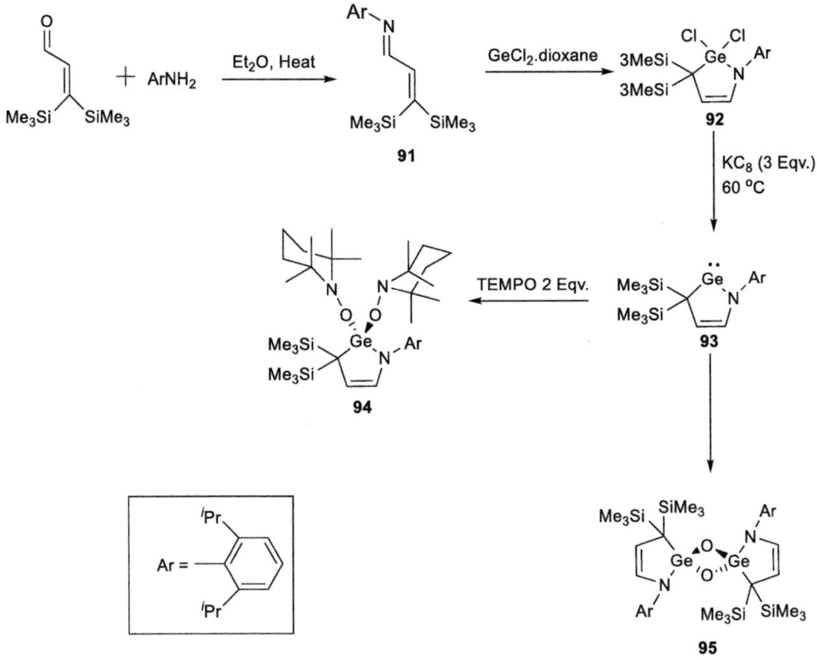

Scheme 12 Synthesis and reactivity of cyclic(alkyl)(amino) germylene (**93**).

increases the electrophilicity at the germylene center. Reactivity of **93** with TEMPO (2,2,6,6-tetramethylpiperidine)-N-oxide and N$_2$O led to the formation of **94** and **95**, respectively.

3. Reactivity of heavier tetrelylenes

Owing to the toxicity, high cost and low earth abundance of transition metals a recent urge prevails for the development of the cost-effective and environmental friendly main group catalysts. A large development in this field has occurred by the implementation of N-heterocyclic carbenes, silylenes and FLP's (frustrated Lewis pairs) as organocatalysts for the hydrogen activation and dehydrogenative coupling reactions.[74] Although there are a large number of reports on heavier tetrelylenes supported transition metal complexes and their application as catalysts,[7] a very handpicked number of reports exists on sole heavier tetrelylenes acting as organocatalysts for various organic reactions or utilizing heavier tetrelylenes for small molecules (H$_2$, CO$_2$, NH$_3$, etc.) activation. Herein, the second part of the chapter will mainly focus on the small molecule activation and catalytic application by germylenes and stannylenes.

3.1 Activation of H_2, NH_3, H_2O and alkynes

Oxidative addition and reductive elimination are two key steps associated with many important catalytic cycles mediated by transition metals. The associated $M^{n+}/M^{(n+2)+}$ redox couples are not well established for main group systems.[75] This highlights the fact that though sub-valent main group system helps in oxidative bond activation but subsequent regeneration of reduced state via reductive elimination is not thermodynamically viable. Recently, keeping reductive elimination in mind many research groups have developed a lot of heavier tetrelylenes based systems which can actively participate in small molecule activation. On a similar line, Power and coworkers developed an acyclic diaryl germylene (**96**) and an acyclic diaryl stannylene (**97**) by using very bulky aryl groups. **96** and **97** were found to maintain thermodynamic balance between reductive elimination and main group facilitated oxidative addition[76] (Scheme 13).

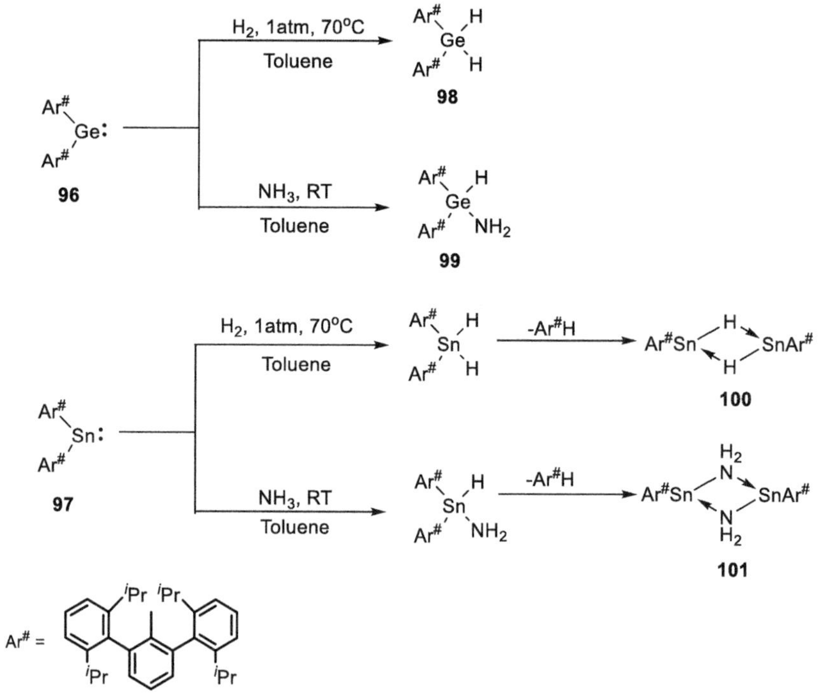

Scheme 13 Oxidative addition of H_2 and NH_3 to **96** and **97**.

Analogous to transition metal complexes, **96** activates ammonia while **97** was used to activate dihydrogen.[77] Similar to carbene complexes reported by Bertrand and coworkers,[78] **96** and **97** upon reaction with dihydrogen and ammonia gave the desired oxidative addition products, **98–101**.

However, for **97** it was observed that a simultaneous elimination of arene took place in a concerted fashion leading to the formation of μ-hydrido complex **100**. But similar is not the case for **96** which gave rise only to tetra-valent complexes (**98, 99**) upon activation of dihydrogen and ammonia, respectively. No facile arene elimination took place. These different results reflect that steric effect are of secondary importance in comparison to the electronic effects and bond strength differences between Ge and Sn. Later Aldridge et al. provided experimental evidences for the oxidative addition of E—H bond (namely dihydrogen) at a mono-nuclear $Sn^{(II)}$ system to yield the corresponding Sn(IV)-dihydride (**103**). This was possible by fine tuning of the Sn^{II}/Sn^{IV} redox couple to promote oxidative addition by the use of strong σ-donating boryl ancillary ligands. Besides, kinetics of E—H oxidative reaction highlights that the singlet-triplet energy difference (ΔE_{st}) is inversely related to its reactivity.[79] The energy of HOMO is raised due to very strong σ-donating boryl substituents, and thus HOMO-LUMO gap is reduced (and thus the related singlet-triplet gap). Aldridge et al. also successfully activated dihydrogen in an oxidative addition pathway and also other protic and hydridic E—H bonds (Scheme 14). The oxidative addition reaction of **102** with dihydrogen at room temperature led to the formation of bisboryltin(IV)dihydride (**103**). This was the first ever report by Aldridge et al. to activate the dihydrogen using any stannylene. **102** was tested for the reaction with hydridic E—H bonds like silanes and an oxidative addition product **104** was formed. Similarly, analogous germylene **102'** with one —N(TMS)$_2$ group (in place of one boryl group) could also activate the dihydrogen to yield a dimeric compound **103'**.[79b] Now moving forward toward the reactivity of protic E—H bonds like H$_2$O and NH$_3$ it was seen that **102** undergoes facile room temperature mediated oxidative addition with both O—H and N—H bonds leading to the formation of **105** and **106**, respectively. The resulting Sn—H linkage in **105** is characterized in solution by large satellite couplings to $^{117/119}Sn$ [$^1J(^{117}Sn-^1H)$ = 1385 Hz; $^1J(^{119}Sn-^1H)$ = 1456 Hz], while the (sharper) Sn—OH resonance can be resolved into a doublet [$^3J(^1H-^1H)$ = 1.4 Hz] with longer range couplings to tin [$^2J(^{117/119}Sn-^1H)$ = 24.6 Hz]. Though activation of water by heavier tetrelylenes is very rare, previous reports do prevail where Porschke and coworkers reported the synthesis of Sn{CH(SiMe$_3$)$_2$}$_2$(H)(OH) from the corresponding stannylene a long time ago in 1998.[80] Lastly oxidative addition of N—H bonds by **102** is currently a high profile synthetic challenge due to its parallel competition provided by transition metals, and also N—H activation is synthetically quite important from the industrial point of view. It was suggested that the reaction of

102 with an excess of NH$_3$ leads to the formation of **106** via an intermediate **102.NH$_3$**. It was proposed that the Sn center of **102** behaves as a Lewis acid. A stepwise mechanism takes place where the extreme Lewis acidity of Sn center abstracts a hydrogen from the second NH$_3$ molecule in turn, which is compensated by hydrogen provided by **102.NH$_3$** leading to the oxidative addition product **106** (Scheme 14). The first ammonia activation by Nacnac

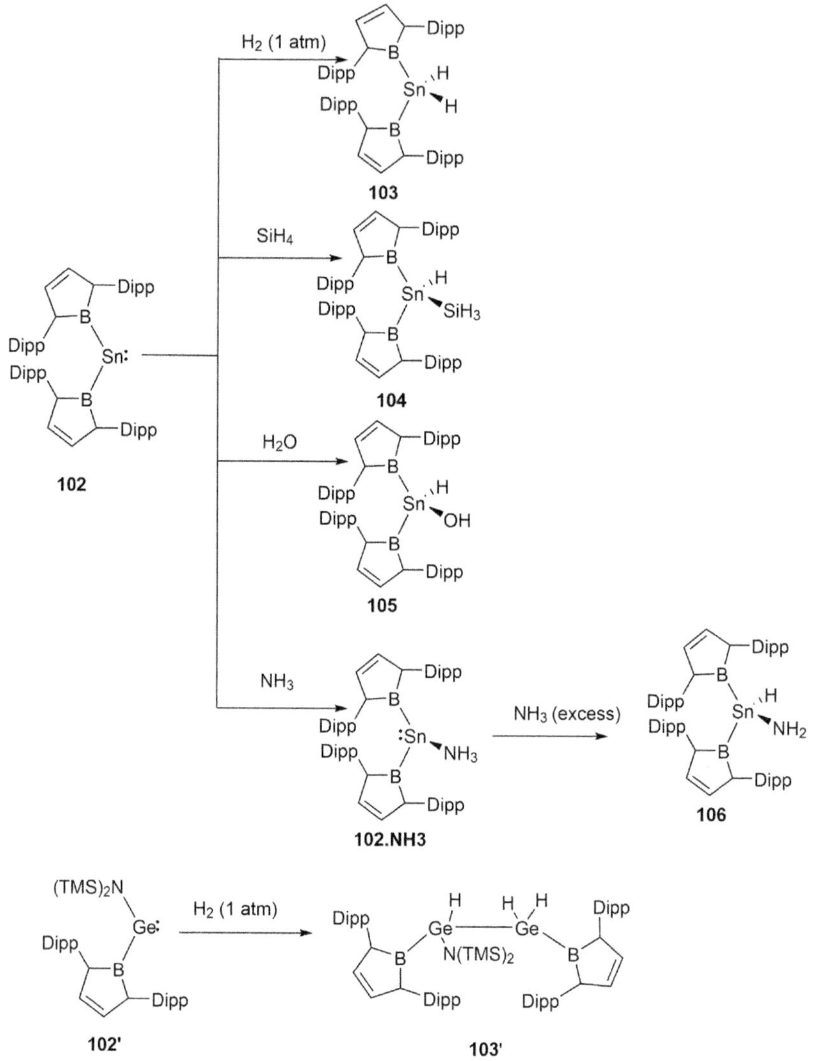

Scheme 14 Small molecules activation using **102** and **102′**.

germylene (**108a**) was demonstrated by Roesky and co-workers. The reaction proceeded via the cleavage of a N—H bond of ammonia to afford a compound with Ge(II)-NH$_2$ (**109a**) moiety.[81] Furthermore, Driess et al. reported the oxidative addition of protic E—H bonds (H$_2$O, NH$_3$) for **108b**. **108b** was synthesized via dehydrochlorination of **107b** with 1,3-di-tert-butylimidazol-2-ylidene as a base.[82] **108b** successfully undergoes ammonolysis and hydrolysis upon reaction with NH$_3$ and H$_2$O, respectively, and give rise to the formation of the corresponding amino-germylene **109b** or germylene hydroxide **110** (Scheme 15). IR spectroscopy for compound **109b** shows two stretching mode of vibration for two N—H bonds at $\nu = 3430$ and $3343\ cm^{-1}$, respectively.[81] A sharp absorption at $\nu = 3643\ cm^{-1}$ for **110** can be attributed to the O—H stretching frequency similar to that of previously reported germylene hydroxide (3571 cm^{-1}).[83]

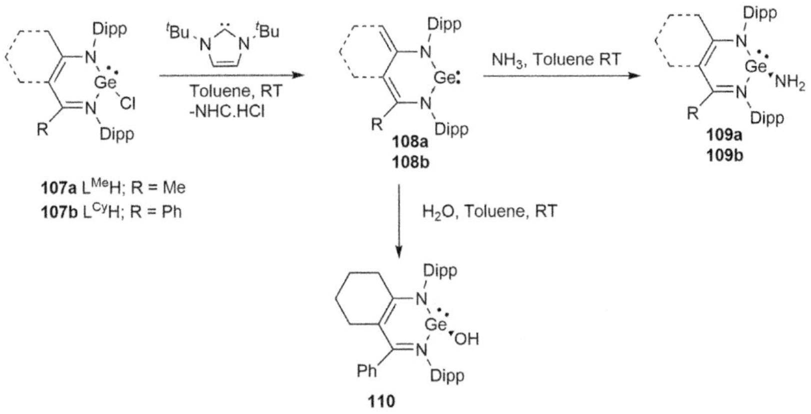

Scheme 15 Synthesis of β-ketiminato germylene (**108**) and its reactivity with NH$_3$ and H$_2$O.

The lighter analogue of germylene, silylene, is well known for the activation of unsaturated bonds like alkene and alkyne to produce stereospecific, stereo-selective and chemo-selective sila-cyclopropenes.[84] Siliranes were first synthesized by Seyfreth et al. in 1972.[85] Siliranes are formed by stereo-specific [2 + 1] cycloadditions which depends on the steric bulk present on the silylene center. However, the heavier tetrelylenes, e.g., germylene and stannylenes, are not so well established in literature to undergo such cycloaddition reactions to form germiranes and stanniranes. The reaction of phenylacetylene and acetylene was first demonstrated with germylene **108a** by Driess and coworkers. Owing to its ylide-like character **108a** displays very distinct reactivities. The reaction of both acetylene and phenylacetylene afforded [4 + 2] cycloaddition products of type **112-I**, whereas in the case of phenylacetylene another C—H activation product **113-I** was also observed.[86a] Jambor et al. also synthesized a cycloaddition product **112-II** by the reaction of diphenyl acetylene with the respective germylene, [(^{i}-Pr)$_2$NB(N-2,6-Me$_2$C$_6$H$_3$)$_2$]Ge: (**111-I**).[86b] A formal [2 + 2 + 2] cycloaddition took place to afford **112-II** (Scheme 16).

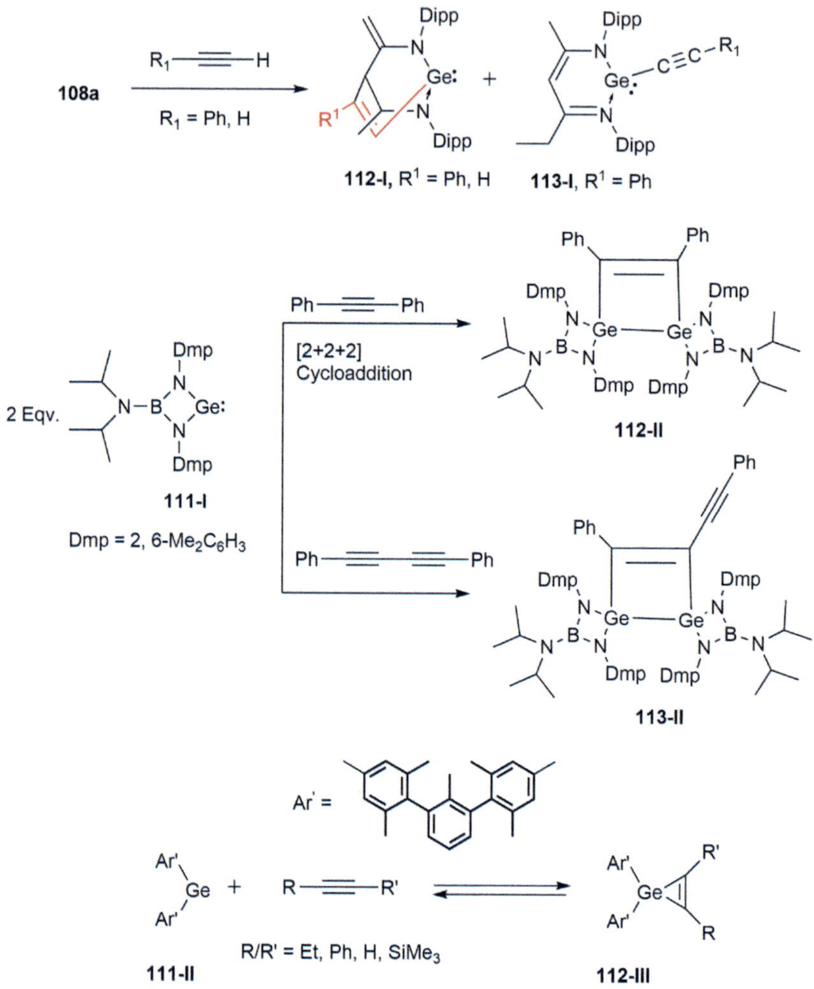

Scheme 16 A cycloaddition reaction of germylenes with alkynes and conjugated diynes.

Unfortunately, the corresponding germirene derivative could not be achieved by unimolar reaction of alkyne with **111-I**. In contrast to reaction of **111-I** with simple internal alkynes only one of the triple bonds of the respective conjugated diynes is attacked by the **111-I** as exemplified by the preparation of 3,4-R,R-1,2-digermacyclobut-3-enes (**113-II**) (Scheme 16). It is noteworthy, that even heating of **113-II** with an excess of the **111-I** did not lead to second cycloaddition reaction to form another four-membered digermacyclobutadiene ring, which can be explained by a significant steric hindrance at the unreacted carbon–carbon triple bond. Although germylenes were successful in giving such cycloaddition reactions, stannylenes could not do the same reactions. The reason for this unreactive nature of stannylenes can be attributed to the greater extent of orbital mismatch (5p-2p overlap). Mostly, these cycloaddition reactions of germylenes are irreversible in nature. However, recently a unique behavior of reversible complexation of alkyne to diarylgermylene has been shown. The carbon-substituted germylene Ge(Ar^{Me6})$_2$ (**111-II**) displays a reversible complexation at ambient temperature with four alkynes 3-hexyne, diphenylacetylene, trimethylsilylacetylene, and phenylacetylene. The reaction proceeds with the formation of germacyclopropene (germirene) products of type **112-III**. The formation of **112-III** was confirmed by X-ray diffraction studies.[86c] The reason of this reversible behavior was attributed to the steric hindrance and ring strain of the **112-III**.

3.2 Activation of CO$_2$ using germanium(II) and tin(II) hydrides

Carbon dioxide is a readily accessible atmospheric gas and can be a valuable raw material for other organic compounds.[87] Organometallic hydrides play an important role in metathesis reactions hence their reactivities are well documented.[88] Roesky et al. utilized a germanium(II) hydride **53** for the activation of CO$_2$.[89a] Reaction of **53** with CO$_2$ lead to the formation of the respective metal formate **115a** (Scheme 16), which are quite well known for transition metal and alkali metal hydrides. The formal oxidation state of germanium center persists upon simultaneous activation of CO$_2$ to form **115a**. Similarly, analogous germylene **114** with a cyclohexyl backbone was also utilized to activate CO$_2$.[89b] Similar to **53** and **114** Roesky et al. also reported an analogous tin(II) hydride (**116**) in the same year.[90] Successful hydrostannylation reaction of **116** occurred with CO$_2$ to form the respective stannylene formate **117**. Similar to **114** the formal oxidation of **116** remains the same upon activation of CO$_2$ to form **117** (Scheme 17).

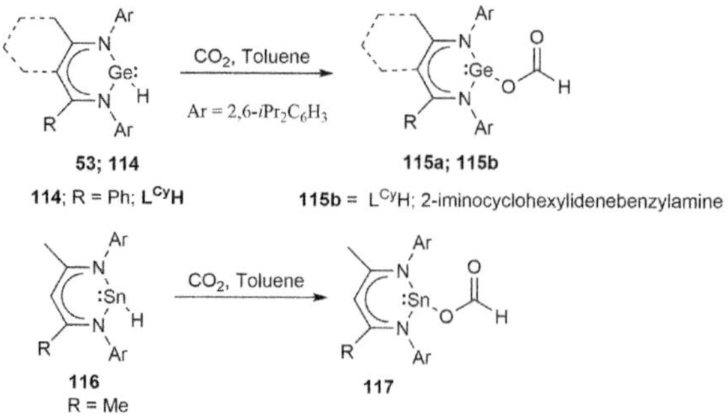

Scheme 17 Activation of CO_2 using N-heterocyclic germylenes (**114**) and stannylene (**116**).

3.3 Activation of white phosphorus by germylene

White phosphorus is the primary raw material for the synthesis of phosphorus containing compounds. Early and late transition metals have been long used to activate phosphorus[91]; however, metal free activation of phosphorus is recently reported by mainly using low coordinate ambiphilic main group compounds.[92] Different variations of carbenes and silylenes have been used for the activation of white phosphorus.[93] But going down the group activation is relatively difficult due to the bond dissociation enthalpy of newly formed bonds, which could reluctantly release phosphorus from the activated compound. Recently, Ragogna et al. demonstrated the first ever reversible phosphorus activation using diaryl substituted germylene **118**[94] (Scheme 18). A room temperature activation of phosphorus takes place within 4 days to afford a GeP_4 complex **119**, which upon irradiation with light reverts to elemental P_4 and **118**. The tin analogue **120** did not produce any phosphorus activated product and only resulted in the decomposition of the diaryl stannylene. This is the only available report for the activation of white phosphorus by any germylene.

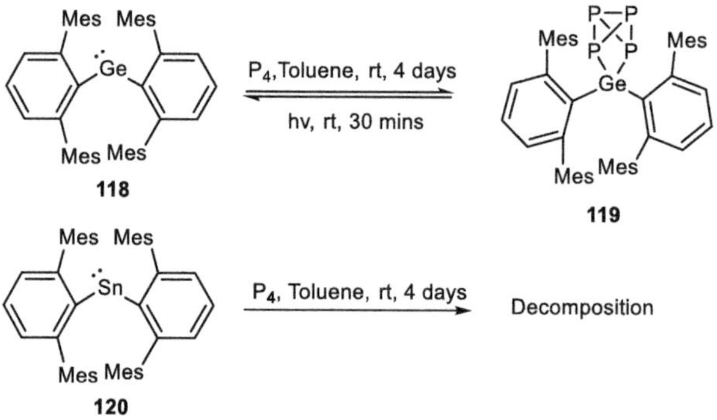

Scheme 18 Activation of white phosphorus using N-heterocyclic germylene (**118**).

3.4 C—F bond activation by heavier tetrelylenes

The chlorides of group 14 elements are well documented in literature as compared to the fluorides which are very few and have huge applications in the laboratory as well as industry. The group 14 based fluoro compounds are mainly known in the +4 oxidation state. Only a handpicked number of organo germanium(II) and tin(II) fluorides are known.[95] Previously, Roesky and coworkers synthesized a lead(II) fluoride, L'PbF (L'=HC(CMeNAr)$_2$, Ar=2,6^{i}Pr$_2$C$_6$H$_3$) via the reaction of L'PbNMe$_2$ with pentafluoropyridine.[96] A substitution reaction of the —NMe$_2$ group at the lead(II) atom by the *para*-fluorine atom of pentafluoropyridine yielded L'PbF. The —NMe$_2$ group of Pb(II) was replaced with fluorine atom. This pattern of reactivity prompted reaction of analogous Ge(II) (**123**) and Sn(II) (**124**) systems,[95] respectively. The dialkyl heavier tetrelylenes (**123, 124**) were successfully isolated via substitution reaction of the respective chloro-tetrelylene (**121, 122**) (Scheme 19). Later Roesky and coworkers reported the reaction of these substituted tetrelylenes, **123** and **124,** with pentafluoropyridine and the fluorination

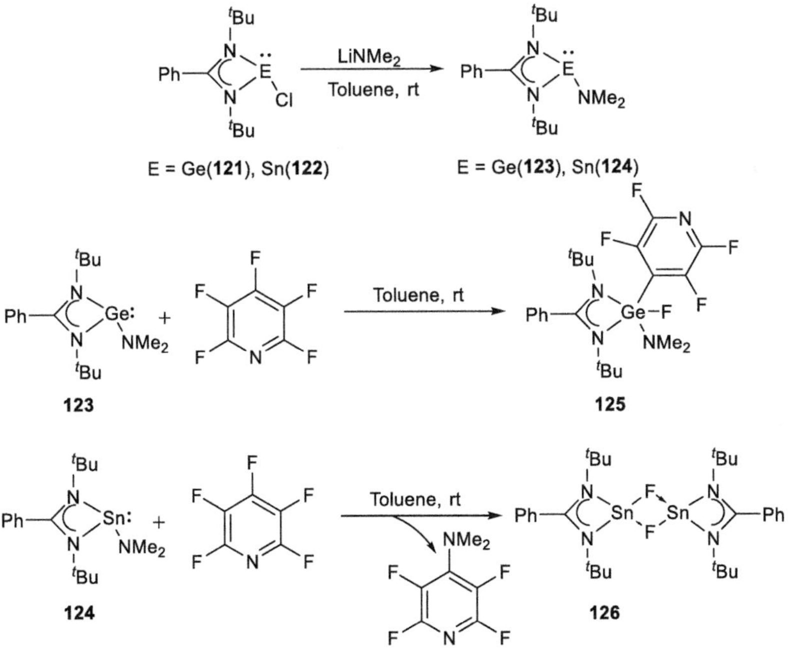

Scheme 19 Fluorination chemistry of benzamidinato germylene (**123**) and stannylene (**124**) with pentafluoropyridine.

chemistry of these heavier tetrelylenes were studied (Scheme 19). These were the first ever reports of C—F bond activation using germylene and stannylene. The activity of the inert lone pair is highly influenced by various functionalities attached to the germanium atom. As an example, even at elevated temperature (60 °C) no reaction takes place for amino functionalized germylenes (LGeN(SiMe$_3$)$_2$) with C$_5$F$_5$N but LGeNMe$_2$ reacts readily at room temperature. X-ray diffraction reveals the molecular structure of **125** (Fig. 12). From these observations it can be concluded that the adjacent atoms bonded to the germanium center tunes up the reactivity of the inert lone pair present in germanium. This can be reasoned out that the increased activity of the lone pair is due to the enhanced basicity of the ancillary ligand. Unlike **123**, the reaction of **124** with pentafluoro pyridine did not proceed in a similar fashion and a substitution product (**126**) was obtained where —NMe$_2$ group was replaced by F-group.

The molecular structure of **126** revealed that it exists as a dimer which is weakly interconnected by intermolecular Sn⋯F bonds and displays two different bond distances (2.0796 and 2.338 Å) (Fig. 13).[97] **126** represents the first example where such an arrangement was observed for a Sn$^{(II)}$—F compound.

Finally, we can conclude that the C—N bond formation balances the energy required for the cleavage of the C—F bond. The C—F bond activation shown by Ge(II) and Sn(II) with pentafluoropyridine is compared to the oxidative addition versus arene elimination reaction of Ar$_2$Ge(II) and Ar$_2$Sn(II)

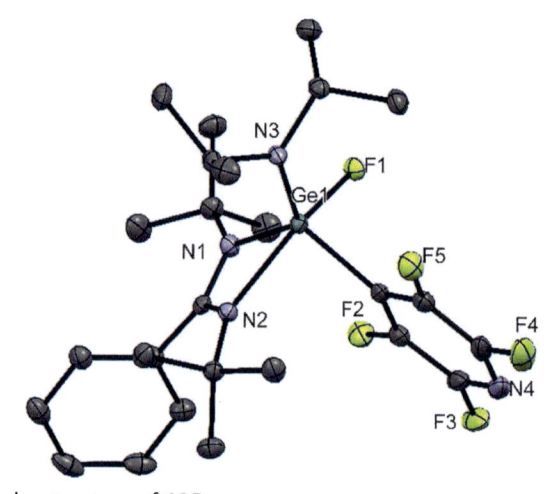

Fig. 12 Molecular structure of **125**.

Fig. 13 Molecular structure of **126**.

[Ar=C$_6$H$_3$-2,6-(C$_6$H$_3$-2,6-iPr$_2$)$_2$] with ammonia and hydrogen as discussed in earlier sections of the chapter reported by Power et al. (Scheme 13).

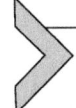

4. N-heterocyclic germylenes and stannylenes as organocatalyst for basic organic transformations

Transition metal catalysis has long overshadowed organocatalysis for various organic transformations. However, owing to the toxicity, less abundance and higher cost expenditure, synthetic chemists are shifting their respective fields of interest to more economical, greener, readily available and abundant catalysts as an alternative to transition metal catalysts for similar organic transformations. There are several reports on N-heterocyclic carbene acting solely as organocatalyst or supporting transition metals as an ancillary ligand.[98] Nevertheless, the catalytic regiments of heavier N-heterocyclic tetrelylenes namely, germylene and stannylene, are in developmental stage. Herein we will be discussing the recent progress of these heavier tetrelylenes as pure organocatalysts for organic transformations, e.g., hydroboration and cyanosilylation reactions.

4.1 Hydroboration reaction catalyzed by heavier N-heterocyclic tetrelylenes

Hydroboration reactions of unsaturated functionalities serve as efficient tools for the preparation of various synthetic organic intermediates.[99a]

A large number of reports on the hydroboration of aldehydes and ketones have come up using transition metal and s-block catalysts.[99b] Recently, a good number of research groups have started using p-block compounds as potential catalysts owing to their cost-effective and non-toxic nature for homogeneous catalytic hydroboration reactions of aldehydes and ketones.[99c] Herein, we will be mainly focusing on heavier low-valent group 14 tetrelylenes and their application as catalysts for hydroboration reactions. N-heterocyclic carbenes display a wide catalytic use as organocatalysts. In 1958 Breslow proposed the activation of an aldehyde by NHC which generates an acyl anion equivalent in the form of an enol tautomer, also known as "Breslow intermediate" today.[100] Nonetheless, there is only a handpicked report on germylene and stannylene as organocatalysts for the hydroboration reactions of unsaturated substrates. The catalytic activities of these heavier tetrelylenes should be enriched owing to the singlet ground state with vacant p-orbital and lone pair of electrons. Thus it can function as an ambiphillic catalyst. Coming to the recent reports for catalyzing hydroboration reaction the most noteworthy mentions are the bulky amine substituted germylene and stannylene hydrides (**57, 62**) reported by Jones et al.[101] (Scheme 20).

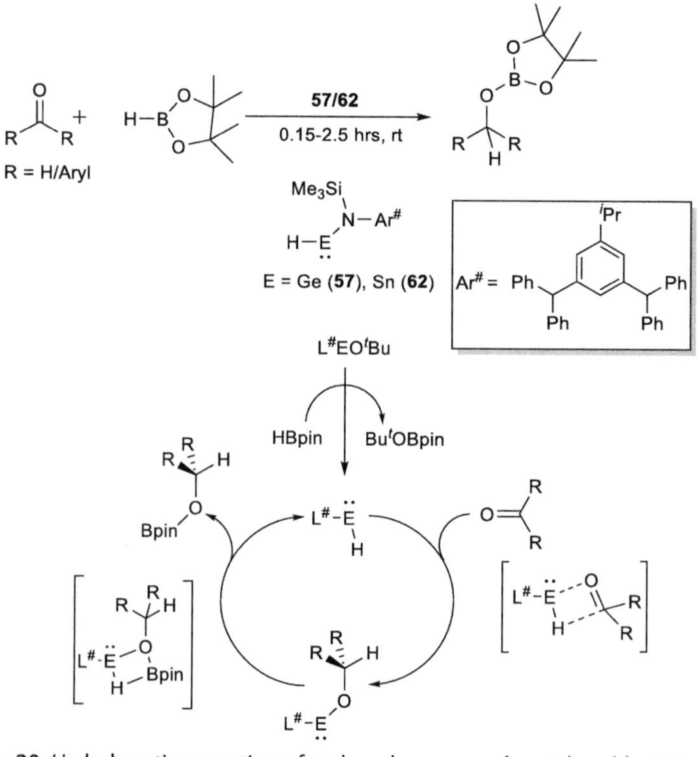

Scheme 20 Hydroboration reaction of carbonyl compounds catalyzed by **57** and **62**.

Interestingly, these catalysts rapidly hydroborate carbonyl compounds and even the sterically demanding substrates can be hydroborated very rapidly at ambient temperature. Undoubtedly, the empty p-orbital available at the tetrel centers of **57** and **62** give rise to their markedly enhanced catalytic activity. The two-coordinate nature of **57** and **62** is also the most plausible reason to explain why they react cleanly with 1equiv. of HBpin at ambient temperature to give the respective borate esters. It was proposed that in the case of **62**, the bulky ligand substituted SnO^{t}Bu acts as the active precatalyst. Which actively dissociates to the respective tin(II) hydride reacting with excess HBpin, hence successfully catalyzing the hydroboration reaction of aldehydes and ketones. Catalysts **57, 62** exhibited a humongous turnover frequency of 13,300 h^{-1}. It was seen that the hydroboration reactions are more efficient using catalyst **62** compared to **57**. This can be explained by the enhanced Lewis acidic metal center in **62** (M=Sn), and more polar $^{\delta+}$M—O$^{\delta-}$ bond present in the hydrostannylated intermediates, thus making these intermediates more prone to react with HBpin than the Ge based catalyst. Jones et al. proposed that the mechanism of the Ge- and Sn-catalyzed reactions initially involves a preliminary attack of the O-center of the substrate present at the two-coordinate metal center, with hydrometalation subsequently taking place via a four-membered transition state (Scheme 20). The two-coordinate, monomeric metal alkoxide intermediate finally undergoes a σ-bond metathesis reaction with HBpin to generate the product and regenerates back the catalyst. Later on in 2016, Zhao et al. showed a cooperative effect of N-heterocyclic germylene **127** in catalyzing the hydroboration reaction of carbonyl compounds[102] (Scheme 21). **127** features a weakly Lewis acidic germanium center, which can interact with the polarized C=O bonds leading to the pre-activation of the substrate. This co-cooperativity effect increases the electrophilicity of the carbon center of the carbonyl compound, thus in the presence of HBpin it undergoes successful conversion to alkoxy pinacolboronate.

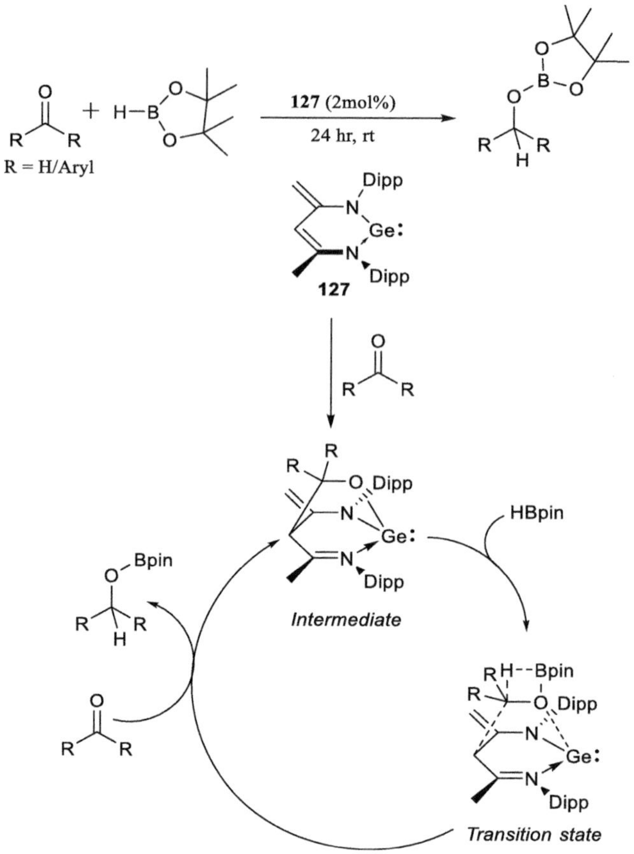

Scheme 21 Hydroboration reaction of carbonyl compounds through cooperativity effect of **127** along with the tentative mechanistic pathway.

Following this, Wesemann et al. reported hydroboration of carbonyl compounds using intramolecular heavier tetrelylenes based Lewis pairs (**128, 129**)[103] (Scheme 22). Both the intramolecular heavier tetrelylene based Lewis pairs gave successful hydroboration of hexanal in satisfactory yields within 30 min at room temperature.

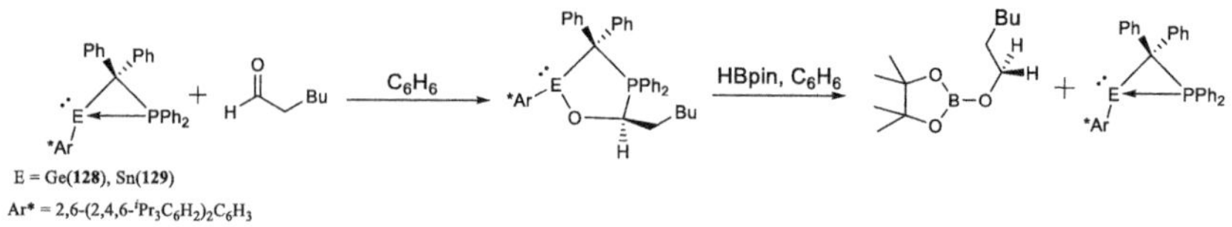

Scheme 22 Hydroboration reaction catalyzed by **128** and **129**.

Lastly N-heterocyclic germylene (**20**) and stannylene (**37**) were utilized as potential catalysts for hydroboration of aldehydes in a chemo-selective way.[104] Selective hydroboration of aldehydes were obtained using these catalysts even in the presence of other unsaturated functionalities. A double anchored mechanism was proposed where the Lewis acidic tetrelylene catalyst anchors to the electronegative oxygen atom of HBpin and then it causes a relative change in electron density and later in the presence of the aldehydic functional group a four-membered transition state is attained. A facile concerted [2 + 2] cycloaddition reaction reverts back the catalyst along with the formation of respective alkoxy pinacolboronate ester (Scheme 23).

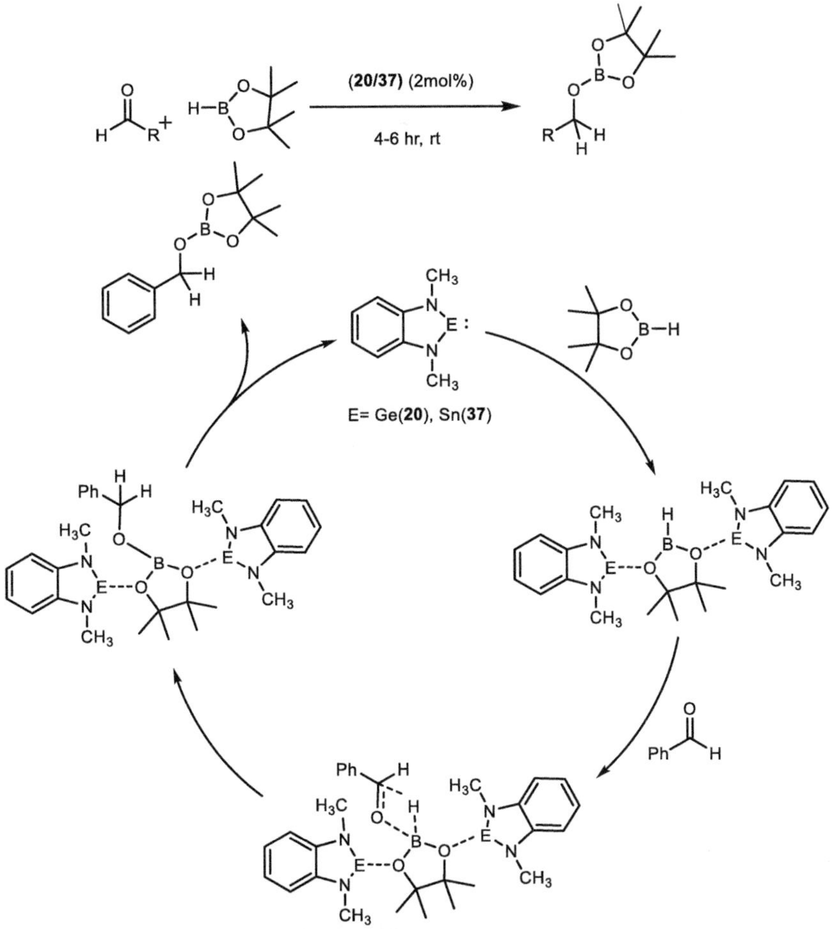

Scheme 23 Selective hydroboration of aldehydes using **20** and **37**.

4.1.1 Hydroboration of CO_2 to methanol

Catalytic conversion of CO_2 into various important fuels is recently catching a lot of research attention.[105] Although transition metals have been vastly used for CO_2 reductions, but non-metallic compounds as a homogeneous catalyst for the reduction of CO_2 has been minimal until 2010. Jones et al. have utilized their amido ligand substituted germanium(II) (**57**) and tin(II) hydrides (**62**) for the successful reduction of CO_2.[106] **62** performs as a better catalyst than compound **57** for catalytic reduction of CO_2. Catechol borane was used as a borane source and it was witnessed that reaction completion using **62** took place within 5 min with a catalyst loading of 1 mol% and showed TOF of 1188 h^{-1}. The tentative mechanism of the reaction is given in Scheme 24. The mechanistic pathway reveals hydrogermylation or hydrostannylation of CO_2 as the preliminary step. Later methanol derivatives are formed utilizing 3 equivalent of catechol borane regenerating catalysts **57** or **62**.

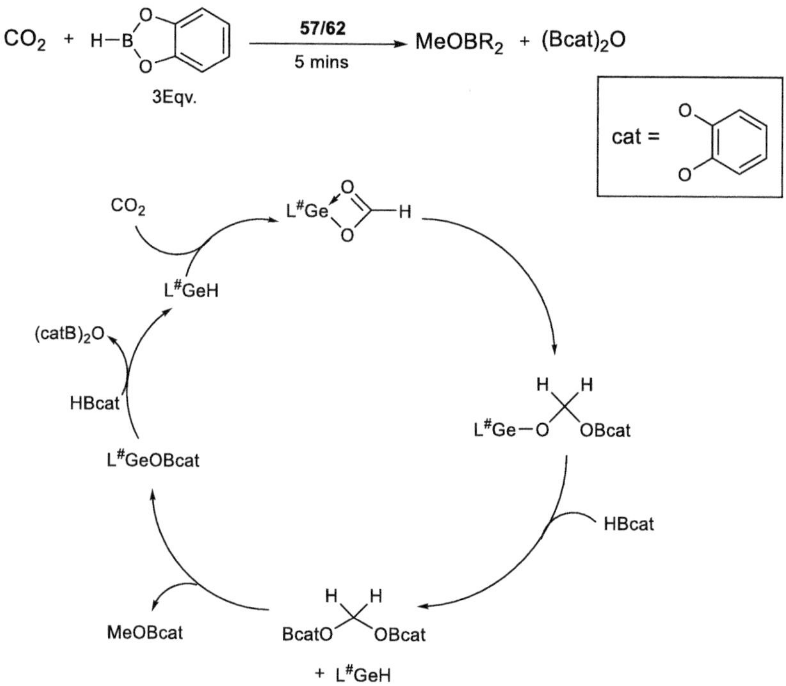

Scheme 24 Hydroboration of CO_2 using catalyst **57** and the proposed mechanistic pathway.

4.2 Cyanosilylation reaction catalyzed by heavier N-heterocyclic tetrelylenes

The cyanosilylation of aldehydes is one of the most important strategies for the C—C bond formation. The cyanohydrin intermediate also serves as a protecting group for further derivatization of alcohols. There has been an increased interest in p-block catalytic systems for cyanosilylation of aldehydes. However, only a handful of well-defined neutral heavier group 14 main group compounds have been devised to act as cyanosilylation catalysts. A couple of reports prevail on silicon based catalysts like Tilley et al. have demonstrated that bis(perfluorocatecholato)silane [Si(cat^F)$_2$] can promote the cyanosilylation of 4-nitro-benzaldehyde.[107] Later Sen et al. showed a benzamidinato framework based tetra-valent silicon based compound (PhC(N^{t}Bu)$_2$SiH(CH$_3$)Cl) could also successfully catalyze cyanosilylation reaction of aldehydes with trimethylsilyl cyanide.[108] Since this chapter preludes on heavier tetrelylenes we will focus on reports of cyanosilylation of unsaturated substrates using germylenes/stannylenes. A brief survey shows that there are only two such reports on neutral heavier tetrelylenes acting as active catalysts for cyanosilylation reaction. The first report was provided by Nagendran et al. where substituted germylene cyanide (**131**) (Scheme 25) could intrinsically act as catalyst owing to the reactive Ge^{II}—CN bond. The isolation of **131** follows a novel route that involves the reaction of digermylene oxide complex [{(^{t}Bu)$_2$ATIGe}$_2$O] (**130**) with TMSCN.[109] This route is based on two fundamental facts (a) the oxophillicity of TMS group in TMS containing reagents and (b) the presence of a reactive oxygen atom within two germanium(II)

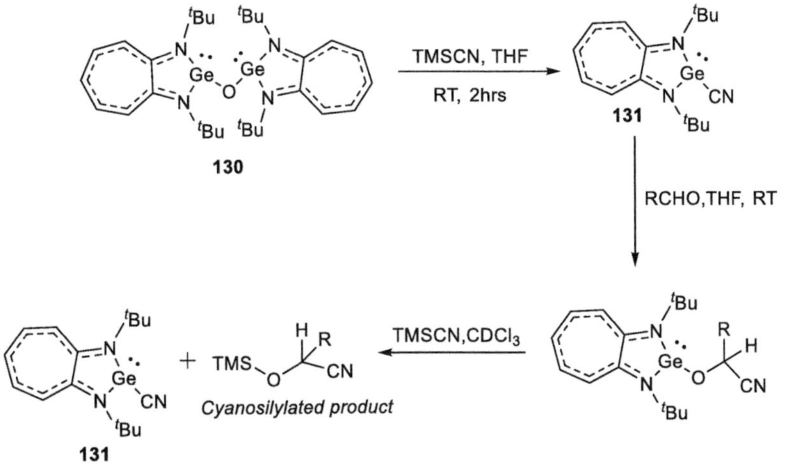

Scheme 25 Germylene cyanide (**131**) as autocatalyst for cyanosilylation of carbonyl compounds.

centers. **131** in situ activates an aldehydic center by forming an alkoxide intermediate with facile transfer of the cyanide to the aldehydic carbon center. Later in the presence of trimethyl silyl cyanide **131** regenerates with the formation of trimethyl silyl cyanohydrin (Scheme 25).

Thus **131** acts as an autocatalyst for the cyanosilylation reactions. The second example includes the use of N-heterocyclic germylene **20** and stannylene **37** as potential catalysts for the cyanosilylation of aldehydes in a chemo-selective way.[104] Selective cyanosilylation of aldehydes was obtained using these catalysts, even in the presence of other unsaturated functionalities. Mechanistically it was seen that the reaction proceeds taking advantage of the vacant p-orbital present in the tetrel center. A coordination mechanism was proposed where the lone pair of nitrogen is donated to the vacant p-orbital of the tetrel center. Further, in the presence of the aldehydic functional group a four-membered transition state is attained and then via a facile concerted [2+2] cycloaddition reaction, catalyst (**20**, **37**) along with the cyanohydrin trimethylsilyl ethers is regenerated (Scheme 26).

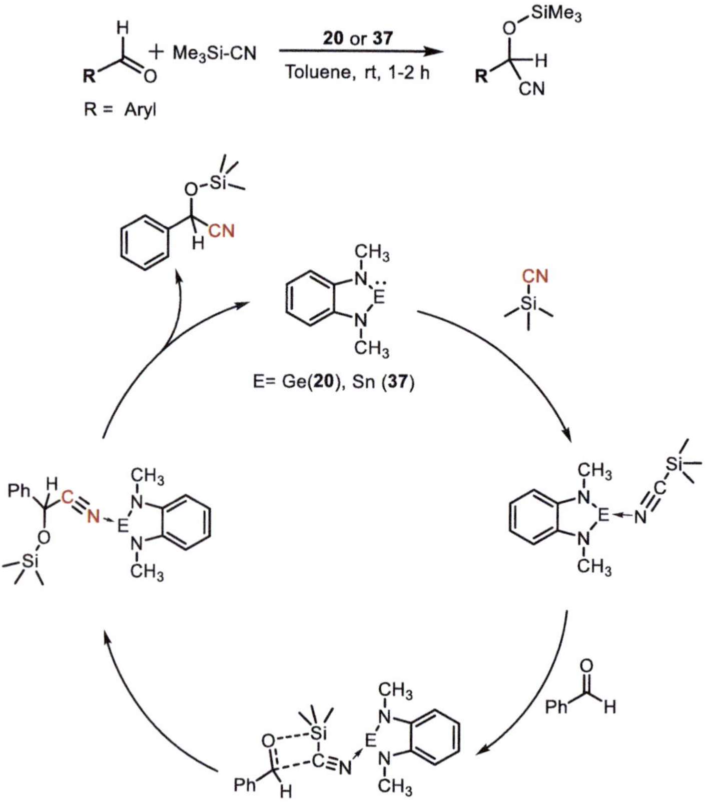

Scheme 26 Selective cyanosilylation of aldehydes using **20** and **37**.

5. Summary at a glance

In summary this chapter provides a detailed discussion on the synthesis, reactivity and catalytic applications of the germylenes and stannylenes. This contribution highlights the unique bonding scenario of the heavier tetrelylenes and how the electronic distribution or availability of vacant p-orbitals leads to the ambiphillic nature of the heavier tetrelylenes. An optimum steric and electronic balance is required for the isolation of the monomeric heavier tetrelylenes and lot of strategies have been employed starting from chelation via pincer ligands to homo-conjugation. These isolated monomeric heavier tetrelylenes have a unique reactivity pattern due to the lower HOMO-LUMO gap and they can perform reactions analogous to transition metals, e.g., small molecule activation via oxidative addition mechanism. Recently, these molecules have also been used as organocatalysts for the hydroboration and cyanosilylation reactions. As a conclusive remark, the fascinating chemistry displayed by the heavier tetrelylenes will open a new domain for them to be used as a better alternative to transition metals for small molecule activation and catalysis.

References

1. Bourissou D, Guerret O, Gabbai FP, Bertrand G. Stable carbenes. *Chem Rev.* 2000;100:39–92. https://doi.org/10.1021/cr940472u.
2. Luke BT, Pople JA, Kroghjespersen MB, et al. A theoretical survey of unsaturated multiply bonded and divalent silicon compounds. Comparison with carbon analogs. *J Am Chem Soc.* 1986;108:270–284. https://doi.org/10.1021/ja00262a014.
3. Liddle ST, Edworthy IS, Arnold PL. Anionic tethered N-heterocyclic carbene chemistry. *Chem Soc Rev.* 2007;36:1732–1744. https://doi.org/10.1039/B611548A.
4. Weidenbruch M. Some recent advances in the chemistry of silicon and its homologues in low co-ordination states. *J Organomet Chem.* 2002;646:39–52. https://doi.org/10.1016/S0022-328X(01)01262-1.
5. Jasinski JM, Becerra R, Walsh R. Direct kinetic studies of silicon hydride radicals in the gas phase. *Chem Rev.* 1995;95:1203–1228. https://doi.org/10.1021/cr00037a004.
6. Moiseev AG, Leigh WJ. The direct detection of diphenylsilylene and tetraphenyl disilene in solution. *Organometallics.* 2007;26:6268–6276. https://doi.org/10.1021/om7006584.
7. Mizuhata Y, Sasamori T, Tokitoh N. Stable heavier carbene analogues. *Chem Rev.* 2009;109:3479–3511. https://doi.org/10.1021/cr900093s.
8. Kira M, Ishida S, Iwamoto T. Comparative chemistry of isolable divalent compounds of silicon germanium and tin. *Chem Rec.* 2004;4:243–253. https://doi.org/10.1002/tcr.20019.
9. Jutzi P, Becker A, Stammler HG, Neumann B. Synthesis and solid state structure of $(Me_3Si)_3CGeCH(SiMe_3)_2$ a monomeric dialkyl germylene. *Organometallics.* 1991;10:1647–1648. https://doi.org/10.1021/om00052a002.

10. Jutzi P, Schmidt H, Neumann B, Stammler HG. Bis(2,4,6-tri-tert-butyl phenyl) germylene reinvestigated: crystal structure, Lewis acidic catalyzed C-H insertion and oxidation to an unstable germanone. *Organometallics*. 1996;15:741–746. https://doi.org/10.1021/om950558t.
11. Bender JE, Holl MMB, Kampf JW. Synthesis and characterisation of a novel diaryl germylene containing electron-withdrawing groups. *Organometallics*. 1997;16:2743–2745. https://doi.org/10.1021/om970200s.
12. Masamune S, Sita LR. Hexakis(2,4,6-triisopropylphenyl) cyclotristannane $(R_2Sn)_3$ and tetrakis(2,4,6-triisopropylphenyl) distannene $(R_2Sn)_2$ their unprecedented thermal conversion and the first solution spectral characterisation of a distannene. *J Am Chem Soc*. 1985;107:6390–6391. https://doi.org/10.1021/ja00308a039.
13. Eaborn C, Ganicz T, Hitchcock PB, Patel D, Smith JD, Zhang SB. Oxidative addition to amonomeric tannylene to give four coordinate tin compounds containing the bulky bidentate ligand $C(SiMe_3)_2SiMe_2CH_2CH_2Me_2Si(Me_3Si)_2C$. Crystal structures of $CH_2Me_2Si(Me_3Si)_2CSnC(SiMe_3)_2SiMe_2CH_2, CH_2Me_2Si(Me_3Si)_2CSnMe(OCOCF_3)$ $(SiMe_3)_2SiMe_2CH_2$ and $(CF_3COO)_2MeSnC(SiMe)_2SiMe_2CH_2CH_2Me_2Si-(Me_3Si)_2$ $CSnMe(OCOCF_3)_2$. *Organometallics*. 2000;19:49–53. https://doi.org/10.1021/om990779p.
14. Weidenbruch M, Schlaefke J, Schafer A, Peters K, Vonschnering HG, Marsmann H. Bis(2,4,6-tri-tert-butylphenyl)stannanediyl: a diaryl stannylene without donor stabilisation. *Angew Chem Int Ed Engl*. 1994;33:1846–1848. https://doi.org/10.1021/anie,199418461.
15. Saito M, Tokitoh N, Okazaki R. A new method for the synthesis of stannylenes: exhaustive desulfurization of tetrathiastannolanes. *Chem Lett*. 1996;25:265–266. https://doi.org/10.1246/cl.1996.265.
16. Kira M, Yauchibara R, Hirano R, Kabuto C, Sakurai H. Chemistry of organosilicon compounds: synthesis and X-ray structure of the first dicoordinate dialkylstannylene that is monomeric in the solid state. *J Am Chem Soc*. 1991;113:7785–7787. https://doi.org/10.1021/ja00020a064.
17. Arduengo III AJ, Harlow RL, Kline M. A stable crystalline carbene. *J Am Chem Soc*. 1991;113:361–363. https://doi.org/10.1021/ja00001a054.
18. Herrmann WA. N-heterocyclic carbene: a new concept in organometallic catalysis. *Angew Chem Int Ed*. 2002;41:1290–1309. https://doi.org/10.1002/1521-3773(20020415)41:8<1290::AID-ANIE1290>3.0.CO;2-Y.
19. Kuhl O, Lonnecke P, Heinicke J. Influence of annelation in unsaturated heterocyclic diaminogermylene. *Polyhedron*. 2001;20:2215–2222. https://doi.org/10.1016/S0277-5387(01)00821-X.
20. Heinecke J, Oprea A. Higher carbene homologues: naphtho[2,3-d]-1,3,2 λ^2-diazagermole, diazastannole, and attempted reduction of 2,2-dichloronaphtho[2,3-d]-1,3,2-diazasilole. *Heteroat Chem*. 1998;9:439–444. https://doi.org/10.1002/(SICI)1098-1071(1998)9:4<439::AID-HC13>3.0.CO;2-S.
21. Veith M, Grosser M. Cyclic diazastannylenes, rings and cages with Ge(II), Sn(II) Pb(II). *Z Naturforsch*. 1982;37b:1375–1381. https://doi.org/10.1016/0022-328X(84)80564.
22. Meller A, Grabe CP. Synthesis and isolation of new germanium(II) compounds and free germylenes. *Chem Ber*. 1985;118:2020–2029. https://doi.org/10.1002/cber.19851180525.
23. (a) Herrmann WA, Denk M, Behm J, et al. Stable cyclic germanediyls ("cyclogermylenes") synthesis structure metal complexes and thermolyses. *Angew Chem Int Ed Engl*. 1992;31:1485–1488. https://doi.org/10.1002/anie.199214851.
(b) Ayers AE, Rasika DHV. Investigation of silver salt metathesis: preparation of cationic germanium(II) and tin(II) complexes and silver adducts containing unsupported silver-germanium and silver-tin bonds. *Inorg Chem*. 2002;41:3259–3268.

https://doi.org/10.1021/ic0200747. (c) Driess M, Yao S, Brym M, Wuellen Cv. A heterofulvene like germylene with a betain reactivity. *Angew Chem Int Ed.* 2006;45:4349–4352. https://doi.org/10.1002/anie.200600237.
24. (a) Piskunov AV, Aivazyan IA, Poddlesky AI, et al. New germanium complexes containing ligands based on 4,6-di-*tert*-butyl-*N*-(2,6-diisopropylphenyl)-*o*-iminobenzoquinone in different redox states. *Eur J Inorg Chem.* 2008;1435–1444. https://doi.org/10.1002/ejic.200701115. (b) Pal S, Dasgupta R, Khan S. Acyclic α-phosphinoamido-germylene: synthesis and characterisation. *Organometallics.* 2016;35:3635–3640. https://doi.org/10.1021/acs.organomet.6b00689.
25. Zabula AV, Hahn FE. Mono- and bidentate benzannulated N-heterocyclic germylenes, stannylenes and plumbylenes. *Eur J Inorg Chem.* 2008;5165–5179. https://doi.org/10.1002/ejic.200800866.
26. Chivers T, Clark TJ, Krahn M, Parvez M, Schatte G. Cubane complexes with two (or more) group 14-group 16 double bonds: synthesis and X-ray structures of $Sn_4Se_2(NtBu)_4$ and $Ge_4Se_3(NtBu)_4$. *Eur J Inorg Chem.* 2003;1857–1860. https://doi.org/10.1002/ejic.200300075.
27. Bartlett RA, Power PP. Synthesis and structural characterization of the cyclic species $[GeN(2,6-iso-Pr_2C_6H_3)]_3$: the first "germanazene" *J Am Chem Soc.* 1990;112:3660–3662. https://doi.org/10.1021/ja00165a062.
28. Kobayashi S, Cao S. The first synthesis of bisgermylene and bisstannylene with acyclic structure. *Chem Lett.* 1994;941–944. https://doi.org/10.1246/cl.1994.941.
29. Braunschweig H, Hitchcock PB, Lappert MF, Pierssens LJM. Synthesis structures and reactions of two bisdiaminostannylenes and a bisdiaminogermylene containing a central C_6 ring. *Angew Chem Int Ed Engl.* 1994;33:1156–1158. https://doi.org/10.1002/anie.119411561.
30. Braunschweig H, Gehrhus B, Hitchcock PB, Lappert MF. Characterisation of N,N' disubstituted 1,2-phenylenebis(Amido)tin(II)compounds. X-ray structures of 1,2-C6H4[N(CH2But)2]Sn and [1,2-C6H4N(SiMe3)2Sn]2(tmeda). *Z Anorg Allg Chem.* 1995;621:1922–1928.
31. Schaeffer CD, Zuckermann JJ. Tin(II) organo-silylamines. *J Am Chem Soc.* 1974;96:7160–7162. https://doi.org/10.1021/ja00829a086.
32. (a) Eichler TG, Gudat D, Niegger M. Tin analogue of Arduengo carbenes: synthesis of 1,3,2 λ2-diazastannoles and transfer of Sn atom between 1,3,2 λ2-diazastannole and diazadiene. *Angew Chem Int Ed.* 2002;41:1888–1891. https://doi.org/10.1002/1521-3773(20020603)41:11<1888::AID-ANIE1888>3.0.CO;2-O. (b) Asay M, Jones C, Driess M. N-heterocyclic carbene analogues with low-valent group 13 and group 14 elements: syntheses, structures, and reactivities of a new generation of multitalented ligands. *Chem Rev.* 2010;111:354–396. https://doi.org/10.1021/cr100216y.
33. Avent AG, Drost C, Gehrhus B, Hitchcock PB, Lappert MF. Synthetic and structural studies on the cyclic bis(amino)stannylenes [Sn(NR2)C10H6-1,8] and their reaction with SnCl2 or Si[(NCH2But)2C6H4-1,2] (R = SiMe3 or CH2But). *Z Anorg Allg Chem.* 2004;630:2090–2096.
34. Babcock JR, Incarvito C, Rheingold AL, Fettinger JC, Sita LR. Double heterocumulene metathesis of cyclic bis(trimethyl silylamido)stannylenesand tethered bimetallic bisamidinates from the resulting α,w-biscarbodiimides. *Organometallics.* 1999;18:5729–5732. https://doi.org/10.1021/om990635a.
35. Ayers AE, Marynick DS, Rasika Dias HV. Azido derivatives of low-valent group 14 elements: synthesis characterisation and electronic structures of [(nPr2)ATl]GeN3 [(nPr2)ATl]SnN3 featuring heterobicyclic 10π electron ring systems. *Inorg Chem.* 2000;39:4147–4151. https://doi.org/10.1021/ic000545u.
36. (a) Henn M, Schurmann M, Mahieu B, Zanello P, Cinquantini A, Jurkschat K. A ferrocenyl bridged intramolecularly coordinated bis(diorganostannylene): synthesis

molecular structure and reactivity of [4-^{t}Bu-2,6-[P(O)(O-^{i}Pr)$_2$C$_6$H$_2$Sn]C$_5$H$_4$]$_2$Fe. *J Organomet Chem.* 2006;691:1560–1572. https://doi.org/10.1016/j.jorganchem.2005. 12.026. (b) Hsu C-Y, Chan L-W, Lee G-H, Peng S-M, Chiu C-W. Triphenylene-based tris-N-heterocyclic stannylenes. *Dalton Trans.* 2015;44:15095–15098. https://doi.org/10.1039/C5DT00694E.

37. (a) Ding Y, Hao H, Roesky HW, Noltemeyer M, Schmidt HG. Synthesis and structures of germanium(II) fluorides and hydrides. *Organometallics.* 2001;20:4806–4811. https://doi.org/10.1021/om010358j. (b) Pineda LW, Jancik V, Starke K, Oswaldand RB, Roesk HW. Stable monomeric germanium(II) and tin(II) compounds with terminal hydrides. *Angew Chem Int Ed.* 2006;2602–2605. https://doi.org/10.1002/anie.200504337.

38. Choong SL, Woodul WD, Schenk C, Stasch A, Richards AF, Jones C. Synthesis characterisation and reactivity of an N-heterocyclic germanium(II) hydride : reversible hydrogermylation of a phosphaalkyne. *Organometallics.* 2011;30:5543–5550. https://doi.org/10.1021/om200823x.

39. Khan S, Samuel PP, Michel R, et al. Monomeric Ge(II) and Sn(II) hydrides supported by a tridented pincer based ligand. *Chem Commun.* 2012;48:4890–4892. https://doi.org/10.1039/C2CC31214J.

40. Xiong Y, Szilvasi T, Yao S, Tan G, Driess M. Synthesis and unexpected reactivity of germyliumylidine hydride [:GeH]$^+$ stabilized by a bis(N-heterocyclic carbene)borate ligand. *J Am Chem Soc.* 2014;136:11300–11303. https://doi.org/10.1021/ja506824s.

41. Al-Rafia SMI, Malcolm AC, Liew SK, Ferguson MJ, Rivard E. Stabilization of the heavy methylene analogues, GeH$_2$ and SnH$_2$, within the coordination sphere of a transition metal. *J Am Chem Soc.* 2011;133:777–779. https://doi.org/10.1021/ja1106223.

42. Hadlington TJ, Schwarze B, Izgorodina I, Jones C. Two-coordinate hydrido-germylene. *Chem Commun.* 2015;51:6854–6857. https://doi.org/10.1039/C5CC01314C.

43. Sindlinger CP, Wesemann L. Hydrogen abstraction from organotin di- and trihydrides by N-heterocyclic carbene: a new method for the preparation of NHC adducts to tin(II) species and observation of an isomer of a hexastannabenzene derivative [R$_6$Sn$_6$]. *Chem Sci.* 2014;5:2739–2746. https://doi.org/10.1039/C4C00365A.

44. Hadlington TJ, Jones C. A singly bonded amido-ditannyne:H$_2$ activation and isocyanide coordination. *Chem Commun.* 2014;50:2321–2323. http://doi.org/10.10339/C3C49651A.

45. Davidson PJ, Lappert MF. Stabilisation of metals in a low co-ordinative environment using the bistrimethylsilylmethyl ligand; coloured Sn(II) and Pb(II) alkyls;M[CH(SiMe3)2]. *J Chem Soc Chem Commun.* 1973;317. https://doi.org/10.1039/C3973000317A.

46. Harris DH, Lappert MF. Monomeric volatile bivalent amides of group IVB elements M(NR1)2, M(NR1R2) [M = Ge, Sn, Pb] [R1 = SiMe3; R2 =CMe3]. *J Chem Soc Chem Commun.* 1974;895–896. https://doi.org/10.1039/C39740000895.

47. Protchenko AV, Birjkumar KH, Dange D, et al. A stable two-coordinate acyclic silylene. *J Am Chem Soc.* 2012;134:6500–6503. https://doi.org/10.1021/ja301042u.

48. Kristinsdottir L, Oldroyd NL, Grabiner R, et al. Synthetic structural and reaction chemistry of N-heterocyclic germylene and stannylene compounds featuring N-boryl substituents. *Dalton Trans.* 2019;48:11951–11960. https://doi.org/10.1039/c9dt02449b.

49. Li L, Fukawa T, Matsuo T, et al. A stable germanone as the first isolated heavy ketone with a terminal oxygen atom. *Nat Chem.* 2012;4:361–365. https://doi.org/10.1038/NCHEM.1305.

50. Del Rio N, Baceiredo A, Merceron SN, et al. A stable heterocyclic amino(phosphanylidene-σ4-phosphorane) germylene. *Angew Chem Int Ed.* 2016;55:4753–4758. http://doi.org/0.1002/anie.201511956.

51. Rit A, Tirfoin R, Aldridge S. Exploiting electrostatics to generate unsaturation: Oxidative Ge=E bond formation using a non-π donor stabilised [RL(Ge):]$^+$ cation. *Angew Chem Int Ed*. 2016;55:378–382. https://doi.org/10.1002/anie.201508940.
52. Dong Z, Reinhold RWC, Schmidtmann M, Muller T. A germylene stabilized by homoconjugation. *Angew Chem Int Ed*. 2016;55:15899–15904. https://doi.org/10.1002/anie.201609576.
53. Kuhl O. Predicting the net donating ability of phosphines—do we need sophisticated theoretical methods. *Coord Chem Rev*. 2005;249:693–704. https://doi.org/10.1016/j.ccr.2004.08.021.
54. Muller C, Vos D, Jutzi P. Results and perspectives in the chemistry of side chain functionalised cyclopentadienyl compounds. *J Organomet Chem*. 2000;600:127–143. https://doi.org/10.1016/S0022-328X(00)00060-7.
55. Siemeling U. Chelate complexes of cyclopentadienyl ligand bearing pendant O-donors. *Chem Rev*. 2000;100:1495–1526. https://doi.org/10.1021/cr990287m.
56. Webb JR, Burgess SA, Cundari TR, Gunnoe TB. Activation of carbon-hydrogen bonds and dihydrogen by 1,2 CH-addition across metal-heteroatom bonds. *Dalton Trans*. 2013;42:16646–16665. https://doi.org/10.1039/C3DT52164H.
57. Krebs KM, Freitag S, Schubert H, Gerke B, Poettgen R, Wesemann L. Chemistry of stannylene based Lewis pairs: dynamic tin coordination switching between donor and acceptor characters. *Chem A Eur J*. 2015;21:4628–4638. https://doi.org/10.1002/chem.201406486.
58. Merino MR, Mansell SM. Synthesis and reactivity of fluorenyl-tethered N-heterocyclic stannylenes. *Dalton Trans*. 2016;45:6282–6293. https://doi.org/10.1039/C5DT04060D.
59. Mansell SM, Russell CA, Wass DF. Synthesis and structural characterisation of tin analogues of N-heterocyclic carbene. *Inorg Chem*. 2008;47:11367–11375. https://doi.org/10.1021/ic8014479g.
60. Aly AA, Brown AB. Asymmetric and fused heterocycles based on [2.2]paracyclophane. *Tetrahedron*. 2009;65:8055–8089. https://doi.org/10.1016/j.tet.2009.06.034.
61. Grimme S. Do special non-covalent π-π stacking really exist? *Angew Chem Int Ed*. 2008;47:3430–3434. https://doi.org/10.1002/anie.200705157.
62. Matsuoka Y, Ishida Y, Sasaki D, Saigo K. Cyclophane type imidazolium salts with planar chirality as a new class of N-heterocyclic carbene precursors. *Chem A Eur J*. 2008;14:9215. https://doi.org/10.1002/chem.200800942.
63. Fürstner A, Alcarazo M, Krause H, Lehmann CW. Effective modulation of the donor properties of N-heterocyclic carbene ligands by "through-space" communication within a planar chiral scaffold. *J Am Chem Soc*. 2007;129:12676. https://doi.org/10.1021/ja076028t.
64. Piel I, Dickschat JV, Pape T, Hahn FE, Glorius F. A planar [2.2]paracyclophane derived N-heterocyclic stannylene. *Dalton Trans*. 2012;41:13788–137909. https://doi.org/10.1039/c2dt31497e.
65. Braunschweig H, Chorley RW, Hitchcock PB, Lappert MF. The first monomeric prochiral tin(II) complexes Sn[N(SiMe3)2]X [X=OC$_6$H$_2$Bu$_2^t$-2,6-Me-4,1 or NCMe$_2$(CH$_2$)$_3$CME$_2$, 2]; the X-ray structure of 1 and oxidative addition reactions of 2. *J Chem Soc Chem Commun*. 1992;1311–1313. https://doi.org/10.1039/C39920001311.
66. Van der Boom ME, Milstein D. Cyclometalated phosphine based pincer complexes: mechanistic insights in catalysis coordination and bond activation. *Chem Rev*. 2013;103:1759–1792. https://doi.org/10.1021/cr960118r.
67. Tomasik AC, Hill NJ, West R. Synthesis and characterisation of three new thermally stable N-heterocyclic germylenes. *J Organomet Chem*. 2009;694:2122–2125. https://doi.org/10.1016/j.jorganchem.2008.12.042.

68. Bestgen S, Rees HN, Goicoechea MJ. Flexible and versatile pincer-type PGeP and PSnP ligand frameworks. *Organometallics*. 2018;37:4147–4155. https://doi.org/10.1021/acs.organomet.8b00698.
69. Kiefer C, Bestgen S, Gamer MT, et al. Coinage metal complexes of bisalkynyl functionalised N-heterocyclic carbenes: reactivity, photo-physical properties and quantum mechanical investigations. *Chem A Eur J*. 2017;23:1591–1603. https://doi.org/10.1002/chem.201604292.
70. Pan B, Evers-Mcgregor DA, Bezpalko MW, Foxman BM, Thomas CM. Multimetallic complexes featuring a bridging N-heterocyclic phosphide/phosphenium ligand: synthesis structure and theoretical investigations. *Inorg Chem*. 2013;52:9583–9589. https://doi.org/10.1021/ic4012873.
71. Hermann WA, Denk M, Behm J, et al. Stable cyclic germanediyls ("cyclogermylenes"). *Angew Chem Intl Ed Engl*. 1992;31:1485–1488. http://doi.org//10.1002/anie.199214851.
72. Gallego D, Bruck A, Irran E, et al. From bis(silylene) and bis(germylene) pincer type nickel(II) complexes to isolable intermediates of the nickel catalyse sonogashira cross-coupling reaction. *J Am Chem Soc*. 2013;135:15617–15626. https://doi.org/10.1021/ja408137t.
73. Wang L, Lim YS, Li Y, Ganguly R, Kinjo R. Isolation of a cyclic(alkyl(amino)germylene. *Molecules*. 2016;21:990–1000. https://doi.org/10.3390/molecules21080990.
74. Stephan DW, Erker G. Frustrated Lewis pair chemistry: development and perspectives. *Angew Chem Int Ed*. 2015;54:6400–6441. http://doi.org./10.1002/anie.201409800.
75. Dunn NL, Ha M. Radosevich AT main group redox catalysis: reversible $P^{(III)}/P^{(V)}$ redox cycling at a phosphorus platform. *J Am Chem Soc*. 2012;134:11330–11333. https://doi.org/10.1021/ja302963p.
76. Peng Y, Ellis BD, Wang X, Power PP. Diarylstannylene activation of hydrogen or ammonia with arene elimination. *J Am Chem Soc*. 2008;130:12268–12269. https://doi.org/10.1021/ja805358u.
77. Peng Y, Guo JD, Ellis BD, et al. Reaction of hydrogen or ammonia with unsaturated germanium or tin molecules under ambient conditions: oxidative addition versus arene elimination. *J Am Chem Soc*. 2009;131:16272–16282. https://doi.org/10.1021/ja9068408.
78. Frey GD, Lavallo V, Donnadieu B, Schoeller WW, Bertrand G. Facile splitting of hydrogen and ammonia by nucleophilic activation at a single carbon center. *Science*. 2007;314:439–441. https://doi.org/10.1126/science.1141474.
79. (a) Protchenko AV, Bates JI, Libah-Saleh MA, et al. Enabling and probing oxidative addition and reductive elimination at a group 14 metal center: cleavage and functionalization of E-H bond by a bis(boryl)stannylene. *J Am Chem Soc*. 2016;138:4555–4564. https://doi.org/10.1021/jacs6b00710. (b) Usher M, Protchenko AV, Rit A, Campos J, Kolychev EL, Tirfoin R, Aldridge S. A systematic study of structure and E-H bond activation chemistry by sterically encumbered germylene complexes. *Chem Eur J*. 2016;22:11685–11698. https://doi.org/10.1002/chem.201601840.
80. Schager F, Goddard R, Seevogel K, Porschke KR. Synthesis structure and properties of {(Me$_3$Si)$_2$CH}$_2$SnH(OH). *Organometallics*. 1998;17:1546–1551. https://doi.org/10.1021/om9700997n.
81. Jana A, Objartel L, Roesky HW, Stalke D. Cleavage of a N-H bond of ammonia at room temperature by a germylene. *Inorg Chem*. 2009;48:798–800. https://doi.org/10.1021/ic801964u.
82. Wang W, Inoue S, Yao S, Driess M. Reactivity of N-heterocyclic germylene towards ammonia and water. *Organometallics*. 2011;30:6490–6494. https://doi.org/10.1021/om200970s.
83. Pineda LW, Jancik V, Roesky HW, Neculai D, Neculai AM. Preparation and structure of the first germanium(II) hydroxide: the congener of an unknown low-valent carbon

analogue. *Angew Chem Int Ed.* 2004;43:1419–1421. https://doi.org/10.1002/anie. 200353205.
84. Lips F, Mansikkamaki A, Fettinger CF, Tuononen HM, Power PP. Reaction of alkenes and alkynes with an acyclic silylene and heavier tetrelylenes under ambient conditions. *Organometallics.* 2014;33:6253–6258. https://doi.org/10.1021/om500947x.
85. Lambert RL, Seyferth DJ. Substituted 7-siladispiro[2.0.2.1] heptanes. The first stable ilacyclopropanes. *J Am Chem Soc.* 1972;94:9246–9248. https://doi.org/10.1021/ja00781a055.
86. (a) Yao S, Wuellen Cv, Driess M. Striking reactivity of ylide-like germylene toward terminal alkynes: [4+2] cycloaddition *versus* C–H bond activation. *Chem Commun.* 2008;5393–5395. https://doi.org/10.1039/B811952J. (b) Boserle J, Zhigulin G, Stepnicka P, et al. Facile activation of alkynes with a boraguanidinato-stabilized germylene: a combined experimental and theoretical study. *Dalton Trans.* 2017;46:12339–12353. https://doi.org/10.1039/c7dt01950e. (c) Kelly TYL, Gullett L, Chen C-Y, Fettinger JC, Power PP. Reversible complexation of alkynes by a germylene. *Organometallics.* 2019;38:1421–1424. https://doi.org/10.1021/acs.organomet.9b00077.
87. Mizuno H, Takaya J, Iwasama N. Rhodium(I) catalysed direct carboxylation of arenes with CO_2 via chelation assisted C-H bond activation. *J Am Chem Soc.* 2011;133:1251–1253. https://doi.org/10.1021/ja109097z.
88. Hoskin AJ, Stephan DW. Early transition metal hydride complexes: synthesis and reactivity. *Coord Chem Rev.* 2002;233:107–129. https://doi.org/10.1016/S0010-8545(02)00030-9.
89. (a) Jana A, Ghosal D, Roesky HW, Objartel I, Schwab G, Stalke D. A germanium(II) hydride as an effective reagent for hydrogermylation reactions. *J Am Chem Soc.* 2009;131:1288–1293. https://doi.org/10.1021/ja808656t. (b) Tan G, Wang W, Blom B, Driess M. Mechanistic studies of CO_2 reduction to methanol mediated by an N-heterocyclic germylene hydride. *Dalton Trans.* 2014;6006–6011. https://doi.org/10.1039/c3dt53321b.
90. Jana A, Roesky HW, Schulzke C. Do4ring A. Reaction of Tin(II) hydride species with unsaturated molecules. *Angew Chem Int Ed.* 2009;48:1106–1109. https://doi.org/10.1002/anie.200805595.
91. Peruzzini M, Gonsalvi L, Romerosa A. Coordination chemistry and functionalisation of white phosphorus via transition metal complexes. *Chem Soc Rev.* 2005;34:1038–1047. https://doi.org/10.1039/B510917E.
92. Scheer M, Balazs G, Seitz A. P4 activation by main group elements and compounds. *Chem Rev.* 2010;110:4236–4256. https://doi.org/10.1021/cr100010e.
93. Back O, Kuchenbeiser G, Donnadieu B, Bertrand G. Nonmetal-mediated fragmentation of P_4: isolation of P_1 and P_2 Bis(carbene) adducts. *Angew Chem Int Ed.* 2009; 48:5530–5533. http://doi.org10.1002/anie.200902344.
94. Dube JW, Graham CME, Macdonald CLB, Brown ZD, Power PP, Ragogna PJ. Reversible, photoinduced activation of P_4 by low-coordinate main group compounds. *Chem A Eur J.* 2014;20:6739–6744. https://doi.org/10.1002/chem.201402031.
95. Jana A, Sarish SP, Roesky HW, Leusser D, Objartel I, Stalke D. Pentafluoropyridine as a fluorinating agent for preparing a hydrocarbon soluble β-diketiminato lead(II) monofluoride. *Chem Commun.* 2011;47:5434–5436. https://doi.org/10.1039/c1cc11310k.
96. Samuel PP, Singh AP, Sarish SP, et al. Oxidative addition versus substitution reactions of group 14 dialkylamino metallylenes with pentafluoropyridine. *Inorg Chem.* 2013;52:1544–1549. https://doi.org/10.1021/ic302344a.
97. Gurnani C, Hector AL, Jager E, Levason W, Pugh D, Reid G. Tin(II) fluoride vs. tin(II) chloride—a comparison of their coordination chemistry with neutral ligands. *Dalton Trans.* 2013;42:8364–8374. https://doi.org/10.1039/C3DT50743B.

98. Peris E. Smart N-heteroyclic carbene ligands in catalysis. *Chem Rev.* 2018;118: 9988–10031. https://doi.org/10.1021/acs.chemrev.6b00695.
99. (a) Brown HC, Rao BCS. A new technique for the conversion of olefins into organoboranes and related alcohols. *J Am Chem Soc.* 1956;78:5694–5695. https://doi.org/10.1021/ja01602a063. (b) Shegavi ML, Bose SK. Recent advances in the catalytic hydroboration of carbonyl compounds. *Cat Sci Technol.* 2019;9:3307–3336. https://doi.org/10.1039/C9CY00807A. (c) Chong CC, Kinjo R. Catalytic hydroboration of carbonyl derivatives, imines and carbon dioxide. *ACS Catal.* 2015;5:3238–3259. https://doi.org/10.1021/acscatal.5b00428.
100. Breslow R. On the mechanism of thiamine action IV[1]. Evidence from studies on model systems. *J Am Chem Soc.* 1958;80:3719–3726. https://doi.org/10.1021/ja01547a064.
101. Hadlington TJ, Hermann M, Frenking G, Jones C. Low co-ordinate germanium(II) and tin(II) hydride complexes: efficient catalysts for the hydroboration of carbonyl compounds. *J Am Chem Soc.* 2014;136:3028–3031. https://doi.org/10.1021/ja5006477.
102. Wu Y, Shan C, Sun Y, et al. Main group metal-ligand cooperation of N-heterocyclic germylene: an efficient catalyst for hydroboration of carbonyl compounds. *Chem Commun.* 2016;52:13799–13802. https://doi.org/10.1039/C6CC08147A.
103. Schneider J, Sindlinger CP, Freitag SM, Schubert H, Wesemann L. Diverse activation modes in the hydroboration of aldehydes and ketones with germanium, tin and lead Lewis pairs. *Angew Chem Int Ed.* 2017;56:333–337. https://doi.org/10.1002/anie.201609155.
104. Dasgupta R, Das S, Hiwase S, Pati SK, Khan S. N-heterocyclic germylene and stannylene catalysed cyanosilylation and hydroboration of aldehydes. *Organometallics.* 2019;38:1429–1435. https://doi.org/10.1021/acs.organomet.8b00673.
105. Aresta M, ed. *Carbon Dioxide as a Chemical Feedstock.* Weinheim, Germany: Wiley-VCH; 2010. ISBN-13 978-3-527-32475-0.
106. Hadlington TJ, Kefalidis CE, Maron L, Jones C. Efficient reduction of carbon dioxide to methanol equivalents catalyzed by two coordinate amido-germanium(II) and tin(II) hydride complexes. *ACS Catal.* 2017;7:1853–1859. https://doi.org/10.1021/acscatal.6b03306.
107. Liberman-Martin AL, Bergman RG, Tilley TD. Lewis acidity of bis(perfluorocatecholato)silane: aldehyde hydrosilylation catalysed by a neutral silicon compound. *J Am Chem Soc.* 2015;137:5328–5331. https://doi.org/10.1021/jacs.5b02807.
108. Swamy VSVSN, Bisai MK, Das T, Sen SS. Metal free mild and selective aldehyde cyanosilylation by a neutral pentaco-ordinated silicon compound. *Chem Commun.* 2017;53:6910–6913. https://doi.org/10.1039/c7cc03948d.
109. Siwatch RK, Nagendran S. Germylene cyanide complex: a reagent for the activation of aldehydes with catalytic significance. *Chem A Eur J.* 2014;20:13551–13556. https://doi.org/10.1002/chem.201404204.

CHAPTER FOUR

Pincer ligands incorporating pyrrolyl units: Versatile platforms for organometallic chemistry and catalysis

C. Vance Thompson, Zachary J. Tonzetich*"
Department of Chemistry, University of Texas at San Antonio (UTSA), San Antonio, TX, United States
*Corresponding author: e-mail address: zachary.tonzetich@utsa.edu

Contents

1. Introduction	153
2. Diaminopyrrolyl pincers (^{R}NNN)	154
3. Diiminopyrrolyl pincers (^{R}DIP)	160
4. Dipyridylpyrrolyl pincers (DPP)	167
5. *Bis*(pyrazolyl)pyrrolyl pincers (BPP)	170
6. *Bis*(oxazolinyl)pyrrolyl pincers (pyrrbox and pyrrmebox)	176
7. Diphosphinopyrrolyl pincers (^{R}PNP)	194
8. Pyridinedipyrrolyl pincers (^{R}PDP$^{R'}$)	210
9. Other pyrrolyl-containing pincers	219
10. Conclusions and outlook	232
Acknowledgment	232
References	232

1. Introduction

Pincers have come to be defined as any tridentate ligand that binds to a metal center in a meridional fashion.[1] Since their inception in the 1970s, pincer-type ligands have seen continual development and they now represent one of the most popular coordination motifs in modern organometallic chemistry. This popularity stems in large part from the kinetic and thermodynamic stabilization afforded by the pincer platform, and the relative ease with which steric and electronic modifications can be made to the ligand framework. Indeed, careful ligand design in the area of pincer complexes has permitted detailed investigations of a variety of organometallic reactions and the development of new catalytic methodologies.[2]

Of the many varieties of pincer ligands reported to date, one class that has emerged in recent years are those which contain a pyrrolyl unit. Outside of their incorporation into porphyrinogens and related macrocycles, pyrroles are not commonly encountered as ligands in organometallic and coordination chemistry when compared with other N-heterocycles such as pyridines and imidazoles. The relative dearth of pyrrolyl-containing ligands is somewhat curious given that N-deprotonated pyrroles (hereafter pyrrolides) offer several desirable features. The pK_a of the pyrrolic NH unit is only 23 in DMSO,[3] rendering it less Brønsted basic than many other deprotonated nitrogen moieties (e.g., dialkylamides). Moreover, the nitrogen lone pair of pyrrole is part of an aromatic system, which mitigates its availability for π-bonding with metal ions. Both these aspects of pyrrole are advantageous to its coordination chemistry as they allow pyrrolide units to bind to both early and late transition metals. One concern surrounding the coordination chemistry of pyrrolide is its potential for η^5-binding.[4,5] This complication, while significant for some systems, is less problematic when considering pyrrolyl groups as components of pincers since the increased denticity of the ligand can help direct η^1-coordination.

In the following review, we aim to describe the chemistry of pincer systems that incorporate one or more pyrrolyl units. These ligands span a variety of architectures and have found use across the periodic table. As a result, pyrrole-based pincer ligands now appear in a variety of subfields of modern organometallic chemistry encompassing catalysis, small molecule activation, and photophysics. We have purposely constrained our coverage to only those systems containing intact pyrrolyl units. Related pincer systems, notably those containing carbazolyl and indolyl-derived donors, will not be explicitly addressed here and the reader is encouraged to consult other accounts for a fuller description of their chemistry.[1,6,7] In addition, several previous reviews have touched upon specific aspects of pyrrole-containing chelates,[8–11] and we have therefore avoided duplicate discussion of those topics whenever possible.

2. Diaminopyrrolyl pincers (^{R}NNN)

Some of the earliest examples of pyrrole-based pincer complexes can be traced to Huang and coworkers initial studies in the early 2000s with the ligand derived from 2,5-*bis*-(dimethylaminomethyl)pyrrole (**1a**, Chart 1). Despite the relatively contemporary nature of pyrrole-based pincer chemistry, the history of pro-ligand **1a** stretches back over 70 years, with the

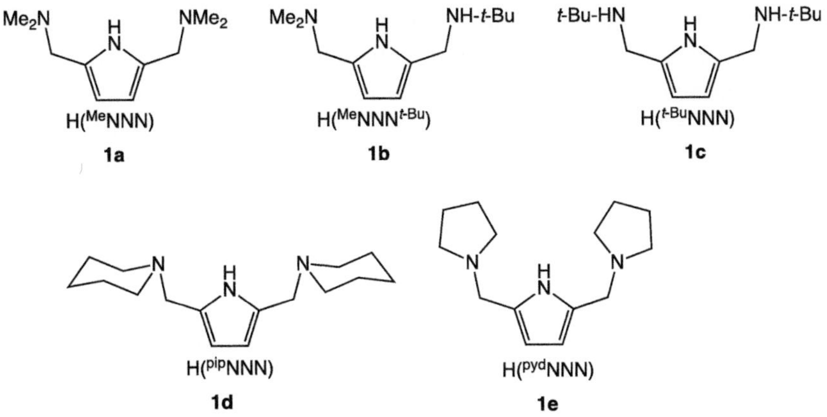

Pincer ligands incorporating pyrrolyl units

Chart 1 Diaminopyrrole proligands.

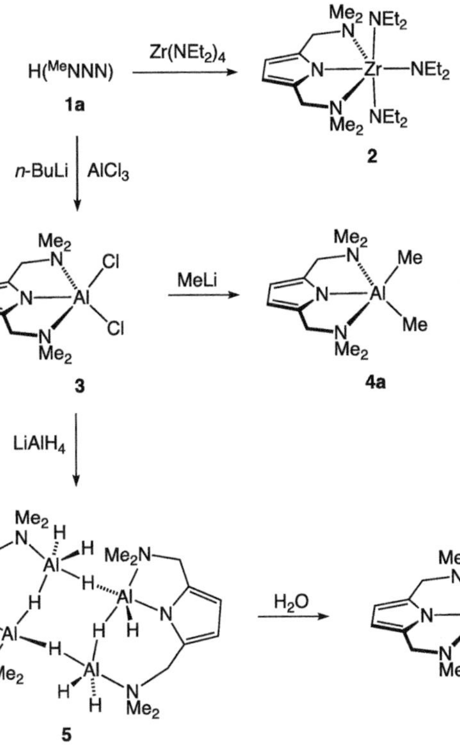

Scheme 1

Mannich reaction used to synthesize it first reported in 1947.[12,13] The first application of **1a** as a ligand was in the preparation of the zirconium complex, [Zr(^{Me}NNN)(NEt$_2$)$_3$] (**2**, Scheme 1).[14] Complex **2** was characterized in solution, although unambiguous evidence for κ^3-coordination of the pincer ligand was not forthcoming. Subsequent work by Huang with Al(III) generated compounds **3** and **4a**, both of which were subjected to crystallographic characterization confirming the expected tridentate coordination mode of the ^{Me}NNN ligand.[15] Further reactivity of **3** with LiAlH$_4$ afforded the tetrametallic cluster **5**, which could be treated with H$_2$O to generate the novel aluminum(III) dihydride, **6a**.[16]

Dihydride **6a** was the subject of a follow-up study examining deprotonation and reductive-addition reactions with a series of phenols, amines, and carbonyl-containing species (Scheme 2).[17] The compound proved effective as a source of Brønsted basic hydride ions, and displayed a propensity to engage in hydroalumination chemistry with unsaturated substrates. Similar reactivity was also observed from a related aluminum dihydride motif featuring the less symmetric pincer ligand derived from **1b** (Chart 1).[18]

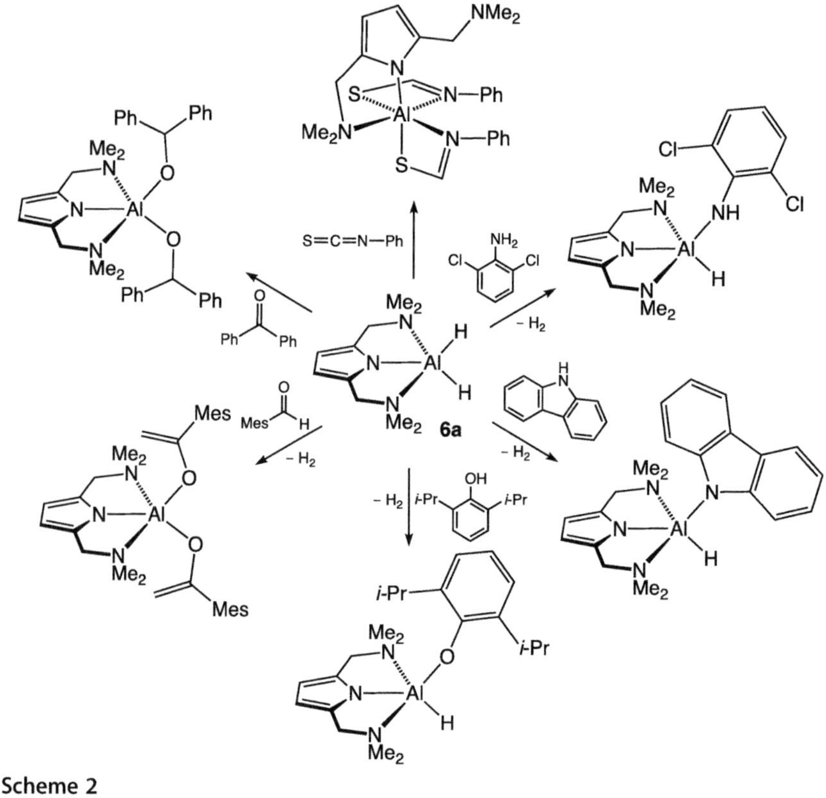

Scheme 2

In another contribution to the area of aluminum chemistry, Huang and coworkers demonstrated that the pincer ligand derived from **1c** could be employed to generate four- and five-coordinate complexes containing

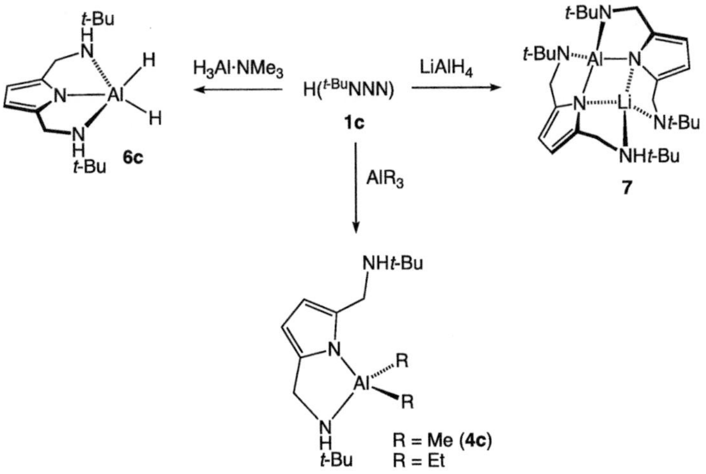

Scheme 3

hydride and alkyl ligands (Scheme 3).[19] Also disclosed was bimetallic Li—Al complex **7**, which resulted from deprotonation of one of the amine donors of the pincer upon treatment with LiAlH$_4$. Notably, compound **7** was found to adopt a structure with bridging pyrrolide nitrogen atoms instead of the more basic amido donors.

Beyond the initial account of compound **2**, several additional Group 4 metal complexes of ^{R}NNN type ligands have been described. Odom and coworkers reported the titanium analog of compound **2** through the reaction of [Ti(NMe$_2$)$_4$] with **1a** (**8**, Scheme 4).[20] The solid-state structure of **8** demonstrated a bidentate mode for the pincer ligand, bringing into question the original assignment of **2** discussed above. Nonetheless, complex **8** proved effective as a catalyst for the iminohydrazination of alkynes, a multicomponent coupling reaction involving alkyne, hydrazine, and an isonitrile. In related work, titanium(IV) imido complexes of ^{R}NNN were prepared by the reaction of the dimeric precursor, [Ti$_2$Cl$_2$(NSiPh$_3$)$_2$(py)$_4$(μ-Cl)$_2$], with the lithium salts of **1a** and **1d** (**9a** and **9d**, Scheme 4).[21] Unlike compound **8**, both **9a** and **9d** were found to coordinate to titanium in a tridentate manner. Crystallographic characterization of **9d** evinced a distorted square-pyramidal geometry with the imido ligand occupying an axial position. In addition to work with **1a** and **1d**, Odom's group has also demonstrated a great deal of Group 4 chemistry with the dianionic *bis*-pyrrolyl ligand derived from *N,N*-bis(pyrrolyl-α-methyl)-*N*-methylamine (H$_2$dpma).[22] However, as this ligand predominantly binds to metal centers in a *fac* fashion, its classification as a pincer is not strictly appropriate and will therefore not be discussed here.

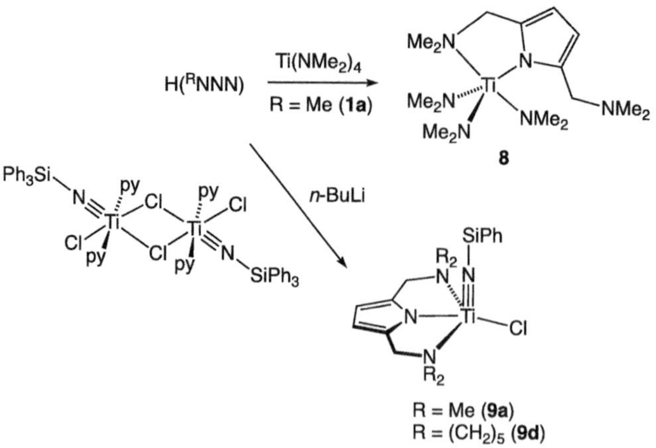

Scheme 4

Other examples of zirconium compounds of ^{R}NNN were reported more recently by Huang and coworkers using pro-ligand **1b**.[23] Reaction with of **1b** with Zr(NR$_2$)$_4$ produced compounds **10** and **11**, which feature tridentate coordination of the pincer ligand. However, in each case additional deprotonation of the *tert*-butylamine group is observed giving rise to a dianionic pincer (Scheme 5). Both **10** and **11** were found to undergo insertion reactions of phenylthiocyanate (**12**) and phenylcyanate (**13**) into the dialkylamido ligands giving rise to higher coordination number species. In one instance, treatment with excess PhN=C(NEt$_2$)OH led to re-protonation of the *tert*-butylamide and reversion to a bidentate binding mode (**14**).

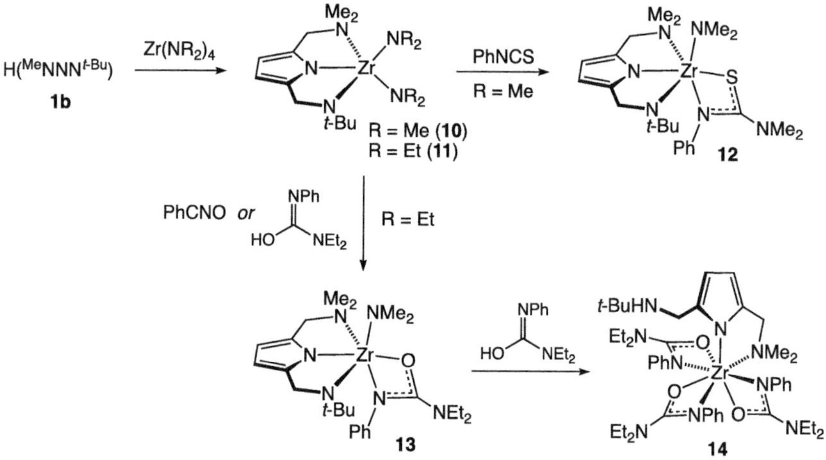

Scheme 5

In the area of rare-earth metal chemistry, Huang and coworkers reported the reaction of Li(^{Me}NNN) with YCl$_3$ to furnish the diyttrium cluster, [Y$_2$Cl$_4$(μ-Cl)$_2$(^{Me}NNN)$_2$·Li$_2$(OEt$_2$)$_4$] (15).[24] Whereas compound 15 was found to display no activity toward polymerization of ε-caprolactone, the lithium salt, Li(^{Me}NNN), actually demonstrated high activity. The resulting polymers showed large polydispersity indices (PDIs), which were attributed to fast initiation by Li(^{Me}NNN) followed by slow propagation of the growing polymer. Also working with rare-earth metals, Cui and coworkers described a series of complexes generated by protonolysis of proligands 1d and 1e with M(CH$_2$SiMe$_3$)$_3$(thf)$_2$ (M = Sc, Y, Lu; Scheme 6).[25] Yttrium and lutetium complexes of the ^{Pip}NN ligand (17 and 18) were found to adopt bimetallic structures featuring η^5-coordintion of two pyrrolide units to a M(CH$_2$SiMe$_3$)$_2$ fragment. When activated by strong Lewis acids, compounds 16–21 all initiated the controlled polymerization of isoprene with those complexes containing ^{Pip}NNN (16–18) displaying higher activity. Different polymer microstructures were found to result depending upon the choice of metal. Scandium complexes produced polymers enriched in 3,4-linkages, yttrium complexes provided polymers with high cis-1,4-selectivity, and lutetium complexes gave atactic polymer.

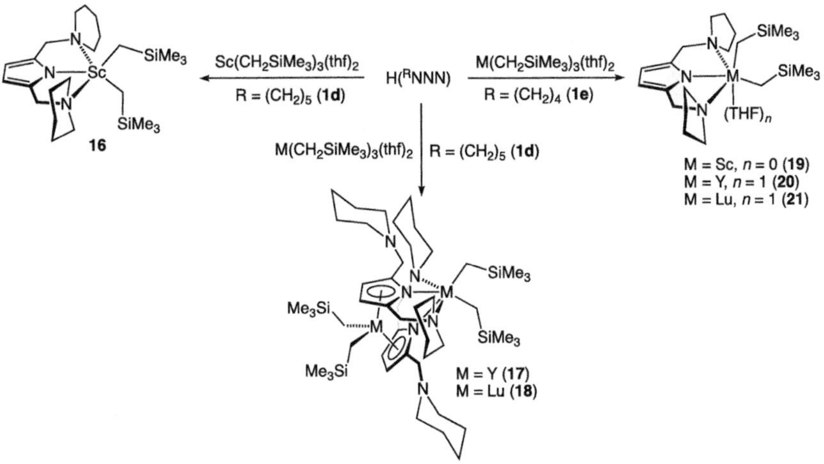

Scheme 6

Main group elements other than Al have also been the subject of studies employing the family of ^{R}NNN ligands. Huang and coworkers described a series of gallium and indium complexes containing ligands derived from 1a–c.[26,27] Perhaps unsurprisingly, each of these compounds demonstrated structures and reactivity patterns akin to the Al analogs discussed above (Schemes 1–3). Moving to Group 14, Stalke and coworkers reported the preparation of Ge, Sn, and Pb complexes **22–24** employing pro-ligand **1e**.[28] Each of the compounds was accessed via the lithium salt, Li(pydNNN), via treatment with the divalent metal chloride, (Scheme 7). Compounds **22–24** exhibited disphenoidal geometries as expected for the presence of a stereochemically-active lone pair on the metal center. Computational studies of the complexes revealed that the extent of π-donation from the pyrrolyl moiety to the metal decreases upon moving from Ge to Pb. This effect was attributed to higher lone pair energy for the heavier congeners, which serves to interfere with π-bonding to the pyrrole. More recently, Dostál and coworkers have taken an analogous approach to extend the chemistry of ^{R}NNN to the Group 15 elements Sb and Bi (Scheme 7).[29] As with compounds **22–24**, the structures of **25a,e** and **26** were consistent with the presence of a lone pair on the metal center.

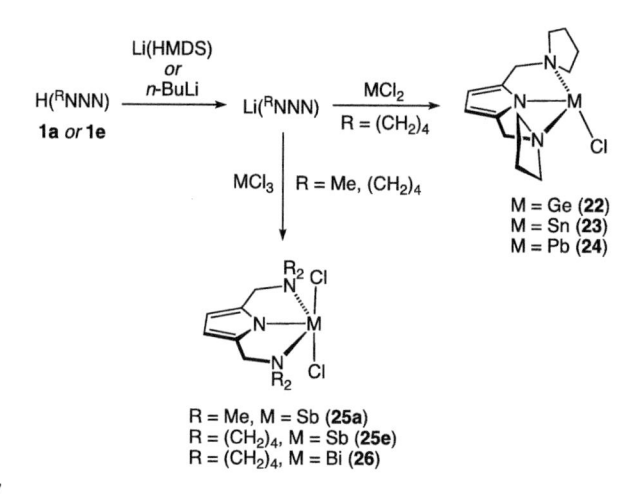

Scheme 7

3. Diiminopyrrolyl pincers (^{R}DIP)

Pyrrole-based pincer ligands containing imine donors (^{R}DIP, Chart 2) were first reported by Bochmann and coworkers slightly before the amine analogs discussed in the previous section.[30] However, this initial report did not contain any examples of the DIP ligand behaving in a true pincer-like manner. Indeed, the larger bite angle subtended by the sp^2 carbon atom of the imine arm renders bona fide tridentate coordination more difficult with ^{R}DIP and often relegates this class of pincers to the chemistry of larger metal

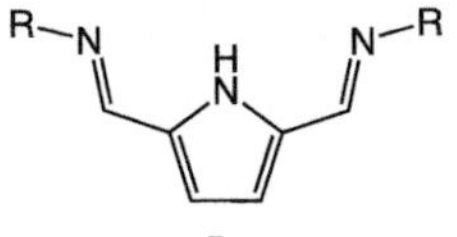

Chart 2 Diiminopyrrole proligand.

ions such as rare-earths and *s* block elements. Moreover, in those compounds that do feature tridentate coordination, the metal-N$_{imine}$ distance is often quite long (>2.5 Å). Nonetheless, several examples of complexes of the ^{R}DIP family of ligands have been reported with a variety of metals.

The first instance of a tridentate diiminopyrrolyl complex was reported by Mashima and coworkers in the area of yttrium chemistry.[31] Protonolysis of several H(^{R}DIP) derivatives with Y(N{SiMe$_3$}$_2$)$_3$ produced a series of new compounds featuring one (**27**), two (**28**), or three (**29**) ^{R}DIP ligands depending upon the reaction stoichiometry and the nature of the imine substituent (Scheme 8). Spectroscopic and structural characterization of the

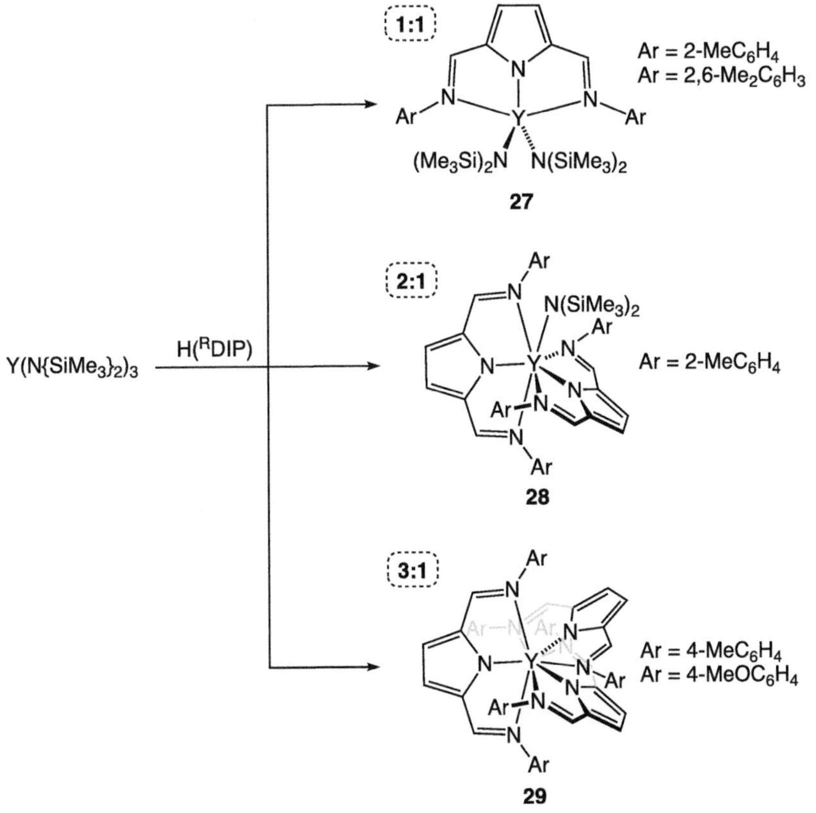

Scheme 8

compounds was consistent with tridentate coordination in each instance with Y-N$_{imine}$ bond distances ranging from 2.65 to 2.78 Å. Much like the yttrium complexes of ^{R}NNN discussed above, compound **27** (Ar=2,6-Me$_2$C$_6$H$_3$) proved competent for the polymerization of ε-caprolactone.

Rare-earth metals have likewise been the subject of several subsequent studies employing ^{R}DIP ligands. Roesky and coworkers described the preparation of several Y and Lu complexes of ^{Ar*}DIP (Ar*=2,6-i-Pr$_2$C$_6$H$_3$) via transmetallation with the Li and K potassium salts of the pincer ligand (Scheme 9).[32] Among the compounds reported was the novel dicyclopentadienyl species **33**. Ensuing transmetallation studies of K(^{Ar*}DIP) with borohydride salts of Sc, La, Nd and Lu resulted in formation of κ^3-BH$_4$ complexes in the case of the larger La^{3+} (**34**) and Nd^{3+} (**35**) ions, but led to generation of the unusual κ^2-iminoborane species, **36** and **37**, in the case of the smaller Sc^{3+} and Lu^{3+} ions.[33–35] Neodymium compound **35** and its dimeric chloride analog, [Nd$_2$Cl$_2$(thf)$_2$(^{Ar*}DIP)$_2$(μ-Cl)$_2$], were further demonstrated to serve as highly active and selective catalysts for the polymerization of butadiene in the presence of co-activators to give poly(cis-1,4-butadiene).

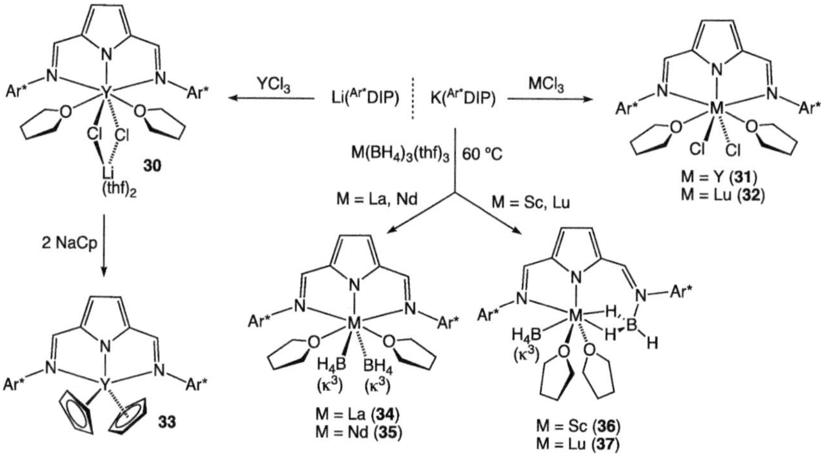

Scheme 9

In addition to the trivalent compounds discussed above, the ^{R}DIP ligand has also been employed to stabilize select examples of divalent rare-earth complexes. Utilizing an analogous transmetallation strategy involving K(^{Ar*}DIP), Roesky and coworkers prepared iodo compounds **38–40** (Scheme 10).[36] Complexes **38** and **39** displayed the expected tridentate binding mode of the pincer ligand but complex **40**, which contains the smaller Yb^{2+} ion, showed inequivalent Yb-N$_{imine}$ distances. The longer of these two distances was measured at 3.20 Å, arguing against true tridentate coordination. Complex **38** was later found to react with the *cyclo*-P$_5$-containing compound, [Cp*Fe(η^5-*cyclo*-P$_5$)], in the presence of reductant to afford monomeric (**41**) and dimeric (**42**) examples of triple-decker Sm—Fe compounds (Fig. 1).[37] In a break with the findings of Roesky, Panda and coworkers found that similar transmetallation reactions of SmI$_2$ with K(^{Bz}DIP) (Bz=CHPh$_2$) led to C—C coupling and formation of the disamarium(III) complex **43**.[38,39]

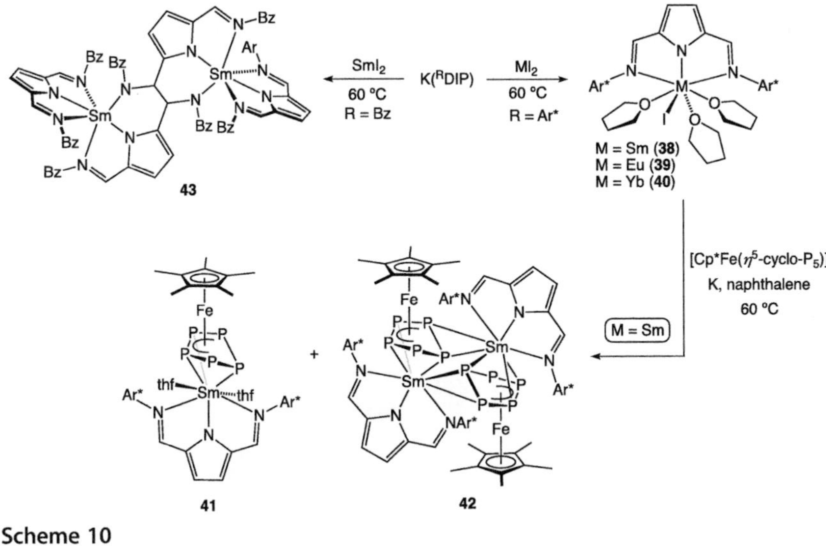

Scheme 10

Group 2 analogs of compounds **38–40** were similarly investigated by Roesky and coworkers through the reaction of K(^{Ar*}DIP) and MI$_2$ (M=Ca, Sr, Ba).[40] Analogous trends in denticity were observed for the pincer ligand with the larger Sr^{2+} and Ba^{2+} ions giving rise to complexes with tridentate coordination and the smaller Ca^{2+} ion resulting in a κ^2-DIP species. Interestingly, treatment of the calcium analog with KN(SiMe$_3$)$_2$ afforded a new complex, [Ca(N{SiMe$_3$}$_2$)(thf)$_2$(^{Ar*}DIP)], which demonstrated κ^3-coordination of the pincer ligand. This compound also proved active for the catalytic intramolecular hydroamination of aminoalkenes.

Use of diiminopyrroles as pincer ligands in transition metal chemistry has met with more limited success on account of the smaller ionic radii of most *d*

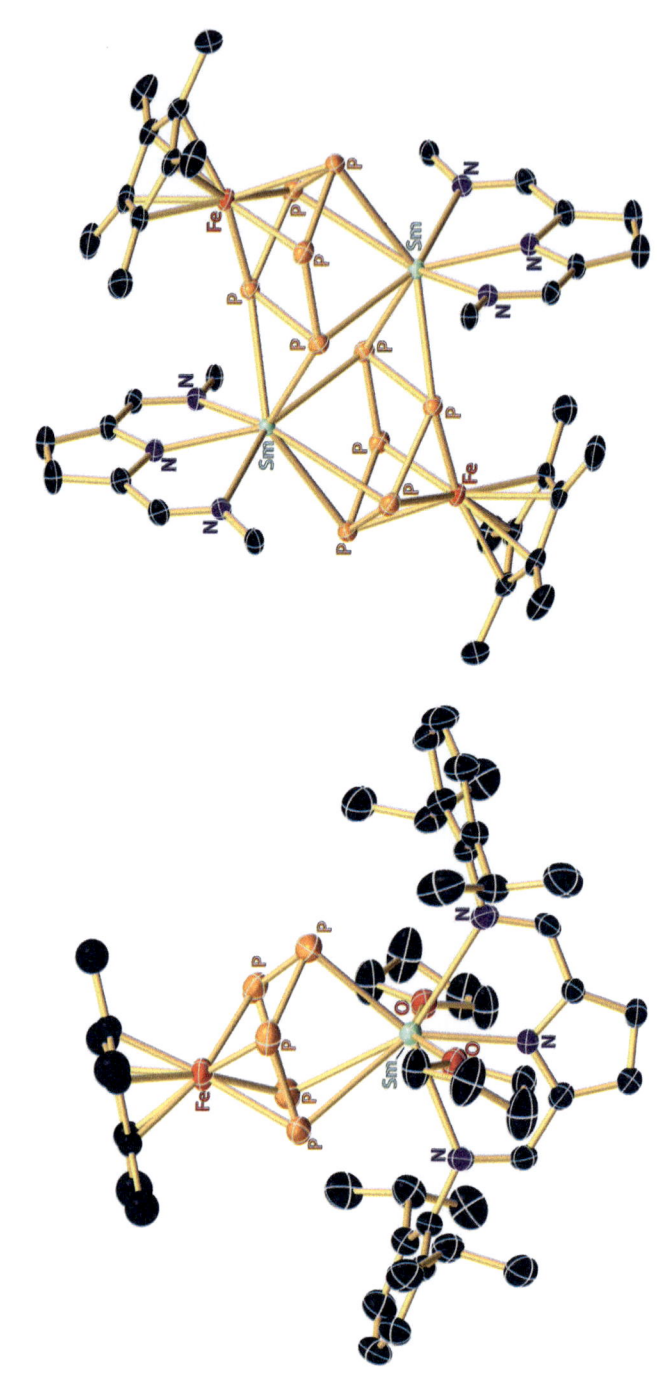

Fig. 1 Solid-state structures of compounds **41** (left) and **42** (right). Diisopropylphenyl groups of **42** omitted for clarity. Images generated from crystallographic data in reference Li T, Wiecko J, Pushkarevsky NA, et al. Mixed-metal lanthanide–iron triple-decker complexes with a cyclo-P5 building block. Angew Chem Int Ed. 2011;50:9491–9495.

block elements, which cannot as easily accommodate the external angles of the ^{R}DIP system. Early work by Mashima and coworkers examined the preparation of zirconium (**44**) and hafnium (**45**) complexes of several different ^{R}DIP derivatives through a protonolysis approach employing M(CH$_2$Ph)$_4$ (M=Zr, Hf).[41] They found that the reactions proceeded smoothly to give new dibenzyl complexes with additional alkylation at one of the imine units (Scheme 11). Alkylide abstraction with trityl salts produced the corresponding monobenzyl cations of the type **46** and **47**. Zirconium dibenzyl complexes (**44**) were found competent for the polymerization of ethylene when activated by modified methylaluminoxane (MMAO). By contrast, cationic complexes such as **46** and **47** failed to produce polyethylene, likely as a consequence of back coordination of the phenyl group after one or two insertion events.

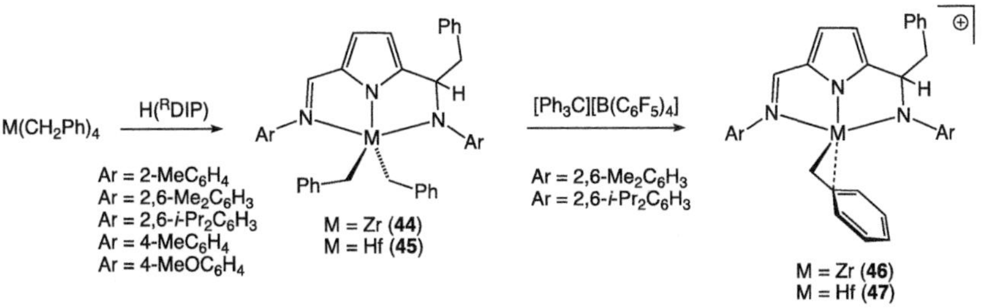

Scheme 11

Also targeting polymerization catalysts, Schaper and coworkers have examined the chemistry of ^{R}DIP ligands with copper(II). In an initial report, chiral bimetallic complexes **48** and **49** were prepared and shown to initiate polymerization of *rac*-lactide (Scheme 12).[42] Compounds **48** and **49** both feature bidentate coordination of the ^{R}DIP ligand reflecting the difficulty in chelating the smaller Cu(II) ion. Of the two initiators, only compound **48** was found to produce istotactically-enriched polylactic acid. Subsequent mechanistic investigations of the reaction were consistent with the observed stereochemistry attributable to chain-end control of the growing polymer.[43] The nature of the bridging-ligand was found to be critical as it must be flexible enough to allow the active site to remain bimetallic during monomer enchainment while still sufficiently rigid to engender a stereoselective active site. In a follow-up study, the authors

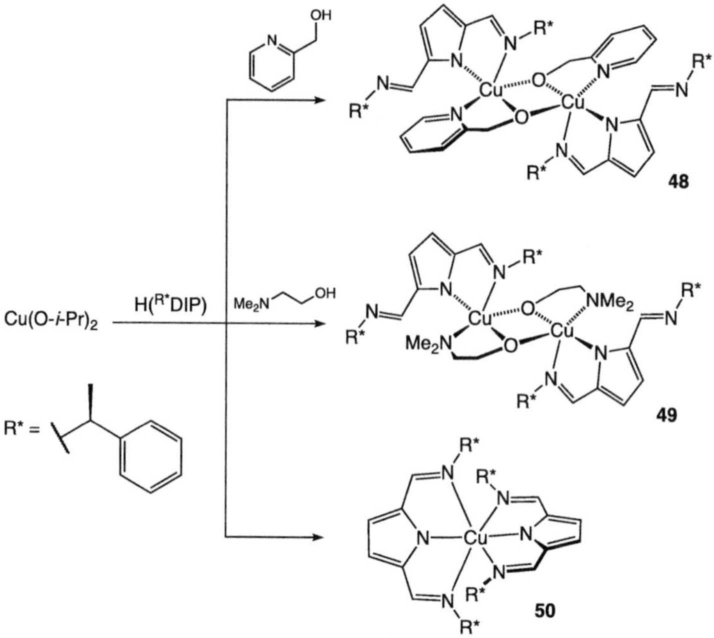

Scheme 12

expanded their investigations to a larger range of ^{R}DIP derivatives akin to **48** and **49**.[42] Homoleptic compounds of the type **50**, which demonstrate tridentate coordination of the pincer, were also synthesized. Despite substantial ligand modification, the nature of the imine substituent was found to have a minor effect on both polymerization activity and stereoselectivity.

In a strategy to foster tridentate coordination of the DIP ligand to transition metal centers, Anderson and coworkers have recently disclosed the synthesis of the ^{Tol,Cy}DIP ligand (Scheme 13). This ligand features enhanced bulk at the imine carbon atom, which was hypothesized to hinder C-N$_{imine}$ rotation and enforce tridentate coordination. Metalation of H(^{Tol,Cy}DIP) with [(COD)PdCl$_2$] generated compound **51**, which indeed displays the desired κ^3 binding mode of the pincer. Moreover, the Pd-N$_{imine}$ bond distances were found to be quite short at 2.15 Å, confirming the efficacy of the design strategy. Reactions of **51** with both proton and reducing equivalents gave rise to compounds **52** and **53**, demonstrating the ability to activate the ^{Tol,Cy}DIP ligand.

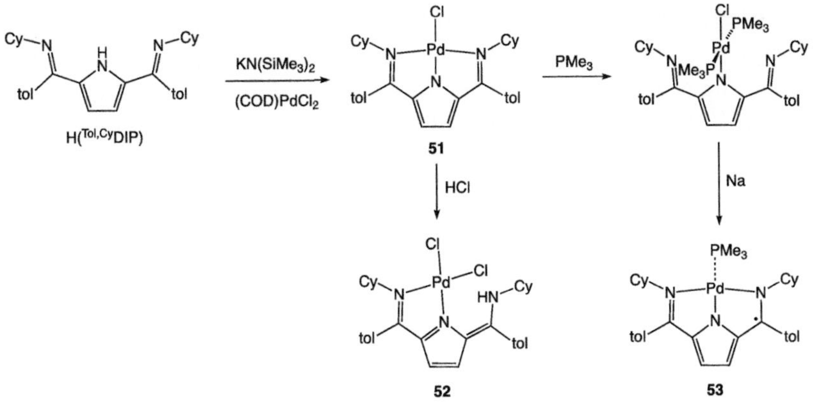

Pincer ligands incorporating pyrrolyl units

Scheme 13

4. Dipyridylpyrrolyl pincers (DPP)

Encapsulation of the imine unit of ^{R}DIP ligands in the form of a pyridine moiety gives rise to the dipyridylpyrrolyl pincers (DPP, Chart 3). These ligands were the earliest pyrrole-based pincers to be reported, first appearing in the mid-1950s.[44,45] Their coordination chemistry, however, was not revisited in earnest until a report in 2006 by Natelson, Tour, and coworkers disclosed the synthesis of a series of homoleptic [M(DPP)$_2$] complexes of the first-row transition metals (M = Fe – Zn).[46] Unlike their ^{R}DIP cousins, DPP ligands have proven more successful in supporting pincer-like complexes of the transition metal ions, likely as a consequence of their more rigid nature. In addition, the enhanced π-conjugation of these ligands has motivated studies into their use as electronic materials. The coordination chemistry of DPP ligands was extensively covered in a recent review by Colbran.[9] Therefore, this section will focus solely upon work published since 2018.

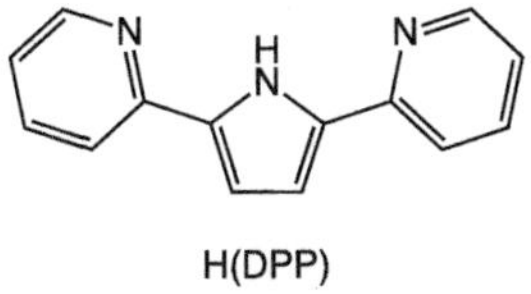

Chart 3 Dipyridylpyrrole proligand.

Comparisons of the DPP ligand to classical terpyridines are logical, and theoretical work by Colbran and coworkers has quantified the strain incurred upon replacement of one or more pyridine groups of terpy with pyrrolyls.[47] This study found that the buildup in strain can be related to a single geometric parameter reflecting the distance between the two *trans* disposed N atoms of the pincer. Their calculations also revealed that this strain energy can correlate with the preferred magnetic ground state (high-spin versus low-spin) of certain complexes. Subsequent experimental work by the authors with a series of homoleptic cobalt(II) complexes of DPP further demonstrated that substitution of the pyrrole backbone can also have a pronounced effect on electronic structure, altering both redox potentials and spin-crossover behavior.[48]

Yi and coworkers have investigated halogenated variants of DPP featuring substitution at both the pyridine and pyrrole moieties. Analogs of H(DPP) featuring brominated pyridyl groups were found to give rise to compounds such as **54** upon treatment with ZnEt$_2$ (Chart 4).[49] Compound **54** catalyzed the coupling of CO$_2$ and epoxides to give cyclic carbonates under solvent-free conditions. In related work, a DPP ligand containing a chlorinated pyrrolyl unit was used to produce a mixture of ruthenium complexes **55** and **56** from [Ru(CO)$_2$Cl$_2$]$_n$ and base.[50]

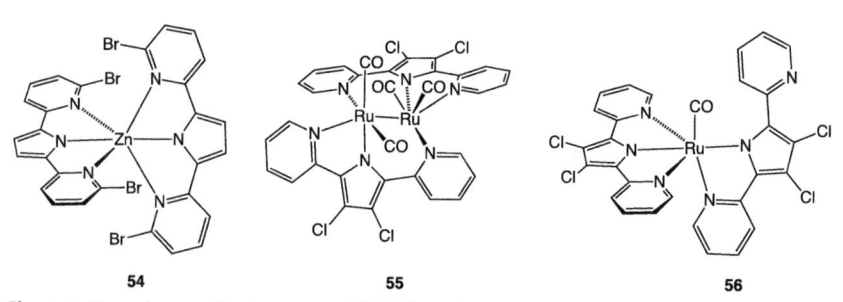

Chart 4 Complexes of halogenated DPP ligands.

Outside of transition metal chemistry, DPP ligands have been employed to prepare several examples of rare-earth complexes. Massi, Colbran, and coworkers reported the synthesis of lanthanide compounds **57–61** (Chart 5) through straightforward ligation with the metal nitrate salts in the presence of Et$_3$N.[51] Each of these complexes features tridentate coordination of three DPP ligands resulting in structures best described as distorted tricapped trigonal prisms. The stability of the complexes toward ligand exchange (hydrolysis) was found to decrease with the size of the lanthanide ion making the Gd (**60**) and Yb (**61**) congeners difficult to handle in solution.

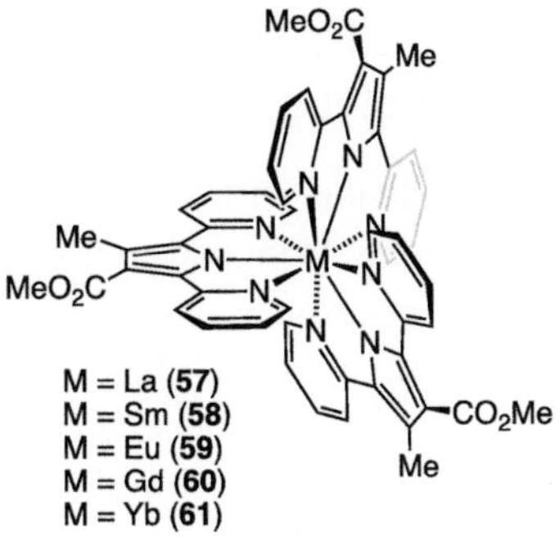

Chart 5 Lanthanide complexes of the DPP ligand.

M = La (**57**)
M = Sm (**58**)
M = Eu (**59**)
M = Gd (**60**)
M = Yb (**61**)

Electrochemical characterization of each complex revealed three closely-spaced reversible cathode processes, attributed to ligand-based oxidation events. Europium and ytterbium compounds **59** and **61** were also found to be emissive at 77 K in the red and near-IR, respectively, as a result of sensitization from the excited triplet state of the DPP ligand.

As a final example of DPP coordination chemistry, Iwasawa and coworkers have recently described the synthesis of iridium complexes bearing gallium and indium derived metalloligands containing a DPP core (Scheme 14).[52] Reaction of **62** and **63** with [Ir$_2$(COE)$_2$(μ-Cl)$_2$] generated the bimetallic species **64** and **65**, each of which features an Ir-M distance within the sum of the respective van der Waals radii. Subsequent reaction of **65** with excess BH$_3$·thf produced the borohydride complex **66** that was shown to serve as a surrogate for an iridium dihydride.

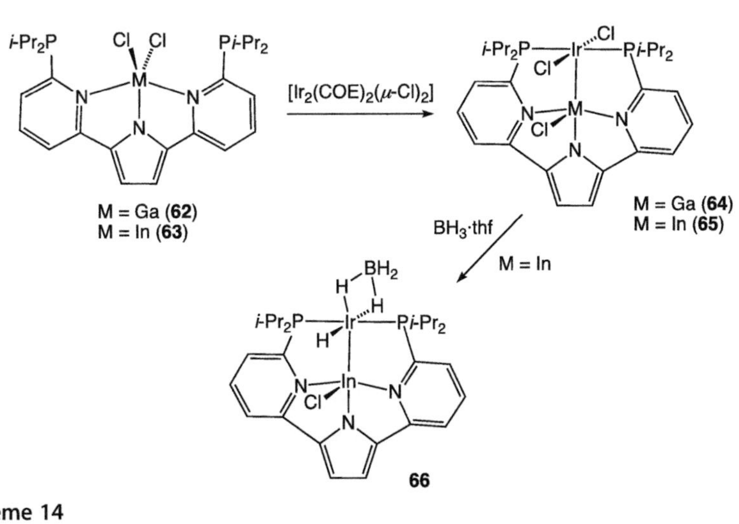

Scheme 14

5. Bis(pyrazolyl)pyrrolyl pincers (BPP)

Related to both the ^{R}DIP and DPP ligands are the *bis*(pyrazolyl)pyrrolyl pincers (^{R}BPP) introduced by Mani and coworkers in 2012 (Chart 6).[53] Much like DPP, these ligands encapsulate an imine-like moiety within a heterocycle to afford a tricyclic framework. Unlike ^{R}DIP and DPP, however, ^{R}BPP ligands possess an extra atom link between the pyrrolyl ring and the flanking donor atoms which helps to accommodate coordination to smaller metal ions.

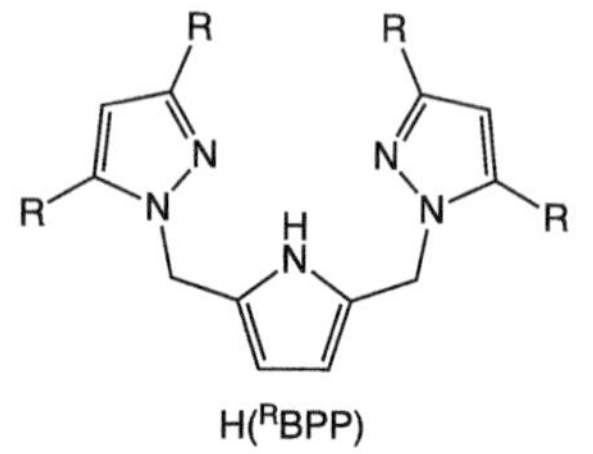

H(^{R}BPP)

Chart 6 Bis(pyrazolyl)pyrrole proligand.

Initial studies with ^{R}BPP employed the tetramethylated version, 2,5-*bis*(3,5-dimethylpyrazolylmethyl)pyrrole (^{Me}BPP). Lithiation of H(^{Me}BPP) was carried out with *n*-BuLi affording the dilithium complex **67** (Scheme 15). Treatment of **67** with half an equivalent of [Pd(NCPh)$_2$Cl$_2$] resulted in the mononuclear Pd complex **68** in moderate yields. Alternatively, **68** can be synthesized by the addition of H(^{Me}BPP) to [Pd(NCPh)$_2$Cl$_2$] in the presence of NEt$_3$ in excellent yields. The same reaction in the absence of base generates the dinuclear helical Pd complex **69**. When the palladium source was changed to Pd(OAc)$_2$, the acetate ions acted as an internal base permitting direct synthesis of **70**. Compound **70** could also serve as a precursor to **68** upon addition of LiCl in acetone/water mixtures. X-ray diffraction data of **68** and **70** revealed a helical twist to the ligand upon coordination to palladium, which gives rise to a pair of enantiomers in the solid state. Variable temperature ^{1}H NMR spectra of **70** revealed a rapid interconversion between the two enantiomers at ambient temperature. Similar studies of **68** demonstrated similar interconversion, however, the rate was slow enough to show a resolved AB quartet for the methylene arm protons at room temperature.

The use of ^{Me}BPP as a ligand has also been expanded to include coordination complexes of Cu and Ag (Scheme 16).[54] Reaction of H(^{Me}BPP) with CuX (X=Cl, Br, I) was found to result in bimetallic Cu complexes of the type **71**, which bare superficial resemble to compound **69** above. Use of lithium salt **67** in place of H(^{Me}BPP) resulted in homoleptic

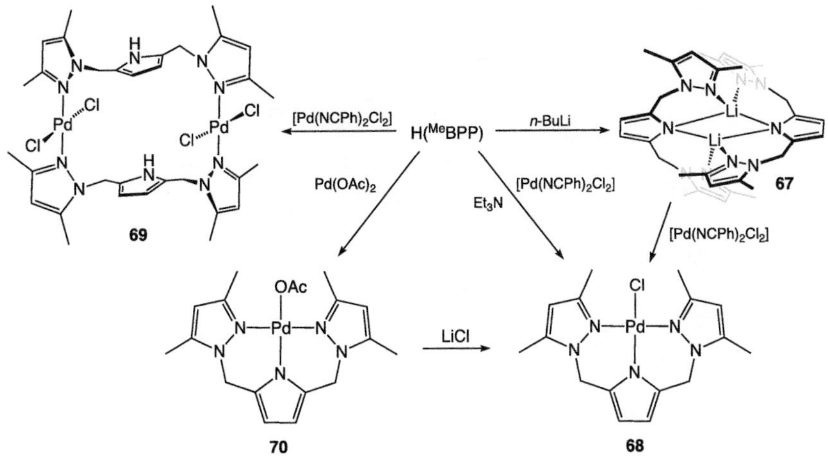

Pincer ligands incorporating pyrrolyl units

Scheme 15

compound **72** upon reaction with CuCl and CuBr in the presence of air. By contrast, analogous reaction of **67** with CuI afforded the novel tetranuclear species **73** (Fig. 2). In the case of silver, treatment of H(^{Me}BPP) with either AgOTf and AgBF$_4$ produced the coordination polymer **74**. The polymeric

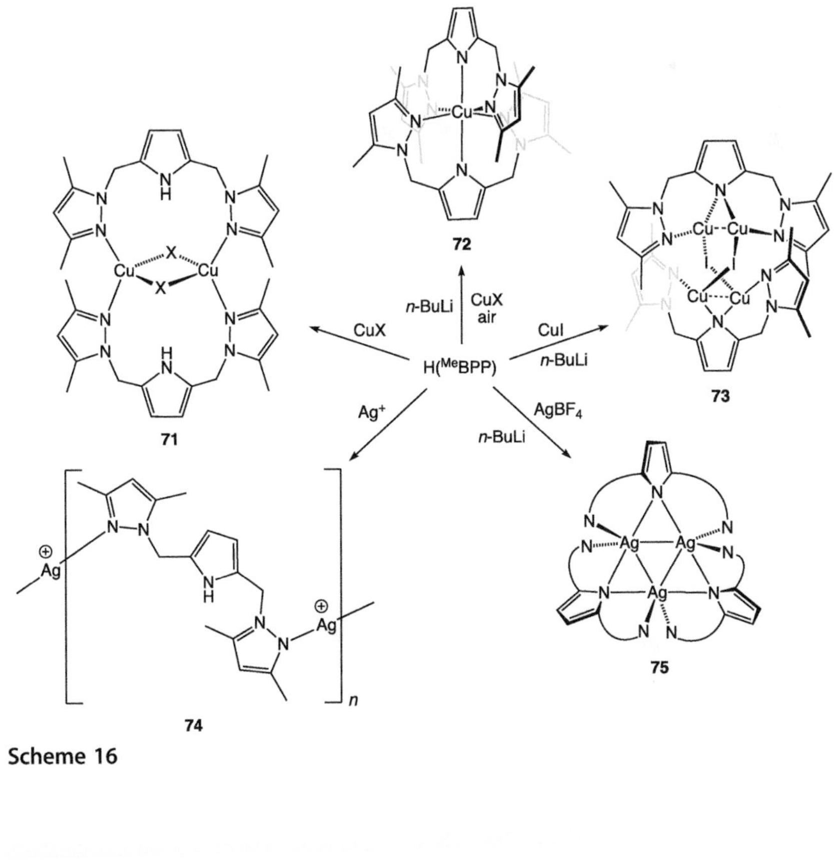

Scheme 16

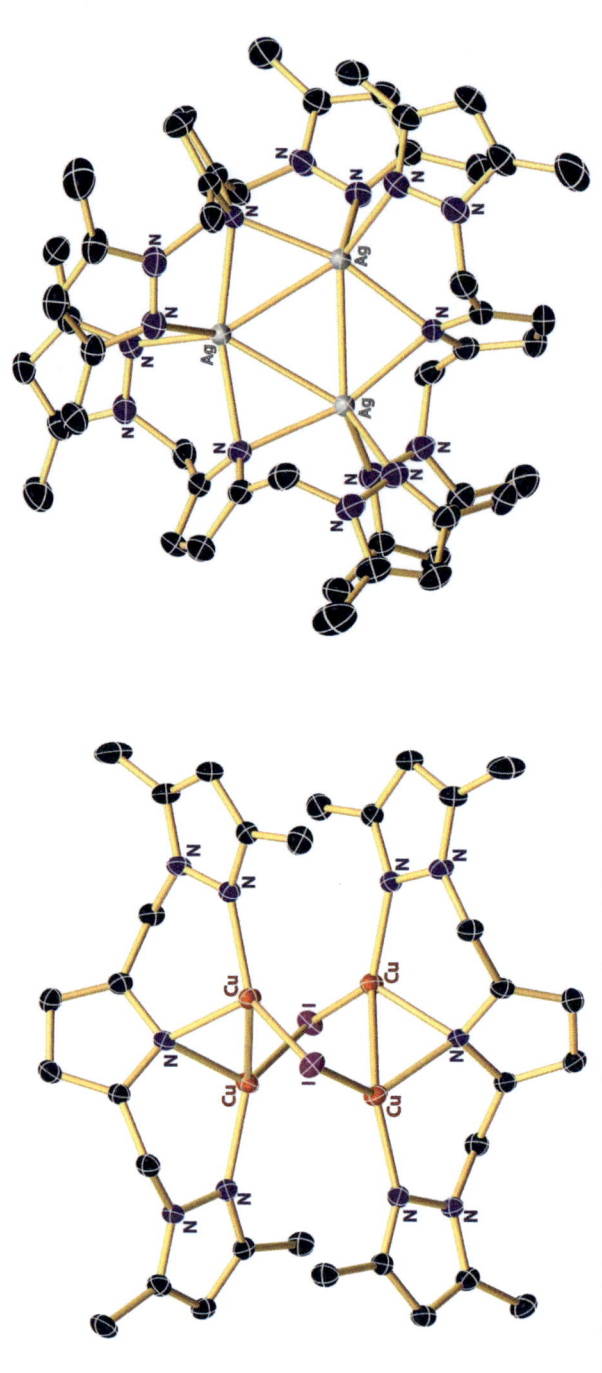

Fig. 2 Solid-state structures of compounds **73** (left) and **75** (right). *Images generated from crystallographic data in reference Jana O, Mani G. New types of Cu and Ag clusters supported by the pyrrole-based NNN-pincer type ligand. New J Chem. 2017;41:9361–9370.*

structure of **74** is likely preferred (vis-à-vis **71**) based on the desire of Ag(I) ions to adopt linear coordination modes. As with Cu, transmetallation reactions of Ag(I) with **67** produced a multinuclear species, **75**, in this case incorporating three metal centers (Fig. 2).

Work by Brothers and coworkers has extended the coordination chemistry of the BPP system to include the unsubstituted variant, ^{H}BPP, and the triazolyl-containing ligand, ^{H}BTP (Scheme 17).[55] Although initially targeted as a precursor to an *N*-heterocyclic carbene type pincer ligand, ^{H}BPP proved unsuccessful in this regard prompting examination of its basic chelating properties. Metalation reactions of proligands H(^{H}BPP) and H(^{H}BTP) with [Pd(COD)Cl$_2$] were performed in the presence of Et$_3$N in similar fashion to the synthesis of **68**. Ligand ^{H}BPP was found to yield a neutral, square-planar Pd complex, **76**, whereas ^{H}BTP afforded the cationic, acetonitrile adduct, **77**.

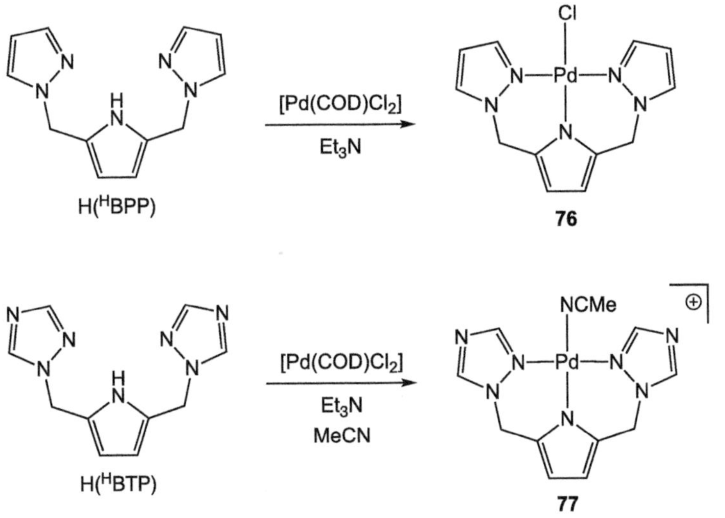

Scheme 17

Modifications to the BPP framework have also been described by Mani and coworkers. In one study, the potentially pentadentate ligand, **78** (Chart 7), was prepared by sequential substitution of each pyrrolic position in **1a** by 3,5-dimethylpyrazole-1-carbinol.[56] The H(^{Me}BPP) proligand was also produced from this reaction, but only in very minor quantities. Several metalation reactions were carried out with **78**, but only Pd was found to bind in a discrete pincer motif.[56,57] Delivery of **78** to Pd without prior deprotonation generated the bimetallic non-pincer complex **79**, which features an S-shaped arrangement of the two N_2PdCl_2 units about the central pyrrole group. By contrast, incorporation of exogenous (*n*-BuLi or NEt$_3$) or internal (AcO$^-$ anion) base during metalation resulted in deprotonation of the pyrrole group and formation of typical tridentate square-planar Pd complexes (**80–82**, Scheme 18).

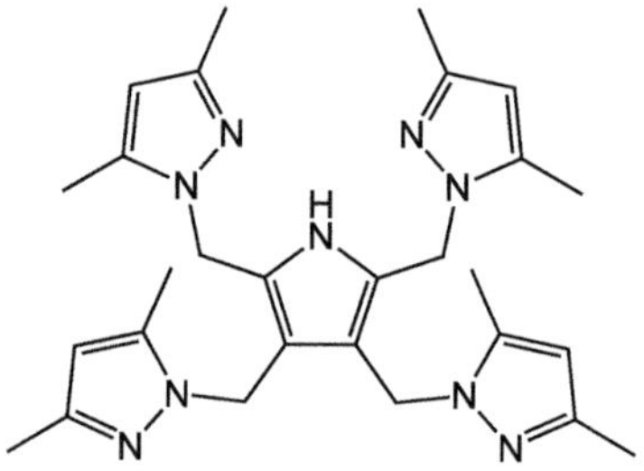

Chart 7 *Tetrakis*(pyrazolyl)pyrrole proligand **78**.

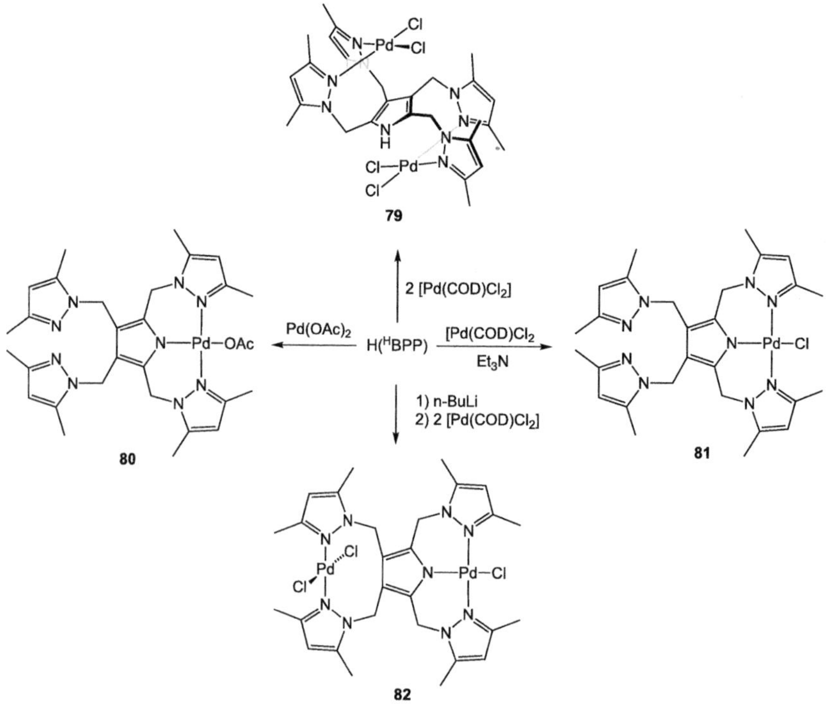

Scheme 18

In an additional modification to the BPP framework, Mani and coworkers have also reported the incorporation of a diphenylphosphine unit in place of one of the pyrazole moieties. The resulting non-palindromic PNN pincer ligand, **83** (Chart 8), was subsequently used to prepare several Ni and Pd complexes.[58]

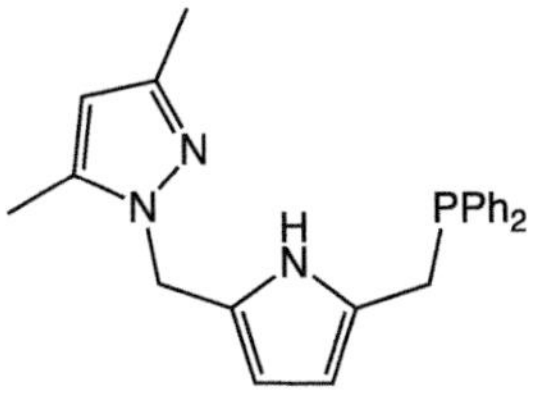

Chart 8 PNN proligand **83**.

In line with the parent ^{Me}BPP system, stoichiometric reaction of **83** with [Pd(NCPh)$_2$Cl$_2$] in the presence of Et$_3$N was found to generate the square-planar Pd complex **84**. Analogous reactions with [NiCl$_2$(DME)] or NiX$_2$ (X=Br, I) in refluxing acetonitrile afforded the Ni(II) congeners **85a–c** (Scheme 19). Norbornene polymerization studies were carried out to compare the activities of **84** and **85a–c** with related complexes bearing the palindromic ^{Me}BPP and ^{Ph}PNP ligands (vide infra). Ni(II) and Pd(II) complexes **84** and **85a–c** proved effective for norbornene polymerization demonstrating activities in the range of 10^7 g of polynorbornene mol^{-1} h^{-1}. At room temperature these activities were comparable to but slightly less than those of complexes bearing the parent ^{Me}BPP ligand, and much higher than complexes of ^{Ph}PNP. Interestingly, upon carrying out the polymerization reactions at 60 °C the relative activities were found to demonstrate a reverse trend with complexes of ^{Ph}PNP performing best. Trends in polymerization behavior among these related pincer frameworks were rationalized in terms of both the steric properties of the pincers and their propensity for dissociation.

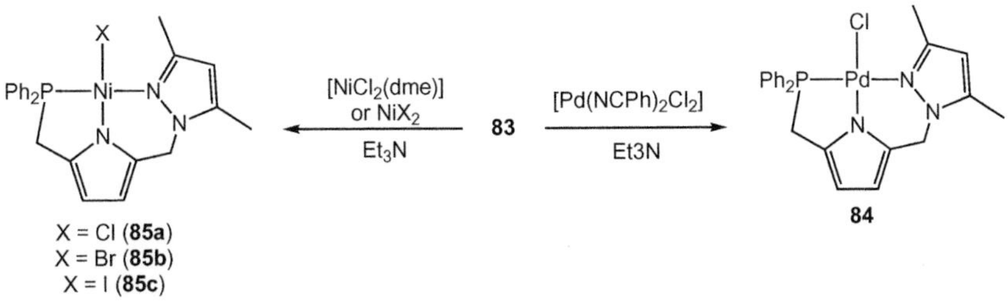

Scheme 19

6. *Bis*(oxazolinyl)pyrrolyl pincers (pyrrbox and pyrrmebox)

In response to the now-commonplace "Pybox" pincer ligand employed in asymmetric catalysis,[59] Gade and Mazet introduced the analogous monoanionic *bis*(oxazolinyl)pyrrole, or "Pyrrbox" class of ligands in 2001 (Chart 9).[60–62] The ligands are synthesized by addition of the corresponding amino alcohol to 2,5-*bis*(carbonitrile)pyrrole in the presence of catalytic amounts of ZnCl$_2$. Much like the related Pybox ligands, the construction of the oxazoline units in Pyrrbox can be varied easily due to the large number of available amino alcohols. However, much like the DIP and DPP ligands discussed above, the constrained bite angle of Pyrrbox renders tridentate coordination challenging.

H($^{R,R'}$Pyrrbox)

Chart 9 Pyrrbox proligand.

Initial studies with the Pyrrbox ligand focused on the dimethyl and isopropyl-substituted variants ^{Me,Me}Pyrrbox and $^{i\text{-}Pr,H}$Pyrrbox and their metalation to Pd (Scheme 20). In situ deprotonation of the proligands by *n*-BuLi followed by the addition of [Pd(COD)Cl$_2$] did not afford monometallic Pd pincer complexes, but instead generated the bimetallic species **86a–c**. Compound **86a** consists of two metal centers in a distorted square-planar geometry linked by two ^{Me,Me}Pyrrbox ligands. The twisting of the second oxazolyl group forms a helical structure that introduces a chiral element into the complex resulting in formation of a pair of enantiomers as shown in Scheme 20. When the enantiomerically pure *R,R*-$^{i\text{-}Pr,H}$Pyrrbox or *S,S*-$^{H,i\text{-}Pr}$Pyrrbox ligands were employed, they gave rise to the diastereomeric complexes **86b** and **86c**, respectively. Each diastereomer forms a pair of enantiomers related by opposing helicity. These species are denoted as ***M*–86b**,**c** *(S,S,S,S)* and ***P*–86b**,**c** *(R,R,R,R)* in Scheme 20. Kinetic analysis of the interconversion between ***M*–86b** and ***P*–86b** revealed a first-order concentration dependence with a large negative activation entropy ($\Delta S^\ddagger = -99 \text{ J} \cdot \text{mol}^{-1} \text{ K}^{-1}$), consistent with a dissociative-type mechanism.

Metalation of Pd with the racemate of R,R-$^{i\text{-Pr,H}}$Pyrrbox and S,S-$^{H,i\text{-Pr}}$Pyrrbox produced a mixture of **86b** and **86c**, and small quantities of the heterochiral species, **86d**. The crystal structure of **86d** was found to contain both helical enantiomers, *M*–**86d** and *P*–**86d** in the unit cell. Similar coordination geometries and helical stereoisomerism were also observed in palladium complexes of a related *bis*(oxazine)pyrrole ligand.[63]

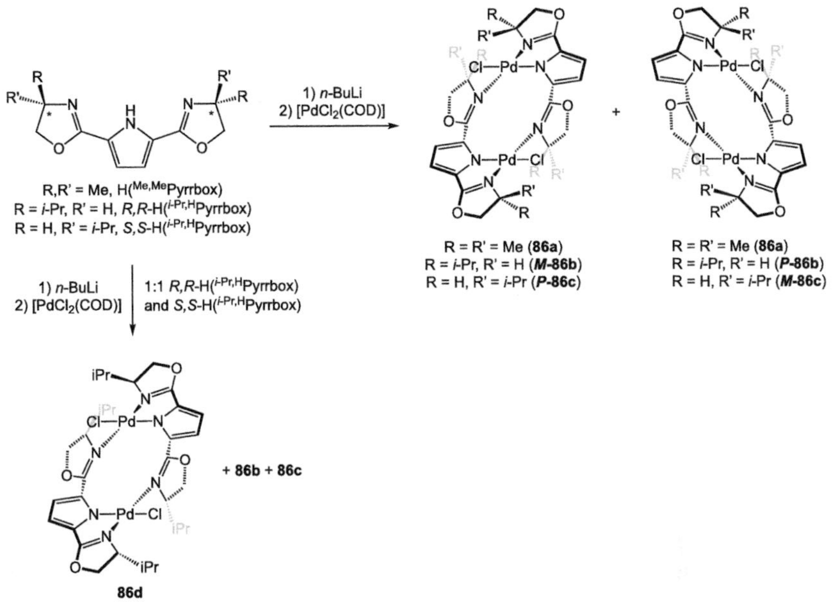

Scheme 20

Compound **86a** and a closely related species containing an ethyl-substituted pyrrole backbone were also employed as catalysts for the Heck and Suzuki couplings of bromoarenes.[60,62] Both complexes proved efficient as catalysts with **86a** displaying a T.O.N. of ca. 9500 for the Heck coupling of styrene and bromobenzene. Catalyst **86a** also demonstrated excellent selectivity for the coupling of methylacrylate and bromoarenes with *E:Z* ratios >99:1. The performance of **86a** in Suzuki couplings was likewise excellent although its activity toward aryl chlorides was limited, and the reaction was found to terminate after only a few cycles at elevated temperatures.

Despite the interesting coordination chemistry of select Pyrrbox derivatives, their constrained geometry renders them less than ideal as pincer ligands. In order to improve the chelating ability of these systems, Gade and coworkers introduced a modified Pyrrbox system based on

2,5-*bis*(oxazolinomethyl)pyrrole (Pyrrmebox, Chart 10).[64] Pyrrmebox ligands feature an extra methylene unit between the pyrrole and oxazoline moieties helping to accommodate tridentate coordination to smaller metal ions. The ligands are produced through a modified synthesis involving the intermediate 2,5-*bis*(cyanomethyl)pyrrole, which can undergo similar reactivity with amino alcohols to generate a series of different oxazoline substituents. The resulting transition metal chemistry with Pyrrmebox related systems encompasses a substantial body of work including both catalysis and small molecule activation.

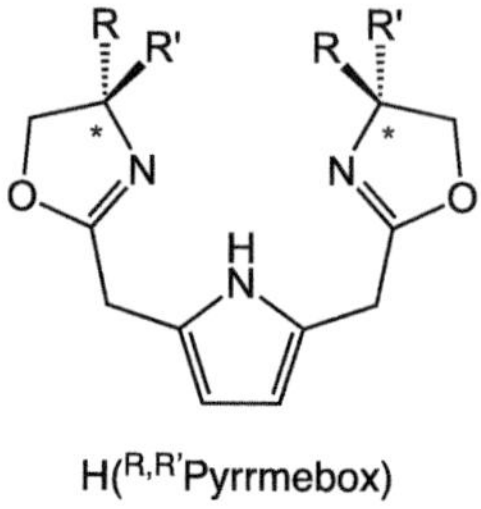

Chart 10 Pyrrmebox proligand.

The first palladium complexes of this ligand, **87a–c**, were readily synthesized by initial deprotonation of H($^{R,R'}$-Pyrrmebox) with *t*-BuLi followed by addition of [Pd(COD)Cl$_2$] (Scheme 21). The solid-state structure of **87a** features a square-planar geometry about the metal center with the expected tridentate coordination. The ligand backbone adopts a twisted arrangement giving rise to a pair of enantiomers (δ and λ) akin to that observed for bimetallic compounds **86a–c**. NMR studies demonstrated that the interconversion barrier between the δ and λ conformers is <10 kcal mol^{-1} leading to rapid isomerization in solution (Scheme 21).

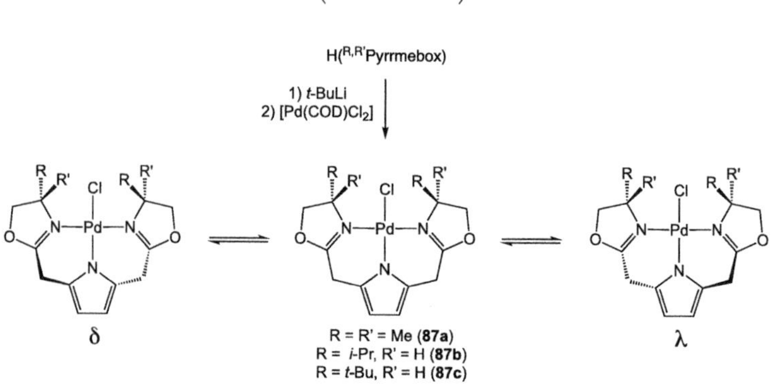

R = R' = Me (**87a**)
R = *i*-Pr, R' = H (**87b**)
R = *t*-Bu, R' = H (**87c**)

Scheme 21

In an effort to avoid the use of lithium reagents, an alternative synthesis of **87a–c** from [PdCl$_2$(NCPh)$_2$] and H($^{R,R'}$-Pyrrmebox) was devised.[64] Yields for this reaction were found to be lower, and in two instances the tautomeric palladium species **88a** and **88b** were produced in tandem (Scheme 22). The quantity of **88** generated was observed to depend on the reaction time suggesting that **87** was the kinetic product whereas **88** represented the thermodynamic product. The enhanced π-delocalization of the ligand backbone in **88** was proposed to provide the thermodynamic driving force for tautomerization from **87**. NMR studies with isolated **87a** and **87c** in the presence of [PdCl$_2$(NCPh)$_2$] confirmed the isomerization process and provided access to the *tert*-butyl derivative **88c**. Modification of the Pyrrmebox backbone was also observed upon autoxidation of **87b,c** to give the new Pd species **89b,c** (Scheme 22). In this case, the ligand is formally oxidized by two electrons with the loss of two protons to afford a planar conjugated backbone.

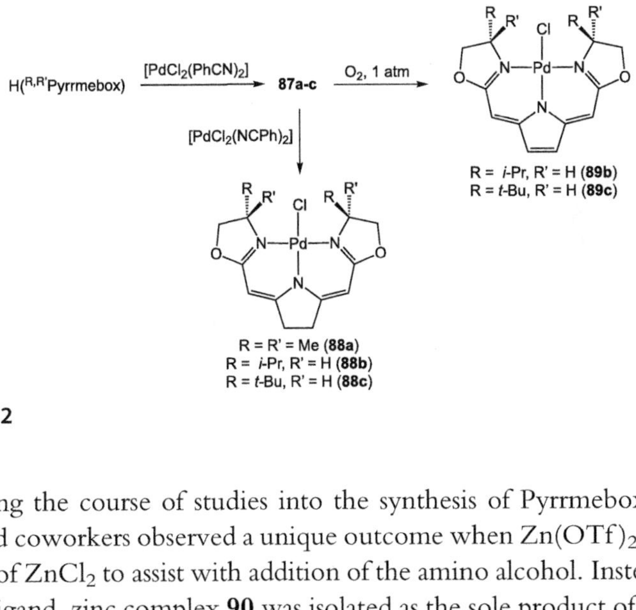

Scheme 22

During the course of studies into the synthesis of Pyrrmebox ligands, Gade and coworkers observed a unique outcome when Zn(OTf)$_2$ was used in place of ZnCl$_2$ to assist with addition of the amino alcohol. Instead of the desired ligand, zinc complex **90** was isolated as the sole product of the reaction (Scheme 23).[65] The increased Lewis acidity of Zn(OTf)$_2$ over ZnCl$_2$ is believed to be responsible for the production of **90**, which requires both metalation and tautomerization. Variable temperature ^{1}H NMR studies of **90** demonstrated a dynamic equilibrium for the complex involving flipping of the chlorine atom through the NNN-plane.

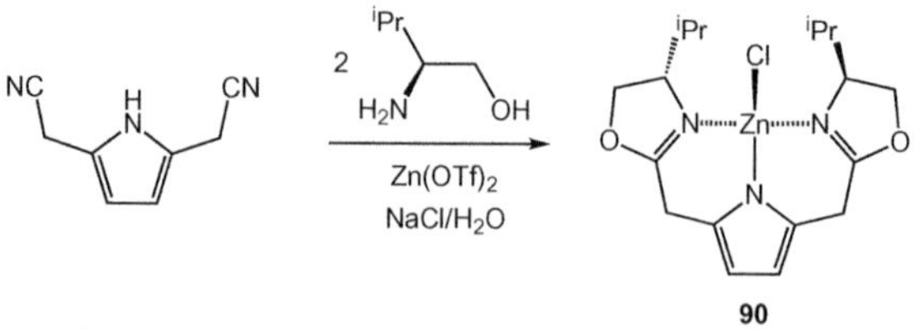

Scheme 23

Synthetic routes to the Pyrrmebox ligand were subsequently expanded to include more extensive substitution patterns on the oxazoline units.[66] Rhodium complexes of these ligands, **91b–f**, were prepared in similar fashion to palladium complexes **87a–c** by initial deprotonation with *t*-BuLi followed by addition of [Rh$_2$(μ-Cl)$_2$(CO)$_4$]. This procedure was found to generate a number of impurities, but a different metalation strategy involving direct reaction of H($^{R,R'}$Pyrrmebox) with [Rh(CO)$_2$(acac)] afforded analytically pure material in excellent yields. Despite the apparent superiority of this method, compound **91c** could not be synthesized in this fashion due to the steric bulk of the *t*-Bu groups. Stoichiometric addition of a Rh source to **91** was attempted to induce the same type of tautomerization event observed for **87** but proved unsuccessful. Instead, elevated temperatures were found to cleanly afford the isomerized complexes **92b,d–f**. Moreover, a one-pot synthesis from H($^{R,R'}$Pyrrmebox) and [Rh(CO)$_2$(acac)] could be adapted to produce **92b–f**, including the *t*-Bu derivative, in nearly quantitative yields (Scheme 24). Oxidative addition of MeI and CsBr$_3$ to **91b** generated the octahedral Rh(III) complexes **93** and **94**, respectively. Compound **93** displayed no ligand tautomerization even at elevated temperatures, but **94** was found to be isomerized by in situ generated Br$_3$ during the course of the reaction.

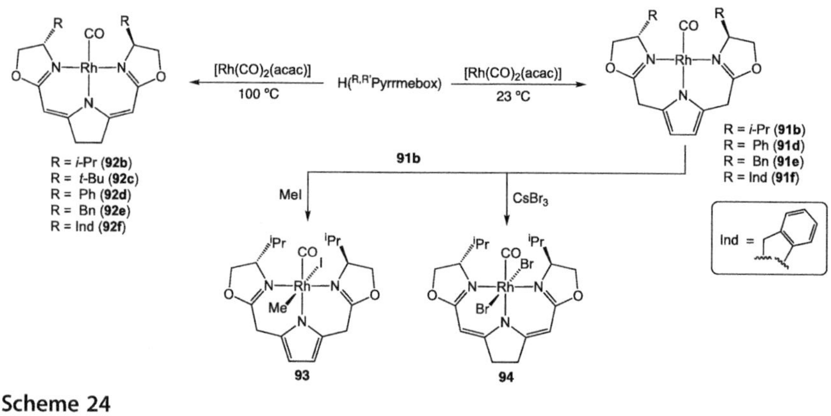

Scheme 24

Nickel complexes bearing $^{R,R'}$Pyrrmebox ligands have also been synthesized and used to investigate the potential mechanisms for intraligand tautomerization.[67] DFT calculations on the proligand, H($^{i\text{-}Pr,H}$Pyrrmebox), demonstrated that the isomerization process is essentially barrierless, yet equilibrium formation of both ligand tautomers could not be detected experimentally even at elevated temperatures. Deuterium scrambling experiments with the proligand in the presence of MeOD and catalytic acetic acid were consistent with a pathway involving the carboxylate unit as a proton transfer relay. To compare the ligand dynamics when bound to a metal center, nickel(II) complexes of $^{i\text{-}Pr,H}$Pyrrmebox were synthesized from both NiCl$_2$ and Ni(OAc)$_2$ generating **95b** and **96b**, respectively (Scheme 25). Notably, tautomerization was found to accompany formation of **96b** but not **95b**. Treatment of **96b** with Me$_3$SiCl generated **97b**, which represents the tautomeric form of **95b**. Attempts to isomerize **95b** directly via thermal means, however, did not proceed to **97b**. These results therefore strengthen the proposal that acetic acid or some other carboxylate source is required for isomerization. Yet this proposal does not account for the isomerization observed both thermally and in the presence metal salts such as [PdCl$_2$(NCPh)$_2$] (vide supra), so other potential mechanisms cannot be ruled out.

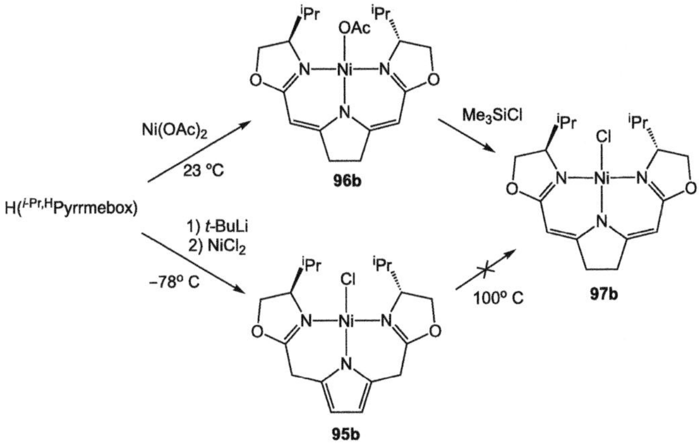

Scheme 25

Outside of their use as model complexes to study ligand tautomerization, compounds **95b** and **97b** have proven to be excellent platforms to explore the organometallic chemistry of Ni pincers.[68,69] Compounds of the type **95** and **97** were expanded to include a variety of oxazoline substituents in similar fashion to the rhodium species discussed above. Reactions of **97b–d,f** with LiHBEt$_3$ produced the square-planar Ni^{II}-H complexes, **98b–d,f**, which exist in a hydrogen-pressure dependent equilibrium with the chiral T-shaped Ni^I species, **99b–d,f** (Scheme 26 and Fig. 3). The reduction of the Ni center in these complexes was corroborated by both DFT calculations and EPR spectroscopy. Furthermore, compounds of the type **99** were found to catalyze the asymmetric hydrodehalogenation of geminal dihalides in the presence of LiHBEt$_3$. Optimization studies identified complex **99d** as the best catalyst in terms of both yield and enantiomeric excess.

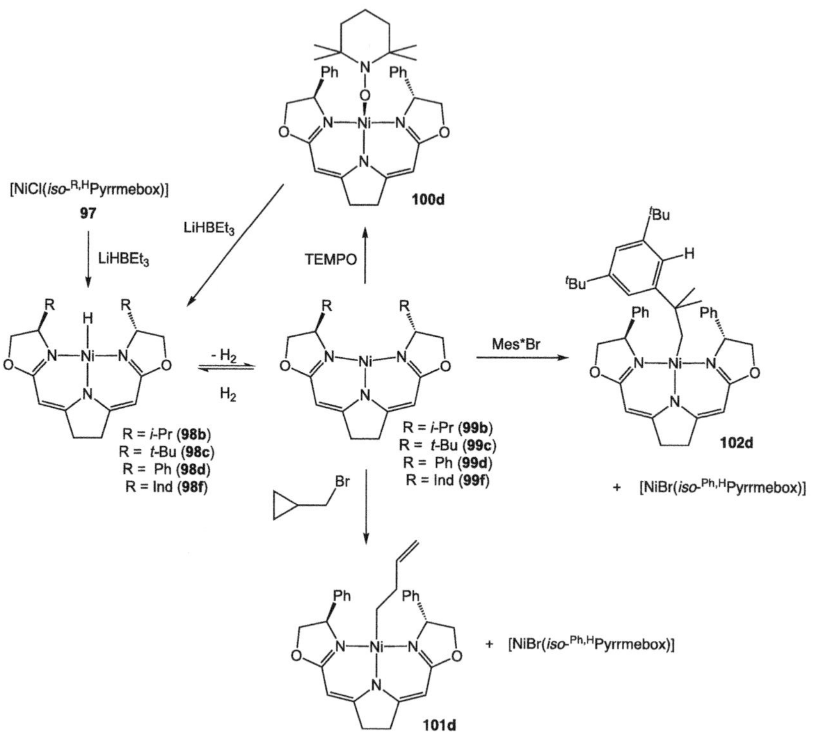

Scheme 26

Owing to the fact that asymmetric hydrodehalogenation reactions have had limited success in the past, mechanistic investigations using **99d** were pursued in order to gain a fuller understanding of the role of Ni(I) species. Dehalogenation in these processes is typically regarded as radical-based, so several stoichiometric reactions were performed to probe this hypothesis. Addition of TEMPO to **99d** generated the distorted square-planar Ni^{II}-TEMPO adduct **100d** indicating facile interception of the Ni(I) center

Pincer ligands incorporating pyrrolyl units

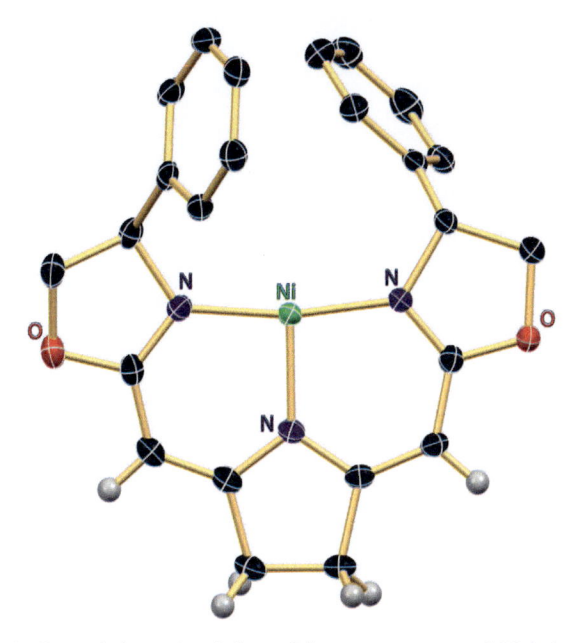

Fig. 3 Thermal ellipsoid drawing of the solid-state structure of **99d**. *Image generated from crystallographic data in reference Rettenmeier C, Wadepohl H, Gade LH. Stereoselective hydrodehalogenation via a radical-based mechanism involving T-shaped chiral nickel(I) pincer complexes. Chem Eur J. 2014;20:9657–9665.*

by free radicals. Similarly, test substrates bromomethylcyclopropane and supermesitylbromide (Mes*Br) afforded the expected addition/rearrangement products **101d** and **102d** plus an equivalent of the Ni bromide complex, further confirming the single-electron transfer reactivity of **99d**. In contrast to **99d**, hydride compound **98d** demonstrated no reactivity toward TEMPO, but could be regenerated from **100d** through addition of LiHBEt$_3$ (Scheme 26).

In line with the stoichiometric reactions depicted in Scheme 26, addition of TEMPO to catalytic hydrodehalogenation trials inhibited product formation through generation of **100**. This dormant state could reenter the catalytic cycle by forming **98** through reaction with LiBEt$_3$H. Accordingly, catalytic reactions inhibited by TEMPO resumed once all the radical had been depleted. Deuterium labeling experiments further established that the hydride transfer component of the hydrodehalogenation reaction is accomplished by **98** and not lithium superhydride. With this evidence in hand, a mechanism for the overall catalytic pathway was proposed as shown

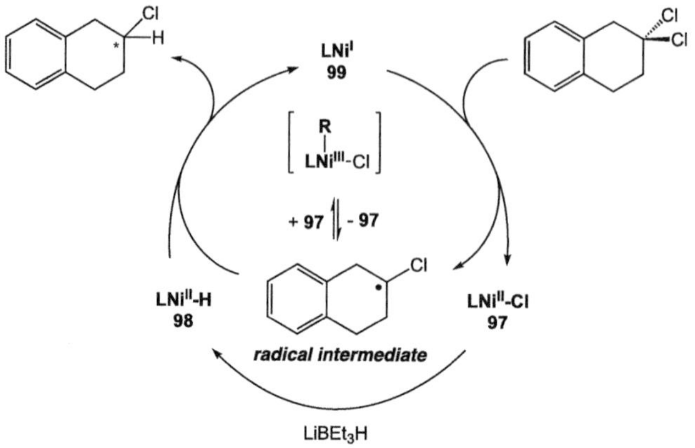

Scheme 27

in Scheme 27. Reversible formation of a nickel(III) adduct between **97** and the newly formed organic radical was postulated as a means of stabilizing this potentially reactive intermediate. Direct synthesis of such a Ni^{III} complex proved unsuccessful, although EPR data from in situ reactions of **97** with aryl and alkyl halides supported its existence.

The hydrodehalogenation chemistry observed with **99d** was extended to include the activation of aryl halides.[69] These reactions demonstrated the same bimolecular oxidative addition processes as their aliphatic counterparts yielding Ni(II) halide complexes alongside aryl species of the type **103d** (Scheme 28). The reactivity trend observed across a series of aryl halides followed the expected ordering I > Br > Cl > F. Carbon–fluorine bond activation was only observed for hexafluorobenzene, with no reactivity seen between fluorobenzene and **99d** even at high temperatures. Hammett correlations between *para*-substituted aryl chlorides showed an increased reaction rate for electron withdrawing groups (CF$_3$ > F > H > Me > OMe) consistent with the preference for systems that can better accommodate the increased electron density from Ni(I) in the transition state. Catalytic hydrodehalogenation of aryl halides was also examined using **99d** but proved unsuccessful. Such a lack of catalytic activity with aryl substrates was ascribed to the enhanced stability of **103d**, which cannot reenter the catalytic cycle easily and therefore serves to sequester Ni from the overall cycle. Although this aspect of aryl halide chemistry was unfavorable from the standpoint of catalysis, it did permit the synthesis of more complex molecules such as the multimetallic species, **104d** and **105d** (Chart 11), that would otherwise be difficult to access from the corresponding Grignard reagents.

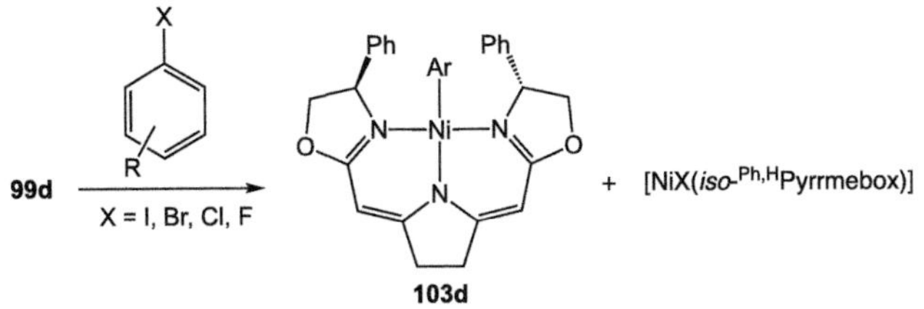

Scheme 28

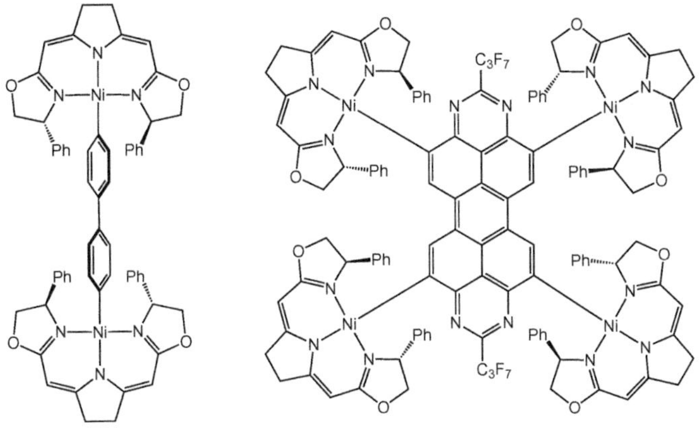

Chart 11 Multimetallic nickel(II) complexes of the iso-Pyrrmebox ligand.

C—F bond activation is still an elusive goal in organometallic chemistry, and the reactivity of **99d** with C—F bonds, albeit limited, prompted a more in-depth investigation of its chemistry with other fluorinated substrates.[70] The activation of hexafluorobenzene mentioned above led to the initial synthesis of the fluoride complex, **106d**. Orthogonal syntheses of related fluoride compounds were subsequently developed from the corresponding bromide and hydroxide compounds via treatment with CsF or NH$_4$F in wet THF (Scheme 29). These reactions produced the desired metal fluorides **106b** and **106d** cleanly in good yields. Compounds **106d** and **106d** were then employed for the catalytic hydrodefluorination of geminal difluorocyclopropanes to afford fluoroalkenes. Silanes and boranes could both be utilized as hydride sources given their ability to convert **106b,d** into the corresponding hydride (Scheme 29). Optimization of the catalytic reaction with 1,1-difluoro-2,2-di(4-fluoro)phenylpropane identified **106d** as the best catalyst as it provided the highest yield and diastereoselectivity for formation of the Z product. A mechanism analogous to the hydrodechlorination process discussed above was put forward for the catalytic cycle.

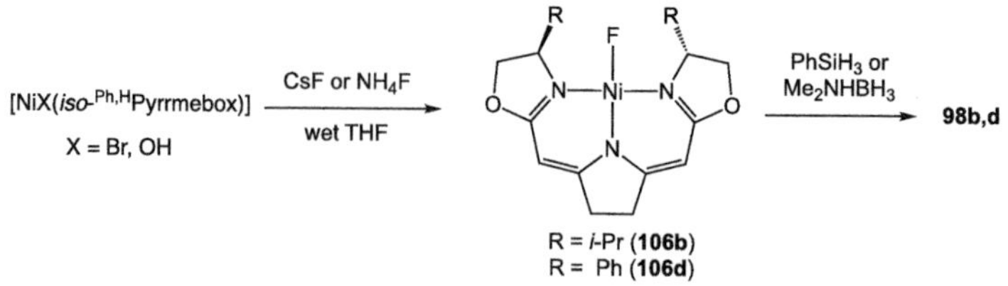

Scheme 29

Stoichiometric defluorination reactions of **99d** with selected substrates were also examined in the absence of a hydride source. Treatment of both 1,1-difluoro-2,2-di-(4-fluoro)phenylpropane and decafluorodecalin with **99d** produced the corresponding defluorinated products shown in Scheme 30. The formation of these products establishes that the second defluorination event from the radical intermediate occurs faster than the initial C—F activation. When the same reaction depicted in Scheme 30A was run in the presence of 5 bar H_2, the hydrodefluorination product was produced instead (Scheme 30C). Given that the presence of H_2 generates a mixture of **98d** and **99d**, the result in Scheme 30C demonstrates that H atom transfer from **98d** is kinetically favored over C—F activation by **99d** in the radical intermediate. Therefore, these stoichiometric reactions highlight the delicate interplay of **98d** and **99d** that is critical to the successful hydrodehalogenation chemistry in this system.

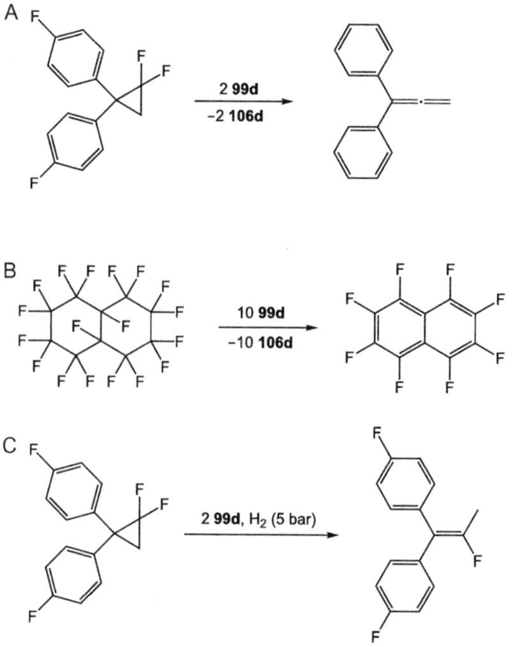

Scheme 30

In addition to dehalogenation, T-shaped Ni(I) complexes have also proved effective in the study of autoxidation and C—H oxygenation processes through the formation of peroxide, superoxide, and hydroperoxide species.[71,72] The low-temperature addition of O_2 to **99b–d** was found to generate the 1,2-μ-peroxo complexes, **107b–d**, which can exist in equilibrium with the end-on Ni^{II} superoxide species (**108b,d**, Scheme 31). Crossover experiments, ^{18}O-labeling, and EPR spectra supported the formation of **107b–d** as occurring through an associative mechanism in which the transition state involves a mixed valent Ni^{II}/Ni^{III} complex where one Ni center has an additional axial superoxo ligand. Further reactions of **107b,d** with hydrogen peroxide and water produced the Ni(II) hydroperoxide complexes **109b,d** and the Ni(II) hydroxide species **110b,d**, respectively. Both **109d** and **110d** were found to decompose at room temperature under aerobic conditions giving rise to a mixture of **111d** and **111d′** (Scheme 31). This decomposition process was proposed to occur through a radical-based C—H activation at the benzylic position of the ligand.

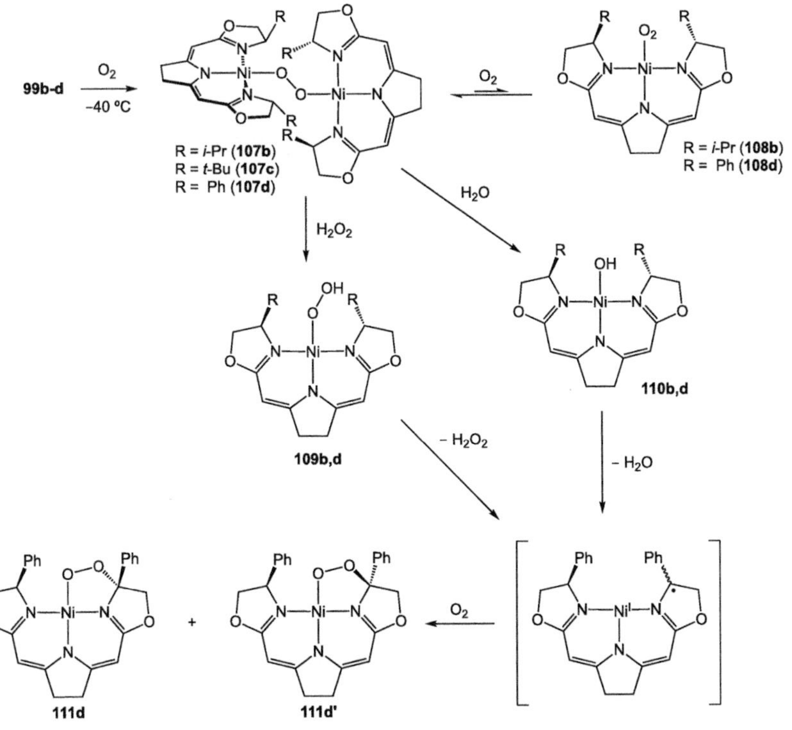

Scheme 31

Decomposition of the peroxo species **107b** and **107c** were further explored to understand the novel oxygenation processes at play in these systems. Decomposition of **107b** in the presence of O_2 was found to generate a 1:1 mixture of **110b** and compound **112b**, which contains a 6-membered peroxo metallacycle (Scheme 32). Notably, both **107b** (tertiary) and **107d** (benzylic) contain reactive C—H bonds in close proximity to the Ni center that facilitates their activation in the presence of O_2. By contrast, **107c** lacks a similar reactive C—H bond such that analogous oxygenation processes were not observed. Instead, autoxidation of **107c** led to a mixture of decomposition products in which the oxazolylcarboxylate compound, **113c**, was the major species. Anaerobic decomposition of **107b** was found to yield a 1:1 mixture of **110b** and the alkoxide species, **114b** (Scheme 32). Detailed investigations on the decomposition processes at play in **107b** supported mechanisms under both aerobic and anaerobic conditions in which the rate determining step is the initial formation of **112b** and the nickel hydride, **98b**. Molecular oxygen can then insert into the Ni—H bond forming peroxide **109b** which subsequently reacts with another equivalent of **98b** to form the hydroxide species **110b**. In the absence of oxygen, **112b** and **98b** react to form **110b** and **114b** by an oxygen atom abstraction pathway.

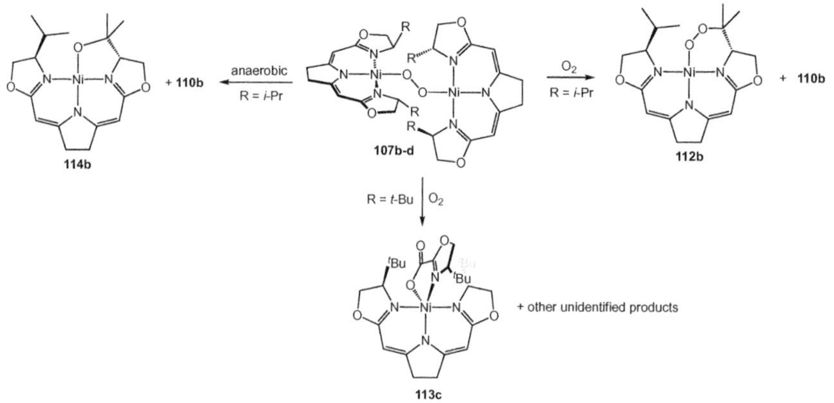

Scheme 32

As a final chapter in the oxidation chemistry of the nickel iso-Pyrrmebox system, alkyl complexes of Ni(II) were investigated for their oxygen reactivity. Exposure of the ethyl complexes **115b,d** and the hexenyl compound **116d** to molecular oxygen generated the peroxide species **109b,d** along with the corresponding alkene de-insertion products (Scheme 33A). To exclude the possibility that the observed oxidation chemistry was occurring via a β-H elimination pathway, the methyl complex, **117d**, was synthesized and exposed to O_2. The reaction produced the methylperoxo species **118d**, thereby establishing that O_2 can react directly with Ni(II) alkyls without the intermediacy of a hydride. Room temperature decomposition of **118d** was found to form equimolar amounts of **110d** and the formate complex, **119d**. An orthogonal synthesis of **119d** was carried out by exposing **110d** to formic acid confirming its composition. In solution at elevated temperatures, **119d** cleanly underwent decarboxylation to generate an equilibrium mixture of **98d** and **99d** (Scheme 33B).

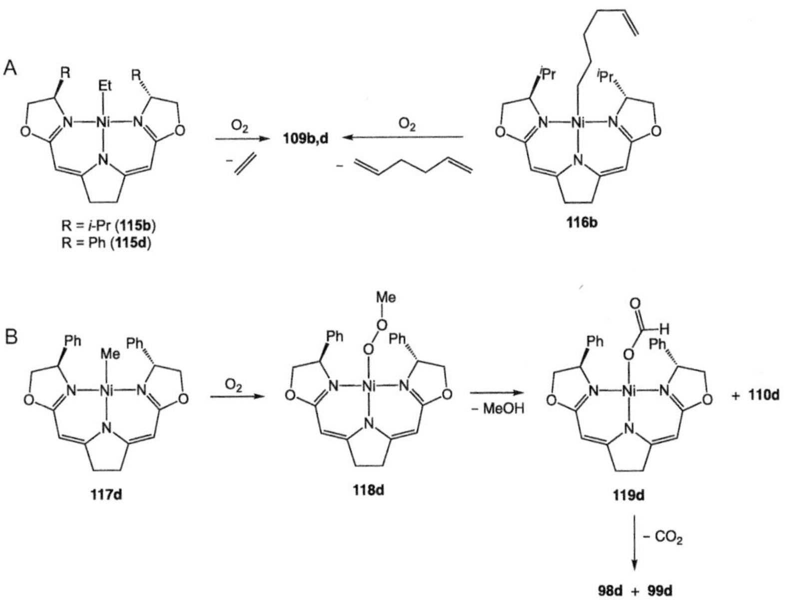

Scheme 33

The chemistry of Pyrrmebox and iso-Pyrrmebox complexes described in the preceding paragraphs highlights the important role of ligand tautomerization in shaping the overall reactivity of these systems. Nowhere was this effect more pronounced than in the stabilization of the T-shaped Ni(I) complexes (**99**). In order to obviate the possibility for ligand isomerization, Gade and coworkers developed the related pincers displayed in Chart 12, which feature dimethyl substitution at the methylene linker.[73]

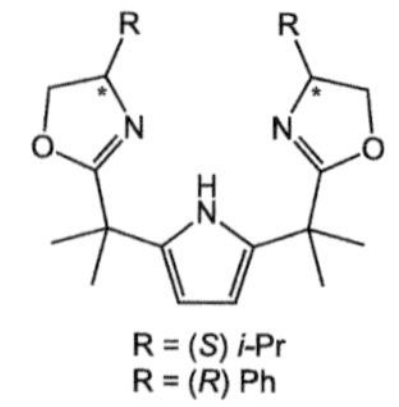

Chart 12 Tetramethylated Pyrrmebox proligand.

R = (S) *i*-Pr
R = (R) Ph

Metalation of the tetramethylated Pyrrmebox ligands to nickel was accomplished by an analogous lithiation protocol as described above for **95**. A wide variety of Ni(II) complexes were subsequently synthesized and characterized using the chloride precursor **120b** via typical halide exchange, salt metathesis, and transmetalation routes (Scheme 34). The solid-state structures of the compounds displayed in Scheme 34 displayed a canted pyrrolic backbone akin to that observed in the un-isomerized parent Pyrrmebox system.

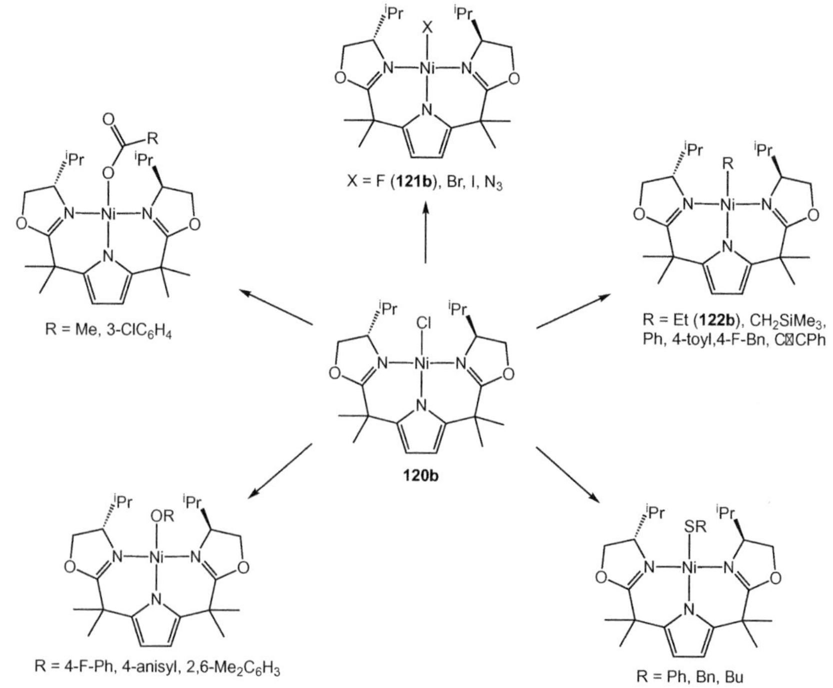

X = F (**121b**), Br, I, N$_3$

R = Me, 3-ClC$_6$H$_4$

R = Et (**122b**), CH$_2$SiMe$_3$, Ph, 4-toyl, 4-F-Bn, C≡CPh

120b

R = 4-F-Ph, 4-anisyl, 2,6-Me$_2$C$_6$H$_3$

R = Ph, Bn, Bu

Scheme 34

The robust chemistry displayed by compounds **98** and **99** described above was attributed in large part to the enhanced stability of the iso-Pyrrmebox ligand, which contains a more planar π-conjugated backbone. The corresponding Ni hydride complex of the tetramethylated Pyrrmebox ligand therefore became an attractive target because of the inability of the ligand to undergo tautomerization.[73] Synthesis of the hydride **123b** was accomplished by addition of PhSiH$_3$ or Ph$_2$SiH$_2$ to the fluoride complex, **121b**. Compound **123b** was found to be stable at room temperature in solution for several days and no equilibrium formation of a Ni(I) species was detected. Bond activation and insertion reactions were performed with **123b** to probe the reactivity of the Ni—H bond. The hydride was found to display no migratory insertion of unsaturated substrates with the exception of ethylene, giving rise to the ethyl complex **123b** after 16 h (Scheme 35). Homolytic bond cleavage of dichalcogen species (R$_2$E$_2$, E=S, Se, Te) by **123b** led to clean formation of the corresponding Ni(II)-chalcogenide species. The thiolate analogs could also be prepared by protonolysis reactions of **123b** with thiols. Addition of elemental sulfur to **123b** generated the novel hydrosulfide complex **124b**, but it was found to decompose readily in solution within 3 h.

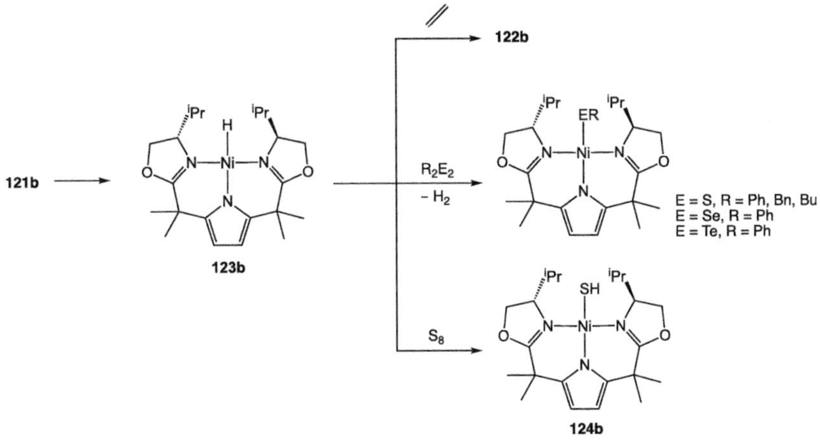

Scheme 35

Given the lack of insertion chemistry observed for aldehydes and ketones, Gade and coworkers next examined the reactivity of **123b** with epoxides.[74] Treatment of **123b** with styrene oxide and 2,2-diphenyloxirane led to smooth formation of the anti-Markovnikov ring-opened species **125b** and **126b** (Scheme 36A). Further reaction of these alkoxide species with silanes regenerated **123b** with concomitant formation of the corresponding silyl ethers. Since the reactions depicted in Scheme 36A represent conceptual steps in the hydrosilylation of epoxides, **123b** was also examined as a catalyst. Indeed moderate to high yields were achieved for the catalytic hydrosilylation of several epoxide substrates at room temperature employing **123b**. 4-Chloro- and 4-bromostyrene oxide failed to undergo hydrosilylation, which was attributed to the likelihood of competing hydrodehalogenation. Detailed mechanistic investigations of the catalytic reaction using 4-fluorostyrene oxide revealed a first-order dependence on both **123b** and epoxide, and a zero-order dependence on silane. A mechanistic proposal was therefore put forward where epoxide inserts into the Ni—H bond in the turnover-limiting step followed by rapid σ-bond metathesis of the resulting alkoxide intermediate with silane (Scheme 36B).

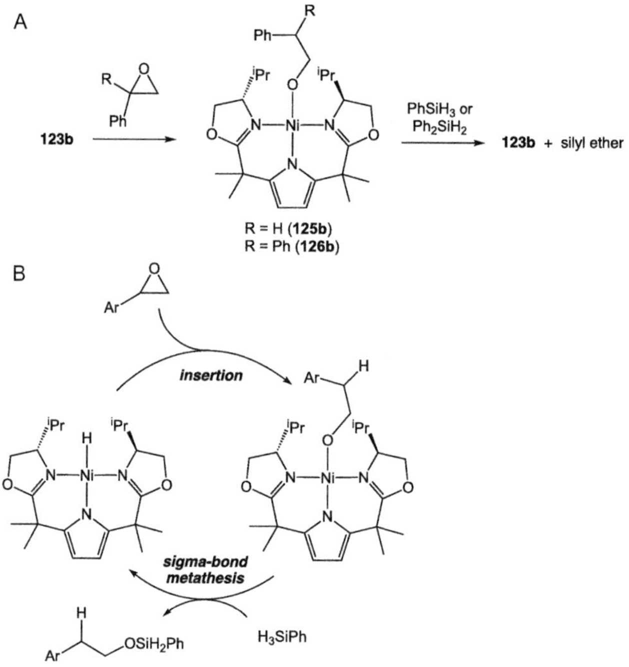

Scheme 36

Extensive coordination chemistry with other transition metals besides nickel has also been investigated with the tetramethylated Pyrrmebox ligands.[75,76] Compounds of Rh(I) and Pd(II) akin to those depicted above in Schemes 21 and 24 could be prepared and were found to adopt

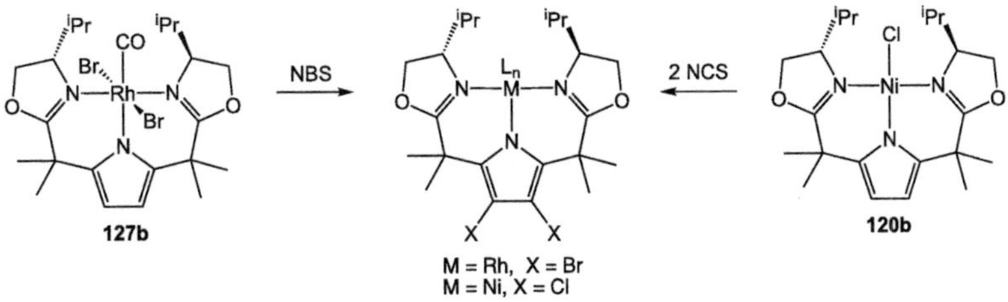

Scheme 37

analogous structures. The Rh(I) compounds were further observed to undergo similar oxidative addition chemistry affording trivalent octahedral species such as **127b** (Scheme 37). Interestingly, halogenation of the pyrrole backbone was observed for both **127b** and **120b** upon reaction with *N*-halosuccinimides.

Successful metalation of tetramethylated Pyrrmebox to Cr, Mn, Fe, and Co was accomplished by in situ deprotonation of the ligand followed by addition of the corresponding divalent metal salt (Scheme 38). Formation of the Fe and Co compounds occurred readily at room temperature, but formation of the Mn analog required extended time under reflux. The structures of both **130b** and **131b** were determined by X-ray crystallography and found to contain metal centers in distorted tetrahedral geometries. A single peak could be observed for the diastereotopic methyl groups in the paramagnetic NMR spectrum of **130b** suggesting that the molecules undergo rapid inversion in solution. Addition of xylyl isocyanide to **130b** afforded the diamagnetic, octahedral ferrous cation **132b**.

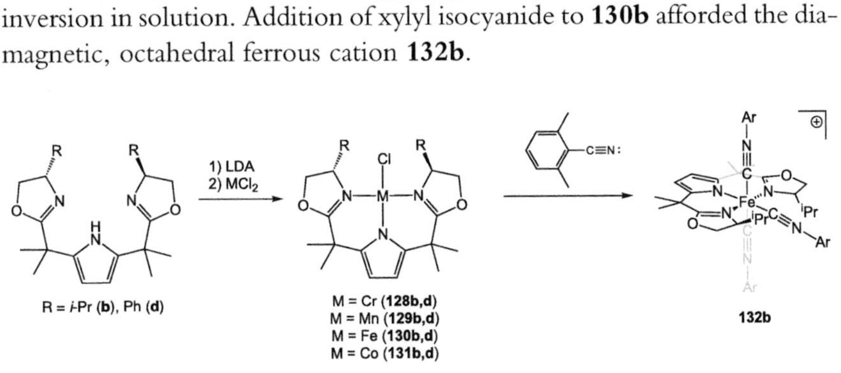

Scheme 38

Complexes **129b–131b** were examined as catalysts for the hydrosilylation and hydroboration of ketones. All three complexes showed high activity toward hydroboration, however, the hydrosilylation results were mixed among the various metals. Unfortunately, the enantioselectivity of the

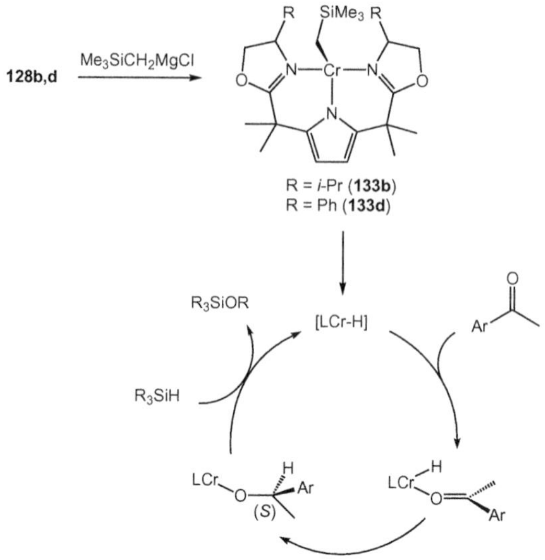

Scheme 39

reactions was low in all cases, but an inversion in the favored enantiomer was observed when changing the oxazoline substituent from *i*-Pr to Ph. Chromium analogs **128b,d** served as precursors to alkylated species **133b,d**, which proved to be highly active and selective for the catalytic asymmetric hydrosilylation of substituted acetophenones. Mechanistic investigations of the reaction were consistent with a modified Chalk-Harrod mechanism in which the rate determining step is association and insertion of the acetophenone for electron-rich substrates, but switches to σ-bond metathesis for electron-poor substrates (Scheme 39).[76]

7. Diphosphinopyrrolyl pincers (^{R}PNP)

Many of the most popular pincer ligand designs in modern coordination chemistry incorporate phosphine donors. Pyrrolyl-containing versions of such pincers (^{R}PNP, Chart 13) were introduced simultaneously in 2012 in a series of papers by the groups of Gade, Mani, and Tonzetich.[77–79] These early studies all considered nickel complexes of the diphenylphosphine derivative, H(^{Ph}PNP). Despite the similarity in the reported nickel chemistry, the synthetic methods used to prepare proligand H(^{Ph}PNP) differed slightly across all three papers. Subsequent work with ^{R}PNP derivatives

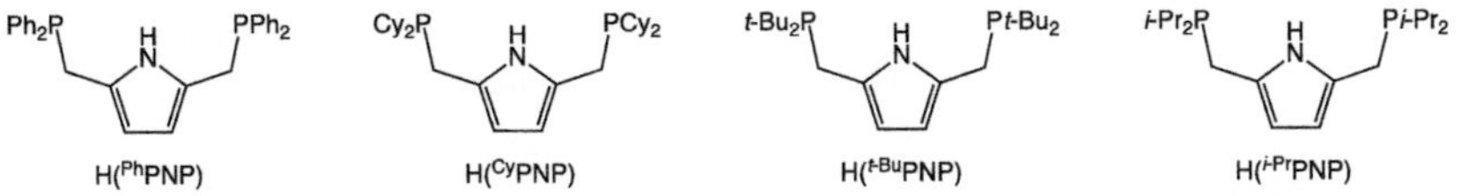

Chart 13 Diphosphino pyrrole proligands.

(R = Cy, i-Pr, t-Bu) has adopted the preparative method first reported by Mani and coworkers,[78] which employs **1a** as a precursor in direct reaction with a secondary phosphine.

Each of the original papers describing the ^{Ph}PNP ligand reported the nickel chloride complex **134a** (Scheme 40). The synthesis of **134a** was accomplished by metalation of H(^{Ph}PNP) with NiCl$_2$(dme) in the presence of base. Subsequent work has extended this synthetic protocol to other ^{R}PNP ligands bearing different phosphine substituents (**134b–d**).[80–82] In their initial report of ^{Ph}PNP, Gade and coworkers also described the direct reaction of H(^{Ph}PNP) with [Ni(COD)$_2$] to generate the bridging Ni(I) species, **135a**.[79] Compound **135a** could alternatively be accessed through treatment of **134a** with hydride sources. The bimetallic palladium analog, **136a**, was prepared by Mani and coworkers through a similar reaction involving Pd$_2$(dba)$_3$.[78] The structures of **135a** and **136a** are unique among PNP species and have only been reported to date with the diphenylphosphine derivative (^{Ph}PNP). The structure of **135a** is depicted in Fig. 4 and features unsymmetrically-bridged phosphine units and a coordination geometry about each nickel center approaching square-planar if the longer M-P contacts are discounted. The palladium analog is similar in structure. Both species **135a** and **136a** react with alky halides and other weak oxidants to regenerate the divalent halide compounds (Scheme 40). Production of **135a** can also be accomplished electrochemically as found by Tonzetich and coworkers. Cyclic voltammetry measurements of **134a** in tetrahydrofuran demonstrated an irreversible cathode process corresponding to formation of **135a**. In the presence of CH$_2$Cl$_2$ as additive or solvent, electrocatalytic dechlorination was observed consistent with the chemical reactivity observed by Gade and Mani. In later work, Walter, Bernskoetter, and coworkers reported the preparation of the mercury-bridged dinickel(I) compound, [Ni$_2$(μ-Hg)($^{t\text{-}Bu}$PNP)$_2$] (**138c**).[83] Compound **138c** was synthesized by the sodium amalgam reduction of the bromide analog of **134c**, and unlike **135a**, features two Ni centers with normal PNP coordination (Fig. 4). Treatment of **138c** with H$_2$ led to homolytic cleavage and generation of nickel hydride **144c** (vide infra).

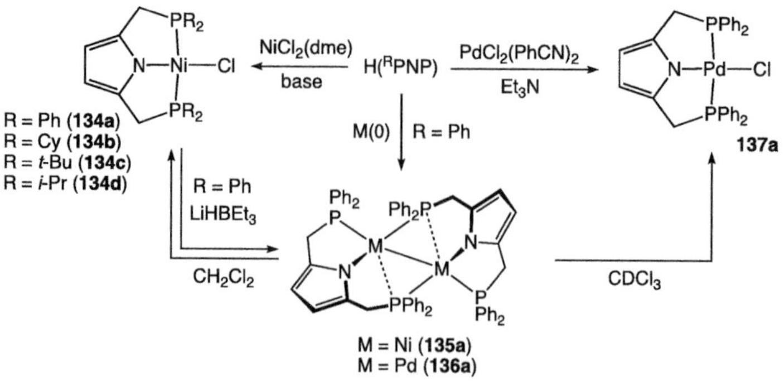

Scheme 40

Subsequent work in the area of nickel chemistry by several different groups utilized compounds **134a–d** as precursors to a wide variety of new complexes (Scheme 41). Tonzetich and coworkers demonstrated the synthesis of several organometallic compounds containing the ^{Ph}PNP and ^{Cy}PNP ligands including a stable ethyl species **140a,b** (Scheme 41).[81] Also reported was nickel hydride **144b**, which was otherwise inaccessible with the ^{Ph}PNP ligand. Compounds **134a** and **134b** both proved active for the Kumada coupling of aryl halides and Grignard reagents at room temperature and **144b** was found to demonstrate facile CO_2 insertion. In later work, compound **134b** was also shown to be an active precatalyst for the C—S cross-coupling of thiols and aryl iodides in the presence of base.[80] Preliminary mechanistic studies of this reaction identified a putative Ni(I) species as a possible intermediate. Synthetic work accompanying this study also disclosed the first instance of the *t*-Bu substituted ligand, $^{t\text{-}Bu}$PNP, as well as compounds **146–150**, which feature S-, N-, and O-bound ligands (Scheme 41). The nickel chemistry of ligands $^{t\text{-}Bu}$PNP and $^{i\text{-}Pr}$PNP was expanded upon by Walter, Bernskoetter and coworkers who also reported the preparation of hydride species (**144c,d**).[83] Much like **144b**, these compounds demonstrated clean insertion of CO_2. It therefore appears that for all phosphine substituents other than Ph reported to date, stable Ni(II) hydrides can be prepared in straightforward fashion from halide precursors.

Cationic complexes of the (^{R}PNP)Ni framework have also been described. In their original report of ^{Ph}PNP chemistry, Tonzetich and coworkers reported the synthesis and structural characterization of the acetonitrile adduct, **151a** (Scheme 41). More recently, Mani and coworkers found that similar preparation of **151a** in the presence of water leads to demetallation and formation of silver complexes of the type **152a**.[84]

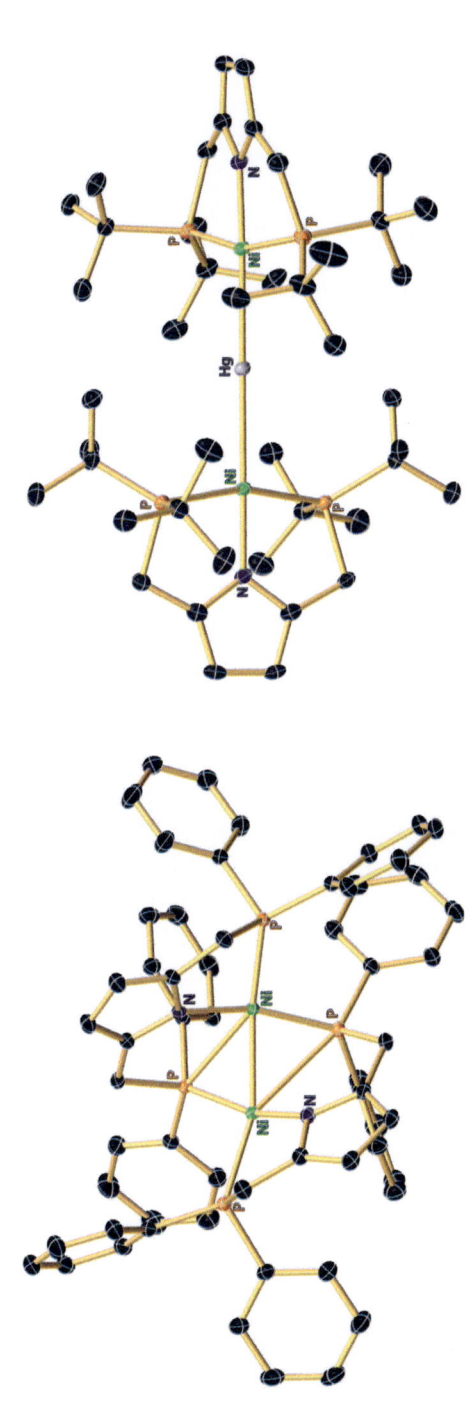

Fig. 4 Thermal ellipsoid depictions of the solid-state structures of Ni(I) compounds **135a** (left) and **138c** (right). *Images generated from crystallographic data in references Grüger N, Wadepohl H, Gade LH. A readily accessible PNP pincer ligand with a pyrrole backbone and its Ni$^{I/II}$ chemistry. Dalton Trans. 2012;41:14028–14030; Kreye M, Freytag M, Jones PG, et al. Homolytic H$_2$ cleavage by a mercury-bridged Ni(I) pincer complex [{(PNP)Ni}$_2${μ-Hg}]. Chem Commun. 2015;51:2946–2949.*

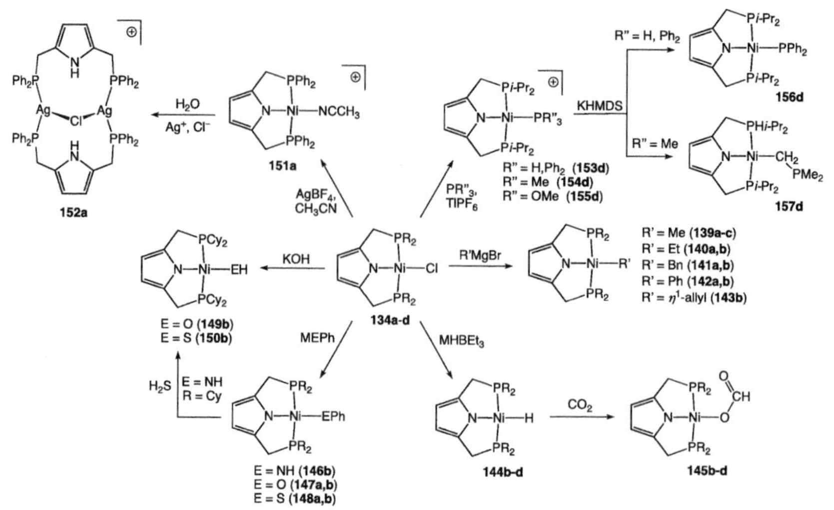

Scheme 41

In related work, Illuc and coworkers reported the preparation of several cationic phosphine derivatives (**153d–155d**) via thallium-assisted chloride abstraction from **134d**.[85] Treatment of **153d** with KHMDS produced the phosphide complex, **156d**, consistent with deprotonation of the secondary phosphine ligand. However, in the case of **154d**, reaction with KHMDS resulted in deprotonation of the methyl group of PMe$_3$ with concomitant rearrangement to a phosphino-alkyl ligand (**157d**, Scheme 41).

As discussed above, nickel precursors **134a–d** can be accessed in a straightforward manner through treatment of the proligands with base in the presence of divalent nickel salts. Discrete deprotonation of H(^{R}PNP) can also be accomplished to generate isolable alkali metal salts, which have found wide use in the coordination chemistry of other transition elements (vide infra). In an alternative approach, Iluc and coworkers have prepared the thallium(I) and silver(I) salts of proligand H($^{i\text{-Pr}}$PNP) and demonstrated their efficacy in transmetallation reactions with Ni, Pd, Pt, Ru, and Ir (Scheme 42).[82]

Outside of nickel chemistry, the metal that has received the most attention with the ^{R}PNP ligand system is iron. Iron compounds of ^{R}PNP were first introduced by Yoshizawa, Nishibayashi and coworkers in 2016 in the context of studies on catalytic nitrogen reduction.[86] The preparation of several new four-coordinate iron species was described beginning from the chloride precursor, **164c** (Scheme 43). Divalent iron compounds **164c–166c**

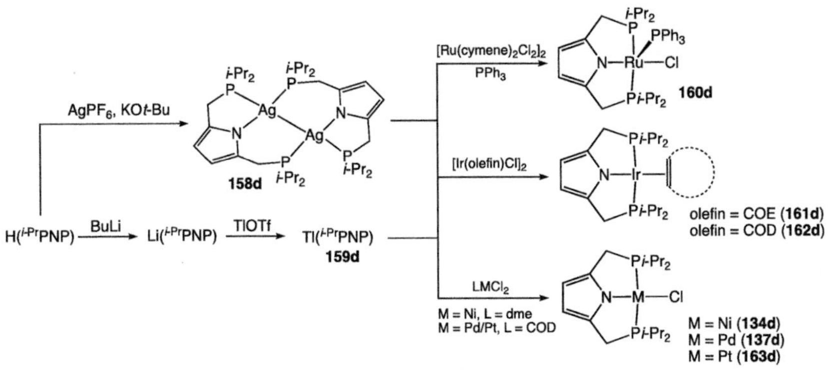

Scheme 42

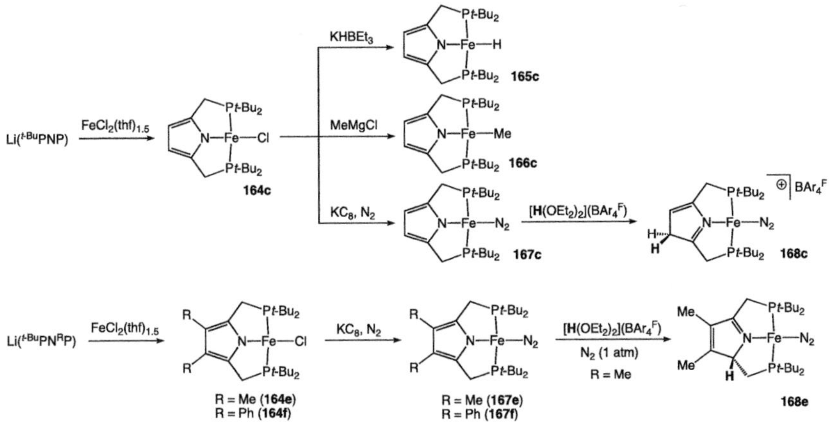

Scheme 43

were found to demonstrate intermediate-spin states ($S=1$), while the iron(I) species, **167c**, was determined to be low-spin ($S=½$). Examination of the compounds as catalysts for N_2 reduction in the presence of excess KC_8 and $[H(OEt_2)_2](BAr_4^F)$ found that the N_2-adduct, **167c**, performed best with up to 14 equivalents of ammonia observed under optimized conditions. Hydrazine was also found as a minor coproduct in the reaction suggesting the intermediacy of possible iron-hydrazine complexes. Mechanistic studies of the catalytic reaction demonstrated that protonation of **167c** at the 3-position of the pyrrolic carbon atom gives rise to the unproductive species, **168c**. Consequently, the authors posited that prior reduction of **167c** to an anionic N_2-adduct is likely the first step in the catalytic cycle. To address issues of protonation during catalytic N_2 reduction, follow-up studies by

Nishibayashi and coworkers examined protection of the pyrrolic backbone by substitution with methyl and phenyl groups (Scheme 43).[87] Indeed, the catalytic performance of methyl-substituted analog **167e** proved superior to that of **167c** generating up to 23 equivalents of ammonia. Treatment of **167e** with [H(OEt$_2$)$_2$](BAr$_4^F$) was still found to result in protonation of the pyrrole unit, but this time at the 2-position. The resulting compound, **168e**, was found to serve as a suitable precatalyst for N$_2$ reduction, which the authors proposed was a result of its facile transformation back to **167e** with concomitant formation of H$_2$ under the catalytic conditions.

Additional early work by Yoshizawa, Nishibayashi and coworkers also investigated a variation of the ^{R}PNP ligand containing an appended ferrocenyl-type unit.[88] Treatment of the lithium salts of ^{R}PNP (R=Ph, Cy, t-Bu) with [Cp*Fe(tmeda)] produced the new Cp*-derivatized ligands, **170a–c** (Scheme 44). In the case of the phenyl-substituted analog, an intermediate iron complex, **169a**, was isolated which was subsequently converted to the new ligand by treatment with MgCl$_2$ at slightly elevated temperatures. Ligands **170a,b** were then utilized to prepare a series of new transition metal complexes of Fe, Cr, and Mo (**171–175**). Dinitrogen adducts **175a,b** performed poorly in catalytic N$_2$ reduction trials, although both species were found to be competent catalysts for N$_2$ silylation in the presence of excess Na and Me$_3$SiCl.

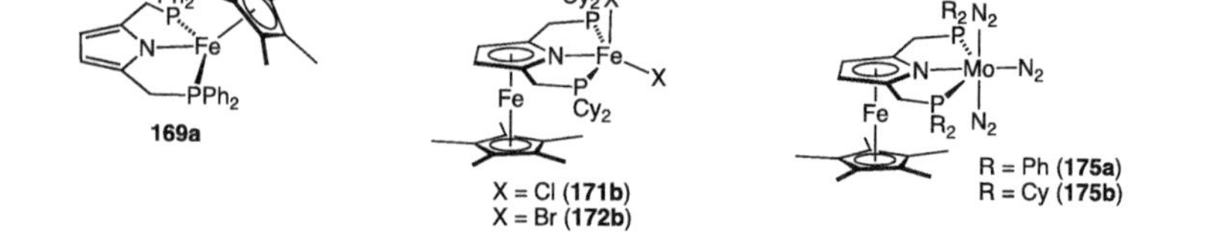

Scheme 44

Outside of nitrogen reduction, several other groups have examined the chemistry of iron complexes containing ^{R}PNP ligands. Also working with $^{t\text{-Bu}}$PNP, Walter and coworkers reported the synthesis and electronic structure of iron(II) complexes **176c–178c** (Chart 14).[89] The modified PNP ligand appearing in compound **178c** was produced from the reaction of

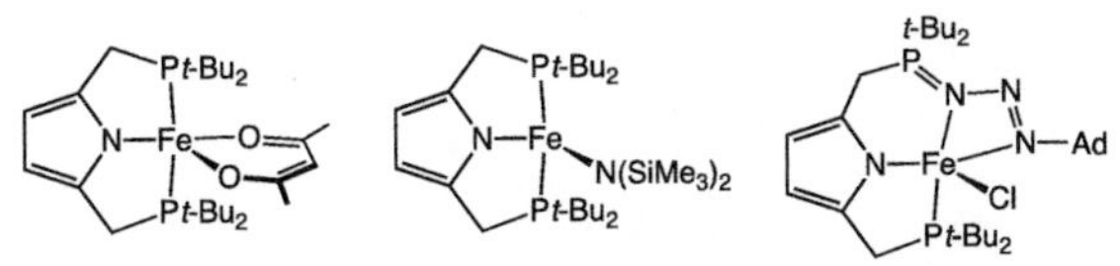

Chart 14 Iron complexes of $^{t\text{-Bu}}$PNP.

1-adamantylazide with **164c**. Notably, **176c–178c** all demonstrated high-spin ground states, which contrasts that found for compounds **164c–166c**.

Employing the ^{Cy}PNP ligand, Tonzetich and coworkers described the preparation of several iron compounds spanning three oxidation states (Scheme 45).[90] In contrast to findings with the $^{t\text{-Bu}}$PNP ligand, four-coordinate square-planar halide complexes of ^{Cy}PNP proved unstable, requiring the presence of additional ligands such as pyridine or CO. The high-spin pyridine adduct, **179b**, proved effective as a precursor to a variety of compounds including intermediate-spin alkyl species of the type [FeR(^{Cy}PNP)] (**181b–183b**). Treatment of **179b** with NaHBEt$_3$ resulted in production of the bimetallic bridging hydride **184b**, which displayed a reorganization of the ^{Cy}PNP ligand about the two iron centers (Fig. 5). By contrast, when the reaction of **179b** with NaHBEt$_3$ was conducted in the presence of CO, the expected six-coordinate monomeric hydride, **185b**, was generated. Reduction of **179b** with KC$_8$ in the presence of CO afforded

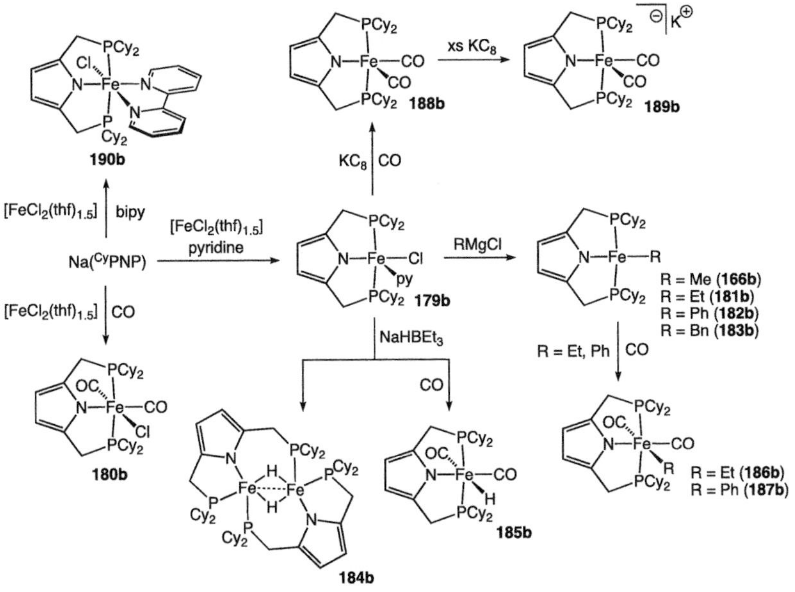

Scheme 45

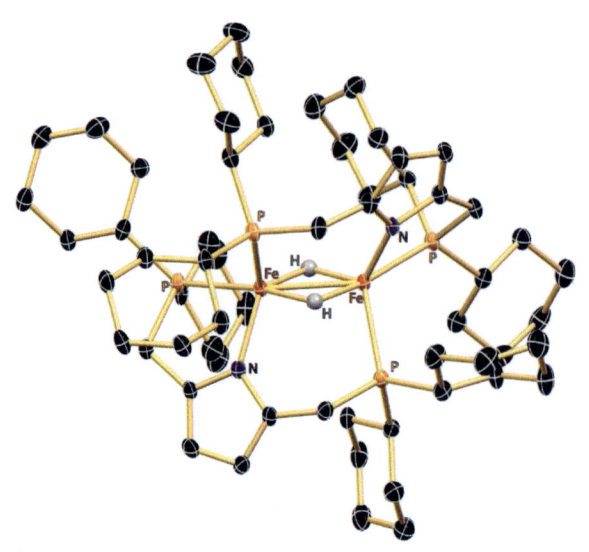

Fig. 5 Thermal ellipsoid drawing of the solid-state structure of **184b**. *Image generated from crystallographic data in reference Thompson CV, Arman HD, Tonzetich ZJ. A pyrrole-based pincer ligand permits access to three oxidation states of Iron in organometallic complexes. Organometallics. 2017;36:1795–1802.*

the dicarbonyl adduct, **188b**, which contains low-spin iron(I). The *t*-Bu analog (**188c**) was subsequently reported by Walter and coworkers in a study re-examining the electronic structure of compounds **164c** and **167c**.[91] Addition of excess KC$_8$ to **188b** produced the rare iron(0) complex, **189b**, which was found to be diamagnetic. Follow-up studies by Tonzetich and coworkers employing bipyridine produced a set of compounds akin to those containing CO ligands (e.g., **190b**).[92] Likewise, Iluc and coworkers reported several analogous compounds containing the $^{i\text{-Pr}}$PNP ligand.[93]

The catalytic potential of iron PNP compounds in reactions other than N$_2$ reduction has also received attention. Nishibayashi, Nakajima and coworkers investigated the catalytic activity of **164c**–**167c** for alkyne hydroboration.[94] Compound **165c** proved most effective demonstrating good yields and high selectivity for the *E*-selective hydroboration of terminal alkynes with pinacolborane (HBpin). Preliminary mechanistic investigations identified boryl complex **191c** from the reaction of **165c** with excess HBpin (Chart 15). Subsequent studies by the same group also demonstrated catalytic C—H borylation using a substituted version of ligand ^{Cy}PNP containing a central tetrahydroisoindolide unit.[95] Much of the chemistry of the isoindolide ligand mirrored that of ^{Cy}PNP shown in Scheme 45. However, borylation of benzene was found to be superior with the methyl complex of the isoindolide ligand (**192**, Chart 15) when compared to **166b**. Indeed **192** proved effective as a catalyst for the C—H borylation of several heteroarenes (sp^2) and anisole derivatives (sp^3) in the presence of pinacol diborane (B$_2$pin$_2$) at 100 °C. Furyl complex **193** was prepared in the course of the study supporting a role for a dual catalytic cycle involving shuttling between hydride and boryl intermediates.

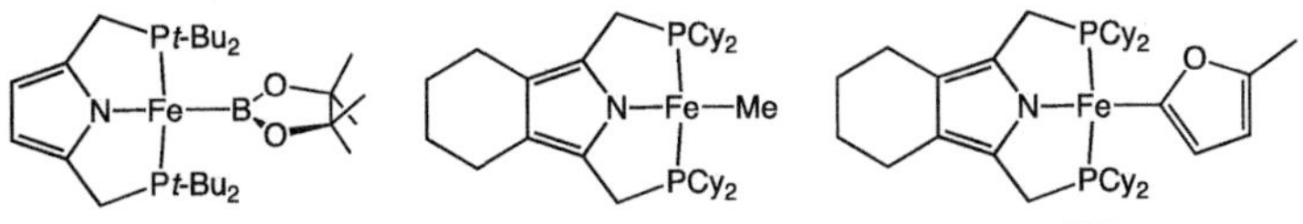

Chart 15 Iron complexes of ^{R}PNP relevant to borylation chemistry.

In work with implications for hydrosilylation chemistry, Tonzetich and coworkers recently described a series of silyl compounds generated by reaction of hydrosilanes with **166b** (Scheme 46).[96] These compounds were found to demonstrate intermediate-spin states, in line with other four-coordinate compounds of ^{Cy}PNP described above. In contrast to other such compounds, however, the silyl species were observed to reversibly bind molecular nitrogen giving rise to low-spin ($S=0$) five-coordinate adducts. Compounds such as **194b** were also found to undergo facile migratory insertion with both aldehydes and alkyne substrates pointing to a potential role for these species in hydrofunctionalization protocols involving silanes. One especially notable finding was the isolation of insertion product **196b**, which was rationalized as originating from tautomerization of phenylacetylene to give a vinylidene intermediate which underwent subsequent 1,1-insertion into the Fe—Si bond.

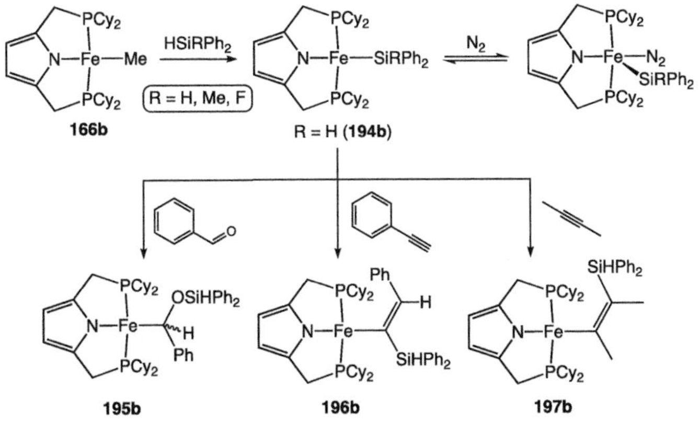

Scheme 46

In addition to iron, several other compounds of the mid-transition elements have been studied with the PNP ligand. Yoshizawa, Nishibayashi, and coworkers employed a series of four-coordinate cobalt(II) compounds bearing both the ^{Cy}PNP and $^{t-Bu}$PNP ligands for catalytic N$_2$ reduction under mild conditions (**198–201**, Scheme 47).[97] Dinitrogen adduct **199c** was found to produce up to four equivalents of NH$_3$ under optimized conditions, representing the first instance of N$_2$ reduction by a molecular cobalt catalyst. Subsequent work by Tonzetich and coworkers re-examined the synthesis of **201c** from a variety of hydride sources and demonstrated its reaction chemistry with a variety of small molecules leading to isolation of several new compounds including the rare hydrido-nitrosyl complex, **203c**.[98] Additional work by Tonzetich and coworkers has also investigated the aldehyde decarbonylation activity of compound **199b**.[99] Owing to the stabilizing nature of the ^{Cy}PNP ligand, examination of the decarbonylation pathway revealed spectroscopic signatures for several key intermediates including an aldehyde adduct. Unfortunately, the strong binding of CO to the reaction product, **202b**, prevented catalytic turnover leading to a purely stoichiometric decarbonylation process.

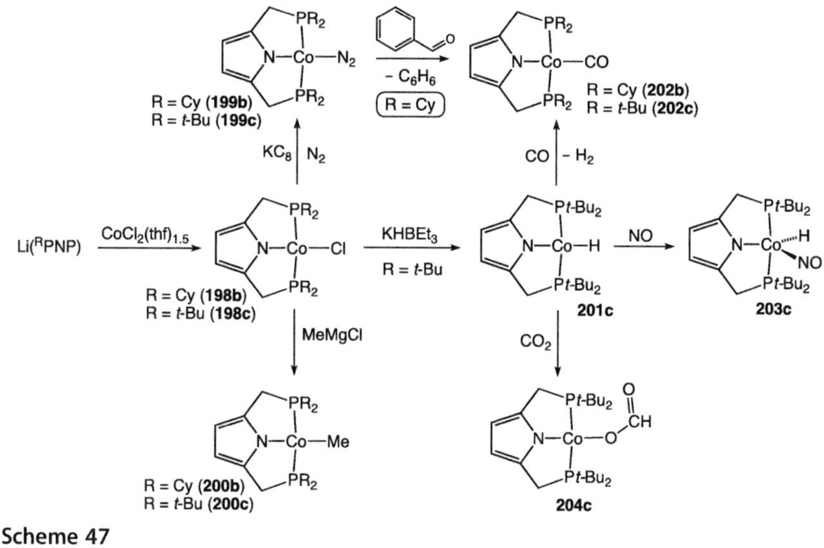

Scheme 47

The chemistry of cobalt's heavier congeners, rhodium and iridium, have also been the subject of investigation with ^{R}PNP ligands. Building upon the preliminary findings of Iluc described above, Yamashita and coworkers reported the synthesis and reactivity of a series of organometallic rhodium and iridium complexes bearing both $^{i\text{-Pr}}$PNP and $^{t\text{-Bu}}$PNP.[100] Entry into the chemistry with both metals was afforded by olefin complexes as shown in Scheme 48. Subsequent reactions with H$_2$ produced a series of hydride species that were found to differ depending upon the nature of the metal ion and supporting ligand employed. The differing structures of compounds **207–210** are especially notable in light of the divergent behavior observed for iron(II) hydrides **167c** and **184b**, and further serves to highlight the dramatic effect of phosphine substituents in these systems. Each of the rhodium and iridium compounds in Scheme 48 was capable of effecting the dehydrogenation of cyclooctane in the presence *tert*-butylethylene, but the catalytic reactions displayed low efficiency even at temperatures above 200 °C.

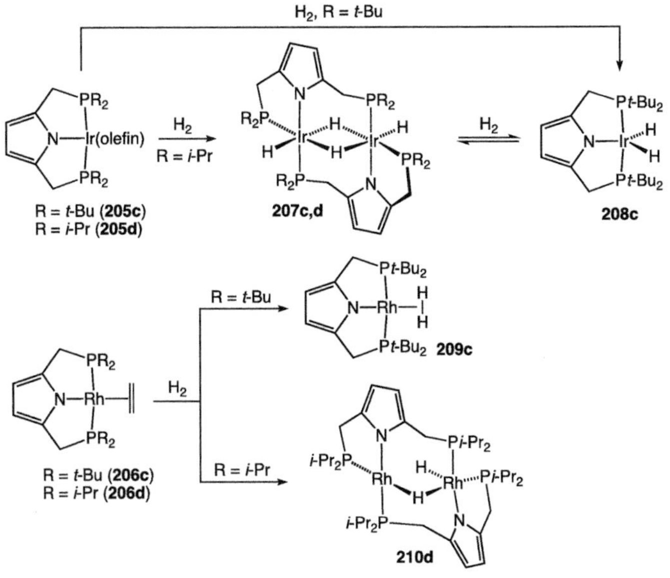

Scheme 48

Rhodium complexes of ^{R}PNP have likewise been examined as catalysts for the reduction of molecular nitrogen. Yoshizawa, Nishibayashi and coworkers reported compounds **211c–214c** (Chart 16) and tested their efficacy for the reduction of N$_2$ to N(SiMe$_3$)$_3$ in the presence of KC$_8$ and Me$_3$SiCl.[101] Each of the compounds along with Yamashita's ethylene adduct, **206c**, demonstrated formation of N(SiMe$_3$)$_3$ from N$_2$, with up to six equivalents observed for reactions initiated with **214c**. Employing **214c** at lower temperature (−40 °C) afforded up to 23 equivalents of the

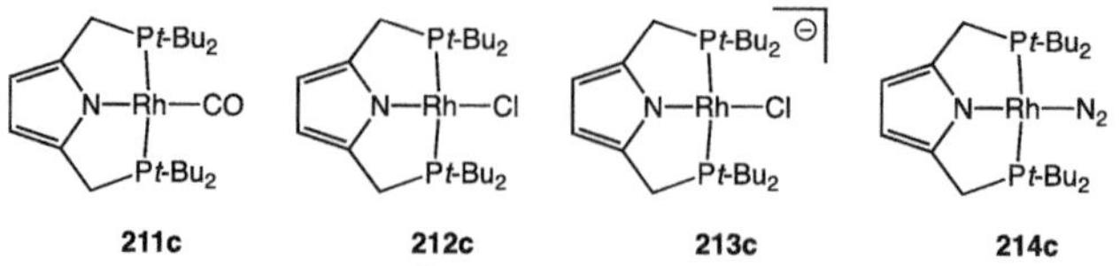

Chart 16 Rhodium complexes of $^{t\text{-}Bu}$PNP.

amine, providing the first instance of efficient N$_2$ reduction by a molecular rhodium catalyst under mild conditions.

The lone application of the ^{R}PNP ligand in Group 7 chemistry was reported by Tonzetich and coworkers and primarily involved a series of low-spin manganese(I) complexes (Scheme 49).[102] Halide compounds of manganese(II) proved inaccessible with either the ^{Cy}PNP or $^{t\text{-}Bu}$PNP ligands, although structural and spectroscopic data on the formally Mn(I) bipyridine adduct, **215c**, demonstrated that it is best regarded as containing a high-spin Mn(II) ion antiferromagnetically coupled to a bipy radical anion. In addition to the Mn(I) compounds, the study also reported the isolation and structural characterization of a rare molecular Mn(0) complex (**218c**, Scheme 49). Despite the unique nature of several of the complexes, none displayed any reactivity toward H$_2$, CO$_2$, or silanes diminishing the prospects for use of the manganese compounds in catalysis.

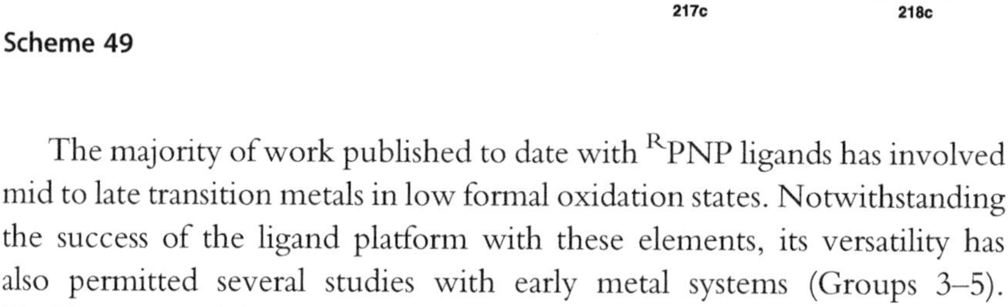

Scheme 49

The majority of work published to date with ^{R}PNP ligands has involved mid to late transition metals in low formal oxidation states. Notwithstanding the success of the ligand platform with these elements, its versatility has also permitted several studies with early metal systems (Groups 3–5). Yoshizawa, Nishibayashi and coworkers reported the synthesis of several $^{t\text{-}Bu}$PNP and $^{i\text{-}Pr}$PNP complexes of vanadium containing aryloxide ligands (Scheme 50).[103] These compounds were accessed through salt metathesis reactions of Li^{R}PNP with VCl$_3$(thf) followed by treatment with LiOAr. Reduction of **220c** with KC$_8$ in the presence and absence of N$_2$ produced the vanadium(II) compounds **221c** and **222c**, respectively. Inclusion of the bulky aryloxide ligand was found to be necessary to stabilize the N$_2$ adduct **221c** as similar reduction of **219c** was found to lead only to decomposition. Alternative treatment of **220c** with NaNH$_2$ afforded the primary amido complex, **223c**. Use of **220c–223c** as catalysts for N$_2$ reduction under conditions similar to those described for **167c** and **199c** resulted in production of up to 14 equivalents of NH$_3$ (**222c**) with small amounts of hydrazine also generated. Computational investigation of several potential stoichiometric steps in the reaction identified a super-reduced V(I) complex, [V(N$_2$)(OAr)($^{t\text{-}Bu}$PNP)]$^-$, as a likely intermediate akin to that proposed for the related iron complexes (vide supra). Subsequent work by Yoshizawa, Nishibayashi and coworkers also reported the N$_2$ reduction activity of a set of related Ti and Zr complexes containing cyclopentadienyl coligands (**225c–228c**, Scheme 50).[104] Unlike the vanadium analogs, however, the Ti and Zr compounds were only capable of generating NH$_3$ in stoichiometric quantities in the presence of KC$_8$ and [H(OEt$_2$)$_2$](BAr$^F_4$).

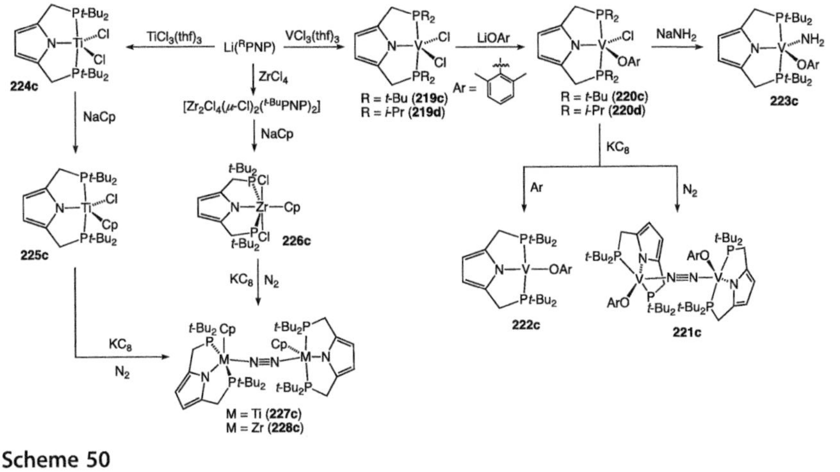

Scheme 50

In the area of Group 3 chemistry, work by Tilley, Anderson and coworkers has described the preparation and reactivity of a series of Sc and Y complexes of the ^{Cy}PNP and $^{t\text{-}Bu}$PNP ligands.[105,106] Entry into Group 3 chemistry can occur through transmetallation reactions of Li(^{R}PNP) or by direct protonolysis with M(CH$_2$EMe$_3$)$_3$(thf)$_2$ precursors (Scheme 51). The resulting scandium neopentyl compounds **230b,c** were found to undergo several subsequent bond activation reactions. Treatment of **230b** with H$_2$ generated a transient dihydride, which could be trapped by 1-hexene to give the dihexyl species **233b**. Both **230b** and **230c** were found to activate N—H and C—H bonds giving rise to a series of new compounds with concomitant loss of neopentane.[105] Thermolysis of **230b** over extended time periods was also found to result in loss of isobutylene and neopentane and formation of a putative bridging methylene species in low yield (**236b**).

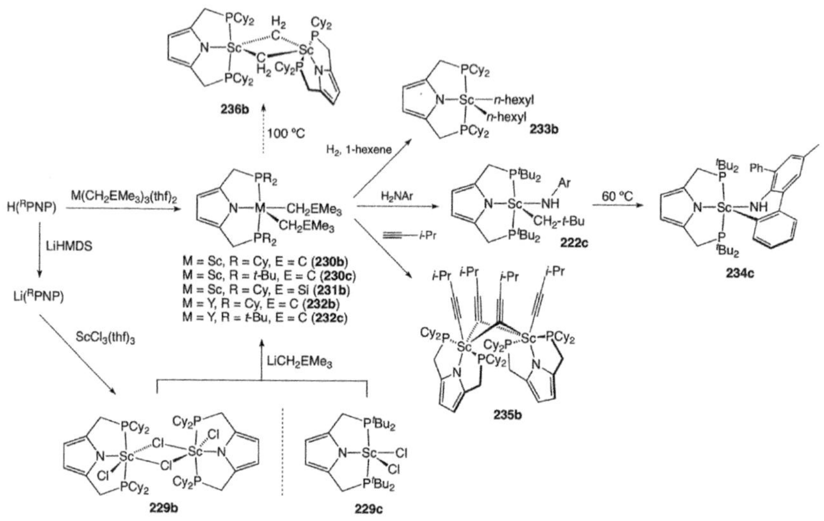

Scheme 51

An additional aspect of the chemistry of the ^{R}PNP ligand involves its dehydrogenation chemistry. Tonzetich and coworkers demonstrated that treatment of either **134c**, **198c**, or **217c** with benzoquinone leads to formal H$_2$ elimination from the backbone of the ligand and formation of a dehydrogenated variant (Scheme 52).[102,107] This procedure is similar to that reported by Schneider and coworkers using an aliphatic diphosphino-amine ligand,[108] and related chemistry was observed earlier by Gade in studies of Pyrrmebox Pd complexes (see Section 6).[64] Ghorai and Mani have also reported analogous oxidation of the H(^{Me}BPP) proligand, although the resulting compound failed to behave as a pincer ligand.[109]

Pincer ligands incorporating pyrrolyl units

Scheme 52

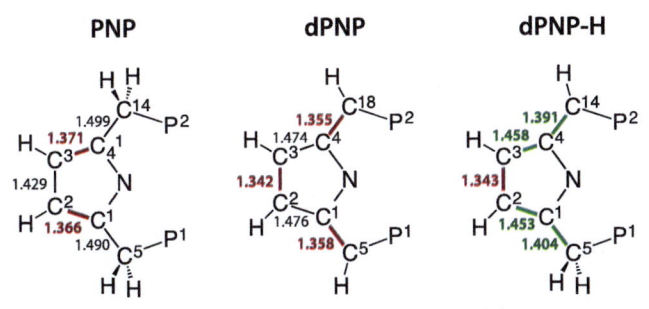

Fig. 6 Comparison of PNP bond metrics as found in the solid-state structures of manganese carbonyl compounds containing PNP, dehydro-PNP (dPNP), and protonated dehydro-PNP (dPNP-H) ligands. Due to a positional disorder of the cation of **243c**, the bond metrics (green) represent an average of formal single and double bonds. Metric data obtained from references Krishnan VM, Davis I, Baker TM, et al. Backbone dehydrogenation in pyrrole-based pincer ligands. Inorg Chem. 2018;57:9544–9553; Narro AL, Arman HD, Tonzetich ZJ. Manganese chemistry of anionic pyrrole-based pincer ligands. Organometallics. 2019;38:1741–1749.

As expected, the dehydrogenated variants **237c–239c** demonstrated metric parameters in line with the formal realignment of pi-bonding within the ligand backbone (Fig. 6). Most surprisingly, however, electrochemical measurements of **238c** demonstrated that dehydrogenation (formal oxidation) of the ligand backbone resulted in a cathodic shift of the metal redox potentials. This apparently contradictory result was rationalized by recognizing that pyrrole dehydrogenation disrupts the aromaticity of the ring resulting in "liberation" of the nitrogen lone pairs for bonding with the metal. Indeed, the more reducing nature of the dehydrogenated ligand permitted isolation of a rare four-coordinate Co(III) species, **240c**, and led to observation of the novel ligand-reduced complex, **242c** (Scheme 52). Using the manganese variant, **237c**, protonation of the methine arm was observed with [H(OEt$_2$)$_2$](BAr$_4^F$) leading to compound **243c**, which represents formal elimination of a hydride equivalent from **217c**.

As a final chapter in the chemistry of ^{R}PNP ligands, we note the report of the phosphaferrocene ligand, ^{Fc}PNP, by Mathey and coworkers (Chart 17).[110] This ligand was introduced the same year as ^{Ph}PNP and its complexation to Rh(I) was described in the form of the compound [Rh(CO)(^{Fc}PNP)]. Unlike the other ^{R}PNP derivatives described above, ^{Fc}PNP exists in both a *rac* and *meso* form by virtue of the planar-chiral nature of the phosphaferrocene units. Both isomers of the corresponding Rh compound were isolated and structural characterization of the *meso* form evinced the expected square-planar geometry about the metal center. Use of the Rh compound for the hydroformylation of several internal alkenes demonstrated modest activity and a preference for the isomerized terminal aldehyde was found for the case of α-pinene. Despite the intriguing nature of this ligand, subsequent studies of its chemistry with other metals have not appeared.

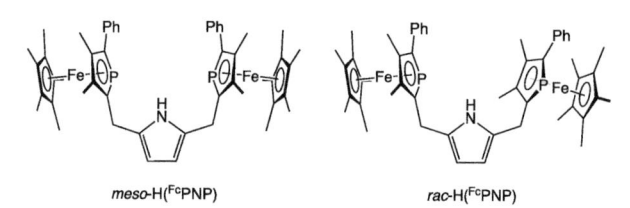

meso-H(^{Fc}PNP)   *rac*-H(^{Fc}PNP)

Chart 17 Phosphaferrocene-substituted PNP ligands.

8. Pyridinedipyrrolyl pincers (^{R}PDP$^{R'}$)

Dianionic pincer ligands of the pyridinedipyrrolyl family, ^{R}PDP$^{R'}$, were first introduced by Caulton and coworkers in 2014 in the form of proligand H$_2$($^{t-Bu}$PDP^{t-Bu}).[111] Subsequent work by several other groups reported the related ligands displayed in Chart 18. Unlike other pincer systems discussed in prior sections, the pyrrolyl units of ^{R}PDP$^{R'}$ are assembled on the ligand framework via two successive ring forming reactions and do not originate with pyrrole itself.

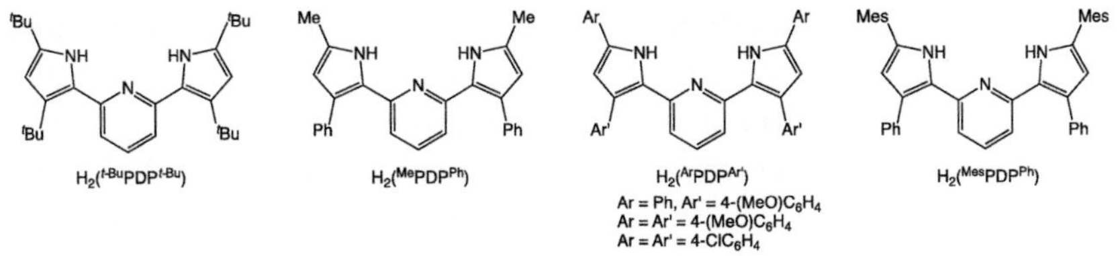

Pincer ligands incorporating pyrrolyl units

Chart 18 Pyridinedipyrrole proligands.

Initial work with $^{t\text{-Bu}}\text{PDP}^{t\text{-Bu}}$ demonstrated its metalation to Pd and Pt to afford benzonitrile adducts **244** and **245** (Scheme 53). Compounds **244** and **245** each demonstrated three anodic events by cyclic voltammetry. In line with the similarity of these events for both metals and the electronic-rich nature of the ligand, the anode processes were assigned as ligand-based oxidations. Later work by Dash and coworkers described similar Pd compounds of the type [Pd(NCCH$_3$)(^{Ar}PDP$^{Ar'}$)].[112] These compounds displayed analogous structures and electrochemical properties to **244** and were also demonstrated to serve as precatalysts for Suzuki-Miyaura cross-coupling reactions of aryl bromides in aqueous alcohol mixtures.

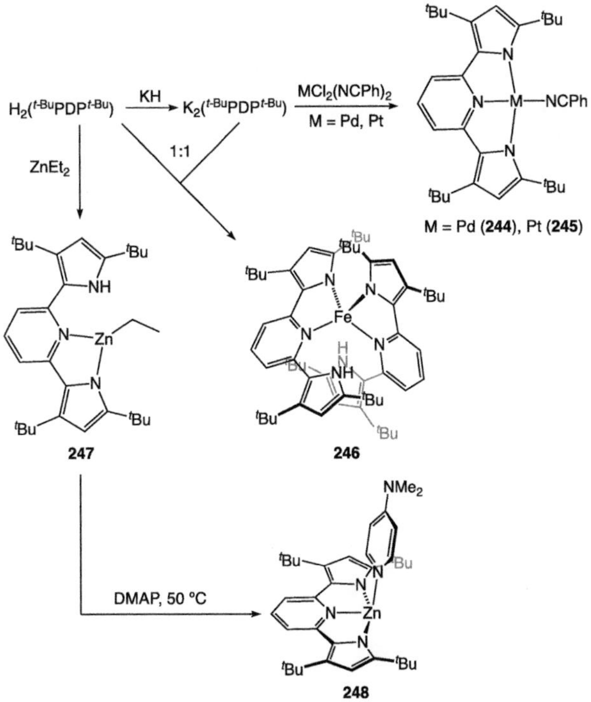

Scheme 53

In addition to Pd and Pt, the initial report of $^{t\text{-Bu}}\text{PDP}^{t\text{-Bu}}$ also disclosed the synthesis of Fe and Zn compounds **246** and **247**. In both instances, the pincer was found to retain a single pyrrolic proton, binding in a bidentate manner to the metal centers. Further treatment of **247** with DMAP and heat, however, generated complex **248**, which contains the desired dianionic form of the ligand. The solid-state structure of **248** demonstrates and unusual *cis*-divacant octahedral geometry about Zn with the DMAP residing perpendicular to the NNN-plane of the pincer (Fig. 7). This unique geometry was ascribed to the steric pressure exerted by the *t*-Bu groups which disfavors additional coordination in the square plane of the complex.

Further work with iron employing $^{t\text{-Bu}}\text{PDP}^{t\text{-Bu}}$ succeeded in generating a 1:1 metal:ligand complex through a protonolysis route employing $Fe(N[SiMe_3]_2)_2$.[113] The resulting iron(II) diethyl ether adduct, **249**, was subsequently found to undergo both one- and two-electron oxidation reactions to generate complexes **250** and **251**, respectively (Scheme 54). Compounds **249–251** all display the *cis*-divalent octahedral geometry observed for **248**, with high-spin ground states observed for both **249** and **250**. By contrast, **251** demonstrated a low-spin ($S=0$) ground state, which was corroborated by both Mössbauer spectroscopy and DFT calculations. Moreover, both the unique *cis*-divacant octahedral geometry and $S=0$ ground state of **251** were reproduced computationally in models where the *t*-Bu groups were replaced by hydrogen atoms. The structure of **251** is therefore most likely a consequence of electronic factors as opposed steric interactions with the *t*-Bu groups.

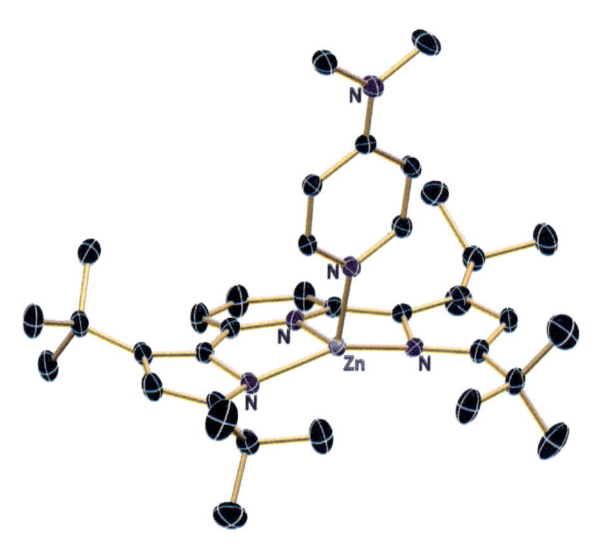

Fig. 7 Solid-state structure of **248** displaying the *cis*-divacant geometry about Zn. *Image generated from crystallographic data appearing in reference Komine N, Buell RW, Chen C-H, Hui AK, Pink M, Caulton KG. Probing the steric and electronic characteristics of a new bispyrrolide pincer ligand.* Inorg Chem. *2014;53:1361–1369.*

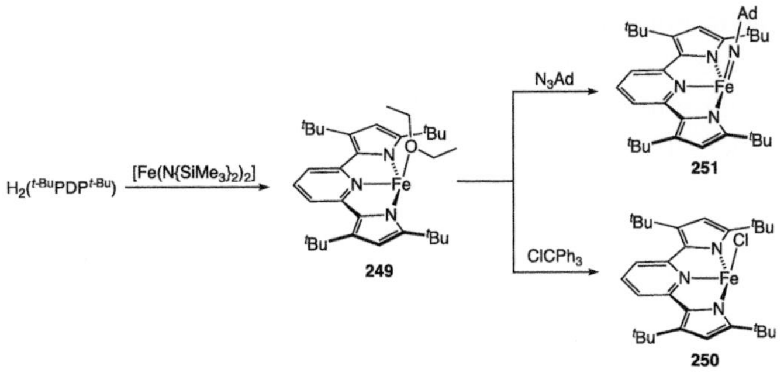

Scheme 54

More recently, work by Milsmann and coworkers described a series of related iron complexes baring the ^{Mes}PDP^{Ph} ligand.[114] Metalation of H(^{Mes}PDP^{Ph}) with [Fe(N{SiMe$_3$}$_2$)$_2$] produced compounds **252** and **253** in similar fashion to **249** (Scheme 55). As a result of the lower steric profile of ^{Mes}PDP^{Ph} versus $^{t\text{-}Bu}$PDP$^{t\text{-}Bu}$, however, both **252** and **253** were found to be square-planar as opposed to adopting a *cis*-divacant octahedral geometry akin to **249**. Five-coordinate square-pyramidal species **254** and **255** were also accessible with ^{Mes}PDP^{Ph} further highlighting the differences in reactivity resulting from the subtle change in sterics. Compounds **252** and **253** are of particular interest because they represent rare examples of neutral, high-spin ($S=2$) iron(II) compounds in square-planar geometries. As already discussed with $^{t\text{-}Bu}$PNP and ^{Cy}PNP (Section 7), square-planar iron(II) compounds typically give rise to intermediate-spin ($S=1$) ground states as a consequence of strong sigma-bonding interactions in the *xy* plane. In the case of **252** and **253**, stabilization of the $S=2$ state was primarily ascribed to a geometric distortion engendered by the smaller bight angle of the ^{Mes}PDP^{Ph} pincer ligand, which moves the pyrrolide nitrogen atoms off axis thereby weakening the sigma-

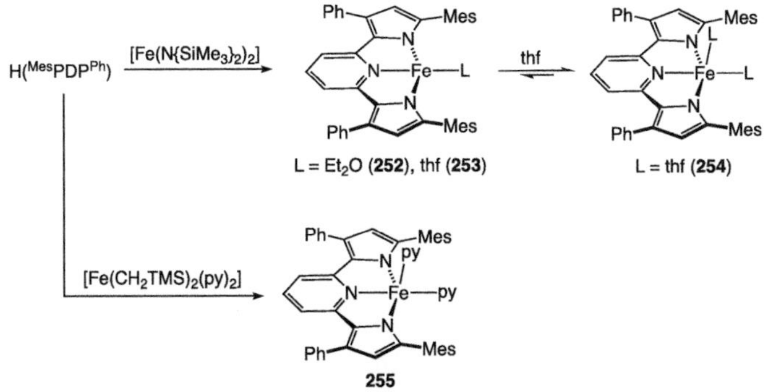

Scheme 55

bonding interactions in the plane. As a result, the σ^* orbital of $d_{x^2-y^2}$ parentage is stabilized more so than in a rigorously D_{4h} ligand field and can better accommodate a high-spin state.

Much like **249**, compound **253** was found to react readily with organic azides to generate species indicative of iron imide formation.[115] In contrast to **251**, however, the putative imido species of $^{Mes}PDP^{Ph}$ proved much more reactive leading to several different amination products depending upon the azide employed (Scheme 56). The diverse reactivity of imido species generated from **253** especially when compared to **251** prompted further theoretical studies. DFT calculations revealed several electronic states for the putative imido species of similar energy consistent with the possibility for two state reactivity. In all cases, however, the lowest energy configuration for [Fe(NR)($^{Mes}PDP^{Ph}$)] was found to be a planar species featuring an intermediate-spin ($S=^3/_2$) iron(III) center antiferromagnetically coupled to an iminyl radical (·NR). Such a description differs markedly from the *cis*-divalent octahedral geometry and singlet ground state found for imido **251** providing a strong rationale for the disparity in reactivity.

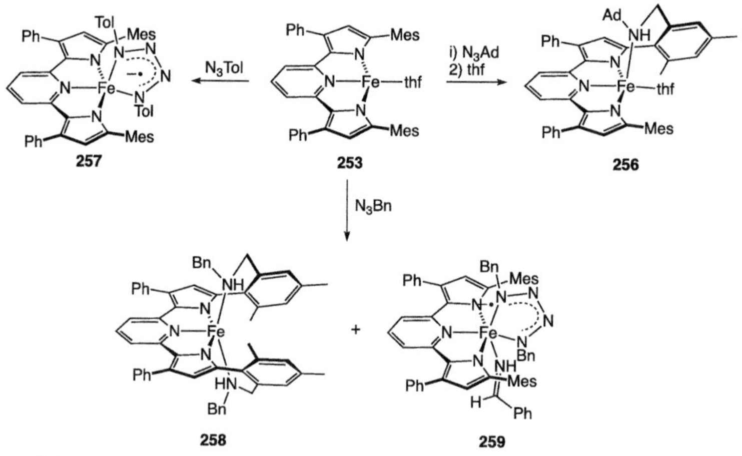

Scheme 56

Imido-type reactivity has also been described for cobalt compounds of the $^{t-Bu}PDP^{t-Bu}$ ligand.[116] Mindiola and coworkers reported Co(II) species **260**, which was prepared through both salt metathesis and protonolysis routes akin to iron species discussed above (Scheme 57). In line with the structure of **249**, compound **260** was also found to adopt a *cis*-divacant octahedral geometry. Treatment of **260** with N₃Ad led to formation of a rare azide adduct, **261**, which demonstrates coordination through the γ-N atom. Photolysis of **261** in benzene generated amination product

262 in similar fashion to the formation of iron analog **256**. One notable feature of **262** is its distorted square-planar geometry, which sets it apart from all other four-coordinate compounds of $^{t\text{-Bu}}\text{PDP}^{t\text{-Bu}}$.

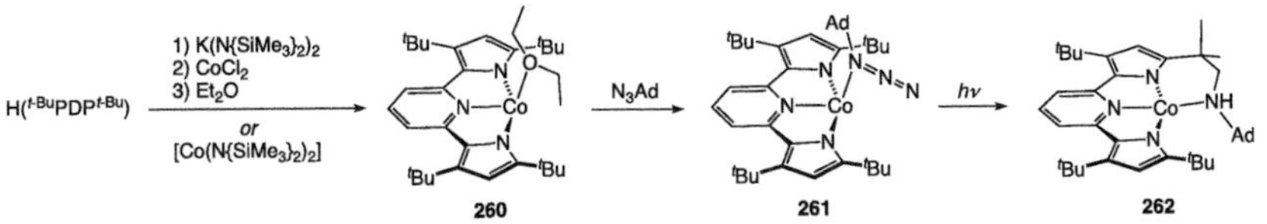

Scheme 57

Additional group transfer chemistry has also been observed for **249** in reactions with elemental sulfur.[117] Meyers, Mindiola, and coworkers reported the preparation of the dimeric iron species, **263**, which features a thiol-functionalized pyrrole moiety (Scheme 58). Activation of the pyrrole unit in **263** proved to be reversible such that treatment with methylenetriphenylphosphorane resulted in *S*-atom transfer and regeneration of the parent ligand in the form of adduct **264**. Synthesis of **264** was also accomplished in an orthogonal manner by treatment of the phosphorane adduct, **265**, with S_8. Mössbauer spectroscopy and magnetic measurements on compounds **263**–**265** demonstrated that each species possesses a high-spin iron(II) center. Accordingly, the geometries of **264** and **265** were found to display the same *cis*-divacant octahedral geometry as other iron(II) compounds of the $^{t\text{-Bu}}\text{PDP}^{t\text{-Bu}}$ ligand.

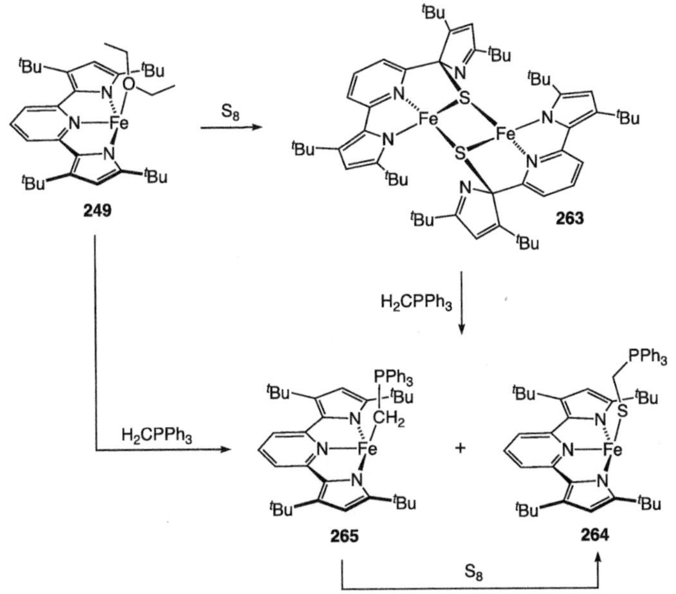

Scheme 58

Replacement of one aryl or *t*-Bu group on the PDP framework with a methyl substituent permits isolation of a series of homoleptic compounds of the type [M(^{Me}PDP^{Ph})$_2$]$^{0/-}$. Work by Milsmann and coworkers described the preparation of several Ti and Zr compounds of ^{Me}PDP^{Ph} (266–270, Scheme 59).[118] Titanium compound 267 and 268 did not display any notable photophysical properties, but the zirconium species, 266, proved to be both highly colored and photoluminescent with a long-lived emission maximum at 594 nm. Electrochemical measurements were used to estimate an excited state reduction potential for 266 of −0.07 V (vs Fc$^{0/+}$) and the compound was demonstrated to serve as a photoredox catalyst in several preliminary reactions.

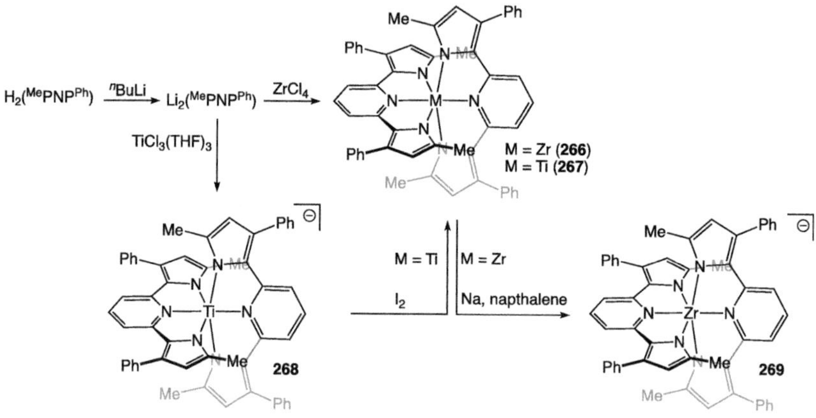

Scheme 59

A follow up study on the photophysical properties of 266 further elaborated upon the nature of the excited state dynamics in the compound.[119] Time-resolved emission spectroscopy identified an exceptionally long excited state lifetime for 266 of 325 μs. DFT calculations supported a mixed ^{3}IL/^{3}LMCT excited state as primarily responsible for the observed photoluminescence. The spin-forbidden nature of this triplet excited state is consistent with the long lifetime, which exceeds that of most known transition metal photosensitizers. Quenching studies of photoexcited 266 established that it serves as a photooxidant leading to formation of the one-electron reduced species, 269. This compound could also be generated independently by reduction with sodium naphthalenide in THF (Scheme 59).

Compound **269** was shown to serve as an intermediate in the photocatalytic homocoupling of benzyl bromide mediated by **266**, thereby establishing its efficacy in single-electron transfer events relevant to photoredox catalysis. Related work by Milsmann and coworkers also described the synthesis of compounds **270** and **271** by treatment of the proligands with ZrBn$_4$ (Scheme 60).[120] Both species demonstrated strong luminescence and long-lived excited states akin to **266**. Compound **271** was likewise active toward the photocatalytic homocoupling of benzyl bromide. By contrast, **270** failed to catalyze the homocoupling reaction due to photoinduced C—C reductive elimination to give **272**.

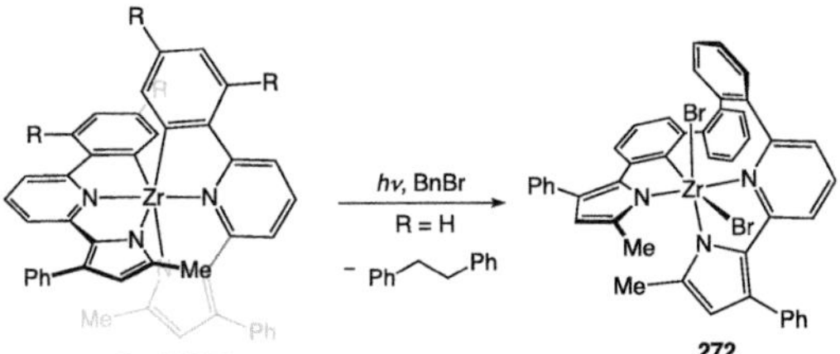

Scheme 60

Additional examples of homoleptic complexes featuring $^{Me}PDP^{Ph}$ were reported by Milsmann and coworkers in the area of Cr and Mo chemistry. Deprotonation/metalation of H$_2$($^{Me}PDP^{Ph}$) with MCl$_3$(thf)$_3$ (M = Cr, Mo) afforded the anionic complexes **273** and **274** (Scheme 61).[121] Each complex demonstrated rich electrochemistry, with multiple reversible electrode processes observed in tetrahydrofuran. Chemical oxidation of **274** afforded the neutral compound, **275**, whereas reduction of both **273** and **274** with one or two equivalents of sodium naphthalenide or potassium graphite generated polyanionic species **276–279** (Scheme 61). Extensive characterization of the redox series by spectroscopic, structural, and theoretical means supported a metal-based oxidation in the case of **275**, but a ligand-based reduction in the case of **276–279**., DFT calculations revealed that the additional electrons were localized on one (**276,277**) or both (**278,279**) of the pyridine moieties of the pincer ligands in the reduced complexes. Thus, as with several of the pincers discussed in this review, the $^{R}PDP^{R'}$ family of ligands is similarly capable of displaying redox non-innocence.

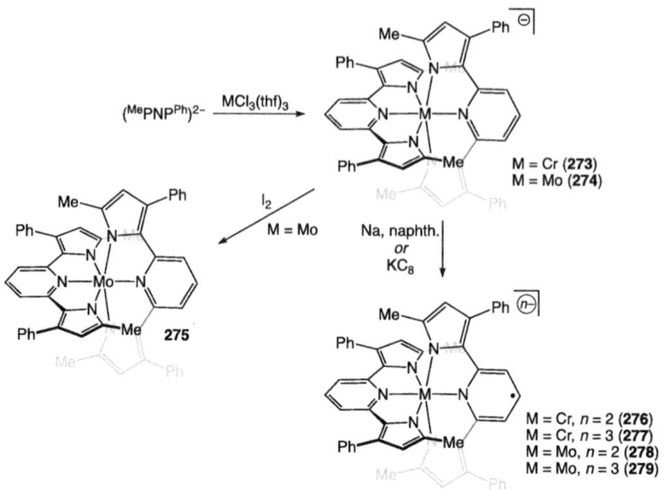

Scheme 61

The related dipyrrolylamine pincer ligand shown in Scheme 62 and its associated copper(II) compounds **280–282** were reported by Mani and coworkers in 2019.[122] Unlike ^{R}PDP$^{R'}$, the more flexible dipyrrolylamine ligand can bind in both a *fac*- and *mer*-fashion depending upon the bulk and hydrogen bonding interactions of the ancillary ligands. Square-pyramidal geometries were preferred in most instances, but enhanced steric encumbrance about the copper center was found to lead to the trigonal bipyramidal species **282**.

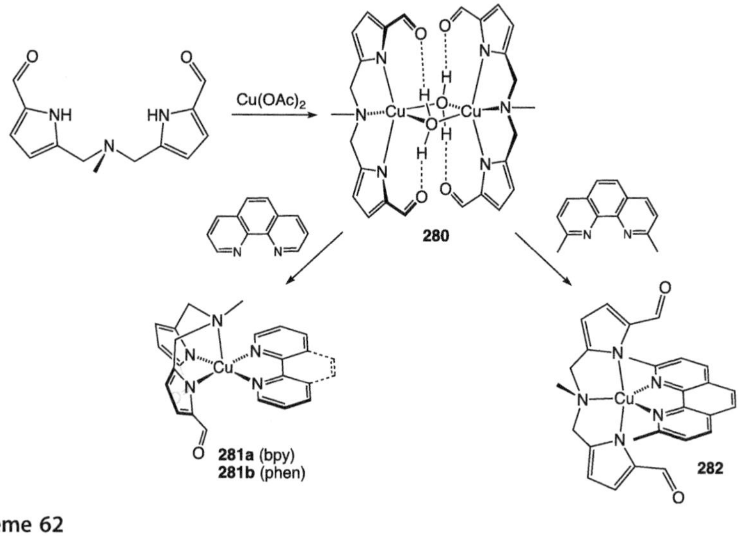

Scheme 62

9. Other pyrrolyl-containing pincers

Outside of the pincer systems already described in this review, a number of other pyrrolyl-containing ligands have been successfully synthesized and applied in transition metal chemistry. Although many of these ligands have not been employed as widely as others discussed in prior sections, their unique attributes contribute to the rich tapestry that is pyrrole-based pincer chemistry. In several instances, these pincers also demonstrate notable parallels with the systems considered above.

The pyrrolidine diester proligand shown in Chart 19 was first introduced by Booker and Li in 2011.[123] Formation of the ligand occurred via tautomerization of the corresponding pyrrole diester in similar fashion to the Pyrrmebox ligands discussed in Section 6. Successful introduction of the ligand to both Cu and Pd was achieved by in situ deprotonation with stoichiometric amounts of Et$_3$N in the presence of the appropriate MX$_2$ precursor (Scheme 63). The synthesis of Pd complex **285** was additionally found to require AgBF$_4$ to generate the [PdCl(CH$_3$CN)$_3$]$^+$ cation prior to metalation. UV–Vis spectroscopy of the Cu(II) complexes **283a,b** and **284** displayed two characteristic absorbances in the visible region corresponding to ligand field transitions of the square-planar d^9 copper centers. Two strong absorbances were also observed in the near-ultraviolet and are believed to originate from metal-ligand charge transfers and/or π-π* transitions of the conjugated ligand. Both **283a** and **284** were characterized crystallographically and found to demonstrate rigorously square-planar coordination geometries. Further interrogation of the bond metrics revealed significant delocalization about the ligand backbone when compared to the free proligand.

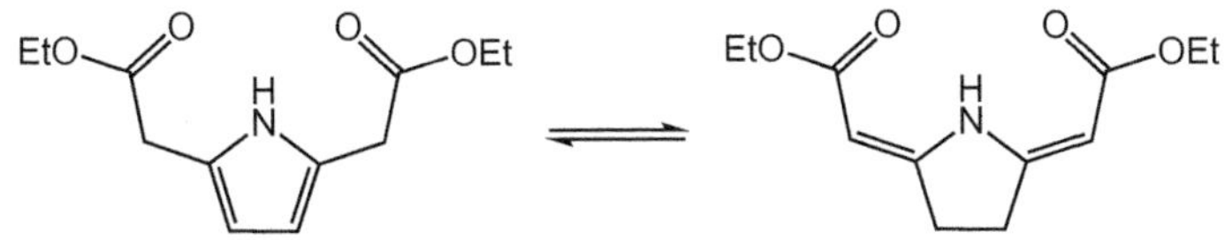

Chart 19 Pyrrole/pyrrolidine diester proligand.

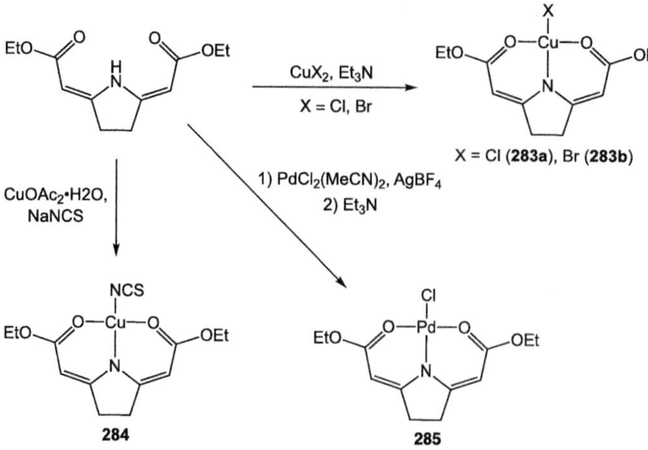

Scheme 63

The alternative ONO ligand based on a pyrrolyldiphenolate motif displayed in Scheme 64 was reported by Veige and coworkers in 2013.[124] The fully deprotonated form of the ligand is trianionic so studies have therefore focused on higher oxidation state metals. Initial reactions of the ligand with the tungsten alkylidyne precursor $(t\text{-BuO})_3\text{W}{\equiv}\text{C-}t\text{-Bu}$ produced the square-pyramidal alkylidene compound **286**, whereas further treatment with the mild phosphorane base, H_2CPPh_3, afforded the anionic alkylidyne **287** (Scheme 64). Both **286** and **287** were characterized crystallographically and single-point DFT calculations on **287** revealed that the nitrogen lone pair of the pyrrolyl unit does not contribute significantly to the HOMO of the molecule. Instead, the HOMO is comprised of the π bonds of the pyrrolyl group, which renders the ligand backbone susceptible to electrophilic attack. Evidence to support this hypothesis was generated experimentally by ligand alkylation and subsequent decomposition of **287** when treated with methyl triflate.

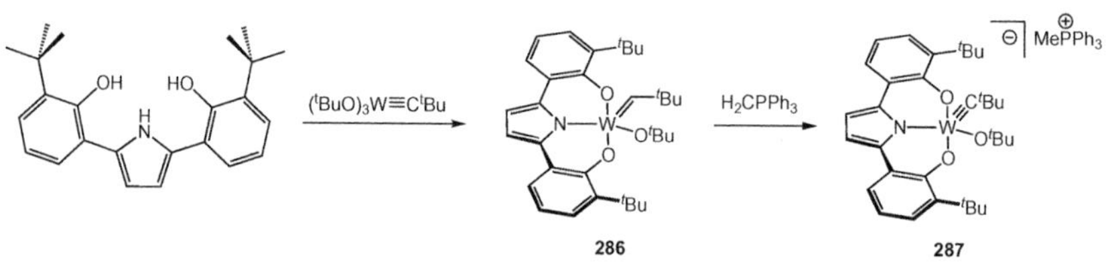

Scheme 64

Further investigations by Veige examined the proton transfer events of the ONO ligand with Group 4 metals.[125] Metalation of the ligand with TiCl$_4$(thf)$_2$ afforded the distorted octahedral Ti(IV) complex **288a**, which features a dianionic form of the ligand generated by proton transfer from N to C within the pyrrolyl unit (Scheme 65). Deprotonation of **288a** by 2,6,-lutidine produced **289**, in which the ligand returns to its normal trianionic form. Beginning from TiCl$_4$ instead of the thf adduct generated a mixture of **288b** and **290** in an 87:13 ratio. Compound **290** is supported by a monoanionic form of the ligand in which both pyrrolic backbone carbons are now protonated. The addition of [H(OEt$_2$)$_2$](BAr$_4^F$) to the mixture of **288b** and **290** led to formation of the single cationic product, **291** (Scheme 65). Alternatively, reactions of the ONO ligand with ZrBn$_4$ produced the homoleptic compound **292** featuring two dianionic ligands. Treatment of **292** with acid generated **293**, which possesses one singly protonated and one doubly protonated ligand backbone. Attempts to create a species with

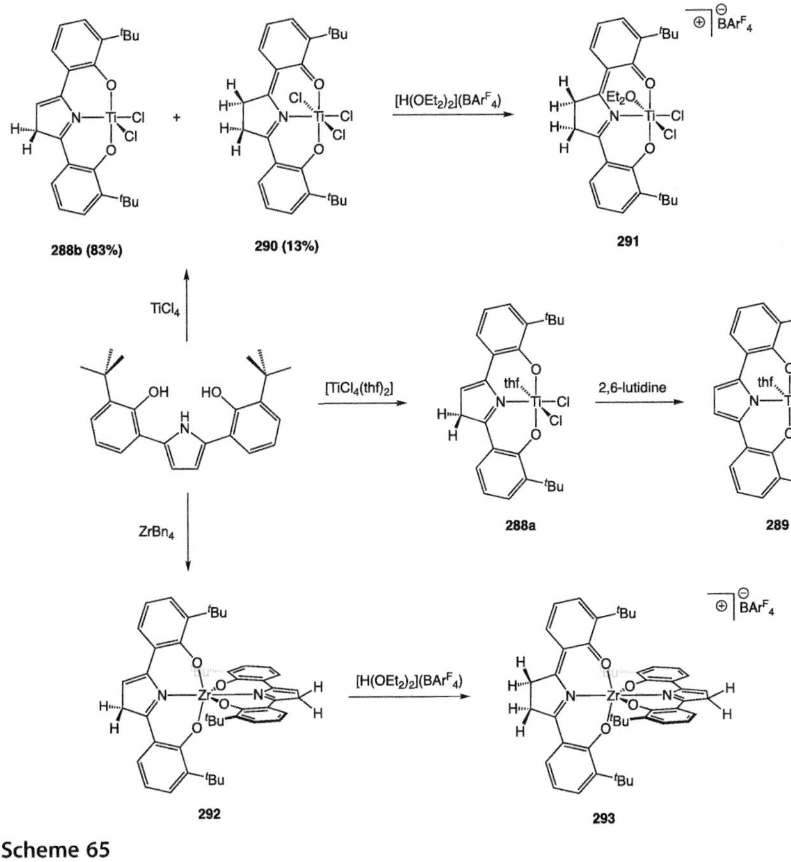

Scheme 65

two doubly protonated ONO ligands was unsuccessful, even in the presence of excess acid.

In a novel variation on the *NNN* platform, Hayes and coworkers designed the *bis*-phosphinimine(pyrrolyl) ligand shown in Scheme 66, which bears resemblance to the diiminopyrrolyl pincers (^RDIP) discussed in Section 3.[126,127] Synthesis of the ligand was carried out by lithium halogen exchange on the *N*-Boc protected 2,5-dibromopyrrole followed by addition of chlorodiphenylphosphine. Subsequent addition of *para*-isopropylphenylazide generated the proligand in moderate yields. Owing to the larger atomic radius of phosphorus versus carbon, the *bis*-phosphinimine(pyrrolyl) ligand is better able to accommodate tridentate coordination than ^RDIP derivatives.

Reactions of the phosphinimine proligand with lanthanide organometallic reagents afforded dialkyl complexes **294–297** via alkane elimination (Scheme 66) Compounds **294–297** did not suffer from intramolecular cyclometallation, which limited the utility of similar lanthanide complexes supported by a carbazole variant of the ligand. Analogous reactions with Sm reagents did not produce compounds akin to **294–297** but instead led to a species, **298**, in which cyclometallation of an *N*-aryl group has occurred (Scheme 66). This compound was only found to be stable at low temperatures and in the solid state. Over time **298** converted to **299** via metalation of one to the phenyl groups of the PPh$_2$ moiety. The addition of a second equivalent of the phosphinimine ligand to **298** can obviate formation of **299** in favor of the *bis*-ligated species, **300**, via a second alkane elimination event. Attempts to form elusive amido and imido complexes from **298** with primary amines were unsuccessful but further reactivity studies are ongoing.

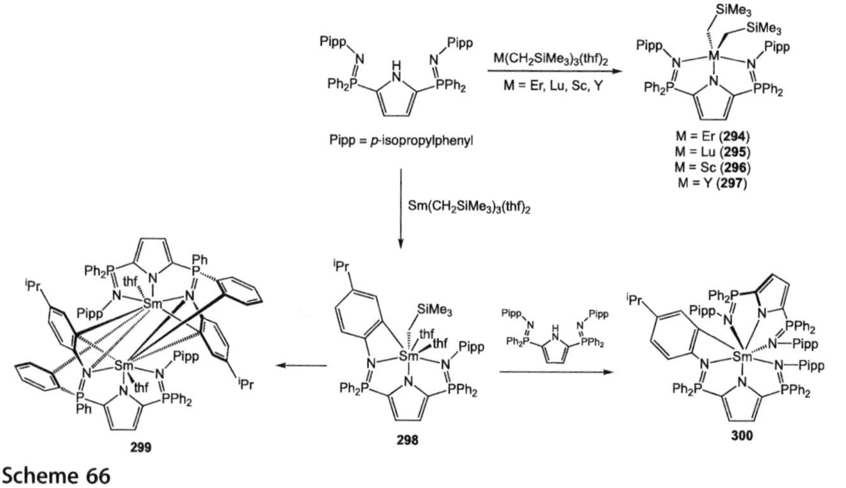

Scheme 66

Hayes and coworkers subsequently expanded the chemistry of the *bis*-phosphinimine(pyrrolyl) ligand to include rhodium.[128] Reactions between the sodium salt of the ligand and [RhCl(COE)$_2$]$_2$ at 50 °C generated the desired Rh(I) complex, **301**, which is best described as having a distorted square-planar geometry. Displacement of cyclooctene by a variety of other π-ligands such as ethylene, diphenylacetylene, and *trans*-stilbene afforded complexes **302–304**, respectively (Scheme 67). Interestingly, the addition of H$_2$ to **303** cleanly formed the stilbene complex **304**. These stoichiometric results suggested the possibility that **301** could serve as an effective hydrogenation catalyst. Accordingly, the addition of 1 atm of H$_2$ to **301** at ambient temperatures was examined and found to produce the new Rh complex **305** (Scheme 67). The ^{1}H NMR spectrum of **305** displayed an exceptionally upfield-shifted doublet of doublets resonance at −35.84 ppm ($^1J_{HRh(1)}$ = $^1J_{HRh(2)}$ = 19.8 Hz) corresponding to a highly shielded hydride interacting with two Rh centers. Compound **305** demonstrated limited reactivity with olefins except for ethylene, where it was observed to generate **302** in near quantitative yields. The addition of cyclooctene to **302** reforms **301** effectively closing a conceptual catalytic cycle for COE hydrogenation. Compound **305** therefore appears to be a resting state for alkene hydrogenation by **301**. Notably, catalytic reactions with the system were actually found to display an improvement in turnover frequency when compared to Wilkinson's catalyst.

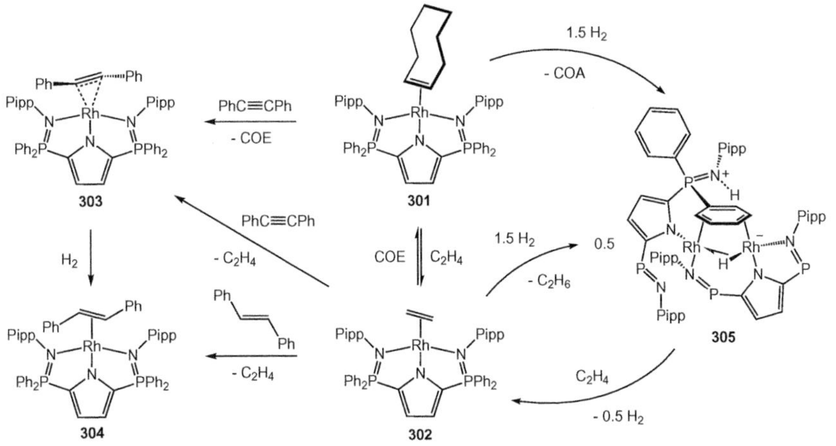

Scheme 67

Switching from an olefinic rhodium precursor to [Rh$_2$(μ-Cl)(CO)$_2$]$_2$ was found to give a 95:5 mixture of the Rh(I) carbonyl species **306** and **307** (Scheme 68).[129] Conversion of **307** to **306** can be achieved by applying a vacuum to a stirred solution of **307** at slightly elevated temperatures. Compound **306** has been proven to be an effective platform for the study of the activation and deoxygenative metathesis of CO upon treatment with Lewis acids. Addition of B(C$_6$F$_5$)$_3$ to **306** produced the Rh μ-CO borane adduct **308**. The nature of **308** closely resembles a borane-stabilized formyl ligand. Titration with thf showed the carbonyl O—B bond to be weak. Consequently, **308** was found to gradually decompose at ambient temperatures overtime. Heating **308** at 50 °C led to complete decomposition and formation of three products, **307**, **310**, and the borane-capped isocyanide (Scheme 68). The formation of these products was proposed to occur by initial dissociation of B(C$_6$F$_5$)$_3$ from **308** followed by metathesis of the C≡O bond to give 4-isopropyl-phenylisocyanide and putative **309**, which contains a phosphine oxide arm. Compound **309** is then believed to react rapidly with another equivalent of **308** to produce the dicarbonyl phosphine oxide compound **310**, an equivalent of **307**, and the isocyanide-borane adduct. It appears that minimal influence comes from the metal center, and the metathesis is likely accomplished through some cooperative action between the phosphinimine and the Lewis acid.

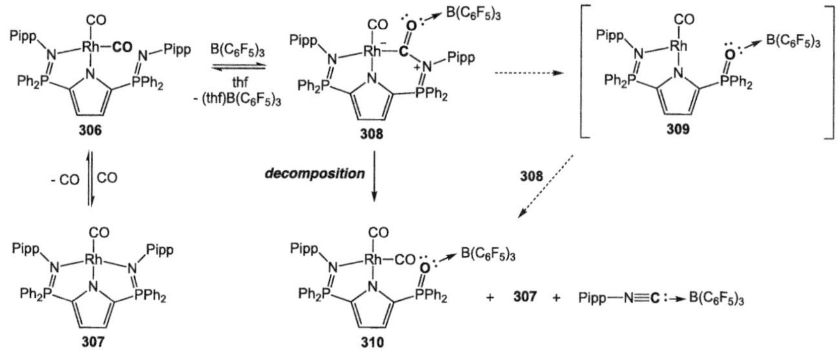

Scheme 68

In a final application of the *bis*-phosphinimine(pyrrolyl) ligand, Hayes and coworkers investigated the synthesis and stability of rhodium silylene complexes (Scheme 69).[130] Reaction of **301** with PhSiH$_3$, Ph$_2$SiH$_2$, and Ph$_3$SiH generated the Rh(III) silyl hydride species **311a-c** with concomitant loss of cyclooctene. Subsequent hydride abstraction reactions of **311a,b** with B(C$_6$F$_5$)$_3$ were attempted but failed to produce the desired Rh silylene cation as judged by NMR. Addition of excess CO gas to **311** triggered reductive elimination of hydrosilane and formation of **306**. The reactivity of hydrosilanes was similarly examined with the carbonyl complex **307** given the differential behavior of the ligand, vis-à-vis phosphinimine dissociation, in the presence of CO. Indeed, these reactions afforded new Rh carbonyl silylene complexes **312a** and **312b** plus an equivalent of molecular hydrogen (Scheme 69). Crossover experiments with a 1:1 mixture of Ph$_2$SiD$_2$ and Ph$_2$SiH$_2$ showed no evidence for production of H-D confirming that silane is the source of H$_2$ generated in the reaction. A pathway to compounds **312a,b** was proposed to involve oxidative addition of silane followed by deprotonation of the Rh—H unit to generate a Zwitterionic intermediate. Subsequent α-elimination within the Zwitterion leads to a Rh silylene hydride, which can combine with the protonated phosphinimine arm to release H$_2$. Migratory insertion of the silylene into the resulting Rh—N bond forms the final product although other mechanisms are possible.

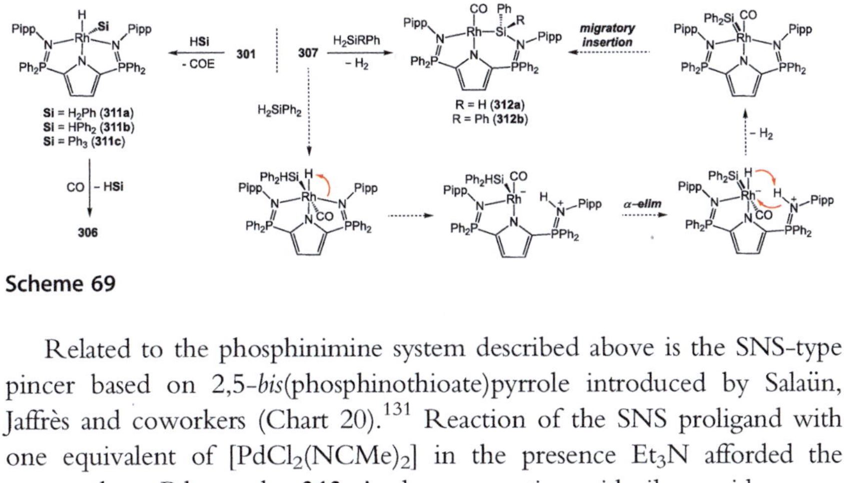

Scheme 69

Related to the phosphinimine system described above is the SNS-type pincer based on 2,5-*bis*(phosphinothioate)pyrrole introduced by Salaün, Jaffrès and coworkers (Chart 20).[131] Reaction of the SNS proligand with one equivalent of [PdCl$_2$(NCMe)$_2$] in the presence Et$_3$N afforded the square-planar Pd complex **313**. Analogous reaction with silver oxide generated the tetranuclear species, **314**. The X-ray crystal structure of **314** revealed that all four Ag atoms align in an almost linear zig-zag arrangement (Fig. 8).

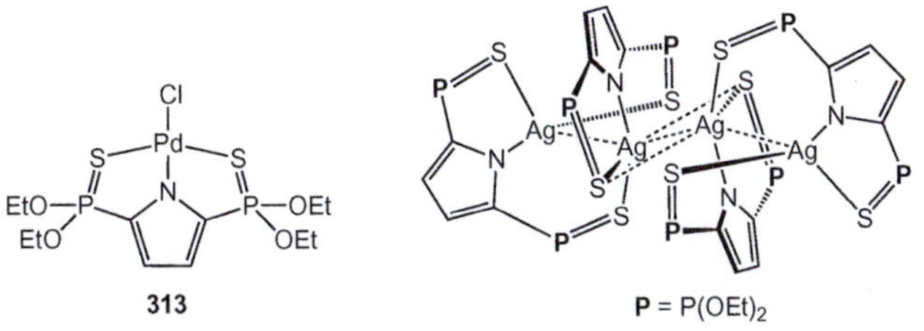

Chart 20 Pd and Ag complexes of an SNS pincer ligand.

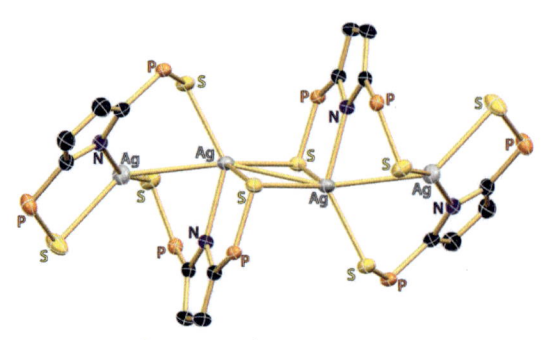

Fig. 8 Solid-state structure of compound **314**. Ethoxy groups on phosphorus omitted for clarity. *Image generated form crystallographic data in reference Fraix A, Lutz M, Spek AL, et al. Construction of a monoanionic S,N,S-pincer ligand with a pyrrole core by sequential [1,2] phospho-fries rearrangement. characterization of palladium and silver coordination complexes. Dalton Trans 2010;39:2942–2946.*

The distance between the central Ag atoms of 3.0903 Å is shorter than the sum of their van der Waals radii indicating a weak bonding interaction. Accordingly, **314** is better regarded as a dimer of bimetallic Ag complexes. Such a formulation was corroborated by both multinuclear NMR and mass spectrometry.

Continuing with the theme of heteroatom-containing arms, Pramanik and coworkers described the synthesis of the *bis*(arylazo)pyrrolyl pincer displayed in Scheme 70.[132] Much like the ^{R}DIP ligands discussed in Section 3, the wide bite angle created by the azo unit was found to prohibit binding of the ligand in the anticipated κ^3-*N* fashion. However, reaction of the proligand with RhCl(PPh$_3$)$_3$ in the presence of air led to oxygenation of one of the aryl groups and formation of the *NNO* Rh(III) complex **315**. Electrochemical measurements on compound **315** identified a reversible anodic process near +0.9 V (vs Ag/AgCl). Chemical oxidation of **315** with ceric ammonium nitrate produced cation **316**, whose EPR spectrum was consistent with a ligand-based radical. In agreement with this finding, DFT calculations on **316** revealed significant unpaired spin density on the phenoxy and pyrrolyl moieties.

Scheme 70

Gale, Ogden and coworkers have reported a unique class of pincers based on 2,5-dicarboxamido- and 2,5-dithioamidopyrroles.[133] The coordination chemistry of these pincers with Ni, Co, and Cu was found to result in variable binding of the donor arms depending upon the ligand employed (Scheme 71). By in large, homoleptic 2:1 complexes were isolated, although in the case of Cu(II), square-planar species **321a** and **321b** could be generated.[134] Interestingly, the thioamide ligand containing N-phenyl substituents was found to undergo oxidative cyclization in the presence of copper to afford benzothiazole species **322**.

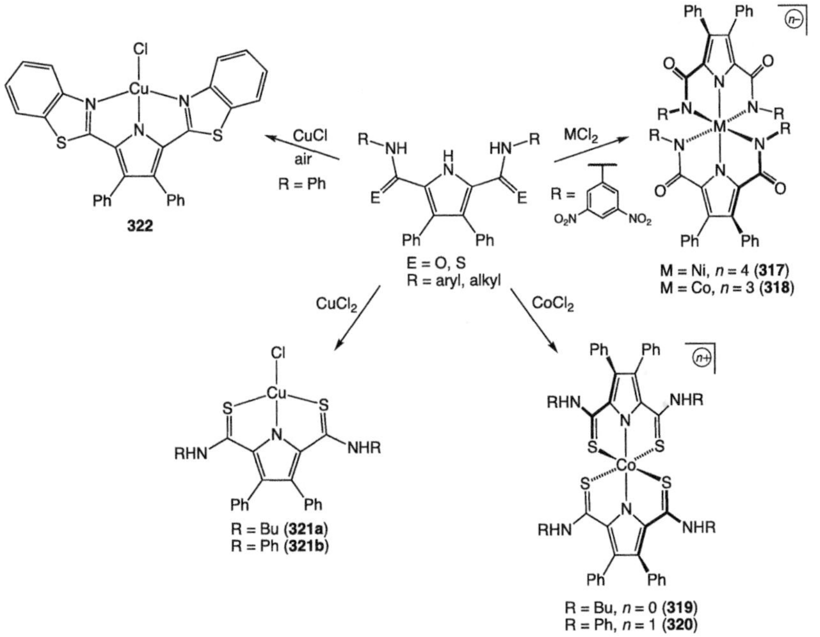

Scheme 71

The final class of pincer ligands to be discussed are those derived from pyrrolyl-pyridylimines, first introduced by Floriani and coworkers in 2000 (Chart 21).[135] The NNN ligands can be prepared in straightforward fashion by simple Schiff base reactions between pyrrole-2-carboxaldehyde and an alkylamino pyridine. Unlike the majority of pincer ligands discussed in this review these ligands feature a more flexible backbone, which has played a prominent role in their chemistry.

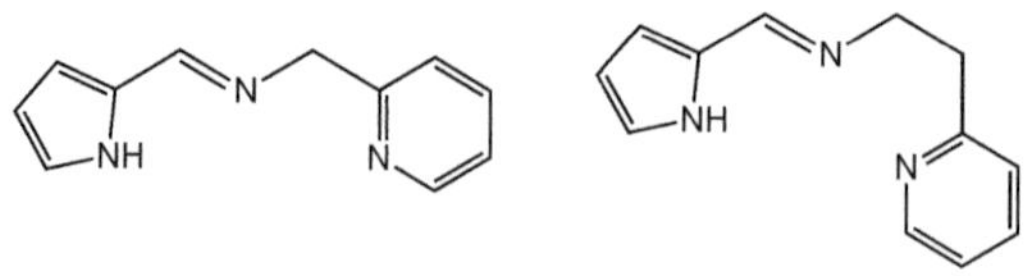

Chart 21 Pyrrole-pyridylimine proligands.

In their initial report, Floriani and coworkers utilized the lithium salt of the ligand to synthesize the series of Ru complexes displayed in Scheme 72. Salt metathesis with [Ru$_2$Cl$_2$(μ-Cl)$_2$(COD)$_2$] produced the octahedral Ru(II) complex **323**, which was used to generate several other compounds by simple displacement reactions. The organometallic compound **326** was found to decompose overtime at room temperature to a novel bimetallic species, **327**, in which the methylene unit of the pincer backbone has undergone C—H activation.

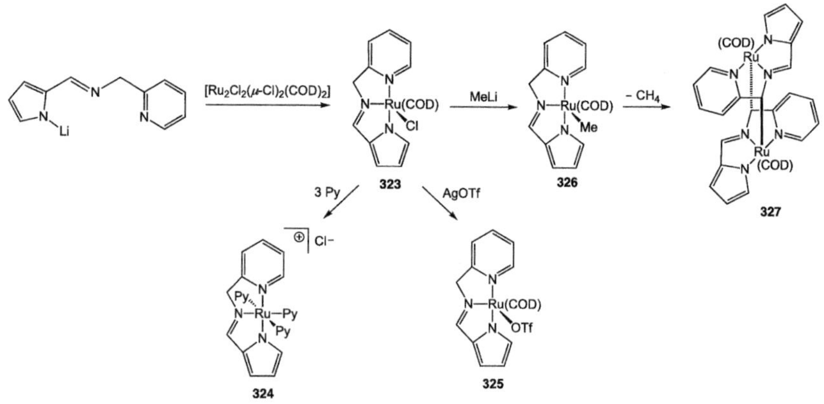

Scheme 72

The research groups of Miller and Brooker have extended the chemistry of the pyrrolyl-pyridylimine pincers to copper. Miller and coworkers reported the synthesis of the asymmetric bimetallic Cu(I) compound **328** (Scheme 73A).[136] Crystallographic analysis of **328** coupled with DFT calculations determined that the most accurate representation of the compound is the amido-imino form **328a** with important resonance contributions from the *bis*-imino form **328b**. Brooker and coworkers prepared copper(II) halide complexes **329a,b** by addition of the proligand to solutions of CuX$_2$ in the presence of Et$_3$N (Scheme 73B).[137] Similar addition of proligand in the absence of Et$_3$N led to formation of **331a,b**. In the solid state, **331a,b** aggregate as dimers stabilized by hydrogen bonding with the pyrrolic and iminyl protons. Introduction of thiocyanate ion following a procedure akin to that used to produce **329a,b** generated the bimetallic complex **330**. The structure of **330** features two distorted square-pyramidal Cu(II) centers bridged by two thiocyanates in a 1,3 fashion. The room temperature magnetic moment of **330** indicated little to no magnetic coupling between the two cupric ions consistent with the three-atom bridge.

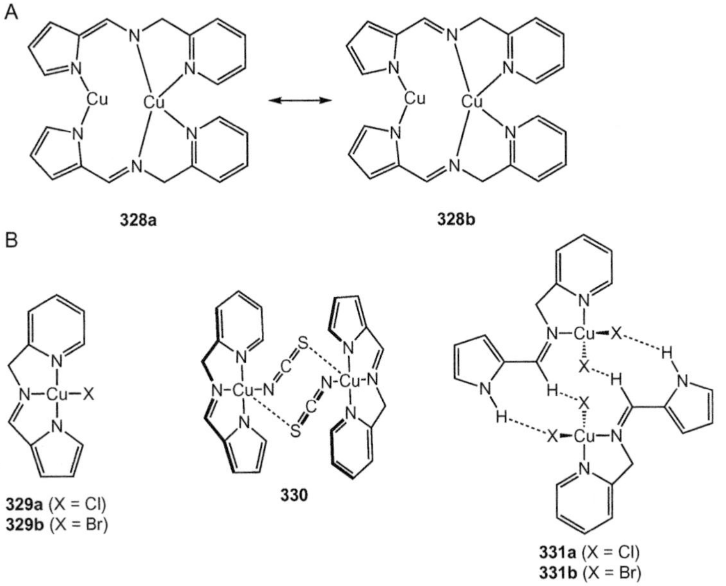

Scheme 73

Similar synthetic methods were employed by Brooker to synthesize Cu(II) compounds supported by the related pyrrolyl-pyridylimine ligand containing an ethylene spacer in the backbone (Chart 21).[137,138] The substitution of a methylene linker by ethylene eliminates the proclivity to form monometallic square-planar compounds resulting in the isolation of only polymetallic species. For example, reaction of the proligand with the appropriate copper(II) salt in the presence of base generated the bimetallic compounds **332a–c** and **334** (Chart 22). The order of addition was found to be critical for the synthesis of **332b**. CuBr$_2$ must be added dropwise to a solution of the proligand and Et$_3$N in order to obtain analytically pure material. If CuBr$_2$ is used in excess, bromination at the 2-position of the pyrrole is observed leading to the coordination polymer **333**. The geometry of the copper centers in compounds **332–334** is best regarded as distorted square-pyramidal similar to that in **330**. Interestingly, **332c** was found to exhibit two μ-N$_3$ anions binding in an end-on or 1,1 fashion to both Cu centers as opposed to the 1,3-bridging mode found for the thiocyanate analog **334**. Accordingly, variable temperature magnetic susceptibility measurements for **332c** demonstrated weak antiferromagnetic coupling as expected for a single atom bridge.

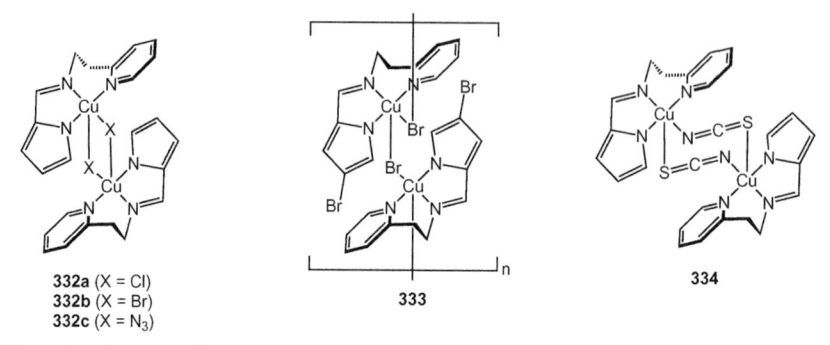

Chart 22 Multimetallic copper(II) complexes of the pyrrolyl-pyridylimine ligand.

In a further demonstration of the chelating properties of the pyrrolyl-pyridylimine ligand Brooker and coworkers disclosed the interesting mixed valent Co(II)/Co(III) salt **335** (Chart 23).[139] The complex was prepared by both stoichiometric and non-stoichiometric means from a combination of proligand, cobalt(II) acetate, sodium thiocyanate, and 2-formylpyrrole. The Co(III) centers of the cations are low-spin as expected for octahedral geometry whereas the Co(II) center of the anion is high-spin. Accordingly, the magnetic moment of the salt was found to be 4.4 μ_B, consistent with the presence of a single paramagnetic Co(II) ion.

Chart 23 Mixed valent Co(II)/Co(III) salt **335**.

The chemistry of the pyrrolyl-pyridylimine pincer was expanded to aluminum by Huang and coworkers in 2014.[140] Dropwise addition of proligand to AlMe$_3$ cleanly generated the trigonal bipyramidal Al complex **336** in high yields (Scheme 74). Reaction of a second equivalent of AlMe$_3$ formed the asymmetric bimetallic complex **337**. Compound **336** also reacted with one or two equivalents of 2,6-diisopropylphenol to produce phenolate complexes **338** and **339**, respectively. Related protonolysis chemistry with dibenzoylmethane afforded the octahedral Al complex **340**. The monometallic Al complexes were tested for potential catalytic activity in the ring-opening polymerization of ε-caprolactone. Both **336** and **338** showed noticeable catalytic activity while **339** displayed a slight decrease in conversion consistent with its added steric bulk. Octahedral compound **340** demonstrated no polymerization activity, which was attributed to the coordinative saturation about the Al center. Interestingly, however, related work by Li and coworkers found that homoleptic octahedral Mg compounds of the pyrrolyl-pyridylimine ligands can serve as successful initiators for the polymerization ε-caprolactone at or above 60 °C.[141] This temperature requirement may reflect the necessity of ligand dissociation from the Mg center prior to interaction with the substrate.

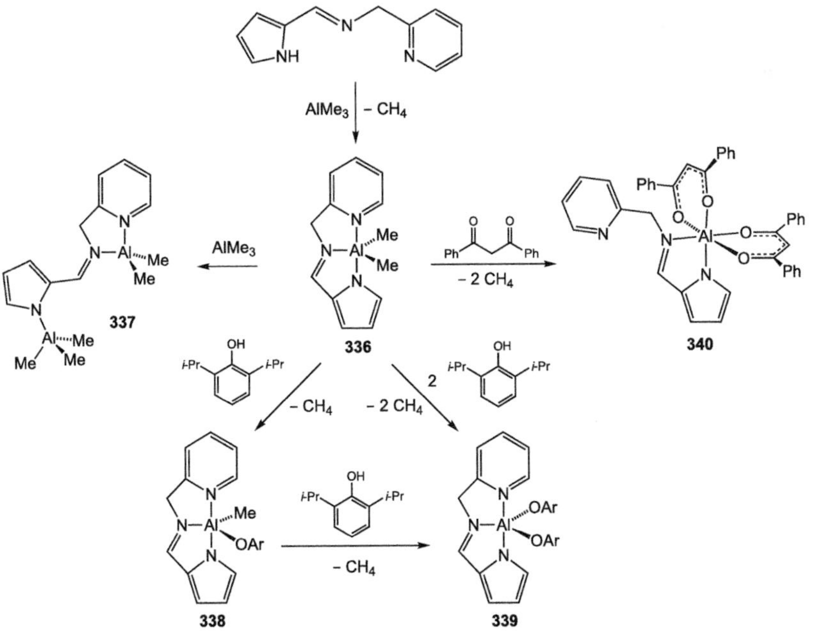

Scheme 74

Huang and coworkers later reported a similar class Al compounds with the related morpholine-containing ligand shown in Chart 24.[142] Similar coordination chemistry was observed to that depicted in Scheme 74. A compound analogous to **336** displayed increased catalytic efficacy for the ring-opening polymerization of ε-caprolactone reaching conversions up to 99% in 6 h.

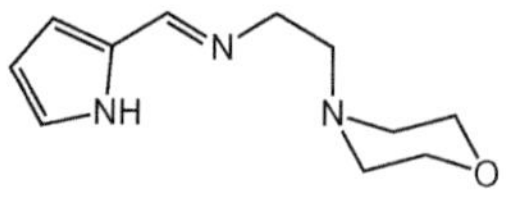

Chart 24 Morpholine-substituted pyrrole-imine proligand.

10. Conclusions and outlook

As is evident from the chemistry described in the previous sections of this review, pyrrolyl-containing pincer ligands have demonstrated remarkable versatility across the periodic table. Although only emerging about 20 years ago, these ligands now occupy an important place in modern inorganic and organometallic chemistry. They have been employed in a number of catalytic reactions from polymerization to N_2 reduction and have given rise to a wealth of new complexes with novel geometries and electronic structures. At the heart of their utility are the unique features of the pyrrolyl moiety, which sets this class of ligands apart from more commonplace pincers containing pyridine, amine, and aryl donors. This fact was evident in the distinctive tautomerization, protonation, and electron transfer events displayed by their various metal complexes. With a firm foundation of success to build upon, the prospects for this class of pincers is bright. There is little doubt that these ligands will find continued use for years to come across many areas of coordination chemistry.

Acknowledgment

The authors thank the Welch Foundation (AX-1772) for their generous support of our work in the area of transition metal chemistry with pyrrole-based pincer ligands.

References

1. Morales-Morales D, ed. *Pincer Compounds: Chemistry and Applications*. Netherlands: Elsevier; 2018.
2. Peris E, Crabtree RH. Key factors in pincer ligand design. *Chem Soc Rev*. 2018;47:1959–1968.

3. Bordwell FG, Drucker GE, Fried HE. Acidities of carbon and nitrogen acids: the aromaticity of the cyclopentadienyl anion. *J Org Chem*. 1981;46:632–635.
4. Rakowski DuBois M. The activation of η5-pyrrole complexes toward nucleophilic attack. *Coord Chem Rev*. 1998;174:191–205.
5. Tanski JM, Parkin G. Synthesis and structures of zirconium–pyrrolyl complexes: computational analysis of the factors that influence the coordination modes of pyrrolyl ligands. *Organometallics*. 2002;21:587–589.
6. Melen RL, Gade LH. New chemistry with anionic NNN pincer ligands. *Top Organomet Chem*. 2016;179–208.
7. Sadimenko A. *Organometallic Chemistry of Five-Membered Heterocycles*. Netherlands: Elsevier; 2020.
8. Nishibayashi Y. Development of catalytic nitrogen fixation using transition metal–dinitrogen complexes under mild reaction conditions. *Dalton Trans*. 2018;47:11290–11297.
9. McPherson JN, Das B, Colbran SB. Tridentate pyridine–pyrrolide chelate ligands: an under-appreciated ligand set with an immensely promising coordination chemistry. *Coord Chem Rev*. 2018;375:285–332.
10. Das B, McPherson JN, Colbran SB. Oligomers and macrocycles with [m]pyridine[n]pyrrole (m + n ≥ 3) domains: formation and applications of anion, guest molecule and metal ion complexes. *Coord Chem Rev*. 2018;363:29–56.
11. Deng Q-H, Melen RL, Gade LH. Anionic chiral tridentate N-donor pincer ligands in asymmetric catalysis. *Acc Chem Res*. 2014;47:3162–3173.
12. Kim IT, Elsenbaumer RL. Convenient synthesis of 1-alkyl-2, 5-bis(thiophenylmethylene) pyrroles using the Mannich reaction. *Tetrahedron Lett*. 1998;39:1087–1090.
13. Herz W, Dittmer K, Cristol SJ. The preparation of some monosubstituted derivatives of pyrrole by the Mannich reaction. *J Am Chem Soc*. 1947;69:1698–1700.
14. Huang J-H, Kuo P-C, Lee G-H, Peng S-M. Synthesis and structure characterization of 2-(dimethylaminomethyl)pyrrolate and 2,5-bis(dimethylaminomethyl)pyrrolate zirconium complexes. *J Chil Chem Soc*. 2000;47:1191–1195.
15. Huang J-H, Chen H-J, Chang J-C, Zhou C-C, Lee G-H, Peng S-M. Synthesis and characterization of organoaluminum complexes containing bi- or tridentate-substituted pyrrole ligands. *Organometallics*. 2001;20:2647–2650.
16. Chang J-C, Hung C-H, Huang J-H. An unusual hydride-bridged aluminum complex with a square-planar tetraaluminum core stabilized by 2,5-bis((dimethylamino)methyl) pyrrole ligands. *Organometallics*. 2001;20:4445–4447.
17. Chen IC, Ho S-M, Chen Y-C, et al. Deprotonation and reductive addition reactions of hypervalent aluminium dihydride compounds containing substituted pyrrolyl ligands with phenols, ketones, and aldehydes. *Dalton Trans*. 2009;8631–8643.
18. Lien Y-L, Chang Y-C, Chuang N-T, et al. A new type of asymmetric tridentate pyrrolyl-linked pincer ligand and its aluminum dihydride complexes. *Inorg Chem*. 2010;49:136–143.
19. Huang W-Y, Chuang S-J, Chunag N-T, et al. Aluminium complexes containing bidentate and symmetrical tridentate pincer type pyrrolyl ligands: synthesis, reactions and ring opening polymerization. *Dalton Trans*. 2011;40:7423–7433.
20. Banerjee S, Shi Y, Cao C, Odom AL. Titanium-catalyzed iminohydrazination of alkynes. *J Organomet Chem*. 2005;690:5066–5077.
21. Li Y, Banerjee S, Odom AL. Synthesis and structure of (triphenylsilyl)imido complexes of titanium and zirconium. *Organometallics*. 2005;24:3272–3278.
22. Harris SA, Ciszewski JT, Odom AL. Titanium η^1-pyrrolyl complexes: electronic and structural characteristics imposed by the N,N-di(pyrrolyl-α-methyl)-N-methylamine (dpma) ligand. *Inorg Chem*. 2001;40:1987–1988.

23. Hsu J-W, Lin Y-C, Hsiao C-S, et al. Zirconium complexes incorporated with asymmetrical tridentate pincer type mono- and di-anionic pyrrolyl ligands: mechanism and reactivity as catalytic precursors. *Dalton Trans*. 2012;41:7700–7707.
24. Kuo P-C, Chang J-C, Lee W-Y, Lee HM, Huang J-H. Synthesis and characterization of lithium and yttrium complexes containing tridentate pyrrolyl ligands. Single-crystal X-ray structures of {Li[C$_4$H$_2$N(CH$_2$NMe$_2$)$_2$-2,5]}$_2$ (**1**) and {[C$_4$H$_2$N(CH$_2$NMe$_2$)$_2$-2,5]YCl$_2$(μ-Cl)·Li(OEt$_2$)$_2$}$_2$ (**2**) and ring-opening polymerization of ε-caprolactone. *J Organomet Chem*. 2005;690:4168–4174.
25. Wang L, Liu D, Cui D. NNN-tridentate pyrrolyl rare-earth metal complexes: structure and catalysis on specific selective living polymerization of isoprene. *Organometallics*. 2012;31:6014–6021.
26. Wang Y-T, Lin Y-C, Hsu S-Y, et al. Synthesis and structural aspects of gallium compounds containing tridentate pincer type pyrrolyl ligands: intramolecular hydrogen bonding of gallium aryloxides. *J Organomet Chem*. 2013;745-746:12–17.
27. Kuo P-C, Huang J-H, Hung C-H, Lee G-H, Peng S-M. Synthesis and characterization of five-coordinate gallium and indium complexes stabilized by tridentate, substituted pyrrole ligands. *Eur J Inorg Chem*. 2003;2003:1440–1444.
28. Maaß C, Andrada DM, Mata RA, Herbst-Irmer R, Stalke D. Effects of metal coordination on the π-system of the 2,5-bis-{(pyrrolidino)-methyl}-pyrrole pincer ligand. *Inorg Chem*. 2013;52:9539–9548.
29. Vránová I, Jambor R, Růžička A, Hoffmann A, Herres-Pawlis S, Dostál L. Antimony(III) and bismuth(III) amides containing pendant N-donor groups—a combined experimental and theoretical study. *Dalton Trans*. 2015;44:395–400.
30. Dawson DM, Walker DA, Thornton-Pett M, Bochmann M. Synthesis and reactivity of sterically hindered iminopyrrolato complexes of zirconium, iron, cobalt and nickel. *J Chem Soc Dalton Trans*. 2000;4:459–466.
31. Matsuo Y, Mashima K, Tani K. Selective formation of homoleptic and heteroleptic 2,5-bis(N-aryliminomethyl)pyrrolyl yttrium complexes and their performance as initiators of ε-caprolactone polymerization. *Organometallics*. 2001;20:3510–3518.
32. Meyer N, Kuzdrowska M, Roesky P. W. (2,5-Bis{[(2,6-diisopropylphenyl)imino]methyl}pyrrolyl)yttrium and -lutetium complexes—synthesis and structures. *Eur J Inorg Chem*. 2008;2008:1475–1479.
33. Meyer N, Jenter J, Roesky PW, Eickerling G, Scherer W. Unusual reactivity of lanthanide borohydride complexes leading to a borane complex. *Chem Commun*. 2009;45:4693–4695.
34. Jenter J, Meyer N, Roesky PW, Thiele SKH, Eickerling G, Scherer W. Borane and borohydride complexes of the rare-earth elements: synthesis, structures, and butadiene polymerization catalysis. *Chem Eur J*. 2010;16:5472–5480.
35. Roesky PW, Jenter J, Köppe R. Mixed borohydride-chloride complexes of the rare earth elements. *C R Chim*. 2010;13:603–607.
36. Jenter J, Gamer MT, Roesky PW. 2,5-Bis{N-(2,6-diisopropylphenyl)iminomethyl} pyrrolyl complexes of the divalent lanthanides: synthesis and structures. *Organometallics*. 2010;29:4410–4413.
37. Li T, Wiecko J, Pushkarevsky NA, et al. Mixed-metal lanthanide–iron triple-decker complexes with a cyclo-P$_5$ building block. *Angew Chem Int Ed*. 2011;50:9491–9495.
38. Das S, Anga S, Harinath A, Pada Nayek H, Panda TK. Synthesis and structure of unprecedented samarium complex with bulky bis-iminopyrrolyl ligand via intramolecular C=N bond activation. *Z Anorg Allg Chem*. 2017;643:2144–2148.
39. Anga S, Banerjee I, Nayek HP, Panda TK. Alkali metal complexes having sterically bulky bis-iminopyrrolyl ligands—control of dimeric to monomeric complexes. *RSC Adv*. 2016;6:80916–80923.

40. Jenter J, Köppe R, Roesky PW. 2,5-Bis{N-(2,6-diisopropylphenyl)iminomethyl}pyrrolyl complexes of the heavy alkaline earth metals: synthesis, structures, and hydroamination catalysis. *Organometallics*. 2011;30:1404–1413.
41. Tsurugi H, Matsuo Y, Yamagata T, Mashima K. Intramolecular benzylation of an imino group of tridentate 2,5-bis(N-aryliminomethyl)pyrrolyl ligands bound to zirconium and hafnium gives amido-pyrrolyl complexes that catalyze ethylene polymerization. *Organometallics*. 2004;23:2797–2805.
42. Daneshmand P, Fortun S, Schaper F. Diiminopyrrolide copper complexes: synthesis, structures, and rac-lactide polymerization activity. *Organometallics*. 2017;36:3860–3877.
43. Daneshmand P, van der Est A, Schaper F. Mechanism and stereocontrol in isotactic rac-lactide polymerization with copper(II) complexes. *ACS Catal*. 2017;7:6289–6301.
44. Hein F, Beierlein U. Preparation and chemistry of the complexes of 2,5-di-(α-pyridyl)pyrrole. *Pharm Zentralhalle Dtschl*. 1957;96:410–421.
45. Hein F, Melichar F. Synthesis of 2,5-di(α-pyridyl)pyrrole and of several complex derivatives of 2,5-di(α-pyridyl)-3,4-dicarbethoxypyrrole. *Pharmazie*. 1954;9:455–460.
46. Ciszek JW, Keane ZK, Cheng L, et al. Neutral complexes of first row transition metals bearing unbound thiocyanates and their assembly on metallic surfaces. *J Am Chem Soc*. 2006;128:3179–3189.
47. McPherson JN, Elton TE, Colbran SB. A strain-deformation nexus within pincer ligands: application to the spin states of iron(II) complexes. *Inorg Chem*. 2018;57:12312–12322.
48. McPherson JN, Hogue RW, Akogun FS, et al. Predictable substituent control of Co$^{III/II}$ redox potential and spin crossover in bis(dipyridylpyrrolide)cobalt complexes. *Inorg Chem*. 2019;58:2218–2228.
49. Chen J-J, Xu Y-C, Gan Z-L, Peng X, Yi X-Y. Zinc complexes with tridentate pyridyl-pyrrole ligands and their use as catalysts in CO2 fixation into cyclic carbonates. *Eur J Inorg Chem*. 2019;2019:1733–1739.
50. Wang Y-M, He P, Peng X, Liu S-Q, Yi X-Y. Chlorogenation of pyrrole-based on di(pyridyl)pyrrolide ligand and synthesis of its ruthenium carbonyl complexes. *Polyhedron*. 2019;168:67–71.
51. McPherson JN, Abad Galan L, Iranmanesh H, Massi M, Colbran SB. Synthesis and structural, redox and photophysical properties of tris-(2,5-di(2-pyridyl)pyrrolide) lanthanide complexes. *Dalton Trans*. 2019;48:9365–9375.
52. Takaya J, Hoshino M, Ueki K, Saito N, Iwasawa N. Synthesis, structure, and reactivity of pincer-type iridium complexes having gallyl- and indyl-metalloligands utilizing 2,5-bis(6-phosphino-2-pyridyl)pyrrolide as a new scaffold for metal–metal bonds. *Dalton Trans*. 2019;48:14606–14610.
53. Ghorai D, Kumar S, Mani G. Mononuclear, helical binuclear palladium and lithium complexes bearing a new pyrrole-based NNN-pincer ligand: fluxional property. *Dalton Trans*. 2012;41:9503–9512.
54. Jana O, Mani G. New types of Cu and Ag clusters supported by the pyrrole-based NNN-pincer type ligand. *New J Chem*. 2017;41:9361–9370.
55. Lin K, Chile L-E, Zhen SC, Boyd PDW, Ware DC, Brothers PJ. Pyrrole pincers containing imidazole, pyrazole and 1,2,4-triazole groups. *Inorg Chim Acta*. 2014;422:95–101.
56. Ghorai D, Mani G. Single-step substitution of all the α, β-positions in pyrrole: choice of binuclear versus multinuclear complex of the novel polydentate ligand. *Inorg Chem*. 2014;53:4117–4129.
57. Jha VK, Mani G, Davuluri YR, Anoop A. The pyrrole ring η2-hapticity bridged binuclear tricarbonyl Mo(0) and W(0) complexes: catalysis of regioselective hydroamination reactions and DFT calculations. *Dalton Trans*. 2017;46:1840–1847.

58. Das S, Subramaniyan V, Mani G. Nickel(II) and palladium(II) complexes bearing an unsymmetrical pyrrole-based PNN pincer and their Norbornene polymerization behaviors versus the symmetrical NNN and PNP pincers. *Inorg Chem.* 2019;58:3444–3456.
59. Nishiyama H, Sakaguchi H, Nakamura T, Horihata M, Kondo M, Itoh K. Chiral and C_2-symmetrical bis(oxazolinylpyridine)rhodium(III) complexes: effective catalysts for asymmetric hydrosilylation of ketones. *Organometallics.* 1989;8:846–848.
60. Mazet C, Gade LH. A bis(oxazolinyl)pyrrole as a new monoanionic tridentate supporting ligand: synthesis of a highly active palladium catalyst for Suzuki-type C–C coupling. *Organometallics.* 2001;20:4144–4146.
61. Mazet C, Gade L. H. Charging and deforming the pybox ligand: enantiomerically pure double helices and their interconversion. *Chem Eur J.* 2002;8:4308–4318.
62. Mazet C, Gade LH. [Bis(oxazolinyl)pyrrole]palladium complexes as catalysts in Heck- and Suzuki-type C-C coupling reactions. *Eur J Inorg Chem.* 2003;2003:1161–1168.
63. Mazet C, Gade LH. Synthesis and coordination chemistry of the first C_2-chiral bis-oxazine ligand. *Inorg Chem.* 2003;42:210–215.
64. Mazet C, Gade LH. Double bonds in motion: bis(oxazolinylmethyl)pyrroles and their metal-induced planarization to a new class of rigid chiral C_2-symmetric complexes. *Chem Eur J.* 2003;9:1759–1767.
65. Capacchione C, Wadepohl H, Gade LH. Synthesis, structure and dynamics of a C_2-symmetrical bis(oxazolinylmethylene)pyrrolinato-zinc complex. *Z Anorg Allg Chem.* 2007;633:2131–2134.
66. Konrad F, Lloret Fillol J, Wadepohl H, Gade LH. Bis(oxazolinylmethyl)pyrrole derivatives and their coordination as chiral "pincer" ligands to rhodium. *Inorg Chem.* 2009;48:8523–8535.
67. Konrad F, Lloret Fillol J, Rettenmeier C, Wadepohl H, Gade LH. Bis(oxazolinylmethyl) derivatives of C4H4E heterocycles (E = NH, O, S) as C_2-chiral meridionally coordinating ligands for nickel and chromium. *Eur J Inorg Chem.* 2009;2009:4950–4961.
68. Rettenmeier C, Wadepohl H, Gade LH. Stereoselective hydrodehalogenation via a radical-based mechanism involving T-shaped chiral nickel(I) pincer complexes. *Chem Eur J.* 2014;20:9657–9665.
69. Rettenmeier CA, Wenz J, Wadepohl H, Gade LH. Activation of aryl halides by nickel(I) pincer complexes: reaction pathways of stoichiometric and catalytic dehalogenations. *Inorg Chem.* 2016;55:8214–8224.
70. Wenz J, Rettenmeier CA, Wadepohl H, Gade LH. Catalytic C-F bond activation of geminal difluorocyclopropanes by nickel(I) complexes via a radical mechanism. *Chem Commun.* 2016;52:202–205.
71. Rettenmeier CA, Wadepohl H, Gade LH. Structural characterization of a hydroperoxo nickel complex and its autoxidation: mechanism of interconversion between peroxo, superoxo, and hydroperoxo species. *Angew Chem Int Ed.* 2015;54:4880–4884.
72. Rettenmeier CA, Wadepohl H, Gade LH. Electronic structure and reactivity of nickel(I) pincer complexes: their aerobic transformation to peroxo species and site selective C-H oxygenation. *Chem Sci.* 2016;7:3533–3542.
73. Wenz J, Kochan A, Wadepohl H, Gade LH. A readily accessible chiral NNN pincer ligand with a pyrrole backbone and its Ni(II) chemistry: syntheses, structural chemistry, and bond activations. *Inorg Chem.* 2017;56:3631–3643.
74. Wenz J, Wadepohl H, Gade LH. Regioselective hydrosilylation of epoxides catalysed by nickel(ii) hydrido complexes. *Chem Commun.* 2017;53:4308–4311.
75. Wenz J, Vasilenko V, Kochan A, Wadepohl H, Gade LH. Coordination chemistry of the PdmBOX pincer ligand: reactivity at the metal and the ligand. *Eur J Inorg Chem.* 2017;2017:5545–5556.

76. Schiwek CH, Vasilenko V, Wadepohl H, Gade LH. The open d-shell enforces the active space in 3d metal catalysis: highly enantioselective chromium(II) pincer catalysed hydrosilylation of ketones. *Chem Commun*. 2018;54:9139–9142.
77. Venkanna GT, Ramos TVM, Arman HD, Tonzetich ZJ. Nickel(II) complexes containing a pyrrole-diphosphine pincer ligand. *Inorg Chem*. 2012;51:12789–12795.
78. Kumar S, Mani G, Mondal S, Chattaraj PK. Pyrrole-based new diphosphines: Pd and Ni complexes bearing the PNP pincer ligand. *Inorg Chem*. 2012;51:12527–12539.
79. Grüger N, Wadepohl H, Gade LH. A readily accessible PNP pincer ligand with a pyrrole backbone and its NiI/II chemistry. *Dalton Trans*. 2012;41:14028–14030.
80. Venkanna GT, Arman HD, Tonzetich ZJ. Catalytic C–S cross-coupling reactions employing Ni complexes of pyrrole-based pincer ligands. *ACS Catal*. 2014;4:2941–2950.
81. Venkanna GT, Tammineni S, Arman HD, Tonzetich ZJ. Synthesis, characterization, and catalytic activity of nickel(II) alkyl complexes supported by pyrrole-diphosphine ligands. *Organometallics*. 2013;32:4656–4663.
82. Kessler JA, Iluc VM. Ag(I) and Tl(I) precursors as transfer agents of a pyrrole-based pincer ligand to late transition metals. *Inorg Chem*. 2014;53:12360–12371.
83. Kreye M, Freytag M, Jones PG, Williard PG, Bernskoetter WH, Walter MD. Homolytic H_2 cleavage by a mercury-bridged Ni(I) pincer complex [{(PNP)Ni}$_2$ {μ-Hg}]. *Chem Commun*. 2015;51:2946–2949.
84. Kumar S, Jana O, Subramaniyan V, Mani G. The 'reverse transmetalation' reaction of the pyrrole-based PNP pincer Ni(II) complexes: X-ray structures of binuclear silver(I) and thiocyanate nickel(II) complexes. *Inorg Chim Acta*. 2018;480:113–119.
85. Kessler JA, Iluc VM. Ni(II) phosphine and phosphide complexes supported by a PNP-pyrrole pincer ligand. *Dalton Trans*. 2017;46:12125–12131.
86. Kuriyama S, Arashiba K, Nakajima K, et al. Catalytic transformation of dinitrogen into ammonia and hydrazine by iron-dinitrogen complexes bearing pincer ligand. *Nat Commun*. 2016;7:12181.
87. Sekiguchi Y, Kuriyama S, Eizawa A, Arashiba K, Nakajima K, Nishibayashi Y. Synthesis and reactivity of iron-dinitrogen complexes bearing anionic methyl- and phenyl-substituted pyrrole-based PNP-type pincer ligands toward catalytic nitrogen fixation. *Chem Commun*. 2017;53:12040–12043.
88. Kuriyama S, Arashiba K, Nakajima K, Tanaka H, Yoshizawa K, Nishibayashi Y. Azaferrocene-based PNP-type pincer ligand: synthesis of molybdenum, chromium, and Iron complexes and reactivity toward nitrogen fixation. *Eur J Inorg Chem*. 2016;2016:4856–4861.
89. Ehrlich N, Baabe D, Freytag M, Jones PG, Walter MD. Pyrrolyl-based pincer complexes of iron—synthesis and electronic structure. *Polyhedron*. 2018;143:83–93.
90. Thompson CV, Arman HD, Tonzetich ZJ. A pyrrole-based pincer ligand permits access to three oxidation states of iron in organometallic complexes. *Organometallics*. 2017;36:1795–1802.
91. Ehrlich N, Kreye M, Baabe D, et al. Synthesis and electronic ground-state properties of pyrrolyl-based iron pincer complexes: revisited. *Inorg Chem*. 2017;56:8415–8422.
92. Thompson CV, Davis I, DeGayner JA, Arman HD, Tonzetich ZJ. Iron pincer complexes incorporating bipyridine: a strategy for stabilization of reactive species. *Organometallics*. 2017;36:4928–4935.
93. Holland AM, Oliver AG, Iluc VM. Iron pyrrole-based PNP pincer ligand complexes as catalyst precursors. *Acta Crystallogr C Struct Chem*. 2017;73:569–574.
94. Nakajima K, Kato T, Nishibayashi Y. Hydroboration of alkynes catalyzed by pyrrolide-based PNP pincer–Iron complexes. *Org Lett*. 2017;19:4323–4326.
95. Kato T, Kuriyama S, Nakajima K, Nishibayashi Y. Catalytic C−H borylation using iron complexes bearing 4,5,6,7-tetrahydroisoindol-2-ide-based PNP-type pincer ligand. *Chem Asian J*. 2019;14:2097–2101.

96. Thompson CV, Arman HD, Tonzetich ZJ. Square-planar iron(II) silyl complexes: synthesis, characterization, and insertion reactivity. *Organometallics*. 2019;38:2979–2989.
97. Kuriyama S, Arashiba K, Tanaka H, et al. Direct transformation of molecular dinitrogen into ammonia catalyzed by cobalt dinitrogen complexes bearing anionic PNP pincer ligands. *Angew Chem Int Ed*. 2016;55:14291–14295.
98. Krishnan VM, Arman HD, Tonzetich ZJ. Preparation and reactivity of a square-planar PNP cobalt(II)–hydrido complex: isolation of the first {Co–NO}^{8}–hydride. *Dalton Trans*. 2018;47:1435–1441.
99. Alawisi H, Al-Afyouni KF, Arman HD, Tonzetich ZJ. Aldehyde decarbonylation by a cobalt(I) pincer complex. *Organometallics*. 2018;37:4128–4135.
100. Nakayama S, Morisako S, Yamashita M. Synthesis and application of pyrrole-based PNP–Ir complexes to catalytic transfer dehydrogenation of cyclooctane. *Organometallics*. 2018;37:1304–1313.
101. Kawakami R, Kuriyama S, Tanaka H, et al. Catalytic reduction of dinitrogen to tris(trimethylsilyl)amine using rhodium complexes with a pyrrole-based PNP-type pincer ligand. *Chem Commun*. 2019;55:14886–14889.
102. Narro AL, Arman HD, Tonzetich ZJ. Manganese chemistry of anionic pyrrole-based pincer ligands. *Organometallics*. 2019;38:1741–1749.
103. Sekiguchi Y, Arashiba K, Tanaka H, et al. Catalytic reduction of molecular dinitrogen to ammonia and hydrazine using vanadium complexes. *Angew Chem Int Ed*. 2018;57:9064–9068.
104. Sekiguchi Y, Meng F, Tanaka H, et al. Synthesis and reactivity of titanium- and zirconium-dinitrogen complexes bearing anionic pyrrole-based PNP-type pincer ligands. *Dalton Trans*. 2018;47:11322–11326.
105. Levine DS, Tilley TD, Andersen RA. C–H bond activations by monoanionic, PNP-supported scandium dialkyl complexes. *Organometallics*. 2015;34:4647–4655.
106. Levine DS, Tilley TD, Andersen RA. Evidence for the existence of group 3 terminal methylidene complexes. *Organometallics*. 2017;36:80–88.
107. Krishnan VM, Davis I, Baker TM, et al. Backbone dehydrogenation in pyrrole-based pincer ligands. *Inorg Chem*. 2018;57:9544–9553.
108. Lagaditis PO, Schluschaß B, Demeshko S, Würtele C, Schneider S. Square-planar cobalt(III) pincer complex. *Inorg Chem*. 2016;55:4529–4536.
109. Ghorai D, Mani G. Unsubstituted quinoidal pyrrole and its reaction with oxygen, charge transfer and palladium(II) complexes via DDQ oxidation. *RSC Adv*. 2014;4:45603–45611.
110. Tian R, Ng Y, Ganguly R, Mathey F. A new type of phosphaferrocene–pyrrole–phosphaferrocene P-N-P pincer ligand. *Organometallics*. 2012;31:2486–2488.
111. Komine N, Buell RW, Chen C-H, Hui AK, Pink M, Caulton KG. Probing the steric and electronic characteristics of a new bis-pyrrolide pincer ligand. *Inorg Chem*. 2014;53:1361–1369.
112. Yadav S, Singh A, Rashid N, et al. Phosphine-free bis(pyrrolyl)pyridine based NNN-pincer palladium(II) complexes as efficient catalysts for Suzuki-Miyaura cross-coupling reactions of aryl bromides in aqueous medium. *ChemistrySelect*. 2018;3:9469–9475.
113. Searles K, Fortier S, Khusniyarov MM, et al. A cis-divacant octahedral and mononuclear iron(IV) imide. *Angew Chem Int Ed*. 2014;53:14139–14143.
114. Hakey BM, Darmon JM, Zhang Y, Petersen JL, Milsmann C. Synthesis and electronic structure of neutral square-planar high-spin iron(II) complexes supported by a dianionic pincer ligand. *Inorg Chem*. 2019;58:1252–1266.
115. Hakey BM, Darmon JM, Akhmedov NG, Petersen JL, Milsmann C. Reactivity of pyridine dipyrrolide iron(II) complexes with organic azides: C–H amination and iron tetrazene formation. *Inorg Chem*. 2019;58:11028–11042.

116. Grant LN, Carroll ME, Carroll PJ, Mindiola DJ. An unusual cobalt azide adduct that produces a nitrene species for carbon–hydrogen insertion chemistry. *Inorg Chem.* 2016;55:7997–8002.
117. Sorsche D, Miehlich ME, Zolnhofer EM, Carroll PJ, Meyer K, Mindiola DJ. Metal–ligand cooperativity promoting sulfur atom transfer in ferrous complexes and isolation of a sulfurmethylenephosphorane adduct. *Inorg Chem.* 2018;57:11552–11559.
118. Zhang Y, Petersen JL, Milsmann C. A luminescent zirconium(IV) complex as a molecular photosensitizer for visible light photoredox catalysis. *J Am Chem Soc.* 2016;138:13115–13118.
119. Zhang Y, Lee TS, Petersen JL, Milsmann C. A zirconium photosensitizer with a long-lived excited state: mechanistic insight into photoinduced single-electron transfer. *J Am Chem Soc.* 2018;140:5934–5947.
120. Zhang Y, Petersen JL, Milsmann C. Photochemical C–C bond formation in luminescent zirconium complexes with CNN pincer ligands. *Organometallics.* 2018;37:4488–4499.
121. Zhang Y, Akhmedov NG, Petersen JL, Milsmann C. Photoluminescence of seven-coordinate zirconium and hafnium complexes with 2,2'-pyridylpyrrolide ligands. *Chem Eur J.* 2019;25:3042–3052.
122. Kumar R, Guchhait T, Subramaniyan V, Mani G. Mixed ligand Cu(II) complexes: square pyramidal vs trigonal bipyramidal with the pyrrole-based dipodal ligand having hydrogen bond acceptors. *J Mol Struct.* 2019;1195:1–9.
123. Li R, Brooker S. Copper(II) and palladium(II) complexes of a terdentate pyrrolidine diester ligand. *Inorg Chim Acta.* 2011;365:246–250.
124. O'Reilly ME, Nadif SS, Ghiviriga I, Abboud KA, Veige AS. Synthesis and characterization of tungsten alkylidene and alkylidyne complexes supported by a new pyrrolide-centered trianionic ONO3- pincer-type ligand. *Organometallics.* 2014;33:836–839.
125. Nadif SS, O'Reilly ME, Ghiviriga I, Abboud KA, Veige AS. Remote multiproton storage within a pyrrolide-pincer-type ligand. *Angew Chem Int Ed.* 2015;54:15138–15142.
126. Johnson KRD, Hannon MA, Ritch JS, Hayes PG. Thermally stable rare earth dialkyl complexes supported by a novel bis(phosphinimine)pyrrole ligand. *Dalton Trans.* 2012;41:7873–7875.
127. Zamora MT, Johnson KRD, Hänninen MM, Hayes PG. Differences in the cyclometalation reactivity of bisphosphinimine-supported organo-rare earth complexes. *Dalton Trans.* 2014;43:10739–10750.
128. Hänninen MM, Zamora MT, MacNeil CS, Knott JP, Hayes PG. Elucidation of the resting state of a rhodium NNN-pincer hydrogenation catalyst that features a remarkably upfield hydride ^{1}H NMR chemical shift. *Chem Commun.* 2016;52:586–589.
129. MacNeil CS, Glynn KE, Hayes PG. Facile activation and deoxygenative metathesis of CO. *Organometallics.* 2018;37:3248–3252.
130. MacNeil CS, Hayes PG. An H-substituted rhodium silylene. *Chem Eur J.* 2019;25:8203–8207.
131. Fraix A, Lutz M, Spek AL, et al. Construction of a monoanionic S,N,S-pincer ligand with a pyrrole core by sequential [1,2] phospho-fries rearrangement. characterization of palladium and silver coordination complexes. *Dalton Trans.* 2010;39:2942–2946.
132. Ghorui T, Roy S, Pramanik S, Pramanik K. RhCl(PPh$_3$)$_3$-mediated C–H oxyfunctionalization of pyrrolido-functionalized bisazoaromatic pincers: a combined experimental and theoretical scrutiny of redox-active and spectroscopic properties. *Dalton Trans.* 2016;45:5720–5729.
133. Bates GW, Gale PA, Light ME, Ogden MI, Warriner CN. Structural diversity in the first metal complexes of 2,5-dicarboxamidopyrroles and 2,5-dicarbothioamidopyrroles. *Dalton Trans.* 2008;37:4106–4112.

134. Karagiannidis LE, Gale PA, Light ME, Massi M, Ogden MI. Further insight into the coordination of 2,5-dicarbothioamidopyrroles: the case of Cu and Co complexes. *Dalton Trans.* 2011;40:12097–12105.
135. Stern C, Franceschi F, Solari E, Floriani C, Re N, Scopelliti R. The use of macrocyclic and polydentate ligands in ruthenium organometallic chemistry. *J Organomet Chem.* 2000;593-594:86–95.
136. Liao Y, Novoa JJ, Arif A, Miller JS. Synthesis and structure of an asymmetric copper(I) dimer with two-coordinate and four-coordinate copper(I) sites. *Chem Commun.* 2002;38:3008–3009.
137. Li R, Moubaraki B, Murray KS, Brooker S. Monomeric and dimeric copper(II) complexes of a pyrrole-containing tridentate Schiff-base ligand. *Eur J Inorg Chem.* 2009;2009:2851–2859.
138. Li R, Moubaraki B, Murray KS, Brooker S. Monomeric, dimeric and 1D chain polymeric copper(II) complexes of a pyrrole-containing tridentate Schiff-base ligand and its 4-brominated analogue. *Dalton Trans.* 2008;37:6014–6022.
139. Li R, Brooker S. An unexpected mixed-valence cobalt(II)/cobalt(III) complex of a pyrrole-containing tridentate Schiff-base ligand. *J Incl Phenom.* 2011;71:303–309.
140. Hsu S-Y, Hu C-H, Tu C-Y, et al. Aluminum compounds containing pyrrole–imine ligand systems—synthesis, characterization, structure elucidation, ring-opening polymerization, and catalytic Meerwein–Ponndorf–Verley reaction. *Eur J Inorg Chem.* 2014;2014:1965–1973.
141. Lin M, Liu W, Chen Z, et al. Magnesium complexes supported by pyrrolyl ligands: syntheses, characterizations, and catalytic activities towards the polymerization of ε-caprolactone. *RSC Adv.* 2012;2:3451–3457.
142. Lin T-H, Cai Y-R, Chang W, et al. Synthesis and characterization of multidentate ethylene bridged pyrrole- and ketoamine-morpholine aluminum compounds. Structure, theoretical calculation and catalytic study. *J Organomet Chem.* 2016;825-826:15–24.

CHAPTER FIVE

Low-coordinate M(0) complexes of group 10 stabilized by phosphorus(III) ligands and N-heterocyclic carbenes

Raquel J. Rama[a], M. Trinidad Martín[a], Riccardo Peloso[b,*], M. Carmen Nicasio[a,*]

[a]Departamento de Química Inorgánica, Universidad de Sevilla, Sevilla, Spain
[b]Instituto de Investigaciones Químicas (IIQ), Departamento de Química Inorgánica and Centro de Innovación en Química Avanzada (ORFEO-CINQA), Consejo Superior de Investigaciones Científicas (CSIC) and Universidad de Sevilla, Sevilla, Spain
*Corresponding authors: e-mail address: rpeloso@us.es; mnicasio@us.es

Contents

1. Introduction — 243
2. Three-coordinate complexes — 244
   2.1 M(0) complexes with phosphorus(III) ligands — 244
   2.2 M(0) complexes with N-heterocyclic carbene ligands — 270
3. Two-coordinate complexes — 289
   3.1 Homoleptic M(0) complexes with phosphorus(III) ligands — 289
   3.2 Homoleptic M(0) complexes with N-heterocyclic ligands — 297
   3.3 Heteroleptic M(0) complexes — 303
4. Conclusions and outlook — 305
Acknowledgments — 305
References — 306

Abbreviations

6-NHC	6-membered ring NHC
7-NHC	7-membered ring NHC
acac	acetylacetonate
Ad	1-adamantyl
BQ	*p*-benzoquinone
Bu	*n*-butyl
Bz	Benzyl
COD	1,5-cyclooctadiene
COT	1,3,5,7-Cyclooctatetraene

CP-MAS	cross polarization magic angle spinning
Cy	cyclohexyl
DAE	diallylether
DAM	diethyl 2,2-allylmalonate
dba	dibenzylideneacetone
DFT	density functional theory
diop	4,5-bis[(diphenylphosphino)-methyl]-2,2-dimethyl-l,3-dioxolane
dippb	1,4-bis(diisopropylphosphino)butane
dippe	diisopropylphosphino
dippp	1,3-bis(diisopropylphosphino)propane
DMFU	dimethylfumarate
dmpe	1,2-bis(dimethylphoshino)ethane
dppm	bis(diphenylphosphine)methane
dppp	1,3-bis(diphenylphosphino)propane
dtbpe	1,2-bis(di-*tert*-butylphosphino)ethane
DVDS	1,1,3,3-tetramethyl-1,3-divinyl-disiloxane
equiv.	equivalents
exc.	excess
FLP	frustrated Lewis pair
IAd	1,3-diadamantylimidazol-2-ylidene
ICy	1,3-dicyclohexylimidazol-2-ylidene
I*i*Pr	1,3-diisopropylimidazol-2-ylidene
IMe	1,3-dimethylimidazol-2-ylidene
IMes	1,3-bis(2,4,6-trimethyl)phenylimidazol-2-ylidene
IPr	1,3-bis(2,6-diisopropylphenyl)imidazol-2-ylidene
IPr*	1,3-bis{2,6-bis(diphenylmethyl)-4-methylphenyl}imidazole-2-ylidene
IPr*^{OMe}	1,3-bis{2,6-bis(diphenylmethyl)-4-methoxyphenyl}imidazole-2-ylidene
***i*Pr**	isopropyl
ITmt	1,3-bis(2,2″,6,6″-tetramethyl-*m*-terphenyl-5′-yl)imidazol-2-ylidene
I*t*Bu	1,3-di-*tert*-butylimidazol-2-ylidene
ITol	1,3-di-*p*-tolyl-imidazol-2-ylidene
MA	maleic anhydride
Me	methyl
Mes	2,4,6-trimethylphenyl
^{Me}IMe	1,3,4,5-tetramethylimidazol-2-ylidene
MMA	methyl methacrylate
MOFLP	metal only frustrated Lewis pair
MVD	metal vapor deposition
NHC	N-heterocyclic carbene
NMR	Nuclear Magnetic Resonance
***n*Pr**	*n*-propyl
NQ	1,4-naphthoquinone
PFAP's	perfluoroalkylphosphines
Ph	Phenyl
RT	room temperature
SIMes	1,3-bis(2,4,6-trimethyl)phenyl-4,5-dihydroimidazol-2-ylidene
SIPr	1,3-bis(2,6-diisopropyl)phenyl-4,5-dihydroimidazol-2-ylidene

SI*t*Bu	1,3-bis(2,6-di-*tert*-butyl)phenyl-4,5-dihydroimidazol-2-ylidene
*t*Bu	*tert*-butyl
*t*BuDAB	bis-*tert*-butyldiazabutadiene
TCNE	tetracyanoethylene
THF	tetrahydrofuran
TIMEN^{tBu}	tris[2-(3-butylimidazol-2-ylidene)ethylamine
TMEDA	tetramethylethylene diamine
TMS	trimethylsilane
TMTBM	3,3'-methylenebis(1-tertbutyl-4,5-dimethylimidazolydene
Vi	vinyl

1. Introduction

The group 10 metals, nickel, palladium and platinum, in the zero oxidation state possess closed-shell nd^{10} electronic configuration. As a result of the availability of only four empty atomic orbitals, i.e., one $(n+1)s$ and three $(n+1)p$ orbitals and the absence of net electrostatic attractive interactions that normally account for great energetic contributions in the formation of transition metal complexes, the coordination number of their complexes is limited up to four. Moreover, the unquestionably "soft" character of these metal centers enhances their affinity toward typical strong-field neutral ligands such CO, alkenes and phosphines. In this context, well-known handbook rules justify the tendency of organometallic chemists to expect a tetrahedral coordination geometry for M(0) complexes of Ni, Pd, and Pt, which would reach this way an electron count of 18.[1]

The first low-oxidation compound, Ni(CO)$_4$, was discovered by Mond[2] as early as 1890. By that time, the nature of the M-CO bond had not yet been established[3] and the compound was considered anomalous owing to its chemical and physical properties. During the 1940s and 1950s, the chemistry of zerovalent metals, particularly those of d^{10}-M(0), experienced a significant rise with the preparation of anionic complexes with the metal in the zero oxidation state, K$_4$[Ni(CN)$_4$][4] and K$_4$[Pd(CN)$_4$][5] and the first examples of tetra-coordinate phosphine complexes of Ni(0), Pd(0) and Pt(0).[6] It soon became apparent that these homoleptic M(PR$_3$)$_4$ complexes displayed a variable stoichiometry, with coordination numbers of four, three and even two.[7]

Two- and three-coordinate species are commonly considered unsaturated and low-coordinate, reaching a 14- or 16-electron count, respectively. Low-coordinate complexes of group 10 metals are well known for their capability to activate unreactive bonds *via* oxidative addition.[8] On this basis, these species are postulated as intermediates in catalytic transformations that proceed through a M(0)/M(II) pathway, including cross-coupling reactions[9] and C-H activation reactions.[8b,10] Although earliest instances of $L_nM(0)$ (n=2, 3) complexes can be traced since the beginning of Organometallic Chemistry, the major impact of palladium catalysis in modern organic synthesis has fostered the interest in the isolation and characterization, in particular, of 14-electron Pd(0) complexes in recent years. In achieving this goal, the development of new ligand platforms that combine strong σ-donor properties and a large steric demand has been instrumental.[11] Notable among them are heteroleptic phosphines (PR_2R') such as PAd_2R introduced by Beller,[12] ferrocenyl-derived phosphines developed by Hartwig,[13] and dialkylbiaryl phosphines put on the stage by Buchwald.[14] To this distinguished group of ligands, N-Heterocyclic carbenes (NHCs), carbon-based σ-donors, must be added. NHCs, long considered as mere phosphine substitutes,[15] currently occupy a remarkable position among common ligands in organometallic chemistry and catalysis.[16]

The aim of this chapter is to review the chemistry of coordinatively unsaturated M(0) complexes of group 10, i.e., two- and three-coordinate species that contain at least one phosphorus(III) or one NHC ligand. Polydentate hybrid ligands with additional oxygen, nitrogen or sulfur donor atoms will not be discussed. The most relevant discoveries are organized in a flexible temporal sequence. Special emphasis will be dedicated to synthetic and reactivity aspects.

2. Three-coordinate complexes
2.1 M(0) complexes with phosphorus(III) ligands
2.1.1 Homoleptic species
In this section, homoleptic complexes of the general formula $M_nL_m^P$ (M=Ni, Pd, Pt; n=1, 2; m=2, 3; L^P=mono- or bidentate P(III) donor ligand, from now on) will be presented in a chronological order, from the first examples of isolated species of this type to the most recent findings in this area. The most common structural motif of $M_nL_m^P$ molecules corresponds to n=1, m=3, that is, trigonal planar mononuclear complexes with monodentate phosphines. However, a few examples of mononuclear and

dinuclear complexes stabilized by bidentate phosphines have been described, which corresponds to n=1, m=2 and n=2, m=3, respectively.

The existence and the reactivity of low-coordinate complexes of group 10 metals bearing phosphines and related ligands attracted the interest of the scientific community since the late 1950s, as witnessed by the pioneering work of Malatesta and Ugo, who described the synthesis of the first homoleptic phosphine and phosphite complexes of palladium(0)[6a] and platinum(0)[6b] with a trigonal planar geometry. The formation of related two-coordinate complexes was also documented (*vide infra*). These early contributions demonstrated that both the synthetic method and the steric hindrance of the ligand affect the stoichiometry of the resulting ML_n^P compounds, i.e., 16-electron complexes (n=3) can be obtained selectively with bulkier ligands or by controlling carefully the molar ratio of the metallic precursor and the P donor ligand.

The first syntheses of tri-coordinate Pd(0) complexes with tertiary phosphines and phosphites was described by Malatesta in 1957,[6a] although the elucidation of the molecular structure of Pd(PPh$_3$)$_3$ was published only in 1987 by Sergienko.[17] Two different reaction strategies were developed (Scheme 1): (a) reduction of Pd(II) precursors in the presence of the P-donor ligand using hydrazine as the reducing agent, (b) substitution/addition reaction of the phosphorated ligand starting from a bis(isocyanide) Pd(0) complex.

A Pd{P(*p*-Tol)$_3$}$_2$I$_2$ + P(*p*-Tol)$_3$ + 2.5 N$_2$H$_4$ ⟶ Pd{P(*p*-Tol)$_3$}$_3$ + 0.5 N$_2$ + 2 N$_2$H$_4$·HI

B Pd{C≡N(*p*-Tol)}$_2$ + 3 P(*p*-Cl-C$_6$H$_4$)$_3$ ⟶ Pd{P(*p*-Cl-C$_6$H$_4$)$_3$}$_3$ + 2 C≡N(*p*-Tol)

Scheme 1 Synthesis of the first tri-coordinate Pd(0) complexes.[6]

The first 16-electron Pt(0) complexes stabilized by phosphorus(III) bases were described 1 year later by the same research group as the result of reduction reactions from Pt(II) complexes (Scheme 2).[6b]

A 4 PtI$_2$L$_2$ + 3 N$_2$H$_4$ ⟶ 2 PtL$_3$ + Pt$_2$I$_4$(N$_2$H$_4$)$_2$L$_2$ + N$_2$ + 4 HI   (L = PPh$_3$, P(OPh)$_3$)

B K$_2$PtCl$_4$ + 4 P(*p*-Cl-C$_6$H$_4$)$_3$ + 2 KOH ⟶ Pt{P(*p*-Cl-C$_6$H$_4$)$_3$}$_3$ + O=P(*p*-Cl-C$_6$H$_4$)$_3$ + 4 KCl + H$_2$O

Scheme 2 Synthesis of the first tri-coordinate Pt(0) complexes.[6b]

Interestingly tris(*p*-clorophenyl)phosphine and triphenylphosphite show a higher tendency to produce tri-coordinate Pt(0) complexes compared to the more basic triphenylphosphine. Some years later the molecular structure of Pt(PPh$_3$)$_3$ was elucidated by X-ray diffraction analyses confirming a trigonal planar geometry around the metal center.[18]

Detailed synthetic procedures for the preparation of several three-coordinate phosphine complexes of Pt(0) were published in 1979 by Yoshida et al.[19]

Di- and three-coordinate Ni(0) complexes stabilized by P-donor ligands proved to be more difficult to isolate and, in spite of some evidence of their existence based on spectroscopic and analytical data,[20] preparation and satisfactory characterization of the tris[tri(o-tolyl)phosphite]nickel(0) complex by reduction of Ni(II) nitrate with NaBH$_4$ in the presence of a phosphite was published only in 1970 by Tolman and Gosser,[21] that is, 13 years after the first palladium analogue. Some years later, two closely related three-coordinate phosphite derivatives were reported.[22] The equilibrium between the unsaturated 16-electron trigonal species and four-coordinate tetrahedral complex in the presence of the phosphite was also studied. Structural characterization of the trigonal planar nickel(0) complex Ni(PPh$_3$)$_3$ was reported only in 1990 by Stephan and co-workers.[23] To the best of our knowledge only one more NiL$_3$ complex, with L = neutral P-donor ligand, has been characterized structurally, namely the tri(o-tolyl)phosphite derivative Ni{P[O(o-tolyl)]$_3$}$_3$ (Fig. 1).[24]

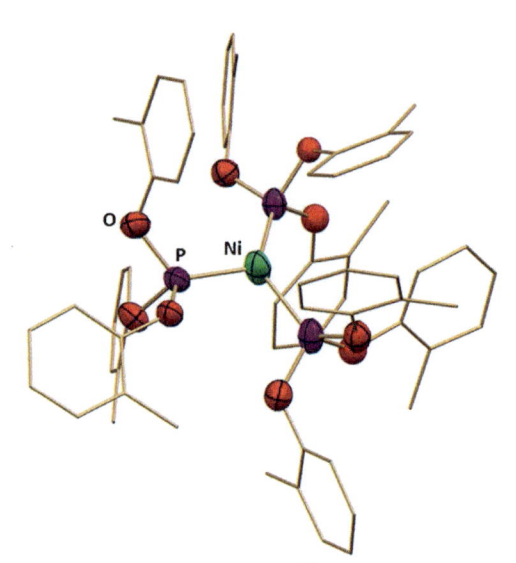

Fig. 1 Molecular structure of Ni{P[O-o-tolyl]$_3$}$_3$.[24]

In addition to successful efforts aimed to accomplish the selective synthesis of homoleptic ML$_3$ complexes of group 10 metals stabilized by phosphines and related ligands, several studies were dedicated to the equilibrium reactions involving two-, three and four-coordinate species of this class (Scheme 3).

$$ML^P_2 \underset{-L^P}{\overset{+L^P}{\rightleftarrows}} ML^P_3 \underset{-L^P}{\overset{+L^P}{\rightleftarrows}} ML^P_4$$

Scheme 3 Equilibrium mixtures of ML^P_n complexes in solution (n = 2, 3 and 4).

By means of NMR spectrometry, cryoscopic molecular weight determination and UV–vis and IR spectroscopies, Tolman provided a deep insight into the right-side equilibrium of Scheme 3 in Ni(0) systems concluding that the corresponding dissociation constant "*is strongly dependent on steric effects*" owing to enthalpic reasons (*i.e.* smaller energies are required for breaking the Ni-P bond of a bulky P-ligand reasonably because of longer bond lengths) or relief of the strain energy which was required to accommodate four ligands in the crowded tetrahedral environment.[22]

Based on ^{13}C{^{1}H} and ^{31}P{^{1}H} NMR studies at variable temperature, Musco and co-workers concluded that the tendency of formation of low-coordinated species of palladium(0), i.e., the thermodynamics of the two reactions represented in Scheme 3, depends on the steric demand of the phosphine, thus increasing in the order: PMe$_3$ ~ PMe$_2$Ph ~ PMePh$_2$ < PPh$_3$ ~ PEt$_3$ < P(CH$_2$Ph)$_3$ ~ P^{i}Pr$_3$ ~ PCy$_3$ ~ P^{t}Bu$_2$Ph.[25,26] So, the fast interconversion of the two-, three-, and four-coordinate complexes in solution generates equilibrium mixtures of the three species PdL^P_n, PdL$^P_{n-1}$, and free L^P (L^P = tertiary phosphine) in variable proportions depending on the temperature and the steric characteristics of L^P. The same research group described in 1975 a convenient access to Pd(0) phosphine complexes,[27] more recently used by Milstein with some modifications,[28,29] consisting in reductive elimination reactions from a Pd(II) allyl derivative promoted by primary amines or phosphines (Scheme 4).

Scheme 4 Synthesis of Pd(0) phosphine complexes described by Milstein.[28,29]

As a further early example of an unsaturated platinum(0) complex with P-donor ligands, King proposed a trigonal planar geometry and κ^1-coordination mode for the diphosphine for a compound with molecular formula Pt(*trans*-Ph$_2$P-CH=CH-PPh$_2$)$_3$, although neither NMR data nor structural characterization were provided.[30]

In 1971 Muetterties and co-workers published a valuable piece of research in which ^{31}P{^{1}H} NMR data were used to deduce the coordination number of PtL$_n^P$ (n=3 or 4) species, taking into account that the $^1J_{PPt}$ coupling constants for PtL$_3^P$ complexes "*are significantly larger than those for PtL$_4$ complexes*", which "*is consistent with the presumed greater s character in Pt-P bonds of tris-phosphine complexes*".[31] In the same contribution, the authors attribute a strong nucleophilic character to the low-coordinate species Pt(PEt$_3$)$_3$ in light of its ability to undergo oxidative addition of H$_2$ and polar bonds such as C-CN, C-Cl, and O-H bonds (Scheme 5):

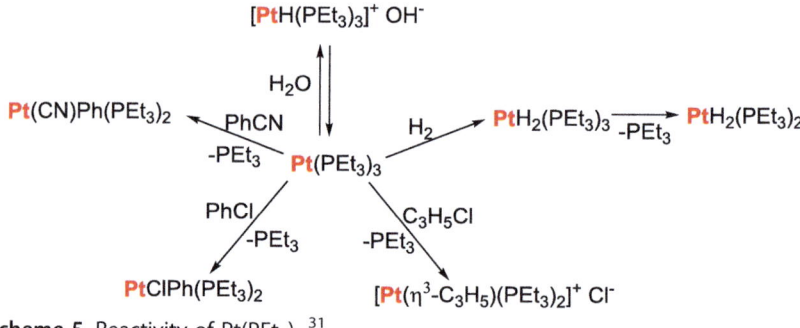

Scheme 5 Reactivity of Pt(PEt$_3$)$_3$.[31]

The nickel congener Ni(PEt$_3$)$_3$ exhibits an analogous reactivity with Ph-X compounds giving rise to Ni(X)Ph(PEt$_3$)$_2$, X=Cl, CN, upon reaction with chlorobenzene and benzonitrile, respectively.[31]

It seems appropriate to include in this section dedicated to homoleptic phosphine complexes with low coordination number some dinuclear species of the general formula MM'(dppm)$_3$ (M, M'=Pd, Pt) stabilized by the bidentate phosphine bis(diphenylphosphine)methane, dppm. As demonstrated by X-ray diffraction analyses for the dinuclear complex Pt$_2$(dppm)$_3$[32,33] and corroborated by numerous spectroscopic data,[34–36] the formal coordination number of the metal center in MM'(dppm)$_3$ (M, M'=Pd, Pt) is three with an additional weaker interaction of the d^{10}-d^{10} type, which compensate partially the coordinative unsaturation (Fig. 2).

Closely related to the aforementioned dinuclear species, the mononuclear palladium complexes Pd(dippp)$_2$ and Pd(dippb)$_2$ (dippp=1,3-bis(diisopropylphosphino)propane, dippb=1,4-bis(diisopropylphosphino)butane), reported by Milstein and co-workers in 1989 and 1993 are claimed to possess an asymmetric trigonal planar geometry at the metal center with the two non-equivalent P,P ligands in a κ^1 and κ^2 coordination mode (Fig. 3A and B).[28,29,37] Moreover, bimetallic species with two Pd$^{(0)}$(κ^2-diphosphine) fragments linked by one further μ^2-dppp or dippb (Fig. 3C and D) are described, although no structural data are provided. Equilibria between mono- and dinuclear species were also studied by means of NMR spectroscopy.

In addition to the asymmetric trigonal planar Pd(0) complexes stabilized by bidentate phosphines reported by Milstein, Fryzuk provided some years later the synthesis and the structural characterization of the analogous 1,2-(diisopropylphosphino)ethane (dippe) derivative, Pd$_2$(dippe)$_3$ (Fig. 3E).[38]

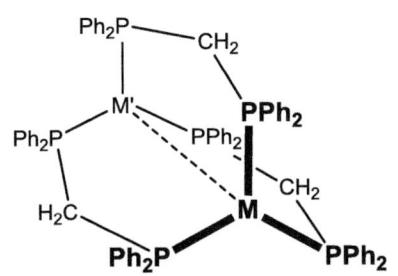

Fig. 2 Dinuclear species of the general formula MM'(dppm)$_3$ showing a d^{10}-d^{10} type interaction (M, M'=Pd, Pt; dppm=bis(diphenylphosphine)methane).

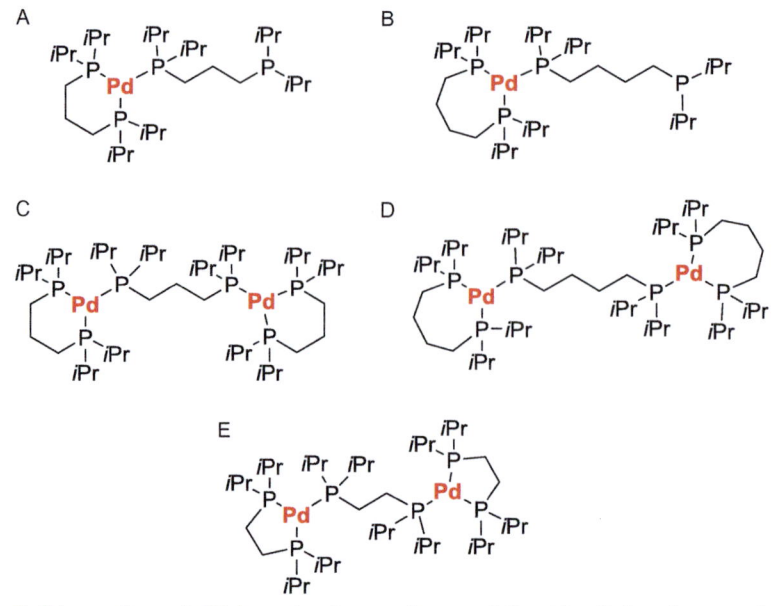

Fig. 3 Tri-coordinate Pd(0) homoleptic complexes stabilized by diphosphines with different coordination modes.[28,29,37,38]

More recently, Braunschweig and co-workers investigated the interaction of highly π-acidic boranes with the Lewis basic three-coordinate Pt(PEt$_3$)$_3$ complex, disclosing its ability to activate C–F bonds in an unprecedented transformation which includes isomerization of the coordinated fluoroborane, oxidation of the metal center, and generation of a zwitterionic Pt(II) complex (Scheme 6).[39]

Scheme 6 C–F bond activation promoted by Pt(PEt$_3$)$_3$.[39]

Perfluoroalkylphosphines (PFAP's) were used by Roddick and collaborators to stabilize low-coordinate platinum(0) complexes and study their reactivity toward dihydrogen, ethylene, and Brønsted acids.[40] Although most of this piece of research published in 2016 was focused on the chemistry of the di-coordinate complex Pt{P(C$_2$F$_5$)tBu}$_2$, the synthesis and molecular structure of the tri-coordinate complex Pt{P(C$_2$F$_5$)Me}$_3$ are described. In particular, phosphine-promoted reductive elimination from the Pt(II) derivative PtPh$_2$(COD) (COD=1,5-cyclooctadiene) (Scheme 7) affords cleanly the trigonal planar complex represented in Fig. 4.

$$\text{PtPh}_2\text{(COD)} \xrightarrow{\text{P(C}_2\text{F}_5)_2\text{Me (exc.)}} \text{Pt\{P(C}_2\text{F}_5)_2\text{Me\}}_3$$

Scheme 7 Synthesis of Pt{P(C$_2$F$_5$)$_2$Me}$_3$.[40]

Reasonably as a consequence of the high π-acidic character of the perfluorinated ligand, i.e., to an increased Pt-P bond order, the metal-phosphorus distances observed in Pt{P(C$_2$F$_5$)Me}$_3$ are significantly shorter than those reported for Pt(PCy$_3$)$_3$, Pt(PtBu$_2$H)$_3$, and Pt(PPh$_3$)$_3$.

2.1.2 Heteroleptic species

The study of low-coordinate heteroleptic species stabilized by phosphines and other phosphorus Lewis bases is closely related to the coordinative reactivity of the corresponding homoleptic ML$_3^P$ and ML$_2^P$ complexes, which are in most cases convenient precursors to produce mixed complexes with CO, alkenes, alkynes, CS$_2$, isocyanates, etc. In the late 1960s, Ugo provided a useful, instructive and admirable review on this topic.[7] In light of the great number of group 10 metals complexes pertaining to this class we considered convenient to separate them into two main categories based on the presence of mono- or diphosphines, respectively.

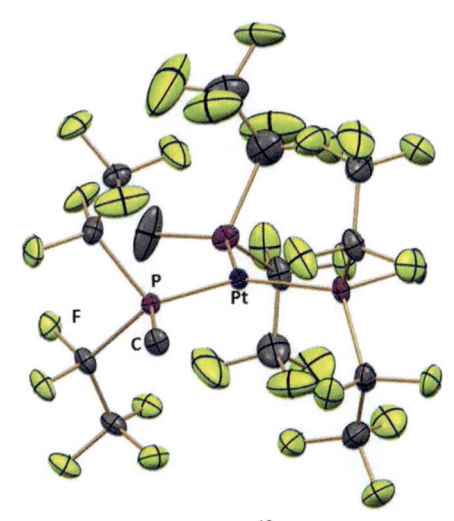

Fig. 4 Molecular structure of Pt{P(C$_2$F$_5$)$_2$Me}$_3$.[40]

2.1.2.1 Monophosphines

This discussion to which this section is dedicated will proceed examining the different classes of complexes according to the type of additional neutral ligands bound to the metal center, e.g., carbon monoxide, alkynes, alkenes, in a flexible chronological order.

In the late 1950s together with the publication of the first three-coordinate complexes of platinum(0) bearing phosphines and phosphites, Malatesta and Carriello showed that carbon monoxide is a convenient partner of P-ligands for the stabilization of low-coordinate species of group 10 metals in the zero oxidation state, owing to its high π-acceptor character.[6b] In particular, the three-coordinate $Pt(CO)\{P(p\text{-}Cl\text{-}C_6H_4)_3\}_2$ complex was prepared by treatment with CO of the homoleptic species $Pt\{P(p\text{-}Cl\text{-}C_6H_4)_3\}_3$ and subsequent dissolution of the reaction product in ethyl ether to promote the loss of carbon monoxide. Two possible pathways to the monocarbonyl bisphoshine derivative were proposed and are outlined in Scheme 8.

A $Pt(CO)_2\{P(p\text{-}Cl\text{-}C_6H_4)_3\}_2 \longrightarrow Pt(CO)\{P(p\text{-}Cl\text{-}C_6H_4)_3\}_2 + CO$

B $Pt(CO)_2\{P(p\text{-}Cl\text{-}C_6H_4)_3\} + \{P(p\text{-}Cl\text{-}C_6H_4)_3\} \longrightarrow Pt(CO)\{P(p\text{-}Cl\text{-}C_6H_4)_3\}_2 + CO$

Scheme 8 Possible pathways for the synthesis of $Pt(CO)\{P(p\text{-}Cl\text{-}C_6H_4)_3\}_2$.

Interestingly, the use of the more donating triphenylphosphine only permitted to access tetrahedral mixed species, i.e., $Pt(CO)_2(PPh_3)_2$ and $Pt(CO)(PPh_3)_3$, under analogous reaction conditions, thus demonstrating that the electronic properties of the phosphorated ligand are crucial in determining the preferred coordination number for Pt(0) species. In spite of these early findings on mixed carbonyl/phosphine species, the high tendency of Pt(0) to form molecular clusters in the presence of CO is responsible for the reduced number of mononuclear species of the general formula $Pt(CO)_n L_m^P$ (n+m=3 or 4) described to date. In recent years, Braunschweig et al. succeeded in the preparation and full characterization of the first 16-electron monocarbonyl platinum complex stabilized by monodentate phosphines using the sterically demanding tris(cyclohexyl)phosphine (Fig. 5).[41]

As far as nickel(0) chemistry is concerned, in spite of the great number of tetrahedral tricarbonyl(monophosphine) or (-phosphite) complexes studied by Tolman as probes for the quantification of the electronic and steric

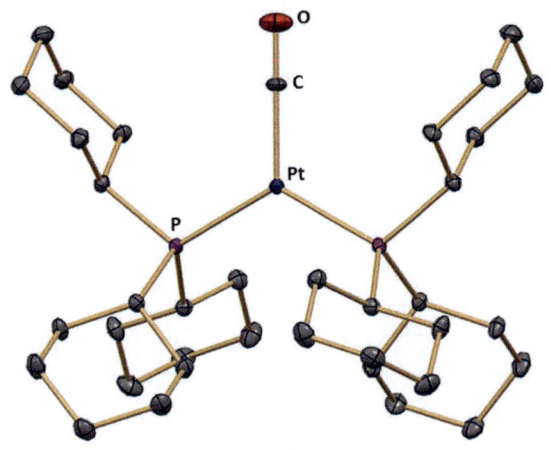

Fig. 5 Molecular structure of Pt(CO)(PCy$_3$)$_2$.[41]

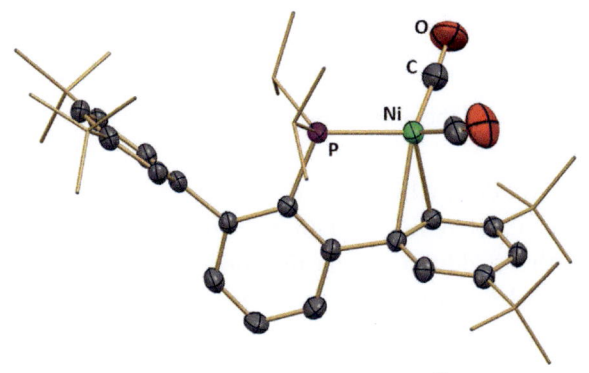

Fig. 6 First Ni(CO)$_2$L^P complex structurally characterized.[43]

properties of L^P ligands,[42] three-coordinate Ni(0) species of the type Ni(CO)$_2$L^P are far less common. As a matter of fact, and to the best of our knowledge, only in 2019 the use of bulky terphenyl substituted phosphines permitted the isolation of the first stable complexes of this class (Fig. 6).[43] It is to specify, anyway, that the presence of non-covalent aryl-metal interactions within the cited molecules put them somehow in the middle between 3- and 4-coordinate species.

Unsaturated hydrocarbons have been used since the 1960s to stabilize heteroleptic unsaturated complexes of group 10 metals in conjunction with L^P ligands, being acetylenes the first class to be investigated in detail.

The influence of electron-withdrawing groups on the relative stability of various three-coordinate alkyne complexes of the general formula Pt(PPh$_3$)$_2$(alkyne), prepared as shown in Scheme 9, was firstly examined by Chat,[44] and later by Allen and Cook.[45]

$$\text{PtCl}_2(\text{PPh}_3)_2 + \text{alkyne} \xrightarrow{\text{NH}_2\text{NH}_2} \text{Pt}(\text{PPh}_3)_2(\text{alkyne})$$

Scheme 9 Synthesis of Pt(PPh$_3$)$_2$(alkyne) complexes.[44,45]

These studies revealed that the introduction of electron-withdrawing groups in the molecular structure of the coordinated acetylene increases the stability of Pt(PPh$_3$)$_2$(alkyne) complexes as a consequence of enhanced π-backdonation. For instance, the equilibrium constants for the displacement of phenylacetylene from Pt(PPh$_3$)$_2$(PhCCH) (Scheme 10) are 125 and 0.5 by using 3-nitrophenylacetylene and 2-methoxyacetylene, respectively:

$$\text{Pt}(\text{PPh}_3)_2(\text{PhC}\equiv\text{CH}) + \text{RC}\equiv\text{CH} \rightleftharpoons \text{Pt}(\text{PPh}_3)_2(\text{RC}\equiv\text{CH}) + \text{PhC}\equiv\text{CH}$$

Scheme 10 Equilibrium of the displacement of phenylacetylene in Pt(PPh$_3$)$_2$(PhCCH).

In accord to IR data, which include absorptions around 1700 cm^{-1} for the acetylenic C-C stretching vibration of the coordinated alkyne, Chatt suggested that these adducts can be regarded as Pt(II) complexes (Fig. 7) or, at least, as a hybrid structure between the two represented below.

Structural data provided by Grim et al. for the closely related diphenyl acetylene adduct Pt(PPh$_3$)$_2$(PhC≡CPh) are consistent with a trigonal planar geometry around the Pt(0) center with a strong contribution of the metal-to-alkyne π-backdonation to the Pt-phenylacetylene bonding, which results in a large C-C bond distance of 1.32 Å, that is typical of a double bond, and a trigonalization of both acetylenic carbons with Pt-C-Ph angles of about 40°.[46] Growing interest in the comprehension of metal-alkyne bonding led also to the structural elucidation of the acetylenedicarboxylate complex Pd(PPh$_3$)$_2$(MeO$_2$CC≡CCO$_2$Me).[47]

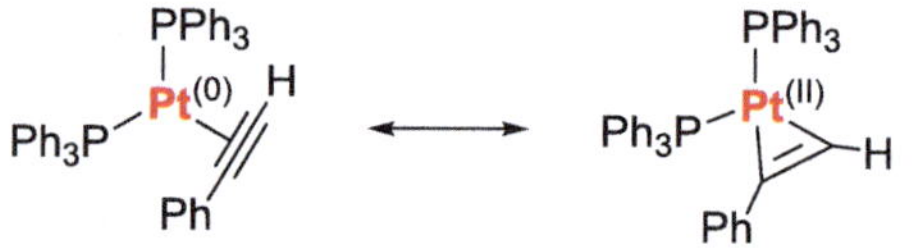

Fig. 7 Hybrid structures for Pt(PPh$_3$)$_2$(PhCCH) proposed by Chatt.[44]

Maitlis and Fitton studied extensively alkyne complexes of palladium(0) and developed a new synthetic procedure based on the simple substitution of triphenylphosphine by the desired alkyne starting from the saturated palladium(0) compound Pd(PPh$_3$)$_4$ (Scheme 11).[48,49]

Pd(PPh$_3$)$_4$ + alkyne ⟶ Pd(PPh$_3$)$_2$(alkyne) + 2 PPh$_3$

Scheme 11 Synthesis of palladium(0) alkyne complexes proposed by Maitlis and Fitton.[48,49]

A variety of analogous Pt(0) compounds either with aryl or alkyl phosphines were also reported by Maitlis in the late 1960s, whereas nickel congeners, described as "very air sensitive and too unstable for elemental analyses,"[50] proved to be more difficult to handle and prepare, and required a specific synthetic methodology based on the displacement of ethylene from the parent alkene complex (Scheme 12). Interestingly, attempts to prepare acetylenebis(trialkylphosphine)nickel(0) complexes were unsuccessful.

[Ni(acac)$_2$]$_3$ + Et$_3$Al + PPh$_3$ + C$_2$H$_4$ ⟶ (Ph$_3$P)$_2$Ni(C$_2$H$_4$) $\xrightarrow{RC\equiv CR'}$ (Ph$_3$P)$_2$Ni(RC≡CR')

R = Ph, CF$_3$
R' = Me, Ph, CF$_3$

Scheme 12 Synthesis of nickel(0) alkyne complexes proposed by Maitlis.[50]

Full characterization of the first phosphine/acetylene complex of Ni(0) stable at room temperature was reported in 1982 by Pörschke, who succeeded in its preparation again by ligand substitution starting from the parent ethene complex.[51] Ni(COD)$_2$ was also used as the metal precursor for the synthesis of a variety of non-symmetrically substituted alkyne derivatives of Ni(0) with phosphines and phosphites.[52]

Mechanistic investigations by Halpern and co-workers concluded that both associative (Scheme 13A) and dissociative (Scheme 13B) pathways take place when reacting Pt(PPh$_3$)$_3$ with prop-1-yn-1-ylbenzene, thus implying

A Pt(PPh$_3$)$_3$ + PhC≡CCH$_3$ ⇌ Pt(PPh$_3$)$_2$(PhC≡CCH$_3$) + PPh$_3$

B Pt(PPh$_3$)$_3$ $\underset{+PPh_3}{\overset{-PPh_3}{\rightleftharpoons}}$ Pt(PPh$_3$)$_2$ $\underset{-PhC\equiv CCH_3}{\overset{+PhC\equiv CCH_3}{\rightleftharpoons}}$ Pt(PPh$_3$)$_2$(PhC≡CCH$_3$)

Scheme 13 Associative (A) and dissociative (B) pathways in the reaction of Pt(PPh$_3$)$_3$ with PhCCMe.[53,54]

that both 14- and 18-electron configurations are accessible for the d^{10} platinum(0) starting material in the formation of the mixed alkyne-phosphine product.[53,54]

Coordination to the basic 14-electron Pt(PPh$_3$)$_2$ fragment permitted the stabilization of the otherwise unstable cycloheptyne thanks to the strain-release effect provided by the metal-to-alkyne π-backdonation, as reported by Bennet in 1971 (Scheme 14).[55]

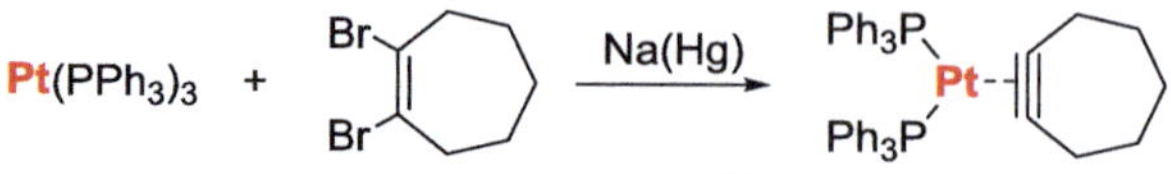

Scheme 14 Synthesis of Pt(PPh$_3$)$_2$(cycloheptyne).[55]

Palenik and collaborators described in 1974 some rare examples of dinuclear alkyne complexes of Pd and Pt in which both the phosphine- and fluorinated alkyne functionalities are located on the same ligand. Syntheses were accomplished by reduction of M(II) precursors (Scheme 15A) or ligand substitution and elimination from the already popular M(PPh$_3$)$_4$ complexes (Scheme 15B).[56]

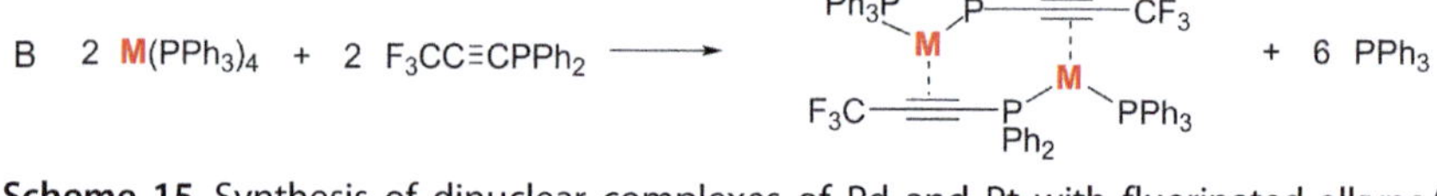

Scheme 15 Synthesis of dinuclear complexes of Pd and Pt with fluorinated alkyne/phosphine hybrid ligands.[56]

Alkenes are undoubtedly the largest class of hydrocarbons to be employed in combination with P(III) ligands to create heteroleptic species of Ni, Pd, and Pt in the zero oxidation state.

Only some years after the isolation of Pt(PPh$_3$)$_4$, Cook and Jauhal[57] declared that treatment of the oxygen adduct Pt(PPh$_3$)$_2$(O$_2$) (to be considered as a Pt(II) peroxide complex in light of the large O-O distance)[58] with sodium borohydride in the presence of ethylene afforded the air stable complex Pt(PPh$_3$)$_2$(C$_2$H$_4$), although satisfactory characterization of this

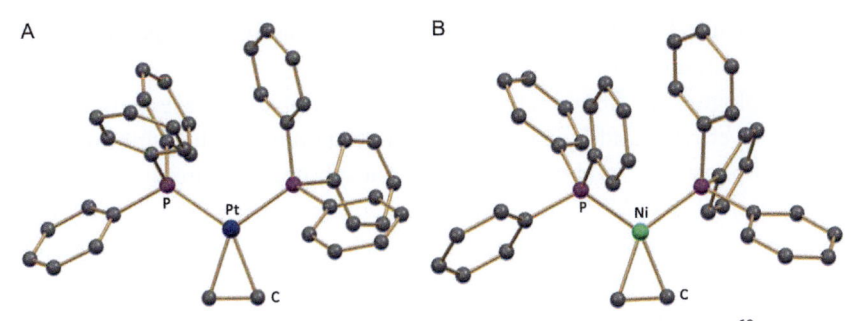

Fig. 8 Molecular structures of: (A) Pt(PPh$_3$)$_2$(C$_2$H$_4$) and (B) Ni(PPh$_3$)$_2$(C$_2$H$_4$).[60]

compound was provided some years later together with the discussion of the nickel analogue, Ni(PPh$_3$)$_2$(C$_2$H$_4$).[59,60] In both complexes, the two P atoms and the C=C double bond are closely coplanar in the solid state and, based on the similarity of the C-C separations within the coordinated alkene, analogous bonding situation is deduced (Fig. 8). Interestingly, NMR studies demonstrated that, whereas the nickel complex dissociate in toluene solution, the platinum congener is largely undissociated and, on the contrary, associates when excess ethylene is present, thus reaching an 18-electron configuration.

Some years later, Pt(PPh$_3$)$_2$(C$_2$H$_4$) was reported to react straightforwardly with cyclopropene and several methyl-substituted cyclopropenes to afford the expected adducts upon loss of ethene.[61]

Reversible coordination of ethylene was documented by Hampden-Smith and co-workers at a Pt(PCy$_3$)$_2$ fragment,[62] thus highlighting the importance of the steric bulk of the P-ligand in the stabilization of di-coordinated species (*vide infra*).

For the sake of completeness, it has to be mentioned in this context the pioneering work of Wilke and Herrmann,[63] who reported in 1962, almost a decade before the publication of the structural details of the compound, the synthesis of Ni(PPh$_3$)$_2$(C$_2$H$_4$) (Scheme 16A) and several alkenes and alkynes

A Ni(acac)$_2$ $\xrightarrow{\text{AlEt}_2(\text{OEt}) \text{ (2 eq)}, \text{PPh}_3 \text{ (2 eq)}, \text{C}_2\text{H}_4 \text{ (exc.)}}$ Ni(PPh$_3$)$_2$(C$_2$H$_4$)

B Ni(PPh$_3$)$_2$(C$_2$H$_4$) + L $\longrightarrow$ Ni(PPh$_3$)$_2$(L) + C$_2$H$_4$

L = styrene, stilbene, α-methylstyrene,
diphenylacetyene, 2-butyne

Scheme 16 Wilke's syntheses of (A) Ni(PPh$_3$)$_2$(C$_2$H$_4$) and (B) Ni(PPh$_3$)$_2$(L) (L=alkene or alkyne).[63]

adducts obtained by simple substitution of ethylene from the parent Ni(0) complex (Scheme 16B).

An analogous synthetic strategy permitted the isolation of a series of three-coordinate ethylene adducts of palladium containing triphenylphosphine, tri(o-tolyl)phosphite, and tri(cyclohexyl)phosphine, as communicated by de Jongh in 1971.[64] Ni(COD)$_2$ was also used as the metal precursor for the high yield preparation of Ni[P(o-tolyl)$_3$]$_2$(stilbene) and Ni(PPh$_3$)$_2$(stilbene) in a one-pot synthesis reported by Ibers.[65]

Triarylphosphites were also used to stabilize Ni(0)-alkene fragments, as documented by the early contributions by Tolman and Seidel,[66] and Guggenberger (Fig. 9).[67]

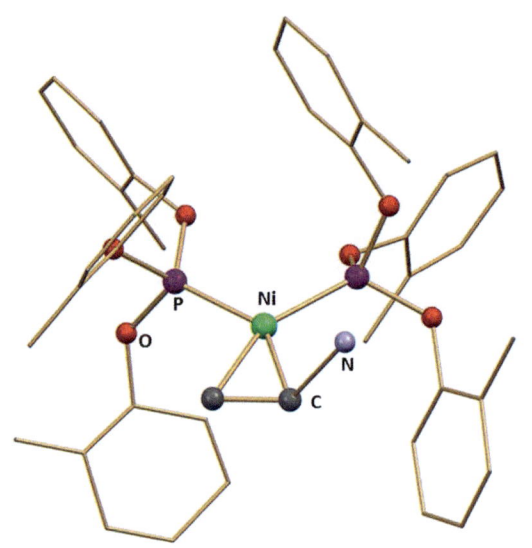

Fig. 9 Molecular structure of Ni{P(O-o-tolyl)$_3$}$_3$(acrylonitrile).[67]

Since the late 1960s, strongly π-acidic olefins such as unsaturated esters,[68] aldehydes, ketones,[69] and anhydrides,[70] quinones,[71] halo-[72] and cyanoalkenes,[73] have been reported to confer exceptional stability to M(0) complexes of the general formula ML$_2^P$(olefin).[49,74] As a representative example, Baddley et al. were able to induce reductive elimination at a hydrido chloride platinum(II) complex by treatment with tetracyanoethylene[75] thus generating a stable three-coordinate complex, whose molecular structure in the solid state is shown in Fig. 10.[76]

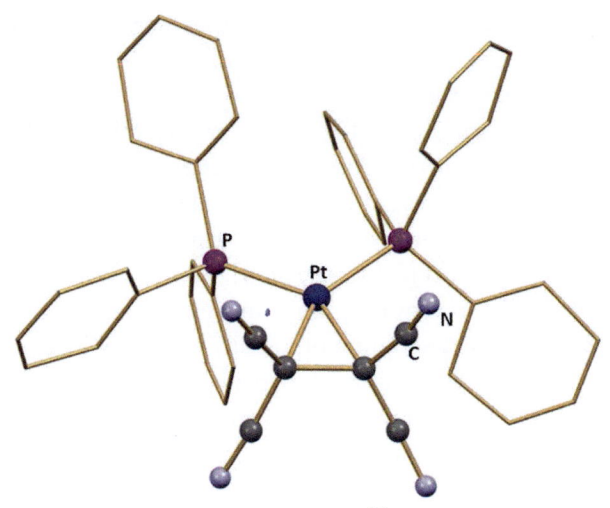

Fig. 10 Molecular structure of Pt(PPh$_3$)$_2${C$_2$(CN)$_4$}.[76]

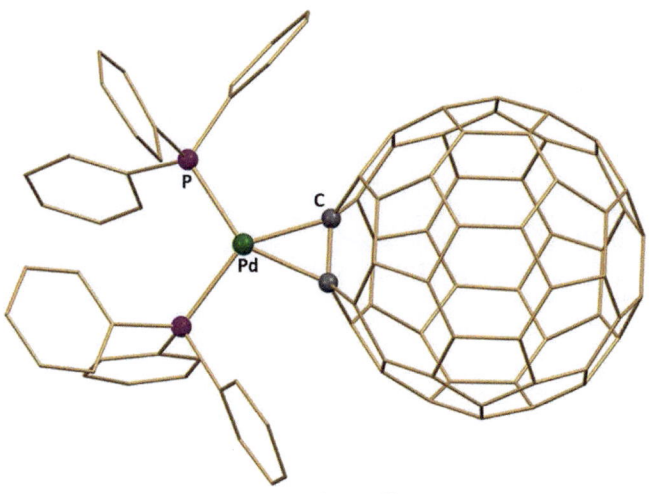

Fig. 11 Molecular structure of Pd(PPh$_3$)$_2$(η^2-C$_{60}$).[78]

Among the plethora of three-coordinate alkene complexes of group 10 metals bearing two L^P ligands characterized in the 80s and 90s,[77] the first fullerene derivative of a transition metal requires a special comment as the first demonstration that this allotropic form of carbon can interact with metal centers in a similar way to simple olefins. Fagan and collaborators prepared the fullerene-bonded compound Pt(PPh$_3$)$_2$(η^2-C$_{60}$) by direct ligand substitution, once again from the parent ethylene complex (Fig. 11).[79] Analogous

Ni and Pd species were also characterized, together with unprecedented hexametallic complexes of the general formula [(PEt$_3$)$_2$M]$_6$(C$_{60}$), M=Pt, Pd, Ni, displaying an idealized T$_d$ symmetry.[80]

The class of heteroleptic complexes of the general formula M(alkene)$_2$L^P is far less populated than that containing related bis(phosphine) mono(olefin) complexes. Nevertheless, the possibility of generating reactive 12-electron-[L^{P}M] fragments by facile olefin extrusion have prompted the scientific community to put effort on the preparation of complexes of this type since the 1970s, as witnessed by the pioneering research of Stone and Spencer which culminated in the description of a family of bis(ethylene) platinum(0) complexes with several alkyl and aryl tertiary phosphines.[81,82] Pt(C$_2$H$_4$)$_3$ was the precursor of choice for those transformations.

Closely related to M(alkene)$_2$L^P species, complexes of the general formula M(diene)L^P were also described owing to their intimate relation with the popular Karstedt's catalyst.[83] In 1999 Pörschke described[84] a great number of compounds of this type obtained in most cases by simple treatment of homoleptic complexes of the general formula M$_2$(diene)$_3$ (M=Pd, Pt) with L^P ligands characterized by diverse electronic and steric properties (Scheme 17A). Dimethyl or diallyl Pd(II) species were also used to generate

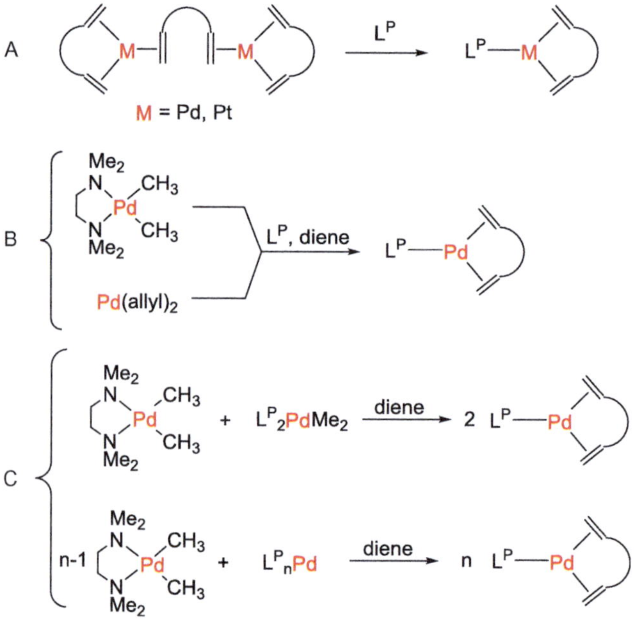

Scheme 17 Syntheses of M(diene)L^P complexes (M=Pd, Pt) reported by Pörschke.[84]

M(alkene)$_2$L^P species by reductive elimination reactions triggered by L^P ligands (Scheme 17B) or by other palladium complexes in which the metal exhibits the 0 or +2 oxidation states (Scheme 17C).

As reported by Markó et al. in 2007,[85] hexachloroplatinic acid is successfully converted in reducing conditions in the presence of diallylether and triphenylphosphine in the tri-coordinate diene phosphine complex represented in Scheme 18A. Noteworthy, the use of the dinuclear complex Pt$_2$(diallylether)$_3$ as the starting material led to the mixture of products showed in Scheme 18B. Several catalytic applications were reported for this and related class of compounds but they will not be discussed here.[86]

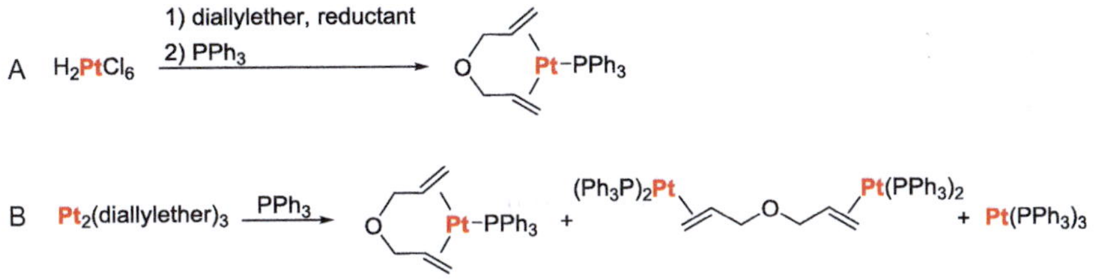

Scheme 18 Synthetic routes for Pt(PPh$_3$)(diallylether) complexes.[85]

A series of bis(alkene) complexes of Pt(0) with the monophosphines PPh$_3$ and PCy$_3$ occupying the third coordination site, were prepared by Hoyte and Spencer by treatment of the corresponding bis(ethene) species, Pt(PR$_3$)(C$_2$H$_4$)$_2$, with methylenecyclopropane or bicyclopropylidene (Scheme 19).[87]

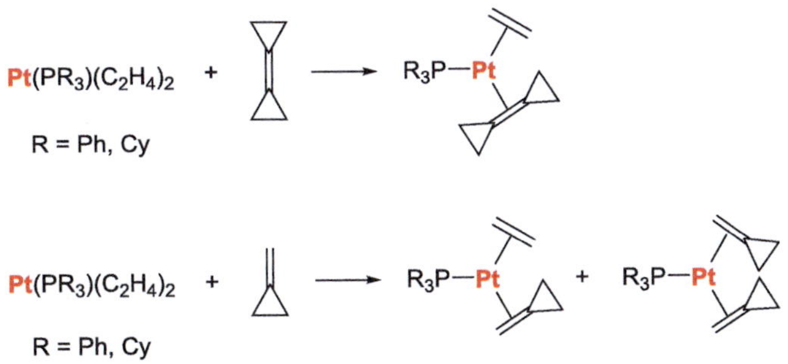

Scheme 19 Synthesis of bis(alkene) complexes of Pt(0) stabilized by PPh$_3$ and PCy$_3$.[87]

More recently Carmona and co-workers communicated that the terphenylphosphine adduct of platinum dichloride represented in

Scheme 20 is conveniently reduced by metallic zinc under ethylene to afford the corresponding bis(ethylene) phosphine complex, which undergoes reversible ethylene dissociation thus generating a formally 14-electron species.[88]

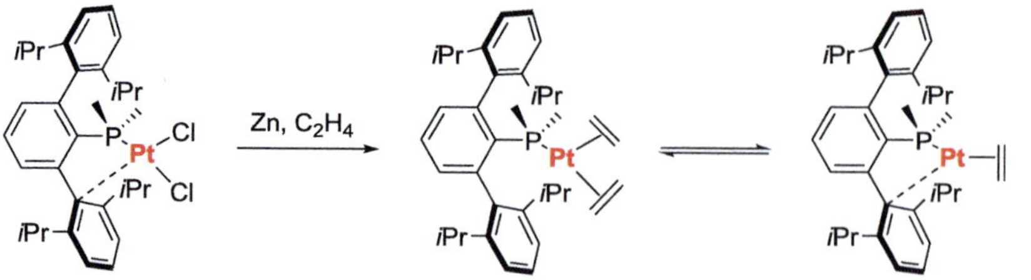

Scheme 20 Synthesis of the bis- and mono-ethylene terphenylphosphine complexes of Pt(0) reported by Carmona.[88]

It is worth mentioning the valuable work of Amatore and collaborators, who provided much information on Pd(0) species stabilized by phosphines and other ligands mainly making use of electrochemical techniques.[89]

2.1.2.2 Diphosphines

Chelating ligands, and particularly diphosphines and related ligands, have been used since the 1980s as crucial components in catalytic processes based on group 10 metals[90] by virtue of their ability to stabilize kinetically and thermodynamically reaction intermediates, which often consist in unsaturated 14-electron species with the metal center in a +2 or 0 oxidation state. Consequently, research contributions focused on heteroleptic complexes of group 10 metals of the general formula M(0)L^{PP}L′ (L^{PP} = bidentate ligand with two P(III) donors, from now on; L′ = neutral 2-electron ligand) are remarkably numerous. In this section, we aim to offer an actualized and concise view on this topic with special emphasis to the most recent findings.

Inspired by the original work of Wilke based on triphenylphosphine,[63] Pörschke and collaborators dedicated great effort to the understanding of alkyne and alkene adducts of group 10 metals stabilized by bidentate phosphines. The combination of 1,2-bis(dimethylphosphino)ethane (dmpe) and Ni(C$_2$H$_4$)$_3$ in conjunction with symmetric and asymmetric alkynes led to several mono- and dinuclear heteroleptic complexes of nickel(0) with a trigonal planar geometry at the metal center,[91] as shown in Scheme 21. Interestingly, the nuclearity of the mixed dmpe/alkyne complexes, i.e., the relative stability of the mononuclear and the dinuclear species, depends strictly on the R and R′ substituents of the alkyne, RC≡CR′.

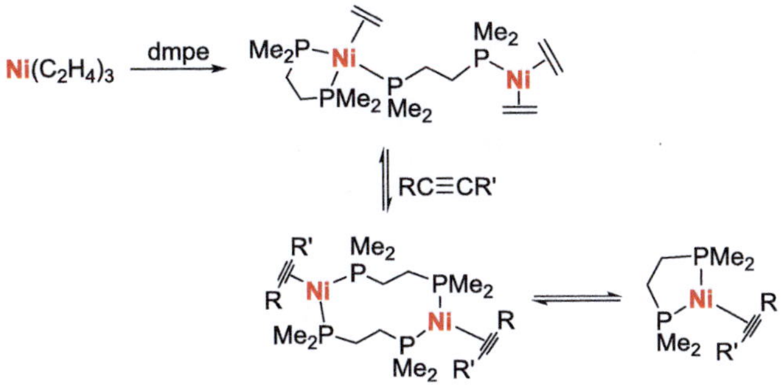

Scheme 21 Synthesis of mono- and dinuclear heteroleptic dmpe/alkyne complexes of Ni(0).[91]

Further contributions of the same research group focused on novel transformations and structures based on 14-electron fragments in which the M(0) center (M=Ni, Pd, Pt) coordinates the bulky diphosphines bis(di-*tert*-butylphosphino)ethane (dtbpe) and bis(diisopropylphosphino)ethane (dippe). In particular, the interaction between the [Ni(dtbpe)] and [Ni(dippe)] moieties with aromatic compounds[92] and cyclic dienes and polyenes[93] permitted the stabilization of unusual species, such as: (a) the μ-η^2:η^2-benzene-bridged dinickel complex showed in Scheme 22A, which is easily converted in a η^2-C$_6$F$_6$ mononuclear complex (Fig. 12) by

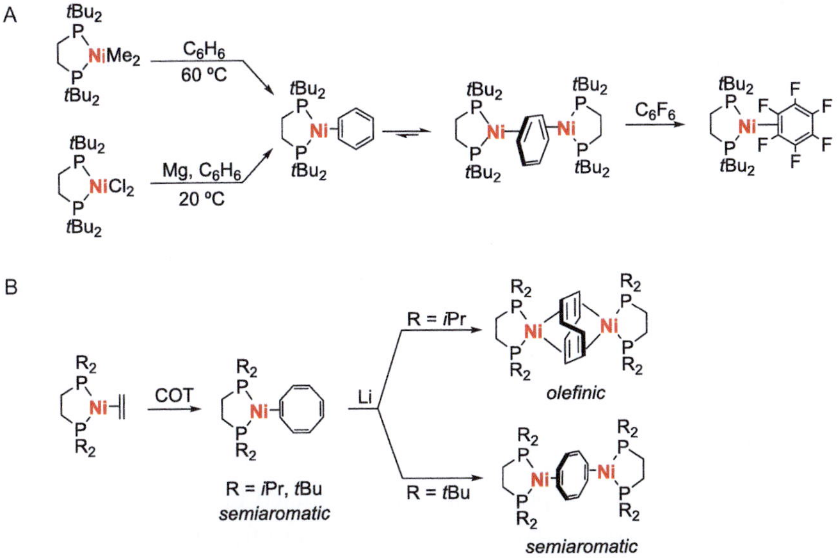

Scheme 22 Synthesis and reactivity of Ni(0) complexes stabilized by diphosphines described by Pörschke.[92,94]

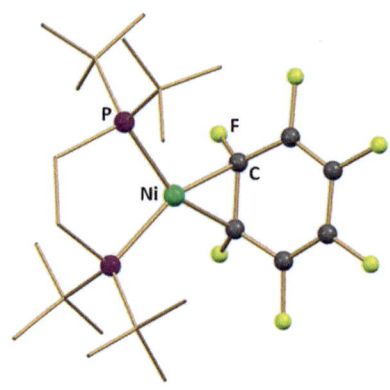

Fig. 12 Molecular structure of Ni(dtbpe)(η^2-C$_6$F$_6$).[92]

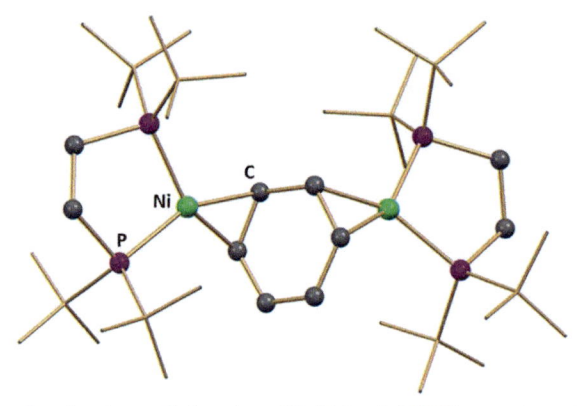

Fig. 13 Molecular structure of dinuclear [Ni$_2$(dtbpe)$_2$(COT)] complex with a bridging semiaromatic COT ligand.[94]

treatment with hexafluorobenzene[92]; (b) mono- and dinuclear Ni(0) complexes with the cyclooctatetraene ligand (COT) (Fig. 13) displaying olefinic or semiaromatic structures (Scheme 22B).[94] Related chemistry of Pd(0)- and Pt(0) diphosphine fragments interacting with cyclooctatetraene and its reduced forms was also explored.[95]

As far as palladium chemistry is concerned,[96,97] Pd(η^3-allyl)$_2$ was conveniently used as the metal precursor to access mixed bis(phosphine)/diene species by straightforward reductive elimination promoted by the bidentate ligand at low temperature (Scheme 23).

Low-coordinate M(0) complexes of group 10   265

Scheme 23 Synthesis of bis(phosphine)/diene Pd(0) species from Pd(η³-allyl)₂.[96,97]

Acetylene displaces easily alkenes and dienes from the [Pd(diphosphine)] fragments to generate the corresponding alkyne complexes, [Pd(diphosphine)(acetylene)], and ethyne-bridged dinuclear species (Scheme 24). As an example, the molecular structure of [Pd(dippe)(phenylacetylene)] is displayed in Fig. 14.

Scheme 24 Synthesis of Pd(diphosphine)(acetylene) complexes *via* diene displacement by acetylene.[96,97]

Ethylene is also displaced by CO from the alkene derivative Pd(dtbpe)(C₂H₄) to afford the corresponding tetra-coordinate dicarbonyl bisphosphine complex, which is easily converted in the CO-bridged

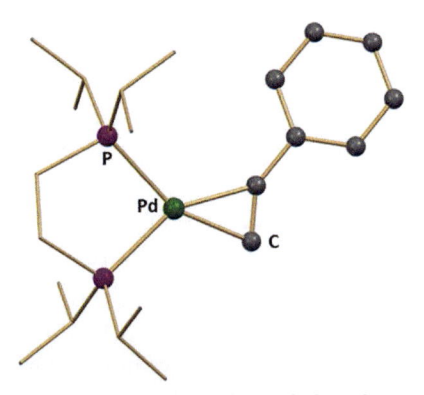

Fig. 14 Molecular structure of phenylacetylene diphosphine complex reported by Pörschke.[97]

dinuclear species [(μ-CO){Pd(dtbpe)}$_2$].[98] More recently, CO exchange at Pt(0) centers stabilized by bidentate phosphines has been studied by Pringle and Wass.[99]

Motivated by the importance of palladium bisphosphine complexes in a variety of catalytic processes, several research groups have worked since the late 1980s on the synthesis and reactivity of species of the general formula PdL^{PP}L′, with L′=neutral 2-electron ligand and L^{PP}=4,5-bis[(diphenylphosphino)-methyl]-2,2-dimethyl-1,3-dioxolane (diop),[100] ferrocenyl-,[101] biphenyl-,[102] cyclohexyl-[103] and isopropyl[104] substituted diphosphines, diphosphines with large bite angles,[105] chiral diphosphines.[106] Theoretical studies were also reported.[107] For the sake of conciseness, these contributions will not be discussed here.

The use of the bulky bidentate P-ligand dtbpe was crucial for the isolation of the first example of a non-heteroatom-stabilized alkylidene complex of Ni(0) described in 2002 by Mindiola and Hillhouse (Scheme 25).[108]

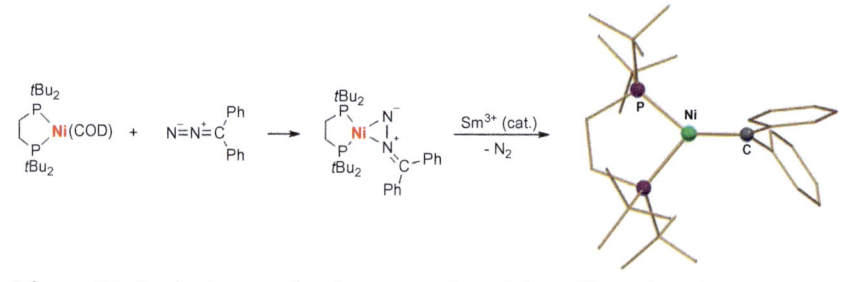

Scheme 25 Synthetic route for the preparation of the Hillhouse's carbene complex, Ni(dtbpe)(=CPh$_2$).[108]

In the original publication the authors disclosed a clear nucleophilic reactivity for the Ni carbene complex [(dtbpe)Ni=CPh$_2$] as proved by its reactions with small acidic molecules such as CO, CO$_2$, SO$_2$, O=C=CPh$_2$, and HNMe$_2$Ph$^+$.

Somehow related to the work of Hillhouse and Mindiola, Iluc and co-workers synthesized and explored the reactivity of a rare example of a nucleophilic palladium carbene generated by C-H activation at a C$_{sp3}$ atom pertaining to a diphosphine ligand with a large bite-angle.[109] Reduction of this alkylidene complex with dihydroanthracene resulted in the tricoordinate Pd(0) species shown in Scheme 26A.[110] Starting from di-coordinate species of Pd(0) and Pt(0) stabilized by a diphosphine ligand,

Scheme 26 Three-coordinated Pd(0) species prepared by Iluc.[110]

mixed diphosphine–carbene complex were also generated by reaction with diazocompounds (*vide infra*).[111]

Two additional ligands with longer linkers, i.e., –CH$_2$CH$_2$–[112] and –CH(Me)CH(Me)– (Fig. 15),[113] between the two diisopropylphenyl moieties were used by the same research group to access tri-coordinate diphosphine/alkene complexes of palladium(0) (see Scheme 26B and C).

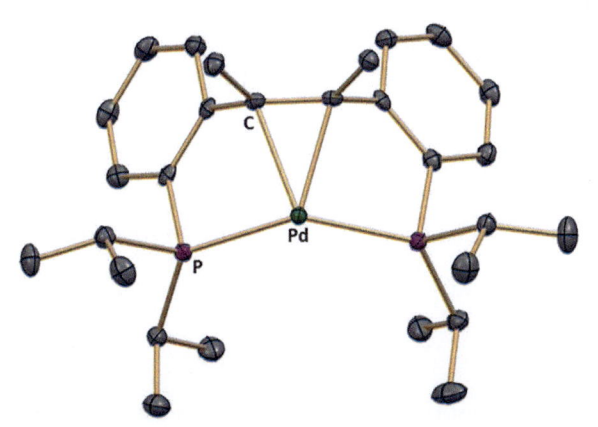

Fig. 15 Molecular structure of Pd(*i*Pr$_2$P-(*o*-C$_6$H$_4$)C(Me)=C(Me)-(*o*-C$_6$H$_4$)-P*i*Pr$_2$).[113]

Another recent and unusual example of a palladium(0) complex stabilized by a large-bite-angle diphosphine has recently been reported by Fairlamb and Lin, who documented that the Pd(II) precursor Pd(η^5-C$_5$H$_5$)(η^3-C$_3$H$_4$Ph) is reduced by a dibenzylideneacetone-derived P,P ligand to afford the dinuclear Pd(0) complex shown in Scheme 27.[105c]

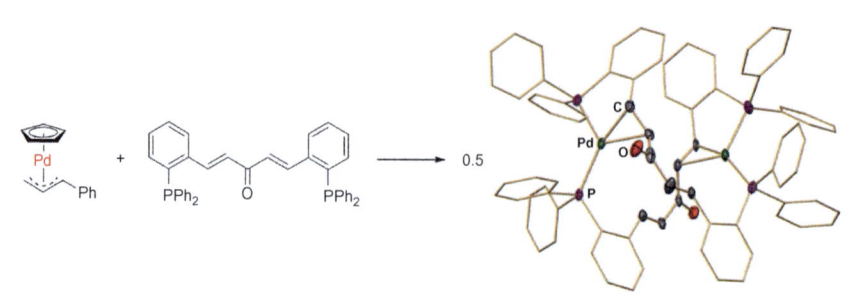

Scheme 27 Synthesis and molecular structure of dinuclear Pd(0) complex stabilized by dibenzylideneacetone-derived P,P ligand.[105c]

Bidentate phosphines of different steric demands and bite angles were also employed to stabilize three-coordinate platinum(0) derivatives of the general formula PtL^{PP}(L′), with L=methylenecyclopropane, bicyclopropylidene, and allylidenecyclopropane (Scheme 28).[87] As a representative species of this series, the molecular structure of the 1,3-bis (diphenylphosphino)propane, dppp, complex is illustrated in Fig. 16.

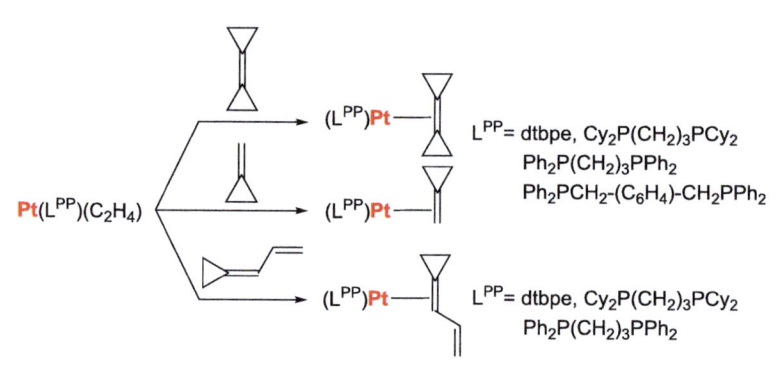

Scheme 28 Synthesis of various tri-coordinate platinum(0) complexes supported by bidentate phosphines.[87]

The growing interest of nickel compounds in catalysis is witnessed by a conspicuous number of articles published since the late 2000s, most of them

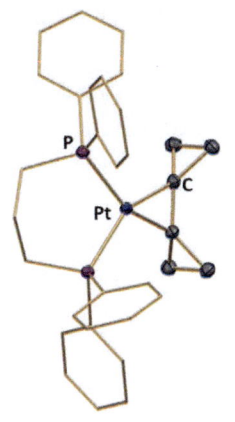

Fig. 16 Molecular structure Pt(bicyclopropylidene)(dppp).[87]

focused on mechanistic investigations, related to a variety of catalytic processes such as isomerization of 2-methyl-3-butenenitrile,[114] hydrogenation of alkynes,[115] hydrocyanation of olefins,[116] Suzuki coupling reactions of N,O-acetals,[117] transfer hydrogenation of unsaturated carbonyl compounds,[118] arene C-H amination,[119] decarbonylative C-H arylation,[120] cycloadditions and cross coupling reactions,[121] carboxylation of olefins with CO_2,[122] reactions involving epoxides,[123] catalyst transfer polymerization,[124] in which low coordinate Ni(0) species are invoked as key intermediates. Among the great number of species reported in the aforementioned publications that should be considered of interest for this review, we decided to present some representative examples exhibiting special structural novelty.

The first one was provided some years ago by Agapie and co-workers, who employed a terphenyl diphosphine to obtain straightforwardly an unusual diphosphine η^2-arene complex of nickel(0) (*vide infra*), which is subsequently reacted with the ylide species $Ph_2S=CH_2$ to promote the insertion of a methylene group into a P-C$_{arene}$ bond (Scheme 29).[119]

Scheme 29 Reaction of NiL^{PP} complex with $Ph_2S=CH_2$ described by Agapie.[119]

Two further examples were reported last year by Kennepohl and co-workers[124] and consist in the coordination of the 12-electron fragment

[Ni(dtbpe)] to thiophene and dithiophene to generate a η^2:η^2 thiophene bridged dinuclear complex and a η^2 dithiophene mononuclear complex, respectively (Scheme 30). Both species are of interest as intermediates in catalyst transfer polymerization of thiophene.

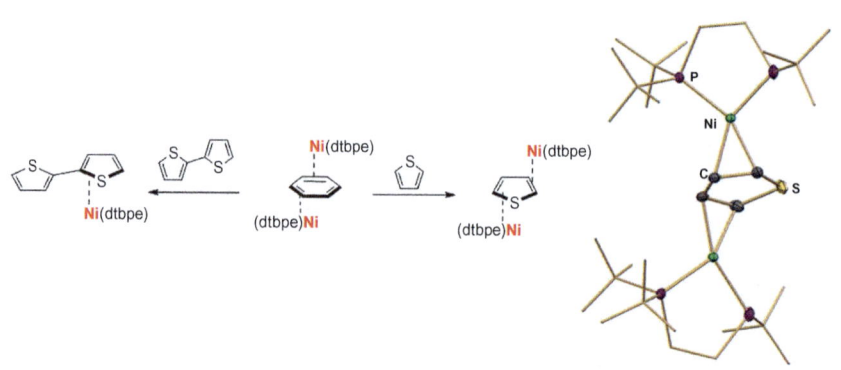

Scheme 30 Synthesis and molecular structure of Ni(dtbpe)(thiophene).[124]

2.2 M(0) complexes with N-heterocyclic carbene ligands

At present, N-heterocyclic carbenes feature prominently as ligands in organometallic chemistry.[125] The rapid growth of NHCs chemistry is a consequence of their unique electronic properties, steric profile and reactivity, which make them tough competitors to phosphines.[126] Most of reports of NHC complexes of group 10 concern with the divalent oxidation state and, in particular with Pd(II), due to their interesting applications in homogenous catalysis.[127] However, the enhanced stability imparted by NHCs ligands to a metal center, as a result of strong metal-carbene binding, has been conveniently exploited by several research groups to prepare stable, low-coordinate NHC-M(0) complexes, as alternative to divalent metal precursors for organic transformations involving oxidative addition steps. While 14-electron two-coordinate (NHC)$_2$M species are well known for all members of the group 10, three-coordinated 16-electron (NHC)$_3$M(0) species are almost unknown. However, a substantial number of examples of 16-electron mixed-NHC/ligand M(0) complexes, particularly with electron-deficient olefin ligands, have been described. Their synthesis, their most striking structural features and their applications, particularly in catalysis, will be presented as follows.

2.2.1 Homoleptic (NHC)₃M(0) complexes

Three-coordinate homoleptic L₃M(0) species with NHC ligands are very uncommon. We are aware of two Ni-based examples. Meyer and co-workers reported the first of them in 2004.[128] The compound, Ni(TIMEN^{tBu}) (TIMEN^{tBu} = tris[2-(3-butylimidazol-2-ylidene)ethylamine) was obtained from the reaction of Ni(COD)₂ with the free tripodal carbene ligand in THF at room temperature (Scheme 31). The structure of this species in solution was consistent with a C3-symmetrical ligand environment. In the solid state, the tripodal NHC ligand was coordinated through the three carbene arms to the metal center, in a trigonal planar arrangement. No interaction with the N atom of the amine functionality was observed. Interestingly, agostic interactions between the Ni(0) center and one hydrogen atom of each of the three *tert*-butyl groups were observed in the solid-state structure. The complex is easily oxidized by benzyl chloride affording the Ni(I) species Ni(TIMEN^{tBu})Cl.

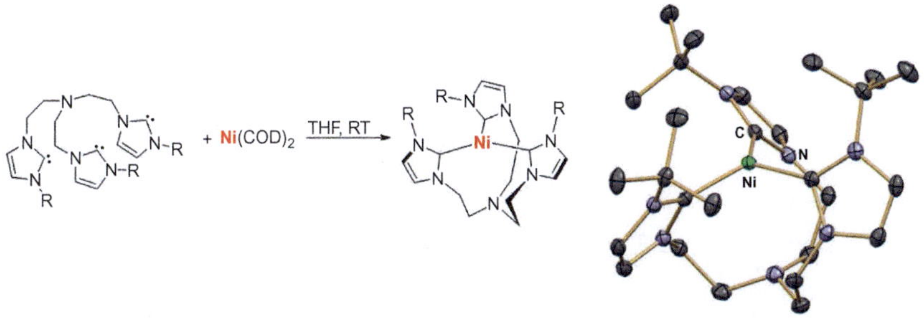

Scheme 31 Synthesis and molecular structure of Ni(TIMEN^{tBu}).[128]

Radius and co-workers studied the reactions of Ni(COD)₂ with a variety of alkyl-substituted NHCs ligands, R₂Im, R = Me, *n*Pr, *i*Pr.[129] With the less bulky IMe carbene (IMe = 1,3-dimethylimidazol-2-ylidene) a three-coordinate Ni(IMe)₃ complex was isolated, regardless the amount of carbene employed. As in the previous case, the compound presents a slightly distorted trigonal planar geometry in the solid state, with the Ni atom aligned in the plane formed by the three carbene carbon atoms. The Ni-C$_{carbene}$ bond lengths falls within the range 1.860–1.888 Å.

2.2.2 Heteroleptic three-coordinate species
2.2.2.1 Three-coordinate mono- and bis-carbene M(0) complexes stabilized by unsaturated organic ligands

Heteroleptic 16-electron complexes of composition M(NHC)$_n$L$_m$ (n = 1, 2; m = 2, 1) bearing π-acid ligands such as electron deficient olefins or alkynes are known for the group 10 triad. As in the case of P$_3$M complexes discussed above, the most favorable geometry for these species in the solid state is the trigonal planar arrangement. Primary examples of 16-electron biscarbene-M(0) compounds supported by alkenes were described by Cavell and co-workers in 1999.[130] Interested in studying the oxidative addition of organic halides to 14-electron (NHC)$_2$M(0) (M=Ni, Pd) species, the group prepared a series of Pd(^{Me}IMe)$_2$(alkene) complexes (^{Me}IMe = 1,3,4,5-tetramethylimidazol-2-ylidene) stabilized with electron-deficient olefins, like tetracyanoethylene (TCNE) and maleic anhydride (MA). The synthesis was accomplished by treating Pd(COD)(alkene) with two equivalents of the carbene ^{Me}IMe in tetrahydrofuran (Scheme 32). Interestingly, Pd(dba)$_2$ could not be used as the Pd(0) source since it reacted with the NHC ligand complicating the workup of the reaction.

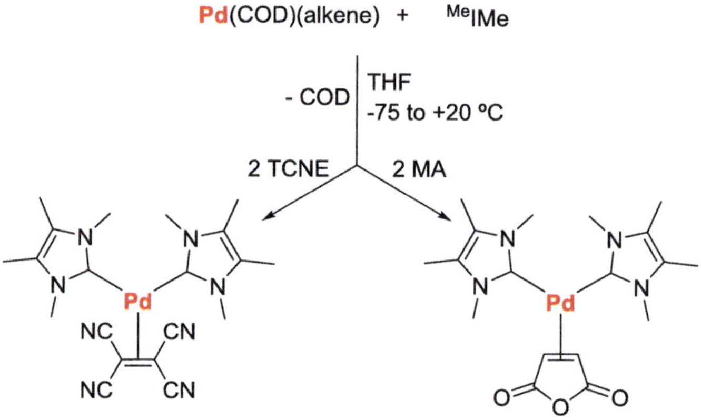

Scheme 32 Synthesis of Pd(^{Me}IMe)$_2$(alkene) devised by Cavell and co-workers.[130]

Both complexes displayed good moisture and air stabilities, but particularly the TCNE derivative appeared to be indefinitely stable. Such behavior is a clear indication of a strong bonding interaction between the electron-deficient olefin and the metal center. As a consequence of the efficient Pd to olefin π-backdonation, a large upfield shift was observed for the olefinic resonances of MAH both in the ^{1}H (δ 3.24) and the ^{13}C{^{1}H} (δ 38.2) NMR spectra with respect to the free ligand (7.05 and 136.6 ppm, respectively). No structural data were provided for any of these complexes. Like other zerovalent compounds of Pd, both derivatives experienced oxidative addition of aryl halides and dihalides affording the corresponding Pd(II) derivatives.

In 2002, Beller and co-workers reported the synthesis and structural characterization of the first 16-electron monocarbene-Pd(0) compound non-bearing phosphine ligands, the complex Pd(IMes)(DVDS) (DVDS=1, 1,3,3-tetramethyl-1,3-divinyl-disiloxane).[131a] The synthesis was attained by reacting the Pd(0) precursor Pd$_2$(DAE)$_3$ (DAE=diallylether) with equimolar amount of the free carbene ligand IMes (1,3-dimesitylimidazol-2-ylidene) in neat DVDS at low temperature (Scheme 33).

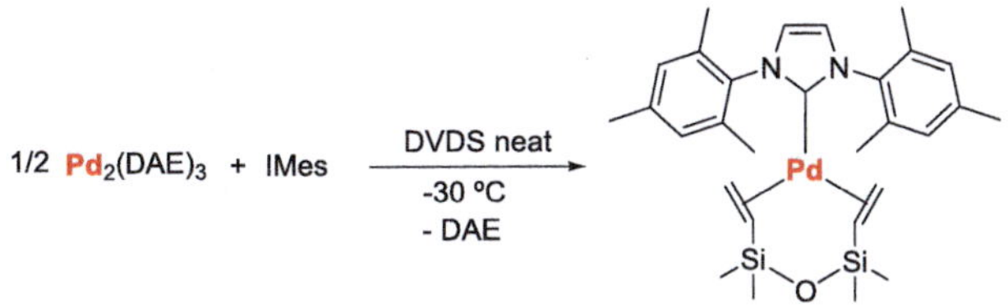

Scheme 33 Synthesis of Pd(IMes)(DVDS) by Beller and co-workers.[131a]

The structure of this compound was elucidated by X-ray diffraction study. It consists of a three-coordinate complex, with the Pd(0) atom bonded to the carbene ligand and to the chelating DVDS ligand in a trigonal-planar arrangement (Fig. 17). The compound exhibited excellent activity and selectivity in telomerization reaction of 1,3-butadiene with methanol. To optimize the catalyst system for said catalytic transformation, the group enlarged the number of DVDS-Pd(0) monocarbene derivatives varying the substituents both on the carbene backbone and on the N-aryl groups[131b] (Fig. 17).

Dinuclear [Pd(IMes)(quinone)]$_2$ complexes were obtained in the reaction of Pd(COD)(quinone) (alkene=p-benzoquinone, BQ and 1,4-naphthoquinone, NQ) with equimolar amounts of free IMes ligand.[131c] The dinuclear nature of these species was elucidated by X-ray crystallography. As shown in Fig. 18, the quinone molecules acted as bridging ligands, binding one Pd center through the olefin moiety and the other through the oxygen atom of one of the carbonyl group. This uncommon coordination mode of the quinone ligands was corroborated by X-ray diffraction studies, which also revealed distorted trigonal-planar coordination geometry at the palladium center. Interestingly, in the IR spectra of these complexes the stretching frequency (νCO) for the two different carbonyl groups of coordinated quinone ligands appeared at lower frequency compared to that of free ligands, the bathochromic shift being larger for the coordinated CO groups (127 and 130 cm^{-1} for the BQ and NQ complexes, respectively)

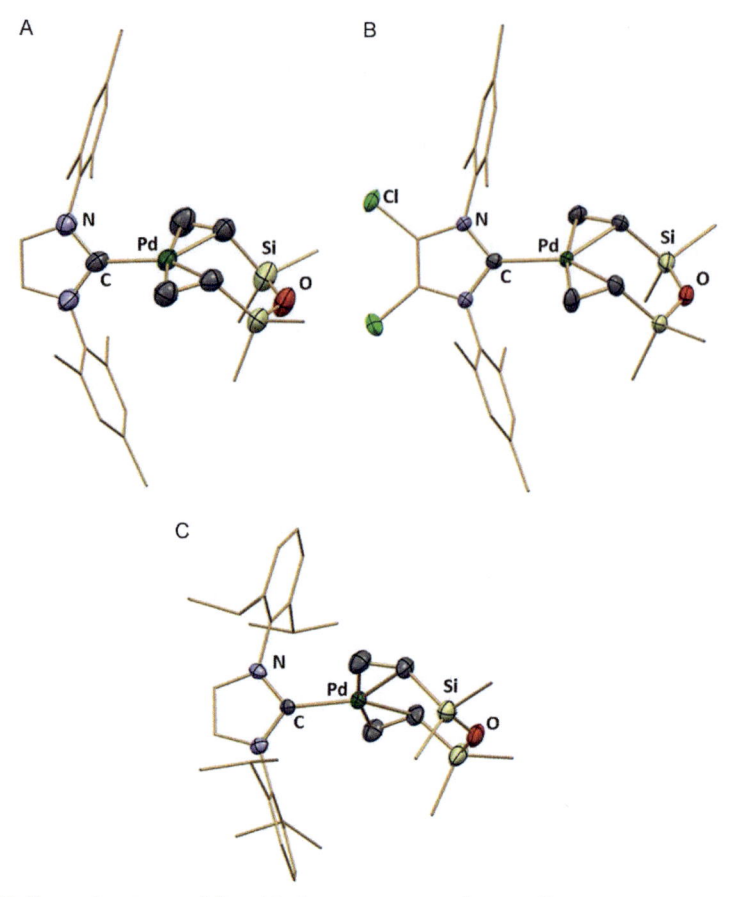

Fig. 17 X-ray structures of first 16-electron monocarbene-Pd(0) compounds Pd(NHC)(DVDS) prepared by Beller and co-workers.[131a,b]

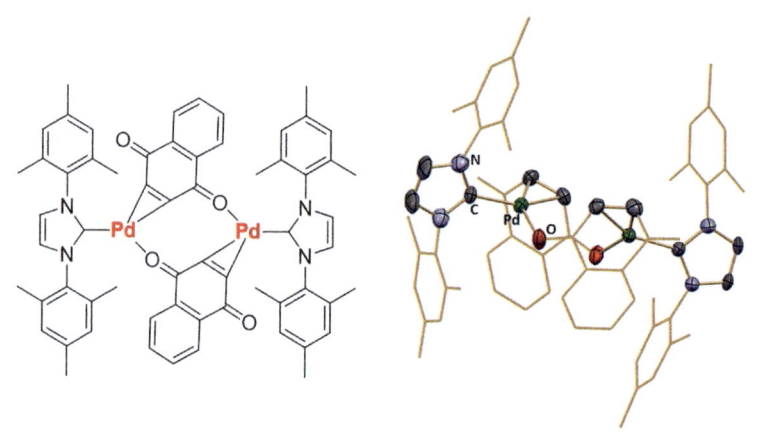

Fig. 18 X-ray structures of [Pd(IMes)(quinone)]$_2$ complexes.[131c]

than for the non-coordinated ones (14 and 28 cm^{-1} for BQ and NQ complexes, respectively). The enhanced stability displayed by these two monocarbene-Pd(0)-monoolefin complexes rely once more on the good π-acceptor behavior of the electron-deficient olefins ligands in the compounds. The reactivity of these complexes as catalysts in various Pd-catalyzed cross-coupling reactions has been described.[132]

To overcome the problems associated with the Karstedt complex Pt$_2$(DVDS)$_3$,[133] the most commonly employed industrial catalysts for the hydrosilylation of alkenes, such as low selectivity and formation of platinum colloids, the group of Markó in 2002 disclosed the preparation of monocarbenes-Pt(0) complexes Pt(NHC)(DVDS)[134a] analogous to that described by Beller for Pd.[131a] In fact, both reports were published almost at the same time.

The platinum(0) complexes were obtained by the room-temperature reaction of commercially available Karstedt catalyst Pd$_2$(DVDS)$_3$ with two equivalents of the NHC ligand, generated in situ by deprotonation of the imidazolium salt with base (NaH or KOtBu), in toluene (Scheme 34).

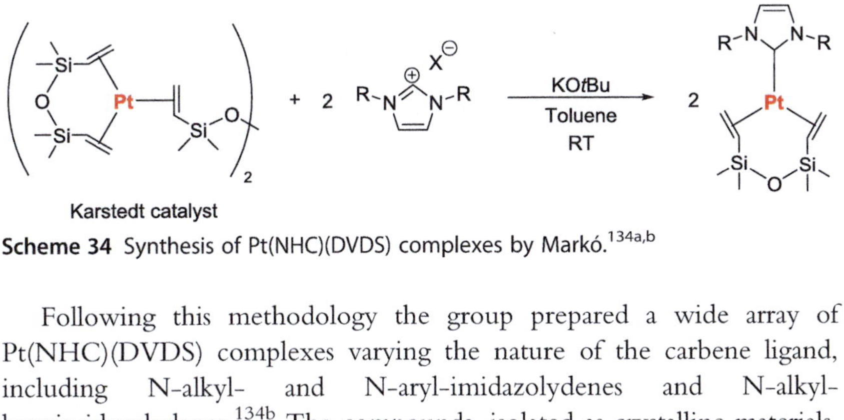

Scheme 34 Synthesis of Pt(NHC)(DVDS) complexes by Markó.[134a,b]

Following this methodology the group prepared a wide array of Pt(NHC)(DVDS) complexes varying the nature of the carbene ligand, including N-alkyl- and N-aryl-imidazolydenes and N-alkyl-benzimidazolydenes.[134b] The compounds, isolated as crystalline materials, were indefinitely stable to air, both as solids and in solution. Structural studies in the solid state reveal the expected trigonal-planar coordination environment around platinum, in which the diene DVDS adopt a pseudo-chair chelating conformation to reduce steric interactions. While the carbene plane is nearly perpendicular to the coordination plane for N-alkyl-substituted NHCs ($\theta=97.6-91.1°$), in the case of N-aryl-substituted NHCs, the carbene plane is tilted away ($\theta=127.9-117.2°$) to prevent destabilizing steric repulsions (Fig. 19). This feature seems to have interesting

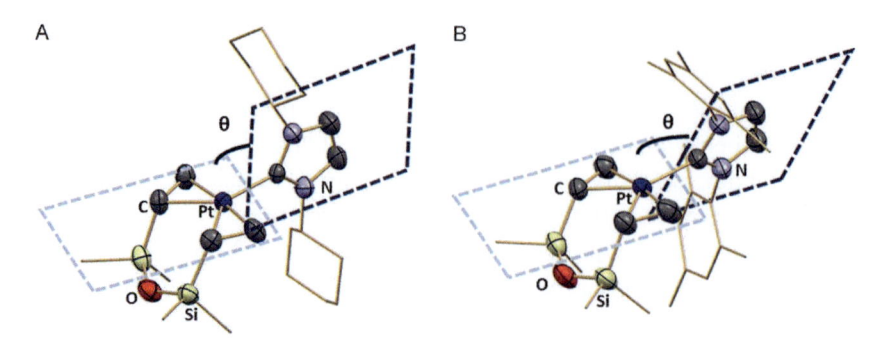

Fig. 19 Structures of IMes and ICy showing the angle between the NHC plane and coordination plane.[134b]

implications in the catalytic performance of the complexes in the hydrosilylation reaction.[134c] Overall, these complexes catalyze the hydrosilylation of alkenes and alkynes with high activities and excellent selectivity, and with a broad tolerance toward reactive functionalities.[135]

To shorten the induction times and lower the reaction temperature, the group reported later the preparation of Pt(NHC)(1,4-hexadiene) bearing more labile 1,4-hexadiene ligands such as diallyl ether, DAE or diethyl 2,2-allylmalonate, DAM.[85] The synthesis was accomplished by direct reduction of $H_2PtCl_6 \cdot xH_2O$ with $(Me_3SiO)_2MeSiVi$ (Vi=vinyl) in *i*PrOH in the presence of excess diene, following by addition of the in situ generated carbene ligand (Scheme 35).

Scheme 35 Synthesis of Pt(NHC)(1,4-hexadiene) complexes.[85]

Recently, Pietraszuk and co-workers have prepared Pt(0)-DVDS derivatives of very bulky, yet flexible NHC ligands IPr* (1,3-bis{2,6-bis(diphenylmethyl)-4-methylphenyl}imidazole-2-ylidene) and IPr*[OMe] (1,3-bis{2,6-bis(diphenylmethyl)-4-methoxyphenyl}imidazole-2-ylidene) and studied the influence of such bulky NHCs ligands on chemo- and regioselectivity in hydrosilylation of terminal olefins and terminal and internal alkynes.[136] Furthermore, de Jesús, Flores and co-workers, using sulfonated–functionalized N-aryl-substituted NHC,[137] have developed a water-soluble version of Markó's type catalyst.

Apart from the excellent performance of Pt(NHC)(DVDS) complexes as catalysts in hydrosilylation reactions, they have been successfully applied in catalytic diboration and tandem hydroboration–cross coupling reactions by Peris and Fernández.[138] The complexes were obtained by a modified synthetic procedure which made use of silver transmetallation to platinum(0) to introduce the carbene functionality.

In 2003, Elsevier and co-workers reported the synthesis of 16-electron monocarbene-Pt(0) stabilized with electron deficient dimethylfumarate, DMFU.[139a] The preparation of Pt(NHC)(DMFU)$_2$ involved the equimolar reaction of Pt(COD)$_2$ with the free carbene IMes or SIMes (1,3-dimesityl-dihydroimidazol-2-ylidene) in the presence of two equivalents of DMFU. In a subsequent work, the synthetic procedure was simplified avoiding the pre-isolation of the carbene ligand and carrying out the reaction in a one-pot fashion by reacting the imidazolium salt, the platinum(0) source and the alkene in the presence of the base NaH[139b] (Scheme 36). However, the method proved unsuccessful with N-alkyl-substituted NHC ligands (IAlk).[139b]

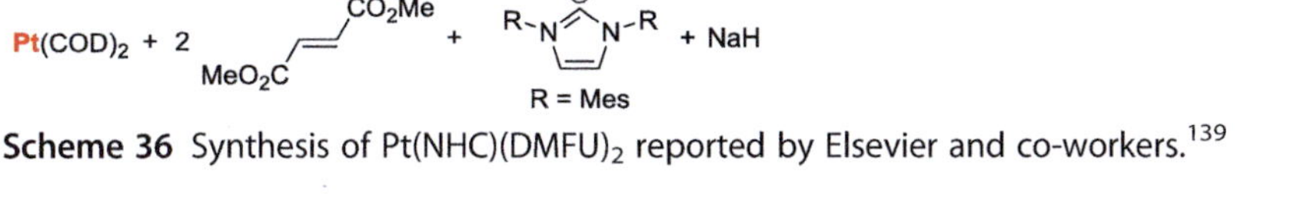

Scheme 36 Synthesis of Pt(NHC)(DMFU)$_2$ reported by Elsevier and co-workers.[139]

^{1}H and ^{13}C{^{1}H} NMR spectra of Pt(NHC)(DMFU)$_2$ complexes revealed C$_2$ symmetry for these species in solution at room temperature. The solid-state structure of Pt(IMes)(DMFU)$_2$ determined by X-ray crystallography unveiled a distorted trigonal-planar arrangement around the metal center[139b] (Fig. 20A). Both DMFU molecules bind the Pt(0) atom in a planar fashion with a dihedral angle between the two Pt-η^2-C=C planes of 10.8(2)°. The latter planes are oriented in angles of 56.05(19)° and 56.7

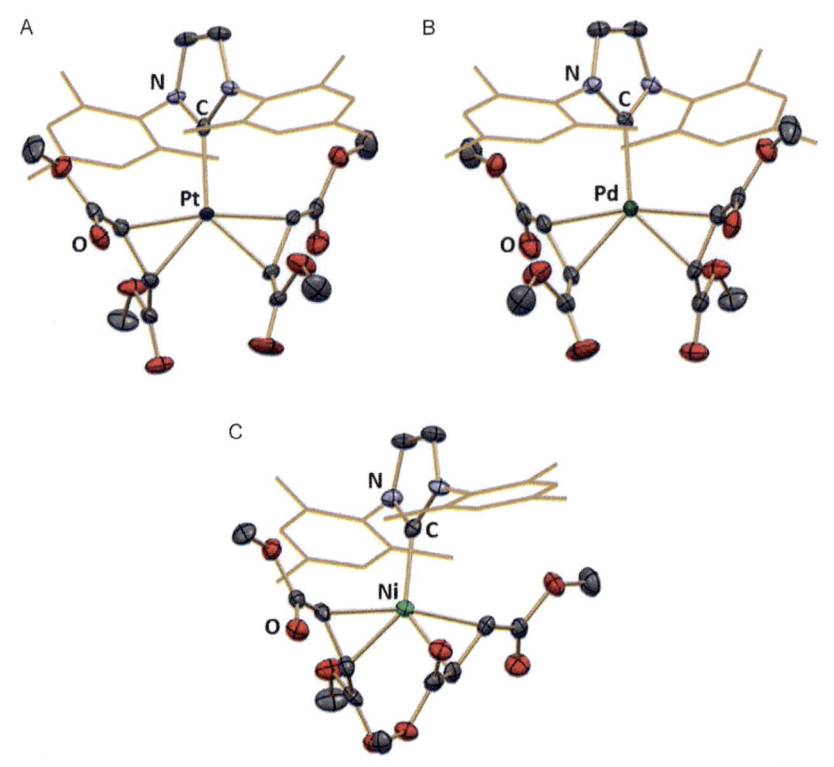

Fig. 20 Molecular structures of complexes M(NHC)(DMFU)$_2$ (M=Pt, Pd and Ni).[140]

(2)° with respect to the IMes plane. To accommodate the steric pressure of DMFU ligands the mesityl substituents are tilted with respect to the carbene plane with dihedral angles of 68.77(14)° and 66.28(14)°.

Pt(NHC)(DMFU)$_2$ complexes were reactive toward C-H oxidative addition of imidazolium salt under mild conditions, generating square-planar hydrido–heterobis(NHC)-Pt(II) derivatives.[139a] In situ generated catalyst derived from Pt$_2$(norbornene)$_3$ and SIMesHCl, in the presence of NaH proved to be competent for the hydrosilylation of styrene with triethylsilane.[141]

In a latter account, the group of Elsevier reported the preparation of a series of Pd(IAlk)(MA)$_2$ complexes.[142a] The synthetic procedure employed depended upon the nature of the alkyl substituents. Thus, for bulky substituents like CH(iPr)$_2$, isopropyl, cyclohexyl or neopentyl, the corresponding complexes were obtained by direct reaction of Pd(0) precursors, Pd(nbd)(MA)[143] or Pd$_2$(DVDS)$_3$,[84] with the imidazolium salt in the presence of a base (KOtBu) and excess of the olefin MA. However, when IMe was used as the ligand, only the transmetallation of the (IMe)$_2$Ag(I) complex to the

Pd(0) source afforded the desired complex. The N-alkyl-substituted Pd(NHC)(MA)$_2$ complexes were thermally less stable, particularly in solution, than their N-aryl counterparts, which precluded their solid-state characterization. They were tested as catalysts in transfer semi-hydrogenation of phenylpropyne.[142b]

Zerovalent palladium and nickel DMFU complexes of composition M(IMes)$_n$(DMFU)$_m$ (for Pd n=1, m=2 and n=2, m=1; for Ni n=1, m=1, 2 and n=2, m=1) were described by Cavell and co-workers.[140] These compounds were prepared from the reaction of a M(0) source (Pd(DAB)(DMFU)[144] or Ni(COD)$_2$) with the free carbene IMes in the presence of DMFU (Scheme 37). The bis-NHC adducts readily exchanged ligand in solution leading to equilibrium mixtures of bis- and mono-NHC-M(0)-DMFU complexes. This exchange was facilitated by the formation of an organic byproduct from NHC-DMFU coupling. These zerovalent complexes showed remarkable long-term stability in the solid state.

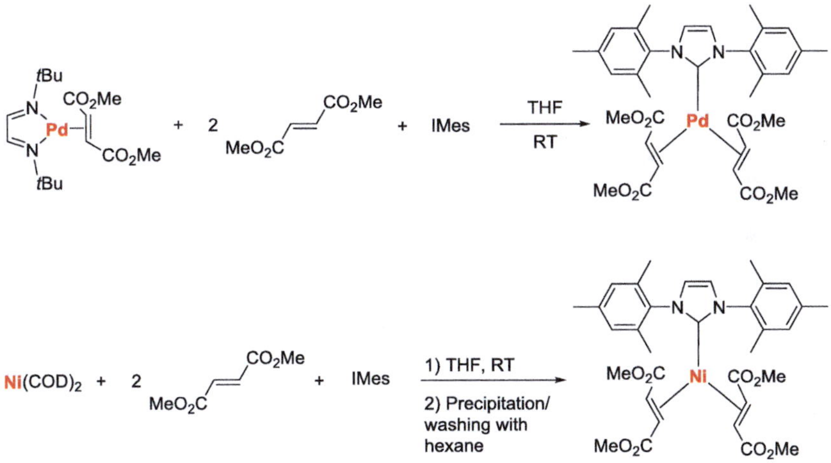

Scheme 37 Synthesis of complexes M(IMes)$_n$(DMFU)$_m$ (M=Pd, Ni).[140]

The X-ray structures of complexes M(IMes)(DMFU)$_2$ (M=Pd and Ni) were obtained and, as expected, they were isostructural with its Pt counterpart (Fig. 20B and C). The most salient features was the longer Pd-C$_{carbene}$ distance of 2.111(3) Å found for the Pd(0) complex, when compared to others Pd(0)-NHC derivatives.[131b] The elongation of this bond was attributed to the steric congestion induced by the two DMFUs molecules in the complex.

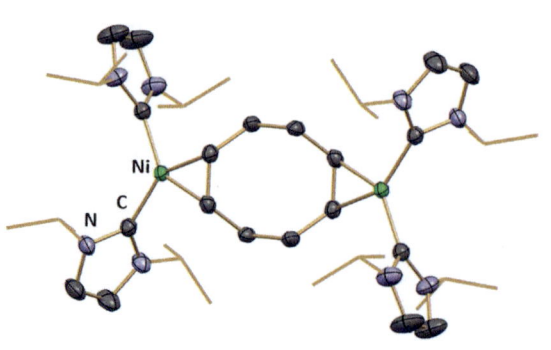

Fig. 21 Molecular structure of [Ni$_2$(I*i*Pr)$_4$(COD)] isolated by Radius and Schaub.[145]

In 2005, Radius and Schaub disclosed the synthesis of the dinuclear complex [Ni$_2$(I*i*Pr)$_4$(COD)] (I*i*Pr = 1,3-diisopropylimidazol-2-ylidene) from the reaction of Ni(COD)$_2$ in toluene with two equivalents of I*i*Pr.[145] The Ni(0) complex was isolated as yellow solid and unlike other instances described in this section, it was extremely sensitive to air and moisture, decomposing around 60 °C both in solution and in the solid-state. X-ray diffraction studies confirmed the dinuclear nature of this compound. As illustrated in Fig. 21, each Ni(0) atom is bonded by two I*i*Pr ligands and by one of the double bond of the bridging COD molecule, in a distorted trigonal-planar geometry. The Ni-C$_{carbene}$ distances, mean value of 1.905 Å, are significantly longer to those reported for homoleptic Ni(IMes)$_2$ (mean value of 1.825 Å)[146a] and Ni(I*t*Bu)$_2$ (I*t*Bu = 1,3-di-*tert*-butyl-imidazol-2-ylidene) (1.874 Å).[146b] Interestingly, the nickel center features agostic interactions to the methine hydrogen atoms of the isopropyl group, with Ni-HC(Me)$_2$ distances in the range of 2.807–2.903 Å. This compound constitutes the first example of COD-bridged nickel complex.

The reactivity of this unprecedented dinuclear Ni(0) complex was explored (Scheme 38), proving to be an excellent precursor of the 14-electron Ni(I*i*Pr)$_2$ fragment in stoichiometric and catalytic transformations. Remarkably, the complex was capable of activating the C-F bonds of C$_6$F$_6$ at room temperature in short reaction time affording the oxidative addition Ni(II) product Ni(I*i*Pr)$_2$(F)(C$_6$F$_6$). The group exploited this reactivity by accomplished Suzuki-Miyaura C-C couplings of challenging perfluorinated arenes with boronic acids in the presence of catalytic amounts of [Ni$_2$(I*i*Pr)$_4$(COD)].[147a] Furthermore, a comprehensive study on the mechanism of these C-F activation reactions by the dinuclear Ni(0) species was accomplished in collaboration with the group of Braun.[147b] These

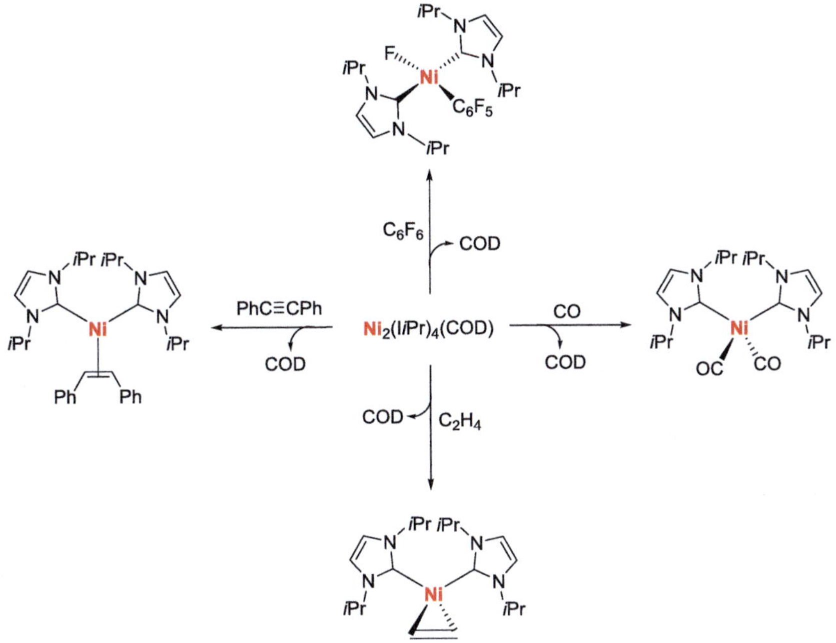

Scheme 38 Reactivity of [Ni$_2$(I*i*Pr)$_4$(COD)].[146]

investigations led to the detection of intermediates with the fluorinated arene coordinated in a η^2-fashion from which the C–F activation products were obtained in a concerted manner.

To enlarge the family of dinuclear COD-bridged Ni(0) species, Radius and co-workers surveyed the applicability of the synthetic protocol to different N-alkyl-substituted NHC ligands.[146b] The outcome of the reaction of Ni(COD)$_2$ with NHCs essentially depended on the steric hindrance of the substituents at the nitrogen atoms of the carbene, a trend similarly observed for the phosphine ligands. Thus, the methyl-substituted carbene IMe (1,3-dimethylimidazol-2-ylidene) produced the aforementioned homoleptic tris-carbene Ni(IMe)$_3$ complex independently of the carbene employed. However, with more sterically demanding *n*Pr- and Me*i*Pr-substituted NHCs, the COD-bridged dinuclear [Ni$_2$(NHC)$_4$(COD)] were obtained. All complexes were active catalysts for the insertion of diphenyl acetylene into the C–C bond of biphenylene.

In recent years, the interest in applying earth-abundant first-row transition metals in homogenous catalysis has increased considerably.[148] They do not only represent a sustainable alternative to precious-metal catalysis, but also offer unique properties that can be exploited to discover novel reactivity patterns not observed with the precious counterparts. Nickel, the lighter member of group 10 metals, is not an exception and the field of nickel catalysis has experienced a remarkable progress during the last decade.[149] Most of these developments have been based on the use of well-defined nickel catalysts. In this regard, the synthetic approach devised by Cavell and illustrated in Scheme 37 for the synthesis of NHC-Ni(0) species stabilized by strong π-acceptor olefins have been applied by different groups to prepare easy-to-handle, well-defined zerovalent Ni complexes for catalytic applications. Thus, Navarro and co-workers isolated in high yield the complex Ni(IPr)(DMFU)$_2$ (Fig. 22A).[150] This compound was used as precatalyst for the anaerobic oxidation of secondary alcohols using 2,4-dichlorotoluene as both the oxidant and the solvent.

We found that styrene was a suitable ligand to stabilize the 12-electron IPr-Ni(0) fragment. The complex Ni(IPr)(styrene)$_2$ was prepared by the reaction of Ni(COD)$_2$ with equimolar amount of the free carbene IPr in the presence of excess styrene[151a] (Fig. 22B). Due to the less accentuated π-acceptor properties of the styrene molecule, this compound could only be handled in air for a short period of time and decomposed rapidly in solution when exposed to an oxygen atmosphere. Ni(IPr)(styrene)$_2$ proved to be a competent precatalyst in the amination of aryl tosylates with cyclic secondary amines and anilines and in challenging N-arylation of indoles.[151b,c]

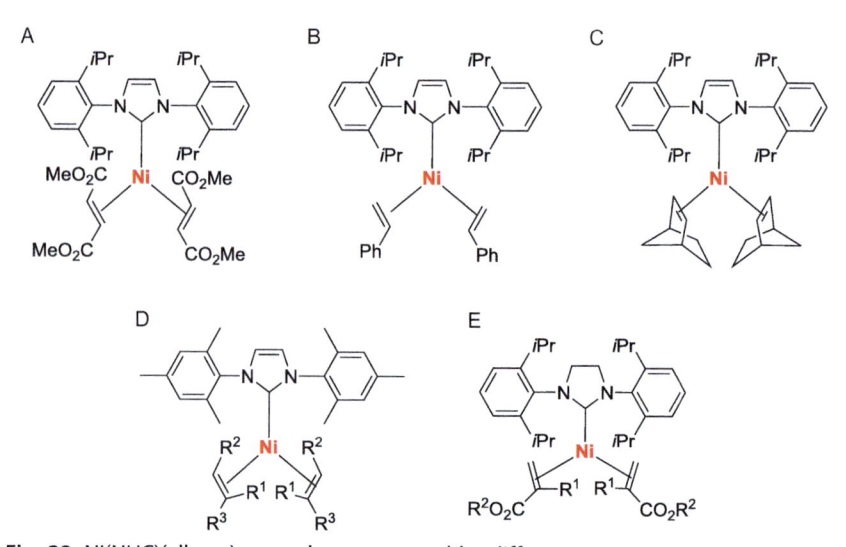

Fig. 22 Ni(NHC)(alkene)$_2$ complexes prepared by different groups.

The complex Ni(IPr)(norbornene)$_2$ was isolated, structurally characterized and identified by Hartwig and co-workers[152] as the resting state of the catalyst in studies aimed at understanding the mechanism of hydroarylation of unactivated olefins with trifluoromethyl-substituted arenes catalyzed by an in situ generated Ni(COD)$_2$/IPr catalyst (Fig. 22C).

The group of Montgomery[153a] unveiled the preparation of Ni(ITol)(MMA)$_2$, (MMA=methyl methacrylate), which contains the unhindered ITol (1,3-di-p-tolyl-imidazol-2-ylidene) carbene ligand. The complex catalyzed the coupling of enones or enal with alkynes in the presence of silane and titanium alkoxide as reductants. In a recent report, the group prepared a series of (NHC)Ni(alkene)$_2$ stabilized with a variety of π-accepting acrylates and fumarate ligands.[153b] With the carbene IMes it was possible to isolate the complete series of fumarates and acrylates derivatives. However, the more sterically demanding IPr*^{OMe} and SIPr ligands formed stable complexes only with methyl and phenyl acrylate (Fig. 22D and E). The complexes proved to be efficient catalysts in C–C (reductive aldehyde-alkyne couplings) and C–N (Buchwald-Hartwig amination) bond forming reactions.

A remarkably stable, 16-electron species Ni(IPr)(η^4-1,5-hexadiene) was obtained by Hazari and co-workers[154a] from the decomposition of Ni(IPr)(η^1-allyl)(η^3-allyl) at room temperature. In a subsequent publication, the group implemented a direct route to prepare a variety of Ni(NHC)(η^4-1,5-hexadiene) from commercially available anhydrous NiCl$_2$[154b] (Scheme 39). The solid-state structures of selected complexes confirmed trigonal planar arrangements at the nickel center in all cases. The Ni–NHC distances were substantially longer, mean value of 1.92 Å, than the values reported for Ni(NHC)$_2$ complexes.[146] These complexes represent uncommon examples of transition-metal species bearing a coordinated 1,5-hexadiene ligand. The complex Ni(IPr)(η^4-1,5-hexadiene) reacted with a wide range of Lewis bases (CO, isonitriles, phosphines) and olefins

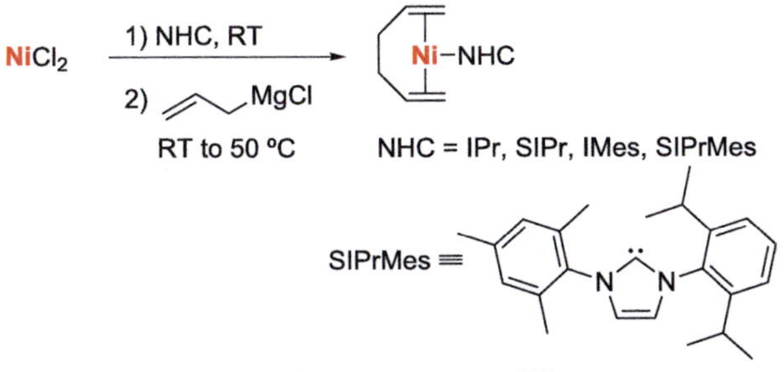

Scheme 39 Synthesis of Ni(NHC)(η^4–1,5-hexadiene).[154b]

(ethylene, DAE, DVDS, DMFU), demonstrating its versatility as a source of the 12-electron NHC-Ni(0) fragment.

Elsby and Johnson[155] described the synthesis of the complex Ni(I*i*Pr)(H$_2$C=CH$_2$SiMe$_3$)$_2$. In the solid state, this molecule exhibited a C$_s$ symmetry as a consequence of the non-symmetric disposition of the SiMe$_3$ substituents of the alkenes. The complex catalyzed the C-H silylation of fluoroarenes with vinyl silanes.

In 2015, Navarro, Spencer and co-workers, reported the stoichiometric bis(silyl)ation of diphenylacetylene, at room temperature, using the Pd(II) complex Pd(^{Me}IMe)$_2$(SiMe$_3$)$_2$. The reaction produced the expected bis(silyl) ated olefin together with the Pd(0) complex Pd(^{Me}IMe)$_2$(PhC≡CPh), in quantitative yields.[156a] This compound was structurally characterized by X-ray crystallography, revealing a Y-shape structure (Scheme 40). The Pd-C$_{carbene}$ bond distances of 2.083(3) and 2.079(3) were within the expected range.[131] The complex Pd(^{Me}IMe)$_2$(PhC≡CPh) represents the first example of biscarbene-Pd(0) complex bonded to a η^2-alkyne ligand. An improved synthetic protocol to this compound using of Pd(^{Me}IMe)(2-methylallyl)Cl as the starting material was devised by this group later[156b] (Scheme 40).

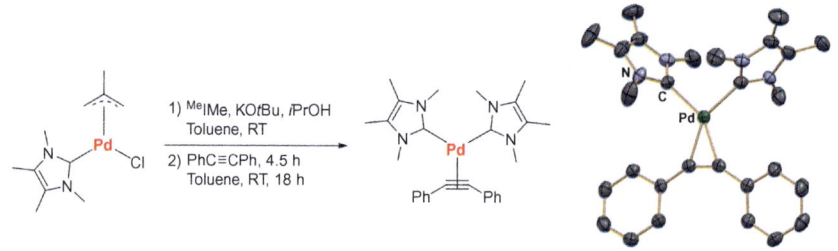

Scheme 40 Improved synthesis of Pd(^{Me}IMe)$_2$(PhC≡CPh).[156b]

The complex Pd(^{Me}IMe)$_2$(PhC≡CPh) proved to be a competent catalyst for bis(silyl)ation diboration and silaboration of terminal and internal alkynes,[156a–c] and diboration and silaboration of azobenzenes.[156d]

Expanded ring N-heterocyclic carbenes are deemed to be stronger σ-donor and more sterically encumbered, due to their larger N-C$_{NHC}$-N angles, than their imidazolium counterparts.[157] However, compared to the latter, the chemistry and catalytic applications of expanded ring N-heterocyclic carbenes-M^0(alkene) complexes are still very limited. So far, only the group of Cavell have reported the synthesis of the complexes M(6-NHC)(DVDS) and M(7-NHC)(DVDS) (M=Pd, Pt).[158] The platinum derivatives were prepared following the Markó's synthesis (see Scheme 34), whereas the Pd analogues were achieved by the reaction of in situ generated free carbene species with a commercial solution of Pd(DVDS) at room temperature. The Pt complexes were tested as catalysts in the hydrosilylation of a variety of unsaturated substrates (e.g., alkenes, alkynes, ketones) and their activity compared to that obtained for the 5-membered NHC derivatives.[158a] The palladium analogues were competent catalysts for the Mizoroki-Heck coupling of 4-bromoacetophenone and *n*-butyl acrylate, the 7-membered NHC palladium derivative outperforming their 6- and 5-membered counterparts.[158b]

In 2013, Elsevier and co-workers[159] described the first examples of Pd⁰(bis-NHC)(MA) complexes supported by chelating bis-NHC ligands (Scheme 41). Compounds with N-alkyl substituents in the chelating NHC framework were more labile than those bearing aryl groups, mainly as the result of the less effective steric protection imparted by the alkyl substituents. The X-ray crystal structure of the complex Pd(bis-(Mes)NHC)(MA) revealed a three-coordinate planar geometry around the palladium center, with the chelating ring adopting a boat conformation.

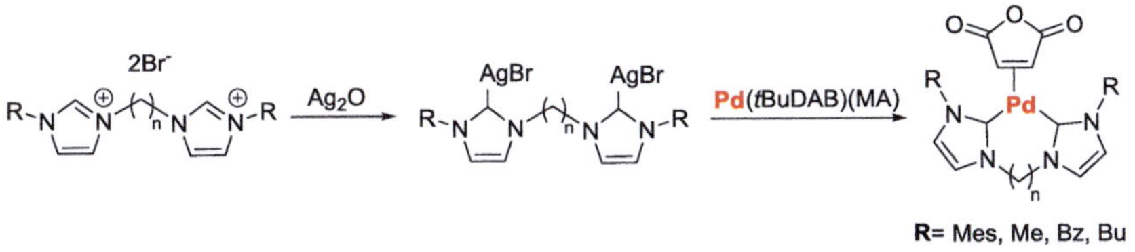

Scheme 41 Synthesis of Pd⁰(bis-NHC)(MA) complexes reported by Elsevier.[159]

Shortly after Elsevier's report, Harrold and Hillhouse[160a] outlined the synthesis of the three-coordinate mononuclear nickel(0) complex Ni(TMTBM)(η²-COD) (TMTBM = 3,3′-methylenebis(1-*tert*-butyl-4,5-dimethylimidazolydene)) stabilized by the non-electron deficient COD ligand. The olefin molecule could be displaced in solution by 2,6-diisopropylphenyl azide (N₃Ar) generating the side-bound organoazide adduct Ni(TMTBM)(η²-N₃Ar) (Scheme 42). The organoazide derivative

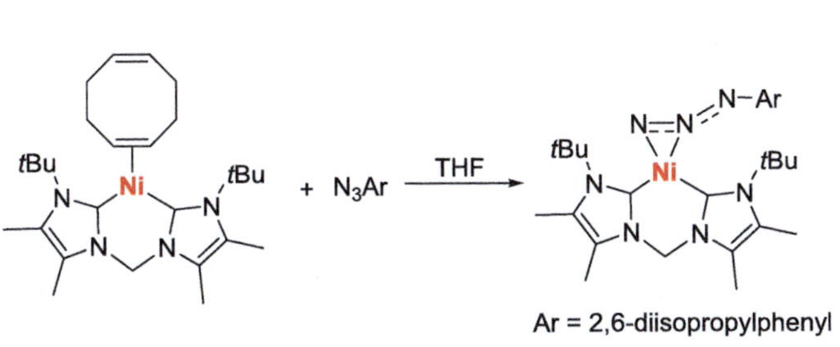

Scheme 42 (TMTBM)-Ni(0) derivatives prepared by the group of Hillhouse.[160a]

was thermally and photochemically robust in contrast to similar trialkylphosphine analogues, which decomposed even in the solid state.[160b]

Related M(bis-NHC)(η^2-COD) analogues of nickel(0) and platinum(0) were prepared by Hofmann and co-workers.[161] Interestingly, the nickel complex underwent C-C activation of benzonitrile in solution. Very recently, Roesler and co-workers[162] reported the synthesis of a Ni(bis-NHC)(η^2-cod) complex bearing a biscarbene ligand with a flexible siloxane linker, which can accommodate a wide range of bite angles. The complex acted as a Lewis base in the reaction with GeCl$_2$, affording a T-shape Ni(0)GeCl$_2$ adduct, which featured a Ni-Ge bond (Scheme 43).

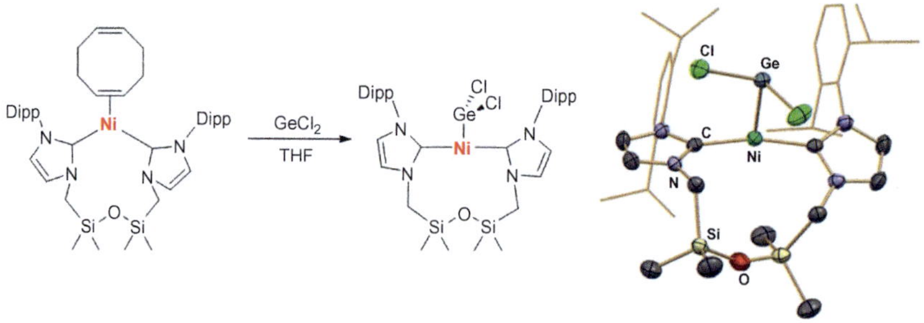

Scheme 43 Example of a T-shape Ni(0) complex.[162]

Heteroditopic NHCs having N-, O-, P- or S-functionalized wingtips are able to coordinate reversibly to a metal center liberating or protecting a vacant coordination site, a valuable attribute for catalytic applications. Most reports on the use of heteroditopic NHCs as ancillary ligands for zerovalent group 10 metals complexes encompass the preparation of Pd(0) complexes, primarily, for applications in C-C bond-forming reactions. Transmetallation of heteroditopic carbene ligands from Ag(I) to zerovalent Pd(0) precursors is usually the method of choice to accomplish their synthesis. A brief description of some representative examples is depicted below.

Elsevier and co-workers[163] focused on the preparation of (η^2-alkene)Pd(0) complexes with N-functionalized NHC ligands (Fig. 23). These molecules exhibit planar Y-shaped geometries in the solid state. The most significant structural features reside in the different lengths of the bonds between palladium and the carbon atoms of the olefin ligand, which gauges the differences in *trans*-influence of the carbene carbon and the nitrogen donors.

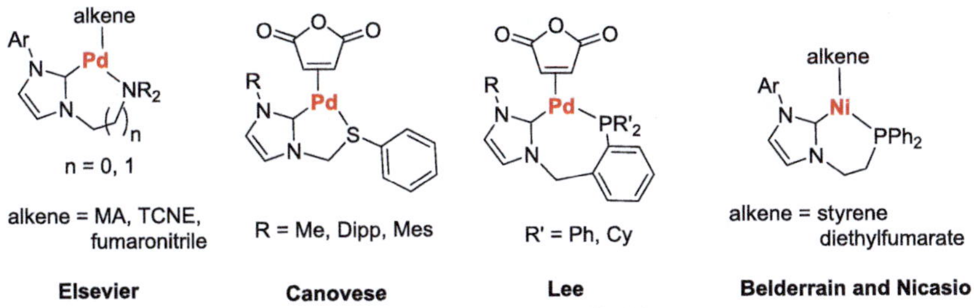

Fig. 23 Pd(X-NHC)(alkene) complexes (X = N, S, P).[163–166]

The complexes were less active and selective than the corresponding mono-carbene Pd(0) complexes counterparts.

Canovese and co-workers[164] described the synthesis of Pd(thioether-NHC)(MA) (Fig. 23). Phosphine-functionalized NHC ligands (NHCP) have been employed to generate Pd(0) complexes by Lee and co-workers.[165] The catalytic activity of these species was examined in the Mizoroki-Heck reaction and in the direct C-H functionalization.

Our group reported the synthesis of first examples of ArNHCP-Ni(0) complexes[166] (Fig. 23). The title compounds were obtained from the one-pot reaction of Ni(COD)$_2$, the imidazolium salt and the base, K[N(SiMe$_3$)$_2$], in the presence of excess alkene (styrene or diethyl fumarate). The precatalyst Ni(ArNHCP)(styrene) catalyzed effectively the coupling of (hetero)aryl chlorides with a range of nucleophiles including Grignard reagents, boronic acids, secondary amines and indoles.

2.2.2.2 Three-coordinate mono- and bis-carbene M(0) complexes stabilized by other Lewis bases

In 2003, the group of Nolan[167] disclosed the formation of stable, three-coordinate Ni(NHC)(CO)$_2$ complexes when reacting a hexane solution of Ni(CO)$_4$ with the sterically hindrance carbene ligands IAd (1,3-di-adamantylimidazol-2-ylidene) or I*t*Bu (1,3-di-*tert*-butylimidazol-2-ylidene) at room temperature. The complexes characterized by X-ray diffraction (Fig. 24), exhibit a distorted trigonal-planar structures with Ni-C$_{carbene}$ bond distances of 1.9528(16) Å and 1.9569(11) Å, slightly longer to those observed for Ni(IPr)(η^4-1,5-hexadiene) (1.916 Å)[154a] and Ni$_2$(I*i*Pr)$_4$(COD) (1.905 Å).[145] The formation of these unsaturated, stable dicarbonyl-Ni(0) species was attributed to the steric pressure imparted by the bulky IAd and I*t*Bu carbenes, which was relieved by dissociation of one of the CO ligand.

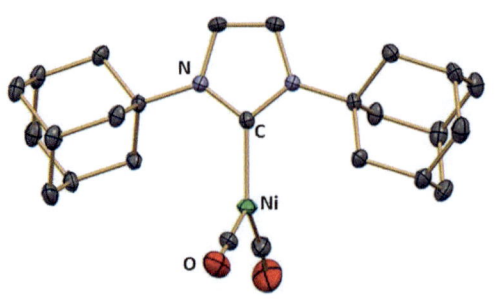

Fig. 24 Molecular structure of Ni(NHC)(CO)$_2$ complexes prepared by Nolan.[167]

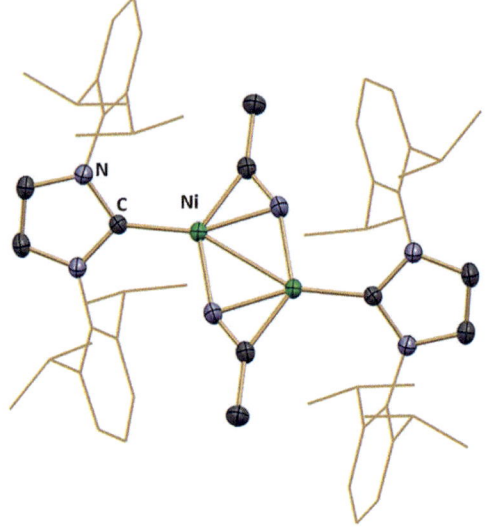

Fig. 25 Molecular structure of [Ni(IPr)(RNC)]$_2$.[168]

Louie and co-workers isolated a series of dinuclear IPr–Ni(0) with nitrile bridging ligands. The compounds [Ni(IPr)(RNC)]$_2$ were formed upon mixing Ni(COD)$_2$ with equimolar amounts of free carbene IPr and the nitrile (acetonitrile, methoxyacetonitrile, benzonitrile) in a hydrocarbon solvent.[168] The reaction was sensitive to steric hindrance of nitriles. The dinuclear nature of these complexes was unambiguously established by single-crystal X-ray diffraction analysis. As shown in Fig. 25, each nickel atoms displays a trigonal-planar coordination environment formed by the IPr ligand and two nitriles molecules in both η^1- and η^2-binding modes. The Ni–C$_{carbene}$ bond lengths, average value of 1.862 Å, compare well to those reported for homoleptic Ni(NHC)$_2$ complexes.[146] The complexes

[Ni(IPr)(RNC)]$_2$ resulted to be competent catalysts for the cycloaddition of diynes and nitriles to yield pyridine derivatives.

3. Two-coordinate complexes
3.1 Homoleptic M(0) complexes with phosphorus(III) ligands

The isolation of ML$_2^P$ complexes is dictated by the size of the ligand. Accordingly, very bulky phosphines (PtBu$_3$, PCy$_3$, P(o-tolyl)$_3$, among others) favor the formation of two-coordinate, 14-electron species that exhibit linear or bent coordination (additional interactions like C-H agostic bonding or M-arene contacts are occasionally observed in the solid state). A variety of ML$_2^P$ complexes have been structurally characterized for Pd and, to a lesser extent, for Pt. In the case of Ni, a few complexes supported by *trans*-spanning chelating diphosphine ligands are the only examples of NiL$_2^P$ reported to date.

As already mentioned, the interest in low coordinate M(0) complexes of group 10 dates back to the beginning of Organometallic Chemistry. In the early 70s, a few 14-electron ML$_2^P$ complexes of Ni, Pd and Pt had been prepared and isolated,[25,169] although some authors questioned the existence of such unsaturated species.[66,170] However, in 1974 the groups of Musco and Otsuka,[171] independently, carried out the first X-ray characterization of two-coordinate Pd(0)P$_2$ complexes. Two different synthetic routes were applied by these groups to generate the 14-electron Pd(0) species (Scheme 44). Musco and co-workers took advantage of the facile reduction of [(2-methylallyl)PdCl]$_2$ with excess phosphine ligand in the presence of a base (sodium methoxide).[25] On the other hand, the group of Otsuka

Musco's synthesis

[(2-methylallyl)Pd(μ-Cl)$_2$Pd(2-methylallyl)] + PR$_3$(exc.) →[NaOMe, MeOH] Pd(PR$_3$)$_2$ PR$_3$= PMe$_3$, PPh$_3$, PEt$_3$, PBu$_3$, PBz$_3$, PiPr$_3$, PCy$_3$, PMe$_2$Ph, PMePh$_2$, PtBu$_2$Ph

Otsuka's synthesis

Pd(cod) + 2 PR$_3$ →[toluene, 70-80 °C] Pd(PR$_3$)$_2$ PR$_3$= PtBu$_3$, PPhtBu$_2$, PCy$_3$

Scheme 44 Synthetic routes to PdL$_2^P$ species devised by Musco and Otsuka.[25,172]

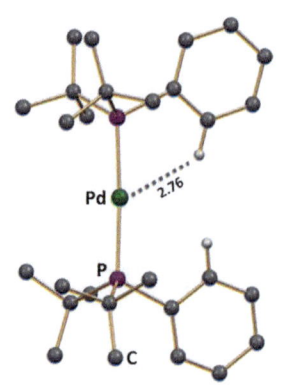

Fig. 26 Solid-state structure of Pd(P*t*Bu₃Ph)₂.[172]

employed Pd(η^5-C₅H₅)(η^3-C₃H₅) as precursor, which reductively eliminates C₅H₅-C₃H₃ upon heating in a hydrocarbon solvent in the presence of the phosphine ligands.[172] To date, both synthetic approaches remain the most common procedures for the preparation of two-coordinate PdL$_2^P$ complexes.

The molecules of Pd(PCy₃)₂ display a bent structure in the solid state with a P-Pd-P angle of 158.4°. Conversely, the complex Pd(P*t*Bu₂Ph)₂ presents an almost linear structure with P-Pd-P angle of 176.8°. In this structure, the substituents of the P atoms adopt an eclipse conformation with nearly parallel phenyl planes (Fig. 26). In this conformation, two of the *ortho*-hydrogen atoms of the phenyl groups are situated in close proximity to the metal center (Pd-H distances of ca. 2.76 Å). Apparently, such interaction is also retained in solution as manifested in the downfield shift observed for the proximal *ortho*-hydrogen atoms in the ¹H NMR spectrum (9.33 ppm in toluene-d_8 at −71 °C).

In the same paper, Otsuka and co-workers[172] also reported the synthesis of PtL$_2^P$ (L^P = PCy₃, P*t*Bu₂Ph, P*i*Pr₃) complexes by the reduction of corresponding *trans*-PtCl₂L$_2^P$ with Na/Hg or sodium naphthalenide in THF at room temperature. However, the synthesis with the bulkier Pt(P*t*Bu₃)₂ was accomplished by the direct reaction of Pt(COD)₂ with P*t*Bu₃ (Scheme 45), a procedure devised for Stone and co-workers to obtain Pt(PCy₃)₂ analogue.[173]

trans-PtCl₂(PR₃)₂ $\xrightarrow[\text{THF, RT}]{\text{Na(Hg) or NaC}_{10}\text{H}_8}$ Pt(PR₃)₂
R = Cy, *t*Bu₂Ph, *i*Pr

Pt(COD)₂ + 2 P(*t*Bu₃) $\xrightarrow[\text{- COD}]{\text{hexane, RT}}$ Pt(*t*Bu₃)₂

Scheme 45 Synthetic routes to Pt(0)P₂ complexes.[172,173]

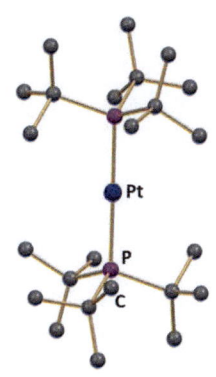

Fig. 27 Solid-state structure of Pt(PtBu₃)₂.[174]

Pt(PtBu₂Ph)₂ exhibits similar structural features as those described above for its Pd counterpart. The solid-state structure of Pt(PtBu₃)₂ was elucidated later by Chieh and co-workers (Fig. 27).[174] Unlike the former, this complex displays a linear geometry around the Pt center (P-Pt-P = 180°) with the substituents of the P in a staggered conformation due to the larger steric demand of the PtBu₃ (cone angle 182°) compared with that of the heteroleptic phosphine PtBu₂Ph (cone angle 170°). These are the firsts, and still among the few, instances of structural studies carried out with a 14-electron PtL$_2^P$ species.

The existence of PdL$_2^P$ complexes in solution was established by Musco and Mann by ^{31}P{^{1}H} NMR in their investigation of the speciation of an array of Pd(PR₃)ₙ complexes in solution.[26] They found that only large size phosphines, such as PtBu₃, PCy₃, PtBu₂Ph and PPiPr₃, formed 14-electron PdL$_2^P$ in solution.

Motivated by the impact of palladium cross-coupling technology in organic chemistry, the isolation of two-coordinate 14-electron PdL$_2^P$ complexes focused the attention of many research groups during the following decades.

Since Pd(PPh₃)₄ complex was the most popular precatalyst in the early cross-coupling chemistry, the preparation of the elusive Pd(PPh₃)₂, postulated as the active species, was targeted by different research groups. Chemical reduction of PdCl₂(PPh₃)₂ with organolithium reagents or electrochemical reduction in the absence of PPh₃ did not produce the expected 14-electron species, but anionic aggregates of the type [Cl$_x$Pd(PPh₃)₂]$_n^{nx-}$ that were observed in solution.[175] Urata and co-workers[176] described the preparation of Pd(PPh₃)₂ from the reaction of [Pd(η³-allyl)(PPh₃)₂]PF₆ with Ph₂MeSiLi. The complex was characterized by means of spectroscopic data

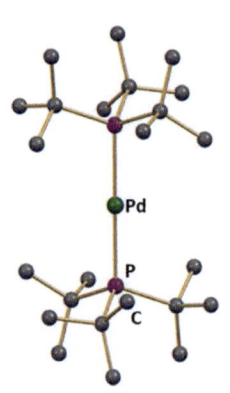

Fig. 28 Molecular structure of Pd(P*t*Bu₃)₂.[177]

and by reactivity studies. No crystallographic data of such species has been reported so far.

The complexes Pd(P*t*Bu₃)₂ and Pd[P(*o*-tolyl)₃]₂ were prepared and crystallographically characterized. The former obtained by Tanaka,[177] following Otsuka's procedure, is isoestructural to the Pt analogue (see Fig. 27). The latter was prepared by the group of Hartwig[178] by the addition of P(*o*-tolyl)₃ to Pd₂(dba)₃ in benzene at room temperature. It shows a linear structure with a staggered conformation of substituents of the P atoms (Fig. 28).

In 2003, Fink and co-workers[179] disclosed the first example of a two-coordinate Pd(0) complex with dialkylbiaryl phosphine ligands, in which the phosphine ligand was adopting a bidentate coordination mode involving the non-functionalized arene ring. The complex was obtained in high yield by the thermal reaction of PdMe₂(TMEDA) with the phosphine PCy₂Ar′ (Ar′ = biphenyl) in stoichiometric ratio (Scheme 46). This molecule in the solid state presents striking structural features, which differentiated it from typical PdL$_2^P$ complexes. The angle P-Pd-P (ca. 155°) is substantially deviated from linearity, facilitating the η^1-coordination of the Pd center to the *ortho*-carbon atom of one of the side aryl ring of the phosphine ligand (Pd-C$_{ortho}$ = 2.676 Å). The CP-MAS ^{31}P{^{1}H} NMR spectrum of a solid sample of this complex confirmed the presence of non-equivalent

Scheme 46 Synthesis and molecular structure of Pd(PCy₂Ar′)₂.[179]

P atoms (AB pattern was observed). However, in solution at room temperature a dynamic process exchanged the environment of two biphenyl moieties. On cooling to −90 °C, the ^{31}P{^{1}H} NMR spectrum revealed the presence of an equilibrium between the π-coordinated conformer (minor species) and the non-coordinated one (mayor species).

Such that Pd-arene interaction, unprecedented at this time, is currently a coordination mode frequently exhibited by biaryldialkyl and related phosphines. In relation with this, the use of potentially hemilabile monophosphine ligands that can enhance the stability of the catalytic active species favoring more rapid catalysis are receiving increasing attention. Heteroleptic phosphines composed of two bulky alkyl groups and a third substituent, which contains the heteroatom (N, O) that can facilitate the hemilabile coordination, have been employed to prepare PdL$_2^P$ complexes (Fig. 29).[180] It is interesting to point out that despite the proximity of the heteroatom to the metal in some of these examples, the phosphine ligand

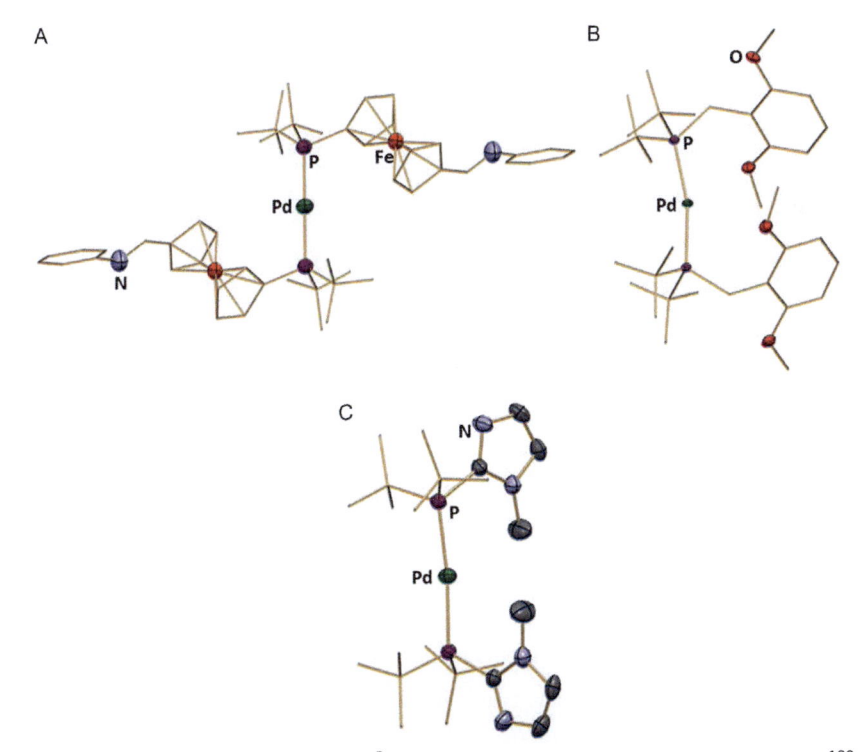

Fig. 29 Solid-state structures of PdL$_2^P$ complexes bearing hybrid monophosphines.[180]

is coordinated to the metal in a classic manner, i.e., through the P atoms, underlining the soft character of the Pd(0) center.

Due to the rising interest in using well-defined Pd-based precursors as catalysts in cross-coupling reactions, improved synthetic methods for the preparation of PdL$_2^P$ complexes have recently appeared (Scheme 47). Baird[181] replaced the precursor Pd(η^5-C$_5$H$_5$)(η^3-C$_3$H$_3$) in the Otsuka's procedure by the complex Pd(η^5-C$_5$H$_5$)(η^3-1-PhC$_3$H$_4$). The latter is easy to handle and store at room temperature due to its enhanced thermal stability, and reacts faster with phosphines to form the Pd(0) complexes. On the other hand, Colacot and co-workers[182] introduced the reduction of PdBr$_2$(COD) with NaOH in MeOH in the presence of phosphine ligands as a suitable protocol for the synthesis of two-coordinate Pd(0) complexes. The detailed examination of this reaction led to the isolation of key intermediates in the reduction of Pd(II) to Pd(0).

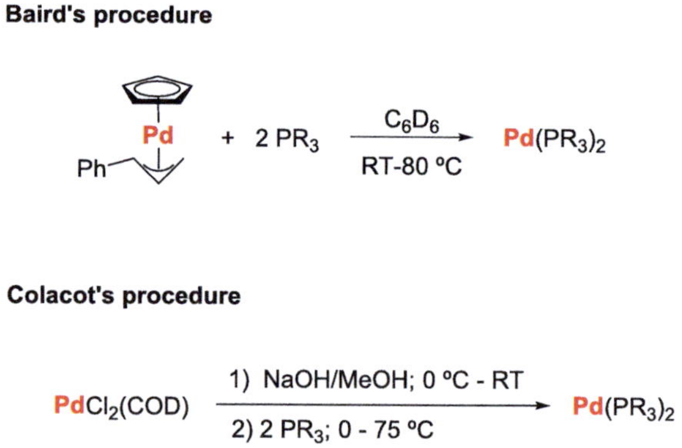

Baird's procedure

Colacot's procedure

Scheme 47 Improved synthetic protocols for the preparation of P$_2$Pd(0).[181,182]

A remarkably reduction of Pd(OAc)$_2$ in the presence of troticenyl phosphines (troticenyl = cycloheptatrienyl-cyclopentadienyl titanium moiety) to generate PdL$_2^P$ complexes has been described by Tamm and co-workers[183] (Scheme 48). Careful examination of this process revealed the participation of the titanium fragment, and not the phosphine, in the Pd(II)/Pd(0) reduction. Interestingly, a stable dianionic 14-electron PdL$_2^P$ complex stabilized by a monoanionic carboranyl phosphine ligand has been published.[184] The synthesis was accomplished by the reaction of Pd(CH$_2$SiMe$_3$)$_2$(COD) with the ligand in benzene-d$_6$ (Scheme 48). The dianionic PdL$_2^P$ complex was more active in Kumada-Tamao-Corriu couplings than the neutral isosteric

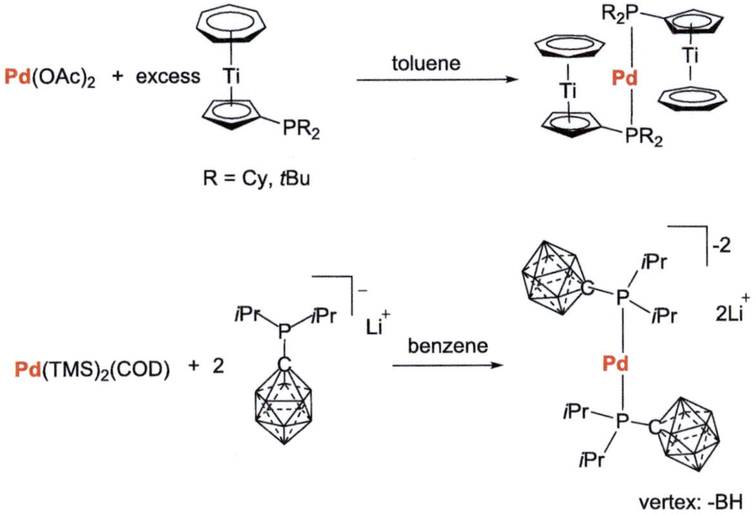

Scheme 48 P$_2$Pd(0) complexes supported by troticenyl phosphines and monoanionic carboranyl phosphine.[183,184]

counterpart. The enhanced reactivity of the former was attributed to electrostatic effects that facilitate both ligand dissociation and oxidative addition steps.

In contrast to the considerable amount of reports on the synthesis and crystallographic characterization of PdL$_2^P$ complexes, the number of corresponding platinum(0) analogues is still scarce. Very recently, Roddick and co-workers[40] published the synthesis of a two-coordinate Pt(0) complex supported by π-acceptor perfluoroalkyl phosphine ligand, P*t*Bu(C$_2$F$_5$)$_2$. The complex was generated by thermolysis of PtPh$_2$(COD) in neat phosphine at 80 °C for 5 days. In the solid state, this molecule shows a linear geometry (P-Pt-P angle 180°) with Pt-P bond lengths of 2.205 Å, slightly shorter than those found in Pt(PCy$_3$)$_2$[173] and Pt(P*t*Bu$_2$Ph)$_2$[174] (2.231 Å and 2.252 Å, respectively), in line with the accentuated π-acceptor character of the perfluoroalkyl phosphine ligand.

In the last few years, electron-rich PtL$_2^P$ complexes have been the focus of attention of those research groups endeavoring to prepare frustrated Lewis pair (FLP), as they can serve as Lewis base partners.[185] In this regard, Campos has described the first examples of metal-only frustrated Lewis pair (MOFLP) built upon the use of Pt(P*t*Bu$_3$)$_2$ as the basic component and a group 11 metal species as the Lewis acid partner.[186] Activation of a variety of X-H bonds (X=H, C, O, N) by these systems have been reported.

It is worth of interest to present in this section the short number of ML_2^P complexes stabilized by *trans*-spanning chelating diphosphine, since these include the only examples of NiL_2^P compounds structurally characterized to date.

In 2008, Hofmann and co-workers[187] reported on the synthesis and characterization of the first monometallic 14-electron PdL^{PP} complex supported by a *trans*-spanning chelating bisphosphine ligand derived from the 9,10-bridge, 9,10-dihydroanthracene scaffold (Scheme 49). X-ray diffraction studies confirmed the expected non-linear, *trans*-spanning coordination imposed by the chelating ligand (P-Pd-P below 150°).

Scheme 49 Synthesis of two-coordinate Pd(0) complexes supported by chelating diphosphine ligands.[187]

Using *p*-terphenyl diphosphine, the group of Agapie isolated the first stable example of a two-coordinate homoleptic Ni(0) stabilized by P-donor ligands.[188] The synthesis was accomplished by reacting $Ni(COD)_2$ with the ligand at room temperature. In the solid state, the Ni(0) unsaturated fragment is stabilized by η^2-coordination to one of the double bonds of the central aryl ring of the terphenyl moiety (Fig. 30). The Ni-C$_{arene}$ distances, close

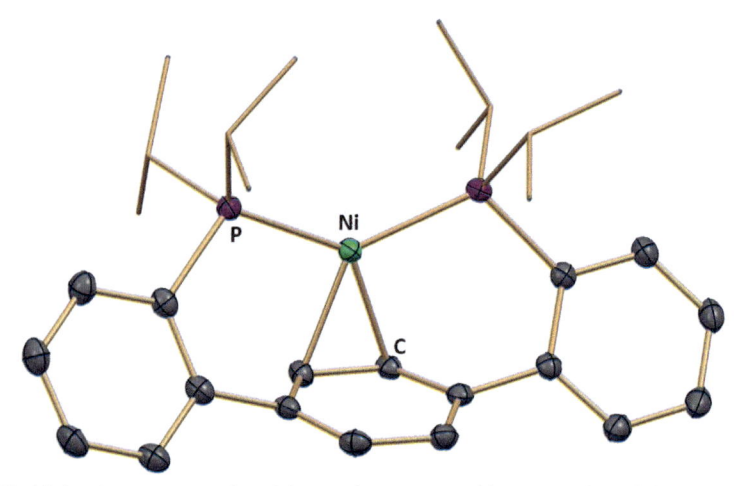

Fig. 30 Molecular structure of a Ni(0) complex supported by a *p*-terphenyl diphosphine.[188]

to 2.0 Å, are almost identical than those found in known Ni(0)-olefin complex Ni(C$_2$H$_4$)$_2$(PPh$_3$).[189] Such strong Ni-arene interaction is also maintained in solution, as emerges from the significantly upfield shift of the four central arene protons (ca. 5.52 ppm) in the ^{1}H NMR spectrum. A variety of two-coordinate complexes of the group 10 triad supported by terphenyl diphosphine ligand platform has been prepared and characterized by Agapie and co-workers during the last few years. In addition to exploring the reactivity of these unsaturated molecules to different Lewis bases, a comprehensive investigation of the hydrogenolysis of aryl ether by the nickel complexes has been undertaken.[190c]

Iluc and co-workers (Scheme 50) took advantage of the geometric flexibility of an *o*-terphenyl diphosphine ligand to stabilize both two-coordinate ML^{PP} species (M=Pd, Pt), with a *trans* distribution of the P donors, and three-coordinate diaryl carbene species).[111]

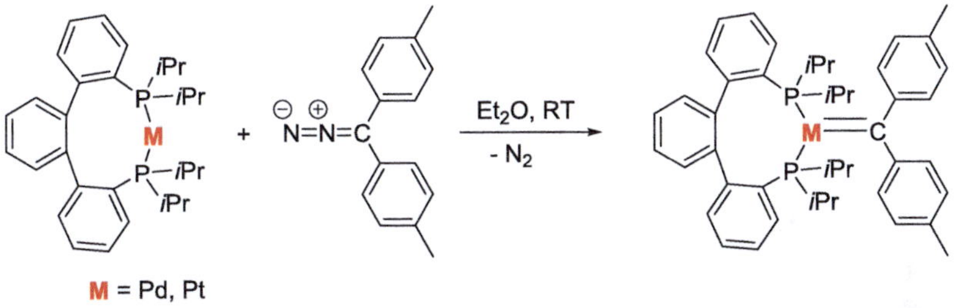

Scheme 50 Synthesis of diarylcarbene complexes of Pt and Pd stabilized by a chelating diphosphine.[111]

3.2 Homoleptic M(0) complexes with N-heterocyclic ligands

Unlike their phosphine stabilized analogues, two-coordinate 14-electron (NHC)$_2$M(0) complexes have been isolated and structurally characterized for all group 10 metals. They exhibit linear geometry with staggered (D$_{2d}$) or eclipsed (D$_{2h}$) arrangements of the carbene rings, depending on their bulkiness.

Arduengo and co-workers reported in the mid-1990s the first homoleptic 14-electron M^0(NHC)$_2$ complexes of nickel and platinum.[146a] The compounds, M(IMes)$_2$, were obtained in good yields by the reaction of zerovalent metal-olefin precursor M(COD)$_2$ (M=Ni, Pt) with two equivalents of free IMes ligand in THF at room temperature (Scheme 51).

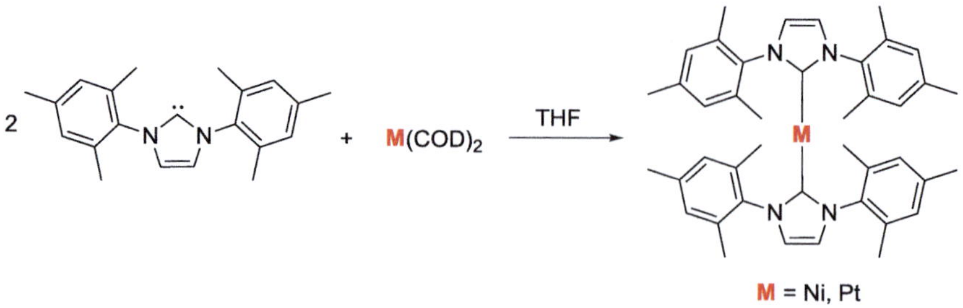

Scheme 51 Synthesis of M(IMes)$_2$ (M=Ni, Pt) by Arduengo.[146a]

The solid-state geometries of these two-coordinate adducts were elucidated by X-ray diffraction studies. Both are almost linear molecules (C$_{carbene}$-M-C$_{carbene}$ angles of 176.4(3)° for the Ni derivative and 177.4(3)° for the Pt one) with very short M-C$_{carbene}$ bond lengths (mean values 1.83 Å and 1.95 Å for Ni(IMes)$_2$ and Pt(IMes)$_2$, respectively).[146a] Such short distances were explained in terms of d_{metal}-to-p_{carbon} π-backdonation and sp hybridization of the linear metal atom. Interestingly, the planes of the imidazole rings are twisted with a deviation of around 50° from coplanarity due to steric interactions between the mesityl groups.

Since the analogous Pd(IMes)$_2$ complex could not be obtained following the route displayed in Scheme 52, the group of Herrmann devised an alternative synthesis involving a ligand exchange procedure.[191] The synthetic strategy encompassed the reaction of Pd[P(o-tolyl)$_3$]$_2$ with excess of free carbene ligand in toluene (Scheme 52). The liberated tri-o-tolylphosphine could be eliminated by washing the crude product with cold hexane. Monitoring the reaction by ^{31}P{^{1}H} NMR revealed a stepwise substitution of the phosphines by the carbene ligand.

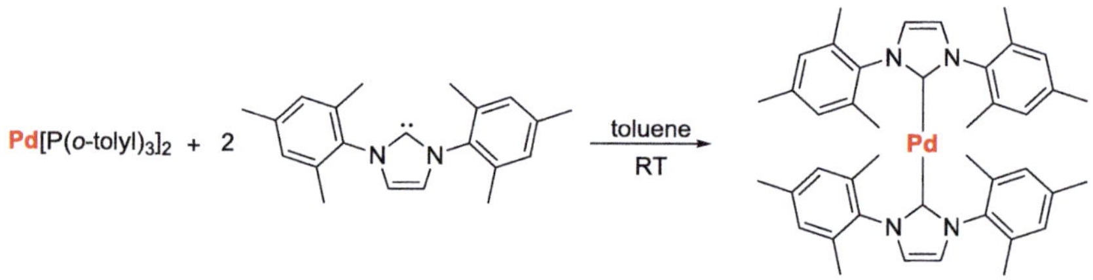

Scheme 52 Synthesis of Pd(NHC)$_2$ by Herrmann and co-workers.[191]

The complex Pd(IMes)$_2$ was only characterized in solution by means of NMR spectroscopy. Its X-ray characterization was accomplished some years later by Stahl and co-workers in their study of the oxidation of Pd(0) by molecular oxygen.[192] The complex is isostructural with their Ni and Pt counterparts.

In 1999, Cloke and co-workers[193] accomplished the preparation of M(NHC)$_2$ complexes for the group 10 triad using metal vapor deposition (MVD) techniques. The synthesis was restricted to the use of the alkyl-substituted carbene I*t*Bu, due to its high vapor pressure at low temperature (the solid sublimes at 40 °C, 10^{-3} bar). The compounds were obtained in low yields (10%–32%) as brown (Ni) or yellow (Pd, Pt) microcrystalline materials. Ni(I*t*Bu)$_2$ was air and moisture unstable, whereas the Pd and Pt analogues were more robust to air oxidation (Scheme 53).

Scheme 53 Synthesis of M(I*t*Bu)$_2$ (M=Ni, Pd and Pt) by MVD.[193]

The heavier complexes M(I*t*Bu)$_2$ (M=Pd, Pt) are linear molecules in the solid state with a geometry approaching a C$_{2v}$ symmetry (the dihedral angle between the two planes of the coordinated carbene ligands is 90°) (Fig. 31). The molecular structure of the Ni(I*t*Bu)$_2$ was described by Caddick and Cloke in a later work.[146b] As expected, the geometry is linear around the Ni center but, the two I*t*Bu carbenes are twisted only 75° one another, not reaching the completed staggered conformation found in its heavier siblings.

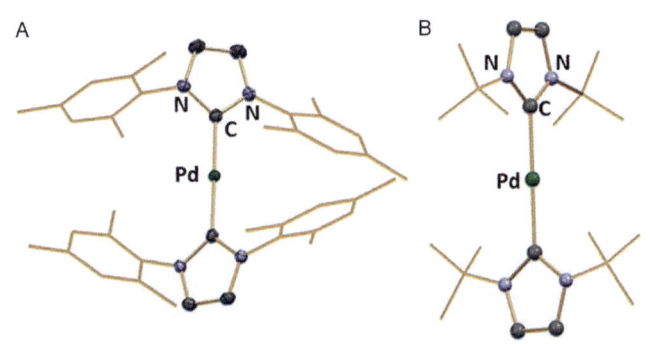

Fig. 31 X-ray structures of (A) Pd(IMes)$_2$[192] and (B) Pd(I*t*Bu)$_2$.[193]

A more facile synthetic protocol for the preparation of Pd(I*t*Bu)$_2$ was published by the same research group shortly thereafter.[194] It involved the addition of a nucleophile to a Pd(II)-allyl complex, typically [Pd(η^3-allyl)Cl]$_2$, in the presence of two equivalents of the free carbene ligand (I*t*Bu) at 90 °C (Scheme 54). This synthetic methodology proved to be versatile and general, allowing the preparation of a series of Pd(NHC)$_2$ complexes (NHC=SIPr, IPr, SI*t*Bu).[195] These complexes were used as precatalysts in C–C and C–N bond forming reactions.

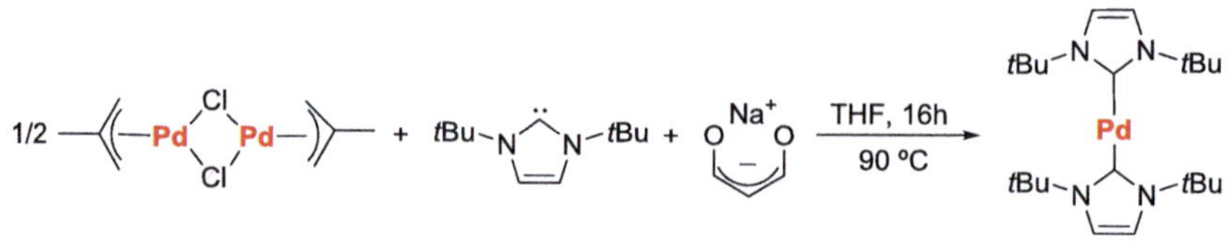

Scheme 54 Improved method for the synthesis of Pd(NHC)$_2$ (NHC=I*t*Bu, SIPr) reported by Caddick and Cloke.[194]

Herrmann and co-workers[196] failed to prepare the homoleptic Pd(IAd)$_2$ complex (IAd=1,3-bis-adamantylimidazol-2-ylidene) by ligand exchange from Pd[P(*o*-tolyl)$_3$]$_2$ with excess of the highly steric demanding IAd ligand. The mixed phosphine/NHC Pd(0) complex was isolated instead (*vide infra*). However, using Pd(P*t*Bu$_3$)$_2$ as the starting material, which bears the poorer π-accepting phosphine P*t*Bu$_3$, ligand exchange was successfully accomplished affording the desired bis-carbene adduct in a 80% yield. Alternatively, Pd(IAd)$_2$ could be obtained using Caddick and Cloke reduction of [Pd(η^3-allyl)Cl]$_2$ in the presence of IAd at high temperatures. The latter method, although affords moderate yields of the Pd(0) complex, is a more convenient route since avoids the use of the expensive Pd(P*t*Bu$_3$)$_2$ starting material. The solid-state structure of Pd(IAd)$_2$ confirms the linear geometry (C$_{carbene}$–Pd–C$_{carbene}$=180°) and the staggered conformation adopted by the two IAd carbene units as a result of their high steric demand. The complex catalyzed efficiently the Suzuki-Miyaura coupling of aryl chlorides at room temperature.

The complex Pd(ITmt)$_2$ (ITmt=1,3-bis(2,2″,6,6″-tetramethyl-*m*-terphenyl-5′-yl)imidazol-2-ylidene) was prepared by Goto and Kawashima[197] by reaction of Pd[P(*o*-tolyl)$_3$]$_2$ with free ITmt ligand. While being bulky enough to facilitate the isolation of low-coordinate 14-electron species Pd(ITmt)$_2$, the less steric congestion of ligand ITmt in the close vicinity of the carbene center accounted for the lower stability of Pd(ITmt)$_2$ when exposed to air in a solid state. A Pd(II) peroxocarbonate complex, in which the two ITmt ligands adopted a mutually *cis* disposition, was generated upon solid-state fixation of oxygen and carbon dioxide at the parent 14-electron complex.

In 2007, the group of Danopoulos[198] devised a new synthetic protocol for the preparation of Pd(0) bis-carbene complexes with less bulky unsymmetrically *ortho* disubstituted NHC ligands using PdMe$_2$(TMEDA) as the metal source (Scheme 55). The complex supported by the unsaturated carbene ligand 1,3-bis-(2-isopropylphenyl)imidazolin-2-ylidene) presents the expected linear geometry but, due to reduced steric congestion around the metal center, the carbene ligands adopt and eclipsed conformation. This methodology was also applied for the preparation of various Ni(NHC)$_2$ (NHC = IMes, IPr, SIPr and SI*t*Bu).[199]

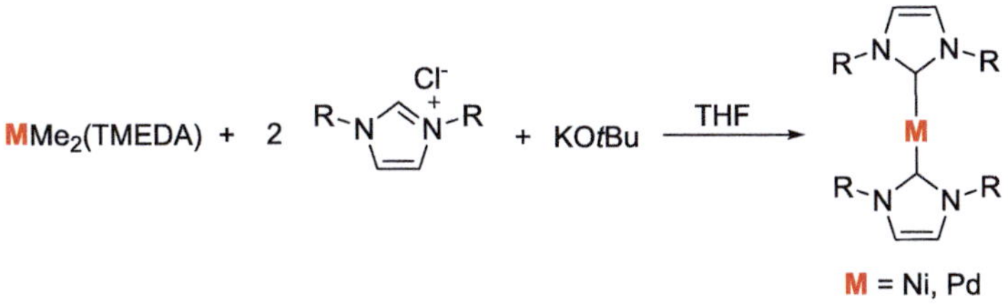

Scheme 55 Danopoulos' synthesis of M(NHC)$_2$ (M = Pd, Ni) complexes.[198,199]

An alternative synthetic route to Ni(NHC)$_2$ complexes was reported shortly after by Matsubara and co-workers[200] involving the use of Ni(acac)$_2$ (acac = acetlyacetonate), as the nickel source, and NaH as the reductant (Scheme 56). The reaction could be performed in one-pot or in a two-step synthesis via isolation of the intermediate Ni(NHC)(acac)$_2$. The complex Ni(IPr)$_2$ was tested as precatalyst in the amination of chlorobenzene with anilines, affording better yields of the coupling products than the in situ made catalyst reported by Fort.[201]

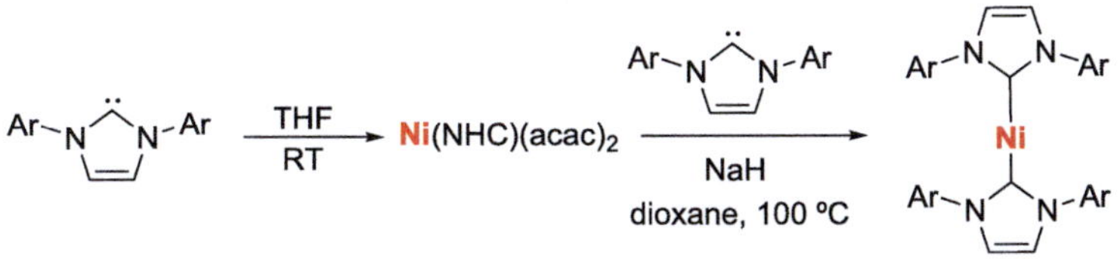

Scheme 56 Alternative synthesis of Ni(NHC)$_2$ complexes devised by Matsubara.[200]

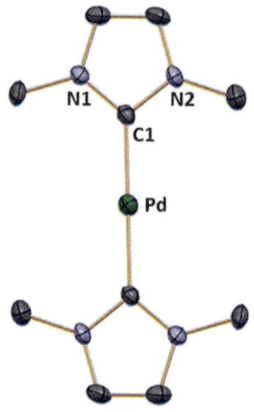

Fig. 32 Solid-state structure of Pd(IMe)₂.[203]

The least sterically encumbered d^{10}-M(NHC)₂ structurally characterized, the complex Pd(IMe)₂ (IMe = 1,3-1,3-dimethylimidazol-2-ylidene), was prepared by Lee and Yandulov[202] by the 2-electron reduction of Pd(IMe)₂Cl₂ with K metal in THF under sonication. The two-coordinate complex shows the usual linear geometry in the solid state, in spite of the reduced steric bulk of the IMe ligand (Fig. 32). Moreover, the two imidazole rings are coplanar, even though DFT calculations predict that the staggered conformer (D_{2d} symmetry) is about 2–3 kcal mol⁻¹ lower in energy that the eclipsed one (D_{2h} symmetry). The conformation adopted by IMe ligands is attributed to packing forces.

As discussed in the previous section,[129,145] Radius and co-workers disclosed the formation of three-coordinate Ni(NHC)₂L (L=olefin) complexes with the sterically modest I*i*Pr ligand. However, two-coordinate M(I*i*Pr)₂ (M=Pd and Pt) were isolated when salt like complexes [M(I*i*Pr)₃Cl]⁺Cl⁻ were reduced with potassium graphite (Scheme 57). In the complex Pd(I*i*Pr)₂ the imidazole rings show an eclipsed alignment (D_{2h} symmetry). As in the previous case, this conformation is also ascribed

Scheme 57 Synthesis of M(I*i*Pr)₂ complexes developed by Radius and co-workers.[203]

to packing forces. Poor reactivity was found when M(I*i*Pr)$_2$ complexes were reacted with I*i*Pr or π-acidic ligands. This behavior, which contrasted to that found for the nickel complexes (see above), appeared to have originated in the high L-M-L bending strain of d^{10}-M(NHC)$_2$ fragments (M=Pd, Pt).

The group of Radius explored other synthetic routes to obtain Pt(I*i*Pr)$_2$.[204] They found that Pt(PPh$_3$)$_2$(η^2-C$_2$H$_4$) was a suitable Pt(0) precursor for ligand substitution, producing Pt(I*i*Pr)$_2$ in good yields when reacted with excess I*i*Pr (3.5–4 equiv.) in hexane at room temperature. However, depending on the reaction conditions (amount of I*i*Pr and solvent) the stepwise exchange of PPh$_3$ by the carbene ligand lead to a variety of different Pt-bases species.

3.3 Heteroleptic M(0) complexes

Mixed NHC/phosphine M(0) complexes are the most common heteroleptic 14-electron complexes of group 10 metals reported to date. As discussed in the previous section, Caddick and Cloke and the group of Herrmann reported the formation of mixed NHC/phosphine Pd(0) complexes in the synthesis of (NHC)$_2$Pd(0) by ligand exchange from Pd[P(*o*-tolyl)$_3$]$_2$ with NHCs,[195a,196] although no structural information of these species was provided. Nolan[205] published the first general synthetic procedure for this type of compounds in 2008 (Scheme 58). The combination of *i*PrOH and basic conditions seems to be crucial for the reduction of Pd(II) to Pd(0) species.

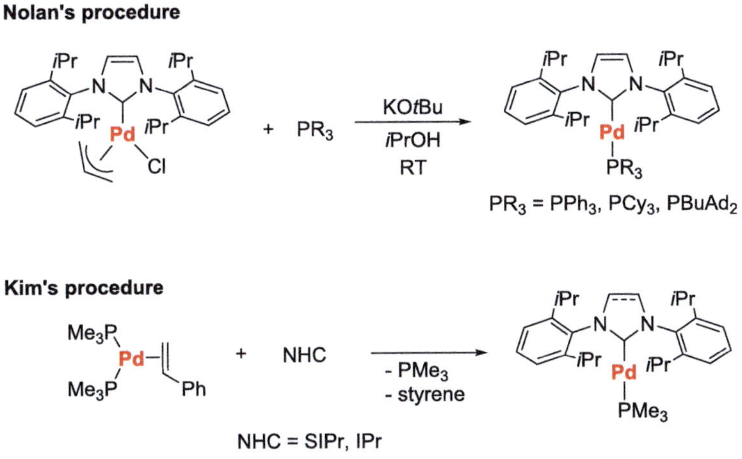

Scheme 58 Synthesis of Pd(NHC)(PR$_3$) devised by Nolan and Kim.[206]

An alternative route was later developed by Kim and co-workers,[206] which involved the reaction of a Pd(0) precursor with the carbene ligand (Scheme 58).

As expected, these complexes are almost linear in the solid state and their structural features are similar to those already described for homoleptic PdL_2^P.

Nolan and Cazin demonstrated the catalytic capability of these mixed NHC/PR$_3$ Pd(0) complexes in oxidation,[207] hydrogenation/dehydrogenation[208] reactions and in Suzuki-Miyaura couplings.[209]

Stradiotto and co-workers[210] reported the formation Pd(IPr)(PR$_2$Cl) complexes by a two-step reaction, including dehydrohalogenation and subsequent P-Cl reductive elimination, triggered by the treatment of Pd(IPr)(Cl)$_2$(PR$_2$H) with NaN(SiMe$_3$)$_2$ (Scheme 59).

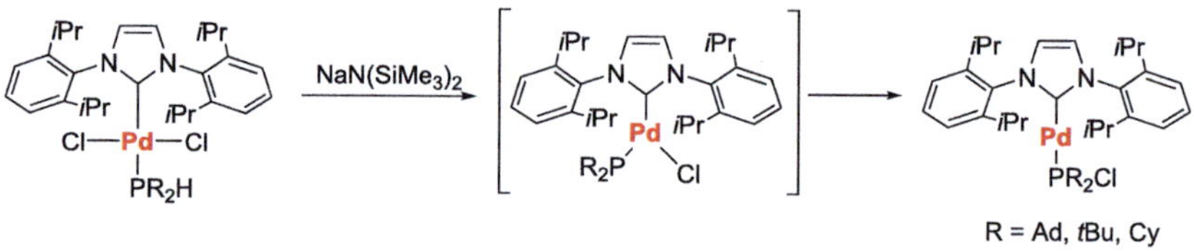

Scheme 59 Synthesis of Pd(IPr)(PR$_2$Cl).[210]

The synthesis of Pt(NHC)(PCy$_3$), the first heteroleptic Pt(0) species of this type, was described by Braunschweig in 2010. The complex was obtained by ligand exchange reaction of Pt(PCy$_3$)$_2$ with NHCs. In a subsequent paper, the same research group targeted the synthesis of the analogous Pd(ItBu)(PCy$_3$) by the stepwise introduction of PCy$_3$ and ItBu to Pd(η^3-allyl)(η^5-C$_5$H$_5$). These complexes were used to prepare metal only Lewis pair by reaction with aluminum chloride.

Finally, it is interesting to include in this section three remarkable examples of unsaturated MP(olefin) complexes. The first two examples come from the group of Buchwald and correspond to Pd(0)-(η^2-dba) complexes supported by dialkyl biaryl phosphine ligands.[211] The complexes were obtained by the direct reaction of Pd$_2$(dba)$_3$ with the ligand and their structure were elucidated by X-ray diffraction studies. As shown in Fig. 33, both structures feature interactions with the non-phosphorous functionalized arene ring, which help to stabilize the unsaturated Pd fragment.

The third example is the complex Pt(PMe$_2$Ar')(C$_2$H$_4$) reported by Carmona and co-workers, in which the Pt(0) center is also stabilized by the interaction with the C$_{ipso}$ of one of the flanking aryl ring of the terphenyl fragment (see Scheme 20).

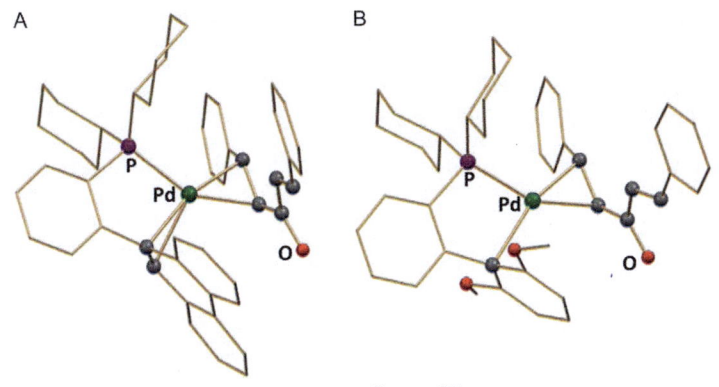

Fig. 33 Molecular structures of Pd(PR$_2$Ar′)(η^2-dba).[211]

4. Conclusions and outlook

In this chapter we have shown that the chemistry of low-coordinate complexes of nickel, palladium and platinum in the zero-oxidation state is still a vivid area of research in organometallic chemistry. Indeed, most efforts have been put on isolating and characterized two- and three-coordinate complexes of Pd(0) owing to the meaningful applications of this metal in organic synthesis. However, driven by sustainability concerns, the chemistry of nickel is flourishing in recent years. The synthesis of a variety of bench-stable 16-electron Ni(0) complexes supported by the combination of phosphine or N-heterocyclic carbene ligands with π-acidic olefins and their application in catalysis illustrates the current interest in the lighter and less expensive member of group 10 metals. One can anticipate further studies in this field as two-coordinate Ni(0) complexes bearing monophosphine ligands or carbenes derived from 1,2,4-triazole, benzimidazole or expanded-NHC systems are presently no available. Finally, concerning to platinum, it becomes evident that the chemistry of 14-electron Pt(0) species is still underexplored. Presumably, the rising interest in the synthesis of organometallic FLPs will help drive its growth, since the enhanced basicity of these Pt(0) species makes them suitable candidates as Lewis base partners.

Acknowledgments

We thank MINECO (Grants CTQ2014-52769-C3-3-R and CTQ2017-82893-C2-2-R) and Junta de Andalucía (Grant US-1262266) for financial support. R.J.R. thanks the Universidad de Sevilla (V Plan Propio de Investigación) for a research fellowship.

References

1. (a) Mitchell PR, Parish RV. The eighteen electron rule. *J Chem Educ*. 1969;46(12):811. https://doi.org/10.1021/ed046p811. (b) Tolman CA. The 16 and 18 electron rule in organometallic chemistry and homogeneous catalysis. *Chem Soc Rev*. 1972;1:337–353. https://doi.org/10.1039/CS9720100337.
2. Bhatt V. Metal carbonyls. In: *Essential of Coordination Chemistry*. Elsevier; 2016:191–236.
3. Pauling L. *The Nature of the Chemical Bond*. Ithaca, NY: Cornel University Press; 1960.
4. Eastes JW, Burgess WM. A study of the products obtained by the reducing action of metals upon salts in liquid ammonia solutions. VII. The reduction of complex nickel cyanides: mono-valent nickel. *J Am Chem Soc*. 1942;64:1187–1189. https://doi.org/10.1021/ja01257a053.
5. Burbage JJ, Fernelius WC. Reduction of potassium cyanopalladate (II) by potassium in liquid ammonia; a zerovalent compound of palladium[1]. *J Am Chem Soc*. 1943;65:1484–1486. https://doi.org/10.1021/ja01248a016.
6. (a) Malatesta L, Angoletta M. Palladium(0) compounds. Part II. Compounds with triarylphosphines, triaryl phosphites, and triarylarsines. *J Chem Soc*. 1957;1186–1188. https://doi.org/10.1039/JR9570001186. (b) Malatesta L, Cariello C. Platinum(0) compounds with triarylphosphines and analogous ligands. *J Chem Soc*. 1958;2323–2328. https://doi.org/10.1039/JR9580002323. (c) Wilke G, Müller EW, Kröner M. Nordwestdeutsche chemiedozenten-tagung hannover, 22. bis 24 September 1960. *Angew Chem*. 1961;73:33–34. https://doi.org/10.1002/ange.19610730114.
7. Ugo R. The coordinative reactivity of phosphine complexes of platinum(o), palladium(o) and nickel(o). *Coord Chem Rev*. 1968;3(3):319–344. https://doi.org/10.1016/S0010-8545(00)80121-6.
8. (a) Labinger JA, Bercaw JE. Understanding and exploiting C-H bond activation. *Nature*. 2002;417:507–514. https://doi.org/10.1038/417507a. (b) Lersch M, Tilset M. Mechanistic aspects of C−H activation by Pt complexes. *Chem Rev*. 2005;105:2471–2526. https://doi.org/10.1021/cr030710y.
9. (a) de Meijere A, Diederich F. *Metal Catalyzed Cross-Coupling Reactions*. Weinheim: Wiley-VCH; 2004. (b) Colacot TJ. *New Trends in Cross-Coupling. Theory and Applications. RSC Catalysis Series*. Cambridge: RSC Publishing; 2015.
10. Dastbaravardeh N, Christakakou M, Haider M, Schnürch M. Recent advances in palladium-catalyzed C(sp^3)–H activation for the formation of carbon–carbon and carbon–heteroatom bonds. *Synthesis*. 2014;46:1421–1439. https://doi.org/10.1055/s-0033-1338625.
11. Stradiotto M, Lundgren RJ. *Ligand Design in Metal Chemistry: Reactivity and Catalysis*. Wiley; 2016.
12. Zapf A, Ehrentraut A, Beller M. A new highly efficient catalyst system for the coupling of nonactivated and deactivated aryl chlorides with arylboronic acids. *Angew Chem Int Ed*. 2000;39:4153–4155. https://doi.org/10.1002/1521-3773(20001117)39:22<4153::AID-ANIE4153>3.0.CO;2-T.
13. (a) Hartwig JF, Kawatsura M, Hauck SI, Shaughnessy KH, Alcazar-Roman LM. Room-temperature palladium-catalyzed amination of aryl bromides and chlorides and extended scope of aromatic C−N bond formation with a commercial ligand. *J Org Chem*. 1999;64:5575–5580. https://doi.org/10.1021/jo990408i. (b) Mann G, Incarvito C, Rheingold AL, Hartwig JF. Palladium-catalyzed C−O coupling involving unactivated aryl halides. Sterically induced reductive elimination to form the C−O bond in diaryl ethers. *J Am Chem Soc*. 1999;121:3224–3225. https://doi.org/10.1021/ja984321a. (c) Shelby Q, Kataoka N, Mann G, Hartwig JF. Unusual in situ ligand modification to generate a catalyst for room temperature aromatic C−O bond formation. *J Am Chem Soc*. 2000;122:10718–10719. https://doi.org/10.1021/ja002543e.

14. (a) Wolfe JP, Singer RA, Yang BH, Buchwald SL. Highly active palladium catalysts for suzuki coupling reactions. *J Am Chem Soc*. 1999;121:9550–9561. https://doi.org/10.1021/ja992130h. (b) Surry DS, Buchwald SL. Biaryl phosphane ligands in palladium-catalyzed amination. *Angew Chem Int Ed*. 2008;47:6338–6361. https://doi.org/10.1002/anie.200800497.
15. Scott NM, Nolan SP. Stabilization of organometallic species achieved by the use of N-heterocyclic carbene (NHC) ligands. *Eur J Inorg Chem*. 2005;2005:1815–1828. https://doi.org/10.1002/ejic.200500030.
16. (a) In: Diez-Gonzalez S, ed. *N-Heterocyclic Carbenes: From Laboratory Curiosity to Efficient Synthetic Tools*. Cambridge: RSC; 2010. (b) Rovis T, Nolan SP. Stable carbenes: from 'laboratory curiosities' to catalysis mainstays. *Synlett*. 2013;24(10):1188–1189. https://doi.org/10.1055/s-0033-1339192.
17. Sergienko VS, Porai-Koshits MA. Crystal structure of tris(triphenylphosphine) palladium. *J Struct Chem*. 1987;28(4):548–552. https://doi.org/10.1007/BF00749589.
18. Albano V, Bellon PL, Scatturin V. Zerovalent metal complexes: crystal and molecular structure of [Pt(PPh$_3$)$_3$]. *Chem Commun*. 1966;57(15):507. https://doi.org/10.1039/C19660000507.
19. Yoshida T, Matsuda T, Otsuka S, et al. Three-coordinate phosphine complexes of platinum(O). *Inorg Synth*. 1979;19:107–110. https://doi.org/10.1002/9780470132500.ch22.
20. Mynott R, Mollbach A, Wilke G. A proof of the existence of Ni[P(C$_6$H$_5$)$_3$]$_4$ in solution. *J Organomet Chem*. 1980;199(1):107–109. https://doi.org/10.1016/S0022-328X(00)84527-1.
21. Gosser LW, Tolman CA. New three-coordinate complex of nickel(0). Tris(tri-o-tolyl phosphite)nickel. *Inorg Chem*. 1970;9(10):2350–2353. https://doi.org/10.1021/ic50092a030.
22. Tolman CA, Seidel WC, Gosser LW. Formation of three-coordinate nickel(0) complexes by phosphorus ligand dissociation from NiL4. *J Am Chem Soc*. 1974;96(1):53–60. https://doi.org/10.1021/ja00808a009.
23. Dick DG, Stephan DW, Campana CF. The crystal and molecular structure of the coordinatively unsaturated Ni(0) species Ni(PPh$_3$)$_3$. *Can J Chem*. 1990;68(4):628–632. https://doi.org/10.1139/v90-096.
24. Kampmann SS, Skelton BW, Wild DA, Koutsantonis GA, Stewart SG. Tris(tri-*o*-tolyl phosphite-κ*P*)nickel: a coordinatively unsaturated nickel(0) complex. *Acta Crystallogr C Struct Chem*. 2015;71:188–190. https://doi.org/10.1107/S2053229615001680.
25. Musco A, Kuran W, Silviani A, Anker MW. Tertiary phosphine palladium(0) complexes. *J Chem Soc Chem Commun*. 1973;938–939. https://doi.org/10.1039/C39730000938.
26. Mann BE, Musco A. Phosphorus-31 nuclear magnetic resonance spectroscopic characterisation of tertiary phosphine palladium(0) complexes: evidence for 14-electron complexes in solution. *J Chem Soc Dalton Trans*. 1975;12:1673–1677. https://doi.org/10.1039/DT9750001673.
27. Kuran W, Musco A. Synthesis and characterization of tertiary phosphine Pd(0) complexes. *Inorg Chim Acta*. 1975;12(1):187–193. https://doi.org/10.1016/S0020-1693(00)89858-8.
28. Portnoy M, Milstein D. Chelate effect on the structure and reactivity of electron-rich palladium complexes and its relevance to catalysis. *Organometallics*. 1993;12(5):1655–1664. https://doi.org/10.1021/om00029a025.
29. Ben-David Y, Portnoy M, Milstein D. Chelate-assisted, palladium-catalyzed efficient carbonylation of aryl chlorides. *J Am Chem Soc*. 1989;111(23):8742–8744. https://doi.org/10.1021/ja00205a039.

30. King RB, Kapoor PN. Polytertiary phosphines and arsines. VII. Zerovalent platinum complexes of arylated polytertiary phosphines and arsines. *Inorg Chem*. 1972;11(7):1524–1527. https://doi.org/10.1021/ic50113a015.
31. Gerlach DH, Kane AR, Parshall GW, Jesson JP, Muetterties EL. Reactivity of trialkylphosphine complexes of platinum(0). *J Am Chem Soc*. 1991;93(14): 3543–3544. https://doi.org/10.1021/ja00743a050.
32. Manoiloviç-Muir L, Muir KW. A diplatium (0) complex with three bridging dppm ligands. The X-ray structure anylysis of [Pt$_2$(μ-dppm)$_3$].2C$_6$H$_6$(dppm = Ph$_2$PCH$_2$PPh$_2$). *J Chem Soc Chem Commun*. 1982;48(20):1155–1156. https://doi.org/10.1039/C39820001155.
33. Manojlović-Muir L, Muir KW, Grossel MC, et al. Tris-μ-[bis(diphenylphosphino)-methane]-diplatinum(0). *J Chem Soc Dalton Trans*. 1986;49(9):1955. https://doi.org/10.1039/DT9860001955.
34. Evrard D, Clément S, Lucas D, et al. Chemistry and electrochemistry of the heterodinuclear complex ClPd(dppm)$_2$PtCl: A M—M′ bond providing site selectivity. *Inorg Chem*. 2006;45(3):1305–1315. https://doi.org/10.1021/ic051102z.
35. Grossel MC, Brown MP, Nelson CD, Yavari A, Kallas E, Moulding RP. Preparation and characterisation of tris-μ-[bis(diphenylphosphino)methane]diplatinum(0). *J Organomet Chem*. 1982;232(1):C13–C16. https://doi.org/10.1016/S0022-328X(00)86857-6.
36. Balch AL, Hunt CT, Lee CL, Olmstead MM, Farr JP. Organo halide addition to tris(bis(diphenylphosphino)methane)dipalladium. Preparation of novel methylene- and phenylene-bridged complexes by two-center, three-fragment oxidative addition. *J Am Chem Soc*. 1981;103(13):3764–3772. https://doi.org/10.1021/ja00403a025.
37. Ben-David Y, Portnoy M, Milstein D. Formylation of aryl chlorides catalysed by a palladium complex. *J Chem Soc Chem Commun*. 1989;1816–1817. https://doi.org/10.1039/C39890001816.
38. Fryzuk MD, Clentsmith GKB, Rettig SJ. Solution and solid-state structures of the binuclear zerovalent palladium complex [(dippe)Pd]$_2$(μ-dippe) (dippe = 1,2-bis (diisopropylphosphino)ethane). *Organometallics*. 1986;15(8):2083–2088. https://doi.org/10.1021/om950825v.
39. Bauer J, Braunschweig H, Dewhurst RB, Radacki K. Reactivity of lewis basic platinum complexes towards fluoroboranes. *Chem A Eur J*. 2013;19(27):8797–8805. https://doi.org/10.1002/chem.201301056.
40. Phelps J, Butikofer JL, Thapaliya B, et al. Structural and reactivity properties of perfluoroalkylphosphine complexes of platinum(0). *Polyhedron*. 2016;116:197–203. https://doi.org/10.1016/j.poly.2016.04.045.
41. Bertsch S, Forster M, Gruss K, Radacki K. Carbonyl complexes of platinum(0): synthesis and structure of [(Cy$_3$P)$_2$Pt(CO)] and [(Cy$_3$P)$_2$Pt(CO)$_2$]. *Inorg Chem*. 2011;50(5):1816–1819. https://doi.org/10.1021/ic102200t.
42. Tolman CA. Steric effects of phosphorus ligands in organometallic chemistry and homogeneous catalysis. *Chem Rev*. 1977;77(3):313–348. and references therein. https://doi.org/10.1021/cr60307a002.
43. Marín M, Moreno JJ, Navarro-Gilabert C, et al. Synthesis, structure and nickel carbonyl complexes of dialkylterphenyl phosphines. *Chem A Eur J*. 2019; 25(1):260–272. https://doi.org/10.1002/chem.201803598.
44. Chatt J, Rowe GA, Williams AA. Series of stable acetylene complexes. *Proc Chem Soc*. 1957;208.
45. Allen AD, Cook CD. Complexes of platinum (0): I. Complexes with triphenylphosphine and substituted phenylacetylenes. *Can J Chem*. 1964;42(5): 1063–1068. https://doi.org/10.1139/v64-162.
46. Glanville JO, Stewart JM, Grim SO. Structure of bis(triphenylphosphine)-diphenylacetyleneplatinum. *J Organomet Chem*. 1967;7(1):P9–P10. https://doi.org/10.1016/S0022-328X(00)90844-1.

47. McGinnety JA. Structure and bonding in (dimethylacetylenedicarboxylate) bis(triphenylphosphine) palladium. *J Chem Soc Dalton Trans*. 1974;1038–1043. https://doi.org/10.1039/DT9740001038.
48. Greaves EO, Maitlis PM. Acetylenic complexes of palladium. *J Organomet Chem*. 1966;6(1):104–106. https://doi.org/10.1016/S0022-328X(00)83359-8.
49. Fitton P, McKeon JE. Reactions of tetrakis(triphenylphosphine)palladium(0) with olefins bearing electron-withdrawing substituents. *Chem Commun*. 1968;4–6. https://doi.org/10.1039/C19680000004.
50. Greaves EO, Lock CJL, Maitlis PM. Metal–acetylene complexes. II. Acetylene complexes of nickel, palladium, and platinum. *Can J Chem*. 1968;46(24):3879–3891. https://doi.org/10.1139/v68-641.
51. Pörschke K-R, Tsay Y-H, Krüger C. Ethynebis(triphenylphosphane)nickel(0). *Angew Chem Int Ed*. 1985;24(4):323–324. https://doi.org/10.1002/anie.198503231.
52. Bartik T, Happ B, Iglewsky M, et al. Synthesis and characterization of bis(phosphine)nickel(0) complexes containing nonsymmetrically substituted acetylenes. *Organometallics*. 1992;11(3):1235–1241. https://doi.org/10.1021/om00039a033.
53. Birk JP, Halper J, Pickard AL. Substitution and oxidative addition reactions of platinum(0) complexes. Evidence for coordinatively unsaturated species in solution and as reactive intermediates. *J Am Chem Soc*. 1968;90(16):4491–4492. https://doi.org/10.1021/ja01018a073.
54. Halpern J, Weil TA. Mechanisms of substitution reactions of platinum(0) complexes. *J Chem Soc Chem Commun*. 1973;631–632. https://doi.org/10.1039/C39730000631.
55. Bennett MA, Robertson GB, Whimp PO, Yoshida T. Stabilization of small-ring acetylenes by complex formation with platinum. *J Am Chem Soc*. 1971;93(15):3797–3798. https://doi.org/10.1021/ja00744a056.
56. Jacobson S, Carty AJ, Mathew M, Palenik GJ. Synthesis and structure of .pi.-phosphinacetylene complexes of zerovalent palladium and platinum. *J Am Chem Soc*. 1974;96(13):4330–4332. https://doi.org/10.1021/ja00820a051.
57. Cook D, Jauhal GS. Oxygen and ethylene complexes of platinum. *Inorg Nucl Chem Lett*. 1967;3(2):31–33. https://doi.org/10.1016/0020-1650(67)80117-X.
58. Cheng PT, Cook CD, Nyburg SC, Wan KY. The crystal and molecular structure of (PPh$_3$)$_2$PtO$_2$·2CHCl$_3$. *Can J Chem*. 1971;49(23):3772–3777. https://doi.org/10.1139/v71-630.
59. (a) Cheng PT, Cook CD, Nyburg SC, Wan KY. Molecular structures and proton magnetic resonance spectra of ethylene complexes of nickel and platinum. *Inorg Chem*. 1971;10(10):2210–2213. https://doi.org/10.1021/ic50104a024. (b) Cheng P-T, Nyburg SC. The crystal and molecular structure of bis(triphenylphosphine)-(ethylene)platinum, (PPh$_3$)$_2$PtC$_2$H$_4$. *Can J Chem*. 1972;50:912. https://doi.org/10.1139/v72-142.
60. Cheng PT, Cook CD, Koo CH, Nynurg SC, Shiomi MT. A refinement of the crystal structure of bis(triphenylphosphine)(ethylene)nickel. *Acta Crystallogr B*. 1971;B27:1904–1908. https://doi.org/10.1107/S0567740871005077.
61. Visser JP, Schipperijn AJ, Lukas J. Platinum(0) complexes of cyclopropenes. *J Organomet Chem*. 1973;47(2):433–438. https://doi.org/10.1016/S0022-328X(00)81755-6.
62. Clark HC, Ferguson G, Hampden-Smith MJ, Kaitner B, Ruegger H. Reversible ethylene coordination in (η2-C$_2$H$_4$)Pt(PCy$_3$)$_2$. *Polyhedron*. 1988;7(15):1349–1353. https://doi.org/10.1016/S0277-5387(00)80384-8.
63. Wilke G, Herrmann G. Bis(triphenylphosphine)-ethylene-nickel and analogous complexes. *Angew Chem Int Ed*. 1962;1(10):549–550. https://doi.org/10.1002/anie.196205492.
64. Van der Linde R, De Jongh RO. Tertiary phosphine–palladium(0)–ethylene complexes. *J Chem Soc D Chem Commun*. 1971;563. https://doi.org/10.1039/C29710000563.

65. Ittel SD, Ibers JA. Nickel(0) stilbene complexes and the structure of bis(tri-*p*-tolylphosphine)(*trans*-stilbene)nickel(0) hemitetrahydrofuranate. *J Organomet Chem*. 1974;74(1):121–134. https://doi.org/10.1016/S0022-328X(00)83769-9.
66. Seidel WC, Tolman CA. Ethylene[bis(tri-o-tolyl phosphite)]nickel(0). *Inorg Chem*. 1970;9(10):2354–2357. https://doi.org/10.1021/ic50092a031.
67. Guggenberger LJ. Crystal structures of (acrylonitrile)bis(tri-o-tolylphosphite)nickel and (ethylene)bis(tri-o-tolylphosphite)nickel. *Inorg Chem*. 1973;12(3):499–508. https://doi.org/10.1021/ic50121a001.
68. (a) Otsuka S, Tani K, Kato I, Teranaka O. Chiral metal complexes. Part III. Nickel(0) complexes containing the bulky chiral ligand tri[(+)-bornan-2-yl] phosphite. *J Chem Soc Dalton Trans*. 1974;2216–2219. https://doi.org/10.1039/DT9740002216. (b) Ishizu J, Yamamoto T, Yamamoto A. Preparation and properties of nickel–phosphine complexes coordinated with α,β-unsaturated ester. *Bull Chem Soc Jpn*. 1978;51(9):2646–2650. https://doi.org/10.1246/bcsj.51.2646. (c) Komiya S, Ishizu J, Yamamoto A, Yamamoto T, Takenaka A, Sasada I. Crystal and molecular structure of (Ethyl methacrylate)bis(triphenylphosphine)nickel(0). *Bull Chem Soc Jpn*. 1980;53(5):1283–1287. https://doi.org/10.1246/bcsj.53.1283.
69. (a) Cherwinski WJ, Johnson BFG, Lewis J. Preparation of platinum π-olefin complexes from bis(l,5-diphenylpenta-1,4-dien-3-one)platinum(0). *J Chem Soc Dalton Trans*. 1974;1405–1409. https://doi.org/10.1039/DT9740001405. (b) Chaloner PA, Davies SE, Hitchcock PB. Synthesis and characterization of platinum(0) alkene complexes; X-ray crystal structure determinations on [Pt(*trans*-PhCH = CHCHO)(PPh$_3$)$_2$] and [Pt(*trans*-PhCH = CHCOMe)(PPh$_3$)$_2$]. *Polyhedron*. 1997;16(7):765–776. https://doi.org/10.1016/S0277-5387(96)00348-8.
70. Minematsu H, Nonaka Y, Takahashi S, Hagihara N. Infrared and nuclear magnetic resonance studies of palladium(0)-maleic anhydride complexes. *J Organomet Chem*. 1973;59:395–401. https://doi.org/10.1016/S0022-328X(00)95056-3.
71. (a) Minematsu H, Takahashi S, Hagihara N. New cyclisation of butadiene with *p*-quinones co-ordinated to palladium. *J Chem Soc Chem Commun*. 1975;466–467. https://doi.org/10.1039/C39750000466. (b) Minematsu H, Takahashi S, Hagihara N. Bonding interaction of *p*-quinones with palladium(0)—phosphine complexes. *J Organomet Chem*. 1975;91(3):389–398. https://doi.org/10.1016/S0022-328X(00)89006-3.
72. Paonessa RS, Trogler WC. Photochemical generation of bis(triethylphosphine)platinum(0) and synthesis of ethylenebis(triethylphosphine)platinum(0). *Organometallics*. 1982;1(5):768–770. https://doi.org/10.1021/om00065a023.
73. McAdam A, Francis JN, Ibers JA. The structure of bis(triphenylphosphine)(1,1-dichloro-2,2-dicyanoethylene)platinum(O), Pt[Cl$_2$C C(CN)$_2$] [P(C$_6$H$_5$)$_3$]$_2$. *J Organomet Chem*. 1971;29(1):149–161. https://doi.org/10.1016/S0022-328X(00)87498-7.
74. (a) Cenini S, Ugo R, La Monica G. Zerovalent platinum chemistry. Part IV. Bistriphenylphosphineplatinum(0) complexes with some activated mono-olefins and monoalkynes. *J Chem Soc A*. 1971;409–415. https://doi.org/10.1039/J19710000409. (b) Ozakawa F, Ito T, Nakamura Y, Yamamoto A. Palladium(0) complexes coordinated with substituted olefins and tertiary phosphine ligands. *J Organomet Chem*. 1979;168(3):375–391. https://doi.org/10.1016/S0022-328X(00)83219-2. (c) Caruso F, Camalli M, Pellizer G, Asaro F, Lenarda M. Further X-ray and NMR investigations on bis-triphenylphosphino-platinum complexes with polysubstituted olefins. *Inorg Chim Acta*. 1991;181(2):167–176. https://doi.org/10.1016/S0020-1693(00)86807-3.
75. Panattoni C, Bombieri G, Belluco U, Baddley WH. Metal complexes of cyanocarbons. V. Crystal and molecular structure of a tetracyanoethylen complex of platinum. *J Am Chem Soc*. 1968;90(3):798–799. https://doi.org/10.1021/ja01005a048.

76. Bombieri G, Forsellini E, Panattoni C, Graziani R, Bandoli G. Crystal and molecular structure of tetracyanoethylenebis(triphenylphosphine)platinum(0). *J Chem Soc A*. 1970;1313–1318. https://doi.org/10.1039/J19700001313.
77. Selected references (a) Christofides A, Howard JAK, Spencer JL, Gordon F, Stone A. Fulvene—platinum complexes: X-ray crystal structure of [Pt(η2-C₅H₄CPh₂)(PPh₃)₂]. *J Organomet Chem*. 1982; 232(1), 279–292. https://doi.org/10.1016/S0022-328X(82)80023-5; (b) Werner H, Crisp GT, Jolly PW, Kraus HJ, Krüger C. Synthesis of (1,2,3,4-tetramethylfulvene)palladium(0) complexes from .eta.5-pentamethylcyclopentadienyl)palladium(II) precursors. The crystal structure of [Pd(PMe3)2(.eta.2-CH2:C5Me4)]. *Organometallics*. 1983;2(10):1369–1377. https://doi.org/10.1021/om50004a020. (c) Stanger A, Boese R. The crystal structures of (R₃P)₂Ni-anthracene (R = Et, Bu). *J Organomet Chem*. 1992;430(2):235–243. https://doi.org/10.1016/0022-328X(92)86010-2. (d) Stanger A, Vollhard KPC. Synthesis and fluxional behavior of [bis(trialkylphosphine)nickelio]anthracene (alkyl = Et, Bu). *Organometallics*. 1992;11(1):317–320. https://doi.org/10.1021/om00037a054. (e) Nicolaides A, Smith JM, Kumar A, Barnhart DM, Borden WT. Synthesis and study of the (Ph₃P)₂Pt complexes of three members of a series of highly pyramidalized alkenes. *Organometallics*. 1995;14(7):3475–3485. https://doi.org/10.1021/om00007a055. (f) Hughes DL, Leigh GJ, McMahona CN. New complexes of platinum(0) with cyclopropenes. *J Chem Soc Dalton Trans*. 1997;1301–1308. https://doi.org/10.1039/A607592D. (g) Chaloner PA, Davies SE, Hitchcock PB. Synthesis and characterization of platinum(0) alkene complexes; X-ray crystal structure determinations on [Pt(trans-PhCH=CHCHO)(PPh₃)₂] and [Pt(trans-PhCH=CHCOMe)(PPh₃)₂]. *Polyhedron*. 1997;16(5):765–776. https://doi.org/10.1016/S0277-5387(96)00348-8. (h) Burrows AD, Choi N, McPartlin M, Mingos DMP, Tarlton SV, Vilar R. Syntheses and structural characterisation of the compounds [Pd(dba)L₂] (where L=PBz₃ and PPh₂Np) and the novel dimer [Pd₂(μ-dba)(μ-SO₂)(PBz₃)₂]. *J Organomet Chem*. 1999;573(1–2):313–322. https://doi.org/10.1016/S0022-328X(98)00877-8.
78. Bashilov VV, Petrovskii PV, Sokolov VI, Lindeman SV, Guzey IA, Struchkov YT. Synthesis, crystal, and molecular structure of the palladium(0)-fullerene derivative (.eta.2-C60)Pd(PPh₃)₂. *Organometallics*. 1993;12(4):991–992. https://doi.org/10.1021/om00028a003.
79. Fagan PJ, Calabrese JC, Malone B. The chemical nature of buckminsterfullerene (C60) and the characterization of a platinum derivative. *Science*. 1991;252(5009):1160–1161. https://science.sciencemag.org/content/252/5009/1160.
80. Fagan PJ, Calabrese JC, Malone B. A multiply-substituted buckminsterfullerene (C60) with an octahedral array of platinum atoms. *J Am Chem Soc*. 1991;113(24):9408–9409. https://doi.org/10.1021/ja00024a079.
81. Harrison NC, Murray M, Spencer JL, Stone FGA. Synthesis and dynamic behaviour of bis(ethylene)(tertiary phosphine)-platinum complexes. *J Chem Soc*. 1978;1337–1342. https://doi.org/10.1039/DT9780001337.
82. Spencer JL, Ittel SD, Cushing Jr MA. Olefin complexes of platinum. *Inorg Synth*. 1979;19:213–219. https://doi.org/10.1002/9780470132500.ch49.
83. (a) Karstedt BD. General Electric Company, US3775452A, 1973; (b) Stein J, Lewis LN, Gao Y, Scott RA. In situ determination of the active catalyst in hydrosilylation reactions using highly reactive Pt(0) catalyst precursors. *J Am Chem Soc*. 1999;121(15):3693–3703. https://doi.org/10.1021/ja9825377
84. Krause J, Cestaric G, Haack K-J, Seevogel K, Storm W, Pörschke K-R. 1,6-diene complexes of palladium(0) and platinum(0): highly reactive sources for the naked metals and [L − M⁰] fragments. *J Am Chem Soc*. 1999;121(42):9807–9823. https://doi.org/10.1021/ja983939h.

85. Berthon-Gelloz G, Schumers J-M, Lucaccioni F, Tinant B, Wouters J, Markó IE. Expedient, direct synthesis of (L)Pt(0)(1,6-diene) complexes from H_2PtCl_6. *Organometallics*. 2007;26:5731. https://doi.org/10.1021/om7007088.
86. (a) Gómez Andreu M, Zapf A, Beller M. Molecularly defined palladium(0) monophosphine complexes as catalysts for efficient cross-coupling of aryl chlorides and phenylboronic acid. *Chem Commun*. 2000;2475–2476. https://doi.org/10.1039/B006791L. (b) Vollmüller F, Mägerlein W, Klein S, Krause J, Beller M. Palladium-catalyzed reactions for the synthesis of fine chemicals, 16—highly efficient palladium-catalyzed telomerization of butadiene with methanol. *Adv Synth Catal*. 2001;343(1):29–33. https://doi.org/10.1002/1615-4169(20010129)343:1<29::AID-ADSC29>3.0.CO;2-I. (c) Littke AF, Fu GC. Palladium-catalyzed coupling reactions of aryl chlorides. *Angew Chem Int Ed*. 2002;41(22):4176–4211. https://doi.org/10.1002/1521-3773(20021115)41:22<4176::AID-ANIE4176>3.0.CO;2-U.
(d) Barder TE, Walker SD, Martinelli JR, Buchwald SL. Catalysts for Suzuki−Miyaura coupling processes: scope and studies of the effect of ligand structure. *J Am Chem Soc*. 2005;127(13):4685–4696. https://doi.org/10.1021/ja042491j. (e) De Pater JJM, Tromp DS, Tooke DM, et al. Palladium(0)-alkene bis(triarylphosphine) complexes as catalyst precursors for the methoxycarbonylation of styrene. *Organometallics*. 2005;24(26):6411–6419. https://doi.org/10.1021/om0506419. (f) Maciejewski H, Sydor A, Marciniec B, Kubicki M, Hitchcock PB. Intermediates in nickel(0)–-phosphine complex catalyzed dehydrogenative silylation of olefins. *Inorg Chim Acta*. 2006;359(9):2989–2997. https://doi.org/10.1016/j.i2005.12.067.
87. Hoyte SA, Spencer JL. Mono- and diphosphine platinum(0) complexes of methylenecyclopropane, bicyclopropylidene, and allylidenecyclopropane. *Organometallics*. 2011;30(20):5415–5423. https://doi.org/10.1021/om200630z.
88. Ortega-Moreno L, Peloso R, Maya C, Suárez A, Carmona E. Platinum(0) olefin complexes of a bulky terphenylphosphine ligand. Synthetic, structural and reactivity studies. *Chem Commun*. 2015;51:17008–17011. https://doi.org/10.1039/C5CC07308A.
89. See for example: (a) Amatore C, Jutand A. Role of dba in the reactivity of palladium(0) complexes generated in situ from mixtures of $Pd(dba)_2$ and phosphines. Coord Chem Rev 1998; 178–180: 511–528. https://doi.org/10.1016/S0010-8545(98)00073-3. (b) Amatore C, Broeker G, Jutand A, Khalil F. Identification of the effective palladium(0) catalytic species generated *in Situ* from mixtures of $Pd(dba)_2$ and bidentate phosphine ligands. Determination of their rates and mechanism in oxidative addition. *J Am Chem Soc*. 1997;119(22):5176–5185. https://doi.org/10.1021/ja9637098. (c) Amatore C, Jutand A, Meyer G, Atmani H, Khalil F, Chahdi FO. Comparative reactivity of palladium(0) complexes generated in situ in mixtures of triphenylphosphine or tri-2-furylphosphine and $Pd(dba)_2$. *Organometallics*. 1998;17(14):2958–2964. https://doi.org/10.1021/om971064u.
90. (a) Hartwig JF. *Organotransition Metal Chemistry. From Bonding to Catalysis*. University Science Books; 2010. (b) Hegedus LS, Söderberg BCG. *Transition Metals in the Synthesis of Complex Organic Molecules*. 3rd ed. University Science Books; 2010.
91. Pörschke K-P, Mynott R, Angermund K, Krüger C. Neue Bis(phosphan)-nickel(0)-alkin-Komplexe. *Z Naturforsch*. 1985;40b:199–209. https://doi.org/10.1515/znb-1985-0210.
92. Bach I, Pörschke K-R, Goddard R, et al. Synthesis, structure, and properties of ${(^tBu_2PC_2H_4P^tBu_2)Ni}_2(\mu-\eta^2:\eta^2-C_6H_6)$ and $(^tBu_2PC_2H_4P^tBu_2)Ni(\eta^2-C_6F_6)$. *Organometallics*. 1996;15(24):4959–4966. https://doi.org/10.1021/om960389s.
93. Pörschke K-R, Pluta C, Proft B, Lutz F, Krüger C. Bis(di-tert-butylphosphino)ethan-nickel(0)-komplexe/Bis(di-tert-butylphosphino)ethane-nickel(0) complexes. *Z Naturforsch B J Chem Sci*. 1993;48(5):608–626. https://doi.org/10.1515/znb-1993-0511.

94. Bach I, Pörschke K-R, Proft B, et al. Novel Ni(0)-COT complexes, displaying semi-aromatic planar COT ligands with alternating C – C and C=C bonds. *J Am Chem Soc.* 1997;119(16):3773–3781. https://doi.org/10.1021/ja964210g.
95. Schager F, Haack K-J, Mynott R, Rufińska A, Pörschke K-R. Novel (R$_2$PC$_2$H$_4$PR$_2$) M^0–COT complexes (M = Pd, Pt) having semiaromatic η^2-COT or dianionic η^2(1,4)-COT ligands. *Organometallics.* 1998;17(5):807–814. https://doi.org/10.1021/om970762b.
96. Krause J, Bonrath W, Pörschke K-R. (R2PC2H4PR2)Pd0 alkene and ethyne complexes. *Organometallics.* 1992;11(3):1158–1167. https://doi.org/10.1021/om00039a023.
97. Schager F, Bonrath W, Pörschke K-R, Kessler M, Krüger C, Seevogel K. (R$_2$PC$_2$H$_4$PR$_2$)Pd^0–1-alkyne complexes. *Organometallics.* 1997;16(20):4276–4286. https://doi.org/10.1021/om9702035.
98. Trebbe R, Goddard R, Rufińska A, Seevogel K, Pörschke K-P. Preparation and structural characterization of the palladium(0)–carbonyl complexes (R$_2$PC$_2$H$_4$PR$_2$)Pd(CO)$_2$ and {(R$_2$PC$_2$H$_4$PR$_2$)Pd}$_2$(μ-CO). *Organometallics.* 1999;18(13):2466–2472. https://doi.org/10.1021/om990239s.
99. Forrest SJK, Pringle PG, Sparkes HA, Wass DF. Reversible CO exchange at platinum(0). An example of similar complex properties produced by ligands with very different stereoelectronic characteristics. *Dalton Trans.* 2014;43:16335–16344. https://doi.org/10.1039/C4DT02303J.
100. (a) Hodgson M, Parker D, Taylor J, Ferguson G. Synthesis, reactions, and X-ray structure of η^2-ethene(diop)palladium: a useful synthetic equivalent for (diop)Pd^0. *J Chem Soc Chem Commun.* 1987;1309–1311. https://doi.org/10.1039/C39870001309. (b) Hodgson M, Parker D, Taylor RJ, Ferguson G. Synthetic and mechanistic aspects of palladium-catalyzed asymmetric hydrocyanation of alkenes. Crystal structure and reactions of (.eta.2-ethene)(diop)palladium. *Organometallics.* 1988;7(8):1761–1766. https://doi.org/10.1021/om00098a011.
101. (a) Baker KV, Brown JM, Cooley NA, Hughes GD, Taylor RJ. Reactive intermediates in asymmetric cross-coupling catalysed by palladium P-N chelates. *J Organomet Chem.* 1989;370(1–3):397–406. https://doi.org/10.1016/0022-328X(89)87301-2. (b) Fong S-WA, Vittal JJ, Hor TSA. Isolation of a stable precursor for [Pt^0(P – P)] and its reductive addition to Ru$_3$(CO)$_9$(μ$_3$-S)$_2$ giving square heterometallic clusters supported by two face-capping μ$_4$-sulfides [P – P = (C$_5$H$_4$PPh$_2$)$_2$M; M = Fe, Ru]. *Organometallics.* 2000;19(5):918–924. https://doi.org/10.1021/om990569c.
102. (a) Herrmann WA, Thiel WR, Broßmer C, Öfele K, Priermeier T, Scherer W. (Dihalogenmethyl)palladium(II)-komplexe aus palladium(O)-vorstufen des dibenzylidenacetons: synthese, strukturchemie und reaktivität ag. *J Organomet Chem.* 1993;461(1–2):51–60. https://doi.org/10.1016/0022-328X(93)83273-X. (b) Tschoerner M, Trabesinger G, Albinati A, Pregosin PS. New chiral complexes of palladium(0) containing P,S- and P,P-bidentate ligands. *Organometallics.* 1997;16(15):3447–3453. https://doi.org/10.1021/om970185r.
103. Pan Y, Mague JT, Fink MJ. Stable bis(silyl)palladium complexes: synthesis, structure, and bis-silylation of acetylenes. *Organometallics.* 1992;11(11):3495–3497. https://doi.org/10.1021/om00059a005.
104. Döhring A, Goddard R, Hopp G, Jolly PW, Kokel N, Krüger C. Intermediates in the palladium-catalysed reaction of 1,3-dienes part 8. The reaction of palladium-butadiene complexes with ethyl methylacetoacetate. *Inorg Chim Acta.* 1994;222(1–2):179–192. https://doi.org/10.1016/0020-1693(94)03907-0.
105. (a) Kranenburg M, Delis JGP, Kamer PCJ, et al. Palladium(0)–tetracyanoethylene complexes of diphosphines and a dipyridine with large bite angles, and their crystal structures. *J Chem Soc Dalton Trans.* 1997;1839–1850. https://doi.org/10.1039/

A607927J. (b) Mora G, Deschamps B, Van Zutphen S, Le Goff XF, Ricard L, Le Floch P. Xanthene-phosphole ligands: synthesis, coordination chemistry, and activity in the palladium-catalyzed amine allylation. *Organometallics*. 2007;26(8):1846–1855. https://doi.org/10.1021/om061172t. (c) Jarvis AG, Sehnal PE, Bajwa SE, et al. A remarkable *cis*- and *trans*-spanning dibenzylidene acetone diphosphine chelating ligand (dbaphos). *Chem A Eur J*. 2013;19(19):6034–6043. https://doi.org/10.1002/chem.201203691.

106. Brunker TJ, Blank NF, Moncarz JR, et al. Chiral palladium(0) *trans*-stilbene complexes: synthesis, structure, and oxidative addition of phenyl iodide. *Organometallics*. 2005;24(11):2730–2746. https://doi.org/10.1021/om050115h.

107. Hofmann P, Heiß H, Müller G. Synthese und Molekülgeometrie von Dichloro[η$_2$-bis(di-t-butylphosphino)methan]platin(II), Pt(dtbpm)Cl$_2$. Die Elektronenstruktur von 1,3-Diphosphaplatinacyclobutan-Fragmenten. *Z Naturforsch*. 1987;42b:395–409.

108. Mindiola DJ, Hillhouse GL. Synthesis, structure, and reactions of a three-coordinate nickel-carbene complex, {1,2-bis(di-*tert*-butylphosphino)ethane}Ni=CPh$_2$. *J Am Chem Soc*. 2002;124(34):9976–9977. https://doi.org/10.1021/ja0269183.

109. Comanescu CC, Iluc VM. Synthesis and reactivity of a nucleophilic palladium(II) carbene. *Organometallics*. 2014;33(21):6059–6064. https://doi.org/10.1021/om500682s.

110. Comanescu CC, Iluc VM. C–H activation reactions of a nucleophilic palladium carbene. *Organometallics*. 2015;34(19):4684–4692. https://doi.org/10.1021/acs.organomet.5b00414.

111. Barrett BJ, Iluc VD. An adaptable chelating diphosphine ligand for the stabilization of palladium and platinum carbenes. *Organometallics*. 2017;36(3):730–741. https://doi.org/10.1021/acs.organomet.6b00924.

112. Comanescu CC, Iluc VM. Flexible coordination of diphosphine ligands leading to cis and trans Pd(0), Pd(II), and Rh(I) complexes. *Inorg Chem*. 2014;53(16):8517–8528. https://doi.org/10.1021/ic5010566.

113. Barrett BJ, Iluc VM. Metal-ligand cooperation between palladium and a diphosphine ligand with an olefinic backbone. *Inorg Chim Acta*. 2017;460:35–42. https://doi.org/10.1016/j.ica.2016.08.035.

114. (a) Acosta-Ramírez A, Flores-Gaspar A, Muñoz-Hernández M, Arévalo A, Jones WD, García JJ. Nickel complexes involved in the isomerization of 2-methyl-3-butenenitrile. *Organometallics*. 2007;26(7):1712–1720. https://doi.org/10.1021/om061037g. (b) Swartz BD, Reinartz NM, Brennessel WW, García JJ, Jones WD. Solvent effects and activation parameters in the competitive cleavage of C−CN and C−H bonds in 2-methyl-3-butenenitrile using [(dippe)NiH]$_2$. *J Am Chem Soc*. 2008;130(26):8548–8554. https://doi.org/10.1021/ja8000216. (c) Tauchert ME, Warth DCM, Braun SM, et al. Highly efficient nickel-catalyzed 2-methyl-3-butenenitrile Isomerization: applications and mechanistic studies employing the TTP ligand family. *Organometallics*. 2011;30(10):2790–2809. https://doi.org/10.1021/om200164f.

115. Barrios-Francisco R, García J. J. stereoselective hydrogenation of aromatic alkynes using water, triethylsilane, or methanol, mediated and catalyzed by Ni(0) complexes. *Inorg Chem*. 2009;48(1):386–393. https://doi.org/10.1021/ic801823x.

116. Göthlich APV, Tensfeldt M, Rothfuss H, et al. Novel chelating phosphonite ligands: syntheses, structures, and nickel-catalyzed hydrocyanation of olefins. *Organometallics*. 2008;27(10):2189–2200. https://doi.org/10.1021/om701140c.

117. Sylvester KT, Wu K, Doyle AG. Mechanistic investigation of the nickel-catalyzed suzuki reaction of *N,O*-acetals: evidence for boronic acid assisted oxidative addition and an iminium activation pathway. *J Am Chem Soc*. 2012;134(41):16967–16970. https://doi.org/10.1021/ja3079362.

118. Castellanos-Blanco N, Flores-Alamo M, García JJ. Nickel-catalyzed alkylation and transfer hydrogenation of α,β-unsaturated enones with methanol. *Organometallics*. 2012;31(2):680–686. https://doi.org/10.1021/om2010222.
119. Herbert DE, Lara NC, Agapie T. Arene C-H amination at nickel in terphenyl–diphosphine complexes with labile metal–arene interactions. *Chem A Eur J*. 2013;19(48):16453–16460. https://doi.org/10.1002/chem.201302539.
120. Kruckenberg A, Wadepohl H, Gade LH. Bis(diisopropylphosphinomethyl) amine nickel(II) and nickel(0) complexes: coordination chemistry, reactivity, and catalytic decarbonylative C-H arylation of benzoxazole. *Organometallics*. 2013;32(18):5153–5170. https://doi.org/10.1021/om400711d.
121. Staudaher ND, Stolley RM, Louie J. Synthesis, mechanism of formation, and catalytic activity of Xantphos nickel π-complexes. *Chem Commun*. 2014;50:15577–15580. https://doi.org/10.1039/C4CC07590K.
122. Jevtovikj I, Manzini S, Hanauer M, Rominger F, Schaub T. Investigations on the catalytic carboxylation of olefins with CO_2 towards α,β-unsaturated carboxylic acid salts: characterization of intermediates and ligands as well as substrate effects. *Dalton Trans*. 2015;44:11083–11094. https://doi.org/10.1039/C5DT01040C.
123. Desnoyer AN, Bowes EG, Patrick BO, Love JA. Synthesis of 2-nickela(II)oxetanes from nickel(0) and epoxides: structure, reactivity, and a new mechanism of formation. *J Am Chem Soc*. 2015;137(40):12748–12751. https://doi.org/10.1021/jacs.5b06735.
124. He W, Patrick BO, Kennepohl P. Identifying the missing link in catalyst transfer polymerization. *Nat Commun*. 2018;9:3866. https://doi.org/10.1038/s41467-018-06324-9.
125. (a) Hopkinson MN, Richter C, Schedler M, Glorius F. An overview of N-heterocyclic carbenes. *Nature*. 2014;510:485–496. https://doi.org/10.1038/nature13384. (b) Hahn FE, Jahnke MC. Heterocyclic carbenes: synthesis and coordination chemistry. *Angew Chem Int Ed*. 2008;47:3122–3172. https://doi.org/10.1002/anie.200703883.
126. Nelson DJ, Nolan SP. Quantifying and understanding the electronic properties of N-heterocyclic carbenes. *Chem Soc Rev*. 2013;42:6723–6753. https://doi.org/10.1039/C3CS60146C.
127. (a) Díez-González S, Marion N, Nolan SP. N-heterocyclic carbenes in late transition metal catalysis. *Chem Rev*. 2009;109:3612–3676. https://doi.org/10.1021/cr900074m. (b) Fortman GC, Nolan SP. N-Heterocyclic carbene (NHC) ligands and palladium in homogeneous cross-coupling catalysis: a perfect union. *Chem Soc Rev*. 2011;40:5151–5169. https://doi.org/10.1039/c1cs15088j. (c) Valenta C, Çalimsiz S, Hoi KH, Mallik D, Sayah M, Organ MG. The development of bulky palladium NHC complexes for the most-challenging cross-coupling reactions. *Angew Chem Int Ed*. 2012;51:3314–3332. https://doi.org/10.1002/anie.201106131.
128. Hu X, Castro-Rodriguez I, Meyer K. Synthesis and characterization of electron-rich nickel tris-carbene complexes. *Chem Commun*. 2004;2164–2165. https://doi.org/10.1039/B409241D.
129. Schaub T, Backes M, Radius U. Nickel(0) complexes of N-alkyl-substituted N-heterocyclic carbenes and their use in the catalytic carbon-carbon bond activation of biphenylene. *Organometallics*. 2006;25:4196–4206. https://doi.org/10.1021/om0604223.
130. McGuiness DS, Cavell KJ, Skelton BW, White AH. Zerovalent palladium and nickel complexes of heterocyclic carbenes: oxidative addition of organic halides, carbon-carbon coupling processes and the Heck reaction. *Organometallics*. 1999;18:1596–1605. https://doi.org/10.1021/om9809771.
131. (a) Jackstell R, Gómez Andreu M, Frisch A, et al. A highly efficient catalyst for the telomerization of 1,3-dienes with alcohols: first synthesis of a monocarbenepalladium(0) olefin complex. *Angew Chem Int Ed*. 2002;41(6):986–989.

https://doi.org/10.1002/1521-3773(20020315)41:6<986::AID-ANIE986>3.0. CO;2-M. (b) Jackstell R, Harkal S, Jiao H, et al. An industrially viable catalyst system for palladium-catalyzed telomerizations of 1,3-butadiene with alcohols. *Chem A Eur J.* 2004;10:3891–3900. https://doi.org/10.1002/chem.200400182. (c) Selvakumar K, Zapf A, Spannenberg A, Beller M. Synthesis of monocarbenepalladium(0) complexes and their catalytic behavior in cross-coupling reactions of aryldiazonium salts. *Chem A Eur J.* 2002;8:3901–3906. https://doi.org/10.1002/1521-3765(20020902) 8:17<3901::AID-CHEM3901>3.0.CO;2-E.

132. (a) Selvakumar K, Zapf A, Beller M. New palladium carbene catalysts for the Heck reaction of aryl chlorides in ionic liquids. *Org Lett.* 2002;4(18):3031–3033. https://doi.org/10.1021/ol020103h. (b) Frisch AC, Rataboul F, Zapf A, Beller M. First Kumada reaction of alkyl chlorides using N-heterocyclic carbene/palladium catalyst systems. *J Organomet Chem.* 2003;687:403–409. https://doi.org/10.1016/ S0022-328X(03)00723-X. (c) Frisch AC, Zapf A, Briel O, Kayser B, Shaikh N, Beller M. Comparison of palladium carbene and palladium phosphine catalysts for catalytic coupling reactions of aryl halides. *J Molec Catal A-Chem.* 2004;214:231–239. https://doi.org/10.1016/j.molcata.2003.12.035.

133. Hitchcock PB, Lappert MF, Warhurst NJW. Synthesis and structure of a rac-tris (divinyldisiloxane) diplatinum(0) complex and its reaction with maleic anhydride. *Angew Chem Int Ed Engl.* 1991;30:438–440. https://doi.org/10.1002/anie.199104381.

134. (a) Markó IE, Stérin S, Buisine O, et al. Selective and efficient platinum(0)-carbene complexes as hydrosilylation catalysts. *Science.* 2002;298:204–206. https://doi.org/10. 1126/science.1073338. (b) Markó IE, Stérin S, Buisine O, et al. Highly active and selective platinum(0)-carbene complexes. Efficient, catalytic hydrosilylation of functionalized olefins. *Adv Synth Catal.* 2004;346:1429–1434. hhttps://doi.org/10.1002/adsc. 200404048. (c) Berthon-Gelloz G, Buisine O, Brière J-F, et al. Synthetic and structural studies of NHC-Pt(dvtms) complexes and their application as alkene hydrosilylation catalysts (NHC=N-heterocyclic carbene, dvtms=divinyltetramethylsiloxane. *J Organomet Chem.* 2005;690:6156–6168. https://doi.org/10.1016/j.jorganchem. 2005.08.020.

135. Dierick S, Markó IE. Chapter 5. NHC platinum(0) complexes: unique catalysts for the hydrosilylation of alkenes and alkynes. In: Nolan SP, ed. *N-Heterocyclic Carbenes: Effective Tools for Organometallic Synthesis.* Wiley-VCH; 2014:111–149.

136. (a) Żak P, Bolt M, Lorkowski J, Kubicki M, Pietraszuk C. Platinum complexes bearing bulky N-heterocyclic carbene ligands as efficient catalysts for the fully selective dimerization of terminal alkynes. *ChemCatChem.* 2017;9:3627–3631. https://doi.org/10. 1002/cctc.201701508. (b) Żak P, Bolt M, Lorkowski J, Kubicki M, Pietraszuk C. Highly selective hydrosylilation of olefins and acetylenes by platinum(0) complexes bearing bulky N-heterocyclic carbene ligands. *Dalton Trans.* 2018;47:1903–1910. https://doi.org/10.1039/C7DT04392A.

137. (a) Silbestri GF, Flores JC, De Jesús E. Water-soluble N-heterocyclic carbene platinum(0) complexes: recyclable catalysts for the hydrosilylation of alkynes in water at room temperature. *Organometallics.* 2012;31(8):3355–3360. https://doi.org/10. 1021/om300148q. (b) Ruiz-Varilla AM, Baquero EA, Silbestri GF, Gonzalez-Arellano C, De Jesús E, Flores JC. Synthesis and behavior of novel sulfonated water-soluble N-heterocyclic carbene (η^4-diene) platinum(0) complexes. *Dalton Trans.* 2015;44:18360–18369. https://doi.org/10.1039/C5DT02622A.

138. (a) Lillo V, Mata J, Ramírez J, Peris E, Fernandez E. Catalytic diboration of unsaturated molecules with platinum(0)-NHC: selective synthesis of 1,2-dihydroxysulfones. *Organometallics.* 2006;25:5829–5831. https://doi.org/10.1021/om060666n. (b) Lillo V, Mata JA, Segarra AM, Peris E, Fernandez E. The active role of NHC ligands in platinum-mediated tandem hydroboration-cross coupling reactions. *Chem Commun.* 2007;2184–2186. https://doi.org/10.1039/b700800g.

139. (a) Duin MA, Clement ND, Cavell KG, Elsevier CJ. C-H activation of imidazolium salts by Pt(0) at ambient temperature: synthesis of hydrido platinum bis(carbene) compounds. *Chem Commun.* 2003;400–401. https://doi.org/10.1039/B211235C. (b) Duin MA, Lutz M, Spek AL, Elsevier CJ. Synthesis of electron-rich platinum centers: platinum^0(carbene)(alkene)$_2$ complexes. *J Organomet Chem.* 2005;690:5804–5815. https://doi.org/10.1016/j.jorganchem.2005.07.059.
140. Clement ND, Cavell KJ, Ooi L-I. Zerovalent N-heterocyclic carbene complexes of palladium and nickel dimethyl fumarate: synthesis, structure and dynamic behavior. *Organometallics.* 2006;25:4155–4165. https://doi.org/10.1021/om0602759.
141. Sprengers JW, Mars MJ, Duin MA, Cavell KA, Elsevier CJ. Selective hydrosilylation of styrene using an in situ formed platinum(1,3-dimesityl-dihydroimidazol-2-ylidene) catalyst. *J Organomet Chem.* 2003;679:149–152. https://doi.org/10.1016/S0022-328X(03)00514-X.
142. (a) Tromp DS, Hauwert P, Elsevier CJ. Synthesis of bis-N-alkyl imidazolium salts and their palladium(0)(η^2-MA)$_2$ complexes. *Appl Organomet Chem.* 2012;16:335–341. https://doi.org/10.1002/aoc.2866. (b) Sprengers JW, Wassenaar J, Clement ND, Cavell KJ, Elsevier CJ. Palladium–(N-heterocyclic carbene) hydrogenation catalysts. *Angew Chem Int Ed.* 2005;44:2026–2029. https://doi.org/10.1002/anie.200462930.
143. Kluwer AM, Elsevier CJ, Buhl M, Lutz M, Spek AL. Zero-valent palladium complexes with monodentate nitrogen σ-donor ligands. *Angew Chem Int Ed.* 2003;42(30):3501–3504. https://doi.org/10.1002/anie.200351189.
144. Cavell KJ, Stufkens DJ, Vrieze K. 1,4-Diazabutadiene olefin complexes of zerovalent palladium: preparation and characterization. *Inorg Chim Acta.* 1981;47:67–76. https://doi.org/10.1016/S0020-1693(00)89309-3.
145. Schaub T, Radius U. Efficient C-F and C-C activation by novel N-heterocyclic carbene-nickel(0) complex. *Chem A Eur J.* 2005;11:5024–5030. https://doi.org/10.1002/chem.200500231.
146. (a) Arduengo III AJ, Gamper SF, Calabrese JC, Davidson F. Low-coordinate carbene complexes of nickel(0) and platinum(0). *J Am Chem Soc.* 1994;116(10):4391–4394. https://doi.org/10.1021/ja00089a029. (b) Caddick S, Cloke GN, Hitchcock PB, Lewis AKK. Unusual reactivity of a nickel N-heterocyclic carbene complex: tert-butyl group cleavage and silicone grease activation. *Angew Chem Int Ed.* 2004;43:5824–5827. https://doi.org/10.1002/anie.200460955.
147. (a) Schaub T, Backes M, Radius U. Catalytic C-C bond formation accomplished by selective C-F activation of perfluorinated arenes. *J Am Chem Soc.* 2006;128(50):15964–15965. https://doi.org/10.1021/ja064068b. (b) Schaub T, Fischer P, Steffen A, Braun T, Radius U, Mix A. C-F activation of fluorinated arenes using NHC-stabilized nickel(0) complexes: selectivity and mechanistic investigations. *J Am Chem Soc.* 2008;130(29):9304–9317. https://doi.org/10.1021/ja074640e.
148. (a) Chirik P, Morris R. Getting down to earth: the renaissance of catalysis with abundant metals. *Acc Chem Res.* 2015;48(9):2495. https://doi.org/10.1021/acs.accounts.5b00385. (b) Rodríguez-Ruiz V, Carlino R, Bezzenine-Lafollée S, et al. Recent development in alkene hydrofunctionalisation promoted by homogeneous catalysts based on earth abundant elements: formation of C-N, C-O and C-P bonds. *Dalton Trans.* 2015;44:12029–12059. https://doi.org/10.1039/C5DT00280J. (c) Docherty JH, Peng J, Dominey AP, Thomas SP. Activation and discovery of earth abundant metal catalysts using sodium ter-butoxide. *Nat Chem.* 2017;9(6):595–600. https://doi.org/10.1038/nchem.2697.
149. (a) Tasker SZ, Standley EA, Jamieson TF. Recent advances in homogeneous nickel catalysis. *Nature.* 2014;509:299–309. https://doi.org/10.1038/nature13274. (b) Ananikov VP. Nickel: the "spirited horse" of transition metal catalysis. *ACS Catal.* 2015;5:1964–1971. https://doi.org/10.1021/acscatal.5b00072. (c) Hazari N, Melvin PR, Beromi MM. Well-defined nickel and palladium precatalysts for cross-coupling. *Nat Rev Chem.* 2017;1:0025. https://doi.org/10.1038/s41570-017-0025.

150. Berini C, Winkelmann OH, Otten J, Vicic DA, Navarro O. Rapid and selective catalytic oxidation of secondary alcohols at room temperature by using (N-heterocyclic carbene)-Ni^0 systems. Chem A Eur J. 2010;16:6857–6860. https://doi.org/10.1002/chem.201000220.
151. (a) Iglesias MJ, Blandez JF, Fructos MR, et al. Synthesis, structural characterization and catalytic activity of IPrNi(styrene)$_2$ in the amination of aryl tosylates. Organometallics. 2012;31:6312–6316. https://doi.org/10.1021/om300566m. (b) Rull SG, Blandez JF, Fructos MR, Belderrain TR, Nicasio MC. C-N coupling of indoles and carbazoles with aromatic chlorides catalyzed by a single component NHC-nickel(0) precursor. Adv Synth Catal. 2015;357:907–911. https://doi.org/10.1002/adsc.201500196. (c) Rull SG, Funes-Ardoiz I, Maya C, et al. Elucidating the mechanism of aryl aminations mediated by NHC-supported nickel complexes: evidence for a nonradical Ni(0)/Ni(II) pathway. ACS Catal. 2018;8:3733–3742. https://doi.org/10.1021/acscatal.8b00856.
152. Bair JS, Schramm Y, Sergeev AG, Clot E, Eisenstein O, Hartwig JF. Linear-selective hydroarylation of unactivated terminal and internal olefins with trifluoromethyl-substituted arenes. J Am Chem Soc. 2014;136:13098–13101. https://doi.org/10.1021/ja505579f.
153. (a) Todd DP, Thompson BB, Nett AJ, Montgomery J. Deoxygenative C-C bond-forming processes via net four-electron reductive coupling. J Am Chem Soc. 2015;137:12788–12791. https://doi.org/10.1021/jacs.5b08448. (b) Nett AJ, Montgomery J, Zimmermann PM. Entrances, traps and rate-controlling factors for nickel-catalyzed C-H functionalization. ACS Catal. 2017;7:7352–7362. https://doi.org/10.1021/acscatal.7b02919.
154. (a) Wu J, Hazari N, Incarvito CD. Synthesis, properties, and reactivity with carbon dioxide of (allyl)$_2$Ni(L) complexes. Organometallics. 2011;30(11):3142–3150. https://doi.org/10.1021/om2002238. (b) Wu J, Faller JW, Hazari N, Schmeier TJ. Stoichiometric and catalytic reactions of thermally stable nickel(0) NHC complexes. Organometallics. 2012;31(3):806–809. https://doi.org/10.1021/om300045t.
155. Elsby MR, Johnson SA. Nickel-catalyzed C-H silylation of arenes with vinylsilanes: rapid and reversible β-Si elimination. J Am Chem Soc. 2017;139(27):9401–9407. https://doi.org/10.1021/jacs.7b05574.
156. (a) Ansell MB, Roberts DE, Cloke GN, Navarro O, Spencer J. Synthesis of an [(NHC)$_2$Pd(SiMe$_3$)$_2$] complex and catalytic cis-bis(silyl)ations of alkynes with unactivated disilanes. Angew Chem Int Ed. 2015;54(19):5578–5582. https://doi.org/10.1002/anie.201501764. (b) Ansell MB, Spencer J, Navarro O. (N-heterocyclic carbene)$_2$-Pd(0)-catalyzed silaboration of internal and terminal alkynes: scope and mechanistic studies. ACS Catal. 2016;6:2192–2196. https://doi.org/10.1021/acscatal.6b00127. (c) Ansell MB, Menezes da Silva VH, Heerdt G, Braga AAC, Navarro O. An experimental and theoretical study into the facile, homogeneous (N-heterocyclic carbene)$_2$-Pd(0) catalyzed diboration of internal and terminal alkynes. Cat Sci Technol. 2016;6:7461–7467. https://doi.org/10.1039/C6CY01266C. (d) Ansell MB, Kostakis GE, Braunschweig H, Navarro O, Spencer J. Synthesis of functionalized hydrazines: facile homogeneous (N-heterocyclic carbene)-palladium(0)-catalyzed diboration and silaboration of azobenzenes. Adv Synth Catal. 2016;358:3765–3769. https://doi.org/10.1002/adsc.201601106.
157. Li J, Shen WX, Li X. Recent developments of expanded ring N-heterocyclic carbenes. Curr Org Chem. 2012;16(23):2879–2891. https://doi.org/10.2174/138527212804546859.
158. (a) Dunsford JJ, Cavell KJ, Kariuki B. Expanded-ring N-heterocyclic carbene complexes of zero valent platinum dvtms (divinyltetramethyldisiloxane): highly efficient hydrosilylation catalysts. J Organomet Chem. 2011;696:188–194.

https://doi.org/10.1016/j.jorganchem.2010.08.045. (b) Dunsford JJ, Cavell JK. Expanded ring N-heterocyclic carbenes: a comparative study of ring size in palladium(0) catalyzed Mizoroki-Heck coupling. *Dalton Trans.* 2011;40:9131–9135. https://doi.org/10.1039/C1DT10596E.

159. Sluijter SN, Warsink S, Lutz M, Elsevier CJ. Synthesis of palladium(0) and –(II) complexes with chelating bis(N-heterocyclic carbene) ligands and their application in semihydrogenation. *Dalton Trans.* 2013;42:7365–7372. https://doi.org/10.1039/C3DT32835J.

160. (a) Harrold ND, Hillhouse GL. Strongly bent nickel imides supported by a chelating bis(N-heterocyclic carbene) ligand. *Chem Sci.* 2013;4:4011–4015. https://doi.org/10.1039/C3SC51517F. (b) Warterman R, Hillhouse GL. η^2-Organoazide complexes of nickel and their conversion to terminal imido complexes via dinitrogen extrusion. *J Am Chem Soc.* 2008;130:12628–12629. https://doi.org/10.1021/ja805530z.

161. Brendel M, Braun C, Rominger F, Hofmann P. Bis-NHC chelate complexes of nickel(0) and platinum(0). *Angew Chem Int Ed.* 2014;53:8741–8745. https://doi.org/10.1002/anie.201401024.

162. Gendy C, Mansikkamäki A, Valjus J, et al. Nickel as a Lewis base in a T-shaped nickel(0) germylene complex incorporating a flexible bis(NHC)ligand. *Angew Chem Int Ed.* 2019;58:154–158. https://doi.org/10.1002/anie.201809889.

163. (a) Warsink S, Hauwert P, Siegler MA, Spek AL, Elsevier CJ. Palladium(0) pre-catalysts with heteroditopic NHC-amine ligands by transmetallation from their silver(I) complexes. *Appl Organomet Chem.* 2009;23:225–228. https://doi.org/10.1002/aoc.1501. (b) Warsink S, Van Aubel CMS, Weigand JJ, Liu S-T, Elsevier CJ. Bulky picolyl substituted NHC ligands and their Pd^0 complexes. *Eur J Inorg Chem.* 2010;5556–5562. https://doi.org/10.1002/ejic.201000768. (c) Warsink S, Chang I-H, Weigand JJ, Hauwert P, Chen J-T, Elsevier CJ. NHC ligands with a secondary pyrimidyl donor for electron-rich palladium(0) complexes. *Organometallics.* 2010;29(20):4555–4561. https://doi.org/10.1021/om100670u. (d) Warsink S, Bosman S, Weigand JJ, Elsevier CJ. Rigid pyridyl substituted NHC ligands, their Pd(0) complexes and their application in selective transfer semihydrogenation of alkynes. *Appl Organomet Chem.* 2011;25:276–282. https://doi.org/10.1002/aoc.1754.

164. Canovese L, Visentin F, Levi C, Santo C, Bertolasi V. The interaction between heteroditopic pyridine-nitrogen NHC with novel sulfur NHC in palladium(0) derivatives: synthesis and structural characterization of a bis-carbene palladium(0) olefin complex and formation in solution of an alkene-alkyne mixed intermediate as a consequence of the ligands hemilability. *Inorg Chim Acta.* 2012;390:105–118. https://doi.org/10.1016/j.ica.2012.04.018.

165. Lee J-Y, Shen J-S, Tzeng R-J, et al. Well-defined palladium(0) complexes bearing N-heterocyclic carbene and phosphine moieties: efficient catalytic applications in the Mizoroki-Heck reaction and direct C-H functionalization. *Dalton Trans.* 2016;45:10375–10388. https://doi.org/10.1039/C6DT01323F.

166. Rull SG, Rama RJ, Álvarez E, Fructos MR, Belderrain TR, Nicasio MC. Phosphine-functionalized NHC Ni(II) and Ni(0) complexes: synthesis, characterization and catalytic properties. *Dalton Trans.* 2017;46:7603–7611. https://doi.org/10.1039/C7DT01805C.

167. (a) Dorta R, Stevens ED, Hoff CD, Nolan SP. Stable, three-coordinate Ni(CO)$_2$(NHC) (NHC = N-heterocyclic carbene) complexes enabling the determination of Ni-NHC bond energies. *J Am Chem Soc.* 2003;125:10490–10491. https://doi.org/10.1021/ja0362151. (b) Dorta R, Stevens ED, Scott NM, et al. Steric and electronic properties of N-hetercyclic carbenes (NHC): a detailed study on their interaction with Ni(CO)$_4$. *J Am Chem Soc.* 2005;127:2485–2495. https://doi.org/10.1021/ja0438821.

168. Stolley RM, Duong HA, Thomas DR, Louie J. The discovery of [Ni(NHC)RCN]$_2$ species and their role as cycloaddition catalysts for the formation of pyridines. *J Am Chem Soc.* 2012;134(36):15154–15162. https://doi.org/10.1021/ja3075924.
169. (a) Ugo R, Cariati F, La Monica G. Bistriphenylphosphineplatinum(0). *Chem Commun.* 1966;868–869. https://doi.org/10.1039/C19660000868. (b) Ugo R, La Monica G, Cariati F, Cenini S, Conti F. Zerovalent platinum chemistry. III. Properties of bistriphenylphosphineplatinum(0). *Inorg Chim Acta.* 1970;4:390–394. https://doi.org/10.1016/S0020-1693(00)93311-5. (c) Englert VM, Jolly PW, Wilke G. Bis(tri-2-biphenylphophit)nickel. *Am Ethnol.* 1971;83:84. (d) De Pascuale RJ. Allene cyclooligomerization and polymerization catalyzed by a nickel(0) complex. *J Organomet Chem.* 1971;32(3):381–393. https://doi.org/10.1016/S0022-328X(00)82649-2.
170. (a) Blake DM, Nyman CJ. Photochemical reactions of oxalatobis (triphenylphosphine) platinum (II) and related complexes. *J Am Chem Soc.* 1970;92(18):5359–5364. https://doi.org/10.1021/ja00721a012. (b) Tolman CA. The 16 and 18 electron rule in organometallic chemistry and homogeneous catalysis. *Chem Soc Rev.* 1972;1:337–353. https://doi.org/10.1039/CS9720100337.
171. (a) Immirzi A, Musco A. Two-coordinate phosphine-palladium(0) complexes: X-ray structure of the triclohexyl- and the di-(t-butyl)phenyl-phosphine derivatives. *J Chem Soc Chem Commun.* 1974;400–401. https://doi.org/10.1021/ja00817a055. (b) Matsumoto M, Yoshioka H, Nakatsu K, Yoshida T, Otsuka S. Two.-coordinate palladium(0) complexes Pd[PPh(t-Bu)$_2$]$_2$ and Pd[P(t-Bu)$_3$]$_2$. *J Am Chem Soc.* 1974;96(10):3322–3324. https://doi.org/10.1021/ja00817a055.
172. Otsuka S, Yoshida T, Matsumoto M, Nakatsu K. Bis(tertiary phosphine)palladium(0) and –platinum(0) complexes: preparation and cristal molecular structures. *J Am Chem Soc.* 1976;98(19):5850–5858. https://doi.org/10.1021/ja00435a017.
173. Green M, Howard JA, Spencer JL, Stone GA. Ligand exchange reactions of bis (cyclo-octa-1,5-diene)platinum; synthesis of trisethyleneplatinum and molecular structure of tris-μ-(t-butyli socyanide)-tris-(t-butyl isocyanide)-*triangulo*-triplatinum, [Pt$_3$(ButNC)$_6$]. *J Chem Soc Chem Commun.* 1975;3–4. https://doi.org/10.1039/C39750000003.
174. Moynihan KJ, Chieh C, Goel RG. Bis(tri-tert-butylphosphine)platinum(0). *Acta Crystallogr B.* 1979;B35:3060–3062. https://doi.org/10.1107/S0567740879011365.
175. (a) Negishi E-i, Takahashi T, Akiyoshi K. Bis(triphenylphosphine)palladium: its generation, characterization and reactions. *J Chem Soc Chem Commun.* 1986;1338–1339. https://doi.org/10.1039/C39860001338. (b) Amatore C, Azzabi M, Jutand A. Stabilization of bis(triphenylphosphine)palladium(0) by chloride ions. Electrochemical generation of highly reactive zerovalent palladium complexes. *J Organomet Chem.* 1989;363:C41–C45. https://doi.org/10.1016/0022-328X(89)87132-3.
176. Urata H, Suzuki H, Moro-oka Y, Ikawa T. Preparation and reactions of bis(triphenylphosphine)palladium(0). *J Organomet Chem.* 1989;364(1–2):235–244. https://doi.org/10.1016/0022-328X(89)85347-1.
177. Tanaka M. Structure of bis-(tri-*tert*-butylphosphine)palladium(0). *Acta Crystallogr C.* 1992;C48:739–740. https://doi.org/10.1107/S0108270191010491.
178. Paul F, Patt J, Hartwig JF. Structural characterization and simple synthesis of {Pd[P(o-Tol)$_3$]$_2$}, dimeric palladium(II) complexes obtained by oxidative addition of aryl bromides, and corresponding monometallic amine complexes. *Organometallics.* 1995;14(6):3010–3039. https://doi.org/10.1021/om00006a053.
179. Reid SM, Boyle RC, Mague J-T, Fink MJ. A dicoordinate palladium(0) complex with unusual intramolecular η^1-arene coordination. *J Am Chem Soc.* 2003;125:7816–7817. https://doi.org/10.1021/ja0361493.

180. (a) Weng Z, Teo S, Koh LL, Horn TSA. Efficient Suzuki coupling of aryl chlorides catalyzed by palladium(0) with a P,N heteroligand and isolation of unsaturated intermediates. *Organometallics*. 2004;23:4342–4345. https://doi.org/10.1021/om0494770. (b) Weissman H, Shimon LJW, Milstein D. Unsaturated Pd(0), Pd(I), and Pd(II) complexes of a new methoxy-substituted benzyl phosphine. Aryl-X (X = Cl, I) oxidative addition, C-O cleavage, and Suzuki-Miyaura coupling of aryl chlorides. *Organometallics*. 2004;23:3931–3940. https://doi.org/10.1021/om0497588. (c) Grotjahn DB, Gong Y, Zakharov L, Golen JA, Rheingold AL. Changes in coordination of sterically demanding hybrid imidazolylphosphine ligands on Pd(0) and Pd(II). *J Am Chem Soc*. 2006;128:438–453. https://doi.org/10.1021/ja054779u.

181. (a) Mitchell EA, Baird MC. Optimization of procedures for the syntheses of bisphosphinepalladium(0) precursors for Suzuki–Miyaura and similar cross-coupling catalysis: identification of 3:1 coordination compounds in catalyst mixtures containing Pd(0), PCy$_3$, and/or PMeBu$_2^t$. *Organometallics*. 2007;26(21):5230–5238. https://doi.org/10.1021/om700580d. (b) Norton DM, Mitchell EA, Botros NR, Jessop PG, Baird MC. A superior precursor for palladium(0)-based cross-coupling and other catalytic reactions. *J Org Chem*. 2009;74(17):6674–6680. https://doi.org/10.1021/jo901121e.

182. Li H, Grassa GA, Colacot TJ. A highly efficient, practical, and general route for the synthesis of (R$_3$P)$_2$Pd(0): structural evidence on the reduction mechanism of Pd(II) to Pd(0). *Org Lett*. 2010;12(15):3332–3335. https://doi.org/10.1021/ol101106z.

183. Tagne Huate AC, Sameni S, Freytag M, Jones PG, Tamm M. Phosphane-functionalized cycloheptatrienyl-cyclopentadienyl titanium sandwich complexes: phosphorus ligands with an integrated reducing agent for palladium(0) catalysts generation. *Angew Chem Int Ed*. 2013;125(33):8800–8804. https://doi.org/10.1002/ange.201304252.

184. Chan AL, Estrada J, Kefalidis CE, Lavallo V. Changing the charge: electrostatic effects in Pd-catalyzed cross-coupling. *Organometallics*. 2016;35:3257. https://doi.org/10.1021/acs.organomet.6b00622.

185. (a) Forrest SJK, Clifton J, Frey N, Pringle PG, Sparkers HA, Wass DF. Cooperative Lewis pairs based on late transition metals: activation of small molecules by platinum(0) and B(C$_6$F$_5$)$_3$. *Angew Chem Int Ed*. 2015;54(7):2223–2227. https://doi.org/10.1002/anie.201409872. (b) Devillard M, Declercq R, Nicolas E, et al. A significant but constrained geometry Pt→Al interaction: fixation of CO$_2$ and CS$_2$, activation of H$_2$ and PhCONH$_2$. *J Am Chem Soc*. 2016;138(14):4917–4926. https://doi.org/10.1021/jacs.6b01320. (c) Barnett RB, Figueroa JS. Zero-valent isocyanides of nickel, palladium and platinum as transition metal r-type Lewis bases. *Chem Commun*. 2016;52:13829–13839. https://doi.org/10.1039/C6CC07863J.

186. (a) Campos J. Dihydrogen and acetylene activation by a gold(I)/platinum(0) transition metal only frustrated Lewis pair. *J Am Chem Soc*. 2017;139:2944–2947. https://doi.org/10.1021/jacs.7b00491. (b) Hidalgo N, Maya C, Campos J. Cooperative activation of X-H (X = H, C, O, N) bonds by a Pt(0)/Ag(I) metal only Lewis pair. *Chem Commun*. 2019;55:8812–8815. https://doi.org/10.1039/c9cc03008e.

187. Schnetz T, Röder M, Rominger F, Hofmann P. Isolation and characterization of stable, distinctly bent *trans*-chelated bisphosphine palladium species. *Dalton Trans*. 2008;2238–2240. https://doi.org/10.1039/B802684J.

188. Velian A, Lin S, Miller AJM, Day MW, Agapie T. Synthesis and C-C coupling reactivity of dinuclear Ni^I-Ni^I complex supported by a terphenyl diphosphine. *J Am Chem Soc*. 2010;132(18):6296–6297. https://doi.org/10.1021/ja101699a.

189. Cook CD, Koo CH, Nyburg SC, Shiomi MT. The structure of bis(triphenylphosphine)(ethylene)nickel. *Chem Commun*. 1967;426b–427. https://doi.org/10.1039/C1967000426B.

190. (a) Kelley P, Lin S, Edouard G, Day MW, Agapie T. Nickel-mediated hydrogenolysis of C-O bonds of aryl ethers: what is the source of the hydrogen? *J Am Chem Soc.* 2012;134(12):5480–5483. https://doi.org/10.1021/ja300326t. (b) Herbert DE, Lara NC, Agapie T. Arene C-H amination at nickel in terphenyl–diphosphine complexes with labile metal–arene interactions. *Chem Eur J* 2013;19(48):16453–16460. https://doi.org/10.1002/chem.201302539. c Edouart GA, Kelley P, Herbert DE, Agapie T. Aryl ether cleavage by group 9 and 10 transition metals: stoichiometric studies of selectivity and mechanism. *Organometallics.* 2015;34(21):5254–5277. https://doi.org/10.1021/acs.organomet.5b00710. (d) Kelley P, Edouart GA, Lin S, Agapie T. Lewis acid accelerated aryl ether bond cleavage with nickel: orders of magnitude rate enhancement using AlMe$_3$. *Chem A Eur J.* 2016;22(48):17173–17176. https://doi.org/10.1002/chem.201604160.

191. Böhm WPW, Gstöttmayr CWK, Weskamp T, Herrmann WA. N-heterocyclic carbenes. Part 26. N-heterocyclic carbene complexes of palladium(0): synthesis and application in the Suzuki cross-coupling reaction. *J Organomet Chem.* 2000;595(2):186–190. https://doi.org/10.1016/S0022-328X(99)00590-2.

192. Konnick MM, Guzei IA, Stahl SS. Characterization of peroxo and hydroperoxo intermediates in the aerobic oxidation of N-heterocyclic-carbene-coordinate palladium(0). *J Am Chem Soc.* 2004;126(33):10212–10213. https://doi.org/10.1021/ja046884u.

193. Arnold PA, Cloke GN, Geldbach T, Hitchcock PB. Metal vapor synthesis as a straightforward route to group 10 homoleptic carbene complexes. *Organometallics.* 1999;18(16):3228–3233. https://doi.org/10.1021/om990224u.

194. Caddick S, Cloke FGN, Clentsmith GKB, et al. An improved synthesis of bis(1,3-di-N-tert-butylimidazol-2-ylidene)palladium(0) and its use in C-C and C-N coupling reactions. *J Organomet Chem.* 2001;617(1):635–639. https://doi.org/10.1016/S0022-328X(00)00724-5.

195. (a) Titcomb LR, Caddick S, Cloke FGN, Wilson DJ, McKerrecher D. Unexpected reactivity of two-coordinate palladium-carbene complexes; synthetic and catalytic implications. *Chem Commun.* 2011;1388–1389. https://doi.org/10.1039/B104297C. (b) Arentsen K, Caddick S, Cloke FGN, Herring AP, Hitchcock PB. Suzuki–Miyaura cross-coupling of aryl and alkyl halides using palladium/imidazolium salt protocols. *Tetrahedron Lett.* 2004;45(17):3511–3515. https://doi.org/10.1016/j.tetlet.2004.02.134. (c) Arentsen K, Caddick S, Cloke FGN. On the efficiency of two-coordinate palladium(0) N-heterocyclic carbene complexes in amination and Suzuki–Miyaura reactions of aryl chlorides. *Tetrahedron.* 2005;61(41):9710–9715. https://doi.org/10.1016/j.tet.2005.06.070.

196. Gstöttmayr CWK, Böhm VPW, Herdtweck E, Grosche M, Herrmann WA. A defined N-heterocyclic carbene complex for the palladium-catalyzed Suzuki cross-coupling of aryl chlorides at ambient temperatures. *Angew Chem Int Ed.* 2002;41(8):1363–1365. https://doi.org/10.1002/1521-3773(20020415)41:8<1363::AID-ANIE1363>3.0.CO;2-G.

197. Yamashita M, Goto K, Kawashima T. Fixation of both O$_2$ and CO$_2$ from air by a crystalline palladium complex bearing N-heterocyclic carbene ligands. *J Am Chem Soc.* 2005;127(20):7294–7295. https://doi.org/10.1021/ja051054h.

198. Styliniades N, Danopoulos AA, Pugh D, Hancock F, Zanotti-Gerosa A. Cyclometalated and alkoxyphenyl-substituted palladium imidazolin-2-ylidene complexes. Synthetic, structural and catalytic studies. *Organometallics.* 2007;26:5627–5635. https://doi.org/10.1021/om700603d.

199. Danopoulos AA, Pugh D. A method for the synthesis of nickel(0) bis(carbene) complexes. *Dalton Trans.* 2008;30–31. https://doi.org/10.1039/B714988C.

200. Matsubara K, Miyakazi S, Koga Y, Nibu Y, Hashimura T, Matsumoto T. An unsaturated nickel(0) NHC catalyst: facile preparation and structure of Ni(0)(NHC)$_2$,

featuring a reduction process from Ni(II)(NHC)(acac)$_2$. *Organometallics*. 2008;27(22):6020–6024. https://doi.org/10.1021/om800488x.
201. Desmarets C, Schneider R, Fort Y. Nickel(0)/dihydroimidazol-2-ylidene complex catalyzed coupling of aryl chlorides and amines. *J Org Chem*. 2002;67(9):3029–3036. https://doi.org/10.1021/jo016352l.
202. Lee E, Yandulov DV. Synthesis and characterization of Pd(IMe)$_2$, and its reactivity by C-S oxidative addition of DMSO. *J Organomet Chem*. 2011;696(25):4095–4103. https://doi.org/10.1016/j.jorganchem.2011.07.006.
203. Hering F, Nitsch J, Paul U, Steffen A, Bickelhaupt FM, Radius U. Bite-angle bending as a key for understanding group-10 metal reactivity of d^{10}-[M(NHC)$_2$] complexes with sterically modest NHC ligands. *Chem Sci*. 2015;6:1426–1432. https://doi.org/10.1039/c4sc02998d.
204. Hering F, Radius U. From NHC to imidazolyl ligands: synthesis of platinum and palladium complexes d^{10}-[M(NHC)$_2$] (M = Pd. Pt) of the NHC 1,3-diisopropylimidazolin-2-ylidene. *Organometallics*. 2015;34(13):3236–3245. https://doi.org/10.1021/acs.organomet.5b00277.
205. Fantasia S, Nolan SP. A general synthetic route to mixed NHC-phosphane palladium(0) complexes (NHC = N-heterocyclic carbene). *Chem A Eur J*. 2008;14:6987–6993. https://doi.org/10.1002/chem.200800815.
206. Lee J-H, Jeon H-T, Kim Y-J, Lee K-E, Jang YO, Lee SW. Facile oxidative addition of organic halides to heteroleptic and homoleptic Pd0-N-heterocyclic carbene complexes. *Eur J Inorg Chem*. 2011;1750–1761. https://doi.org/10.1002/ejic.201001257.
207. (a) Fantasia S, Egbert JD, Jurčik V, et al. Activation of hydrogen by palladium(0): formation of the mononuclear dihydride complex trans-[Pd(H)$_2$(IPr)(PCy$_3$)]. *Angew Chem Int Ed*. 2009;48:5182–5186. https://doi.org/10.1002/anie.200900463. (b) Jurčik V, Schmid TE, Dumont Q, Slawin AMZ, Cazin CSJ. [Pd(NHC)(PR$_3$)] (NHC = N-heterocyclic carbene) catalyzed alcohol oxidation using molecular oxygen. *Dalton Trans*. 2012;41:12619–12623. https://doi.org/10.1039/C2DT30133D.
208. (a) Broggi J, Jurčik V, Songis O, et al. The isolation of [Pd{OC(O)H}(H)(NHC)(PR$_3$)] (NHC = N-heterocyclic carbene) and its role in alkene and alkyne reductions using formic acids. *J Am Chem Soc*. 2013;135:4588–4591. https://doi.org/10.1021/ja311087c. (b) Hartmann CE, Jurčik V, Songis O, Cazin CSJ. Tandem ammonia borane dehydrogenation/alkene hydrogenation mediated by [Pd(NHC)(PR$_3$)] (NHC = N-heterocyclic carbene) catalyst. *Chem Commun*. 2013;49:1005–1007. https://doi.org/10.1039/C2CC38145A.
209. Schmid TE, Jones DC, Songis O, et al. Mixed phosphine/N-hetrocyclic carbene palladium complexes: synthesis, characterization and catalytic use in aqueous Suzuki-Miyaura reactions. *Dalton Trans*. 2013;42:7345–7353. https://doi.org/10.1039/C2DT32858E.
210. Tardiff BJ, Hesp KD, Ferguson MJ, Stradiotto M. Generation of [(IPr)Pd(PR2Cl)] complexes via P-Cl reductive elimination. *Dalton Trans*. 2012;41:7883–7885. https://doi.org/10.1039/C1DT11910A.
211. (a) Yin J, Rainka MP, Zhang X-X, Buchwald SL. A highly active Suzuki catalyst for the synthesis of sterically hindered biaryls: novel ligand coordination. *J Am Chem Soc*. 2002;124(7):1162–1163. https://doi.org/10.1021/ja017082r. (b) Walker SD, Barder TE, Martinelli JR, Buchwald SL. A rationally designed universal catalysts for Suzuki-Miyaura coupling processes. *Angew Chem Int Ed*. 2004;43:1871–1876. https://doi.org/10.1002/anie.200353615.

CHAPTER SIX

Recent advances in Pd-catalyzed asymmetric addition reactions

Wenbo Li[a], Junliang Zhang[b],*

[a]Shanghai Key Laboratory of Green Chemistry and Chemical Processes, School of Chemistry and Molecular Engineering, East China Normal University, Shanghai, P.R. China
[b]Department of Chemistry, Fudan University, Shanghai, P.R. China
*Corresponding author: e-mail address: junliangzhang@fudan.edu.cn

Contents

1. Introduction	325
2. Palladium-catalyzed asymmetric conjugate addition	326
2.1 Palladium-catalyzed asymmetric conjugate addition to cyclic electrophiles	326
2.2 Palladium-catalyzed asymmetric conjugate addition to linear electrophiles	335
3. Palladium-catalyzed asymmetric 1,2-addition	356
3.1 Palladium-catalyzed asymmetric 1,2-addition to aldehydes/ketones	356
3.2 Palladium-catalyzed asymmetric 1,2-addition to imines	362
3.3 Palladium-catalyzed asymmetric 1,2-addition to olefins/allenes/cumulenes	374
4. Conclusion and outlook	393
Acknowledgments	394
References	394

1. Introduction

Palladium catalysts are very attractive in the academic and industrial areas, which have been extensively studied for the formation of biologically active molecules and natural products. Over the last several decades, synthetically significant methods via Pd-catalyzed asymmetric addition of various nucleophiles to C=X bonds (X=C, O, N) have been widely applied in organic synthesis, which provide one of the most effective strategies to construct carbon–carbon and carbon–heteroatom bonds. Many new chiral palladium catalysts such as NHC-Pd, palladacycle, pincer-Pd, etc., and chiral ligands such as bidentate N-(sp^2)-based ligands, phosphinoimine ligands, spirophosphite ligands, etc., have been developed, which allowed various asymmetric addition reactions to reach high yields and excellent enantioselectivities.[1] These new palladium catalysts allowed to expand the scope of the asymmetric addition reaction to various electrophiles including conjugated alkenes, aldehydes, ketones, imines, unactivated olefins, allenes and cumulenes.

Generally, Pd catalysts are able to promote the asymmetric addition to a variety of types of substrates via two different catalytic reaction modes. On one side they may act as a pure organometallic catalyst in the catalytic asymmetric reaction, such as 1,4-addition, promoting a transmetallation/insertion/hydrolysis process resulting the corresponding products. On the other, they may act as a Lewis acid to activate unsaturated bonds through coordination with the coordination group in the electrophiles, and then nucleophiles attack the Pd-activated substrates prior to dissociation of the addition products and catalyst regeneration. In both catalytic cycles, the chiral ligands employed have a major impact on the stereochemical outcome.

This chapter mainly refers to the last 10 years advances in the asymmetric conjugate addition and 1,2-addition reactions catalyzed by chiral palladium complexes. Additionally, Pd-catalyzed formal asymmetric addition reactions including Friedel-Crafts, umpolung allylation, Wacker-type cyclization and alkene difunctionalization reactions initiated by addition step are also discussed in this context.[2]

2. Palladium-catalyzed asymmetric conjugate addition

The Pd-catalyzed asymmetric additions of electron-deficient alkenes and various nucleophiles including boronic acids, diphenylphosphine, aryltriethylsiloxanes and amines constitute very efficient methodologies in organic synthesis, providing products in highly enantioselective manner under mild conditions.[3] Both cyclic and acyclic electrophiles have been successfully employed within this strategy.

2.1 Palladium-catalyzed asymmetric conjugate addition to cyclic electrophiles

Chiral (hetero)cyclic rings are incorporated in many biologically active and naturally occurring substances. The Pd-catalyzed asymmetric addition of cyclic conjugate acceptors is an efficient synthetic tool for the formation of enantioselective cyclic compounds. Due to the restricted configuration of cyclic substrates, these types of the reactions are easily accessible with high reactivity and stereoselectivity. In addition, the asymmetric conjugated addition reaction always tolerates atomspheric oxygen and moisture when the air-stable and functional group-tolerant boron nucleophiles were used.

The Shi group developed the axially chiral bis(NHC)-Pd(II) catalytic system for the enantioselective conjugate addition of arylboronic acids to cyclic enones **4** or N-substituted 2,3-dihydro-4-pyridones **7**, furnishing the desired adducts **6** or **8** in high yields and excellent *ee* values (Scheme 1).[4] Performing the addition reaction of 5,6-dihydropyran-2-one or 2-cyclopentenone with arylboronic acids, the corresponding adducts were achieved in 32–38% *ee* under the standard condition. The presence of an electron-withdrawing group or *ortho*-substituents in arylboronic acids for the asymmetric conjugate additions was not reported in both systems. Mechanistic studies suggest that the hydroxopalladium catalytic species generated from complex $(NHC)_2\text{-}PdX_2$ and KOH undergoes transmetallation to produce the arylpalladium intermediate. The sequential process involves migratory insertion to generate π-oxaallylpalladium species or the palladium enolate species. After hydrolysis by water, the adducts were produced and the hydroxopalladium catalytic species was regenerated.

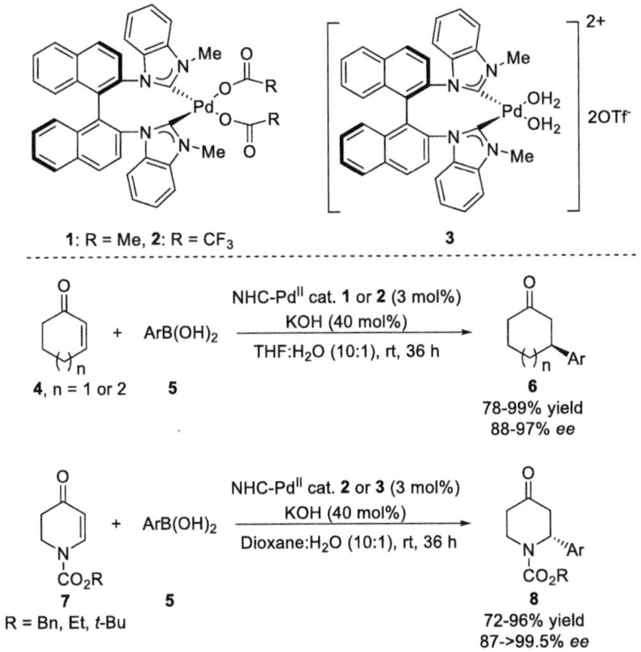

Scheme 1 Pd/bis(NHC)-catalyzed asymmetric addition of arylboronic acids to cyclic enones or N-substituted 2,3-dihydro-4-pyridones.

An enantioselective 1,4-addition of arylboronic acids **5** and cyclopentenone **9** was accomplished working with a chiral crown ether ligand comprising P-stereogenic phosphine and ethyleneoxy units in the ring (Scheme 2).[5] The best enantioselectivities were observed by using the Pd/P-stereogenic diphosphacrown complex **10**.

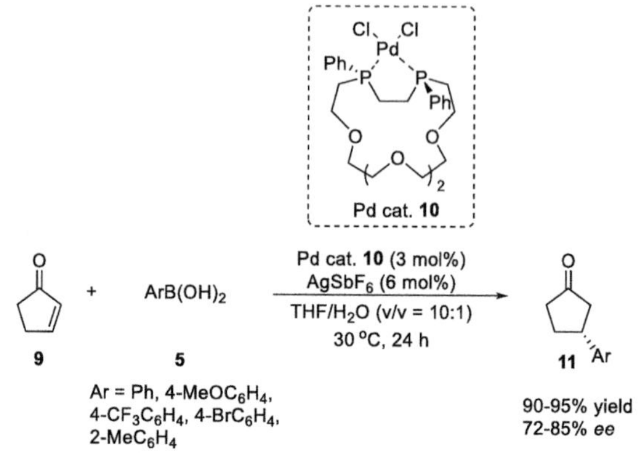

Scheme 2 Pd/diphosphacrown-catalyzed asymmetric addition of arylboronic acids to 2-cyclopentenone.

In 2014, Pullarkat and co-workers synthesized a range of metallacycles, which have been applied in the asymmetric conjugate addition of enones and arylboronic acids.[6] Among those chiral palladacycles, catalyst **12** with the chiral naphthyl backbone proved to be the most efficient catalyst, affording adducts **6** in 64–89% yield with 50–92% *ee* (Scheme 3).

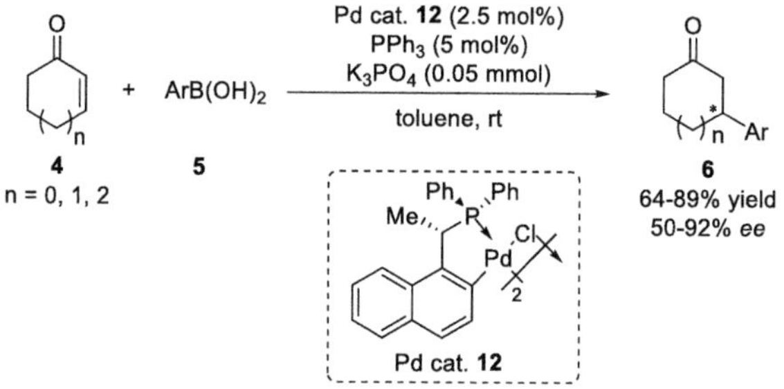

Scheme 3 Phosphapalladacycle-catalyzed asymmetric addition of arylboronic acids to cyclic enones.

As reported by the Tamura group, chiral phenanthroline ligand **13** along with Pd(OTf)$_2$ induced the asymmetric 1,4-addition of arylboronic acid to cyclic enones to give the corresponding adducts **6** or **15** in good yields with excellent *ee*s (Scheme 4).[7]

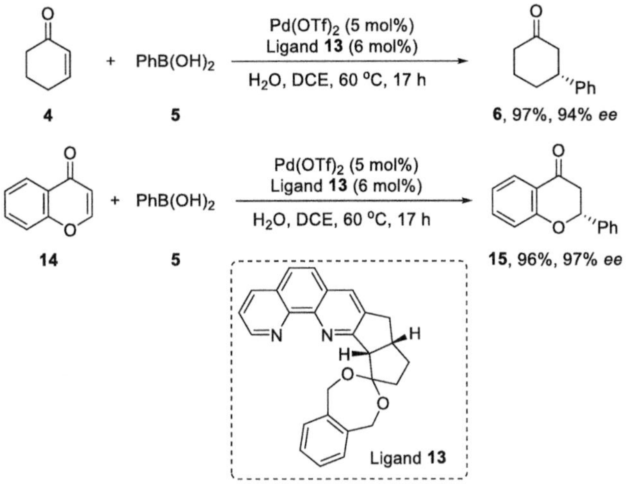

Scheme 4 Pd/chiral 1,10-phenanthroline ligand-catalyzed asymmetric addition of arylboronic acids to cyclic enones.

Constructing all-carbon quaternary stereocenters via palladium-catalyzed asymmetric conjugate addition remains a challenging problem in synthetic chemistry. Lu and co-workers have described the cationic Pd(II)-catalyzed conjugated addition of arylboronic acids to β-substituted cyclic enones.[8] In 2011, the Stoltz group promoted the enantioselective conjugated addition of arylboronic acids to β-alkyl-substituted cyclic enones to generate β,β-disubstituted cyclic ketones (Scheme 5A).[9] They found that the asymmetric addition process proceeded with high yield and enantioselectivity by using a catalyst formed in situ from the combination of Pd(OCOCF$_3$)$_2$ and (*S*)-*t*-BuPyOx in DCE at 40–80 °C. A broad range of arylboronic acids and 5-, 6-, 7-membered enones with various β-alkyl-substituted are well-tolerated. However, some arylboronic acids (such as 4-OBn, 4-OMe, 3-Cl, 3-NO$_2$) furnished lower yields or moderate *ee*s while *ortho*-substituted arylboronic acids did not verify the reaction.

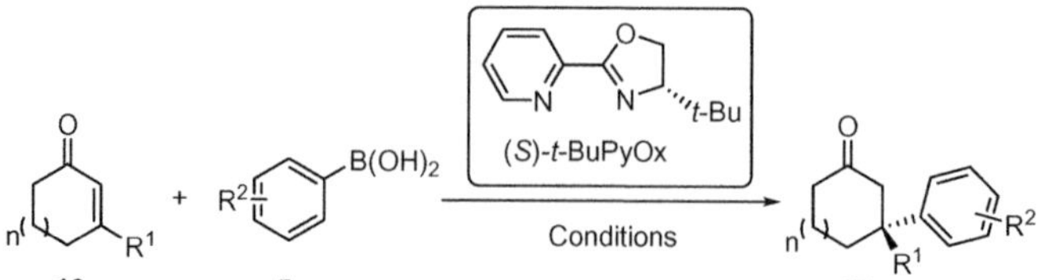

A Stoltz 2011

Conditions: Pd(OCOCF$_3$)$_2$ (5 mol%), (S)-t-BuPyOx (6 mol%)
DCE, 40-80 °C, 12-24 h

n = 0,1,2

R^1 = Me, Et, Bu, Bn, iPr, cyclopropyl, cyclohexyl, (CH$_2$)$_3$OBn

R^2 = H, 4-Me, 4-Et, 4-MeO, 4-OBn, 4-OTBS, 4-Ac, 4-Cl, 4-F, 4-CF$_3$, 3-Me, 3-Cl, 3-Br, 3-CO$_2$Me, 3-NO$_2$

40-99% yield
69-96% ee

B Stoltz 2013

Conditions: Pd(OCOCF$_3$)$_2$ (5 mol%), (S)-t-BuPyOx (6 mol%)
NH$_4$PF$_6$, H$_2$O, DCE, 40-60 °C, 12 h

n = 0,1,2

R^1 = Me, Et, Bu, Bn, iPr, cyclopropyl, cyclohexyl, (CH$_2$)$_3$OBn, COMe

R^2 = H, 3-Cl, 3-Br, 3-NO$_2$, 2-F, 4-Cl, 4-F, 3-Me, 3-NHCOCF$_3$-4-Me,

57-96% yield
77-96% ee

C Stanley 2017

Conditions: Pd(OCOCF$_3$)$_2$ (10 mol%), (S)-t-BuPyOx (12 mol%)
DCE, 80 °C, 9 h

n = 0,1

R^1 = 4-MeOC$_6$H$_4$, Ph, 4-Me$_2$NC$_6$H$_4$, 4-FC$_6$H$_4$, 4-CF$_3$C$_6$H$_4$, 3-MeOC$_6$H$_4$, 2-MeOC$_6$H$_4$, 4-MeC$_6$H$_4$, 3-indolyl

R^2 = H, 4-Me, 4-Ph, 4-Cl, 4-F, 4-CO$_2$CH$_3$, 4-CF$_3$, 2-F, 3-Me, 3-MeO, 3-Cl, 3-F, 3-F-4-MeO, 3,4-Me$_2$, 3,4-(MeO)$_2$, 3,5-Me$_2$

18-92% yield
78-93% ee

Scheme 5 Pd/PyOx-catalyzed asymmetric addition of arylboronic acids to β-substituted cyclic enones.

Later, they modified the Pd/PyOx system using NH$_4$PF$_6$ and water as additives to accelerate the rate of the reaction at low temperatures (Scheme 5B).[10] It was interesting noting that β-acyl cyclic enones **16** and arylboronic acids **5** containing nitrogen and other heteroatoms also provided the conjugated addition products **17** in modest to good yields and ee. This Pd/PyOx catalyst system is not applicable to the asymmetric addition reactions of arylboronic acids to cyclic enones without the β-substituent, whereas this methodology is compatible with chromones and 4-quinolones, but with no β-substituent.[11]

Computational and experimental studies of the reaction mechanism indicate a cationic palladium (II) hydroxide species, which undergoes rapid transmetallation with arylboronic acids (Scheme 6). The subsequent

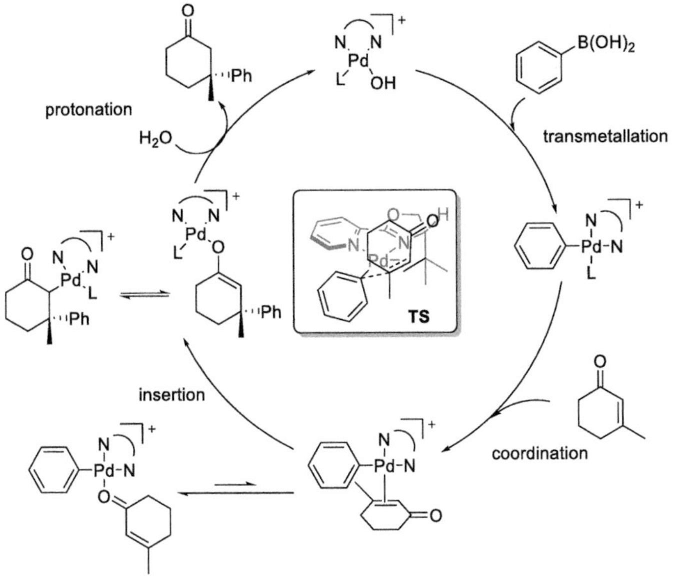

Scheme 6 Mechanistic pathway of Stoltz's catalyst system.

ligand coordination with ketones generated an equilibrium mixture of olefin-bound complex and carbonyl-bound complex. DFT calculations suggested that the olefin insertion is the enantioselectivity-determining step. The stereoselectivity-determining transition state (TS) is shown in Scheme 6, which leads to the formation of the (R)-ketones. The hydrolysis of cationic palladium enolate provides the addition product and regenerates the palladium (II) species.

The Stoltz group further applied the Pd/PyOx-catalyzed asymmetric conjugate addition reaction as the key step in the total synthesis of (+)-dichroanone and (+)-taiwaniaquinone H (Scheme 7).[12] In 2017, the Stanley group established the enantioselective Pd/PyOx-catalyzed conjugate addition of arylboronic acids to β-aryl cyclic enones **16** for the synthesis of cyclic ketone products containing bis-benzylic quaternary carbon stereocenters in 18–92% yield and 78–93% ee (Scheme 5, C).[13] In contrast to electron-rich arylboronic acids, electron-deficient arylboronic acids provided the desired products with similar levels of ee, but in lower yields. The ortho-substituted arylboronic acids or heteroarylboronic acids cannot give the corresponding β,β-disubstituted ketones.

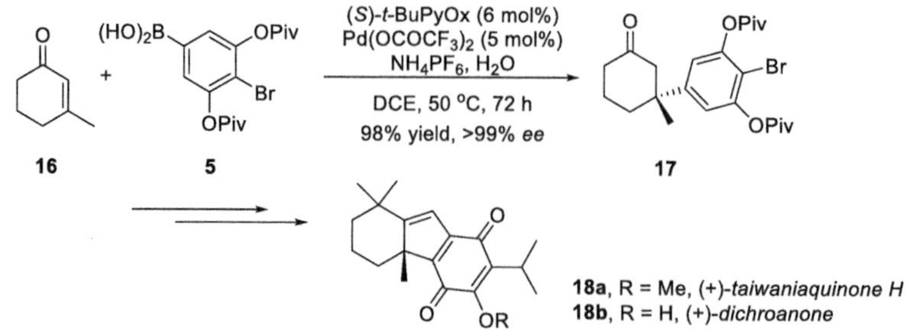

Scheme 7 Total synthesis of (+)-taiwaniaquinone H and (+)-dichroanone.

In 2012, the Minnard laboratory reported a similar asymmetric conjugate addition of arylboronic acids to β-methyl cyclic enones and lactones using PdCl$_2$/(R,R-PhBOX) as catalyst and silver salt as an additive (Scheme 8).[14] The β,β-disubstituted cyclic ketones with a quaternary stereocenter **19** were obtained in 44–98% yields and 88–99% ee. This catalytic system could be further employed to more challenging acyclic enones with an allylic ether moiety, but the ee values of the adducts bearing a quaternary stereocenter are low to moderate (23–60% ee).

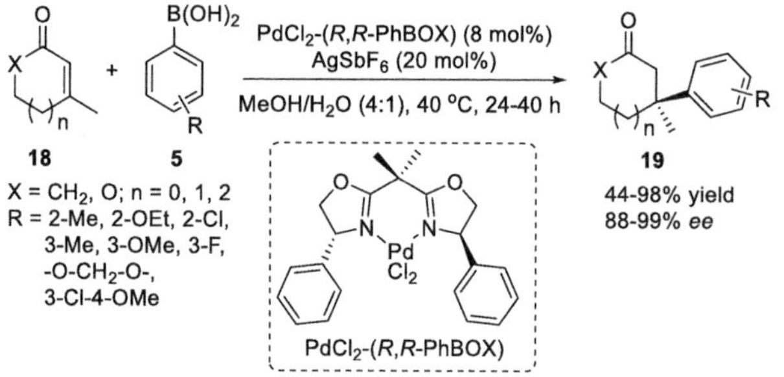

Scheme 8 Pd/BOX-catalyzed asymmetric addition of arylboronic acids to β-methyl cyclic enones/lactones.

A modified catalytic system also employing NH$_4$PF$_6$ and H$_2$O as additives was successfully applied by Minnard and co-worker to the stereoselective addition of *ortho*-substituted arylboronic acids and β-methyl cyclic enones (Scheme 9).[15] Good to excellent enantioselectivities of the desired products are obtained, albeit the yields are low to moderate. This method was applied in the total synthesis of Herbertenediol and Enokipodine A and B in shorter routes compared with previous reports.

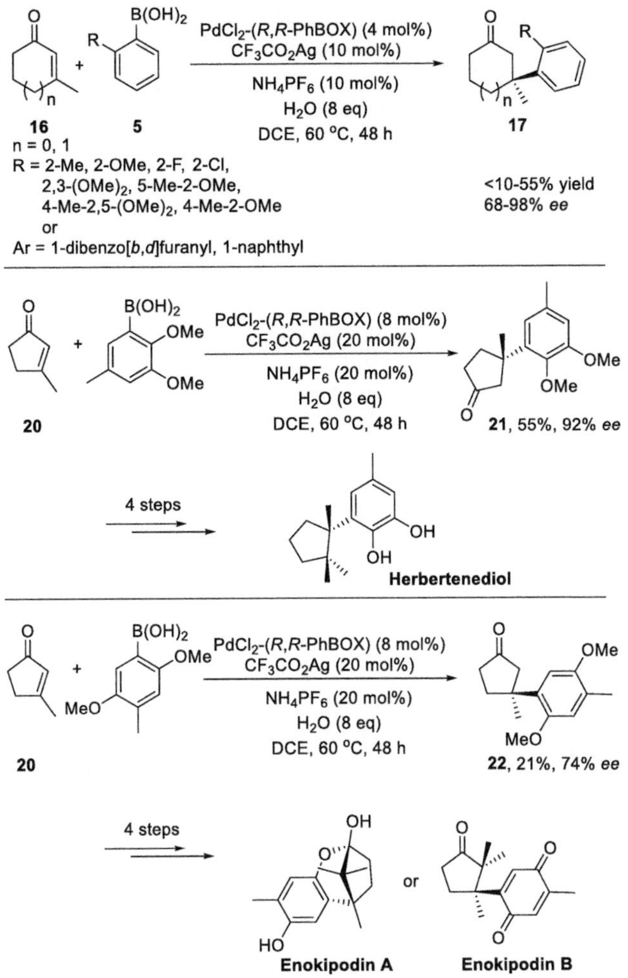

Advances in Pd-catalyzed asymmetric addition reactions 333

Scheme 9 Pd/BOX-catalyzed asymmetric addition of *ortho*-substituted arylboronic acids to β-methyl cyclic enones.

Very recently, pyridine–hydrazone ligand **23** in combination with Pd(TFA)$_2$ has been proved to be an useful catalyst in the asymmetric 1,4-addition of arylboronic acids to β-substituted cyclic enones **16**, affording the 1,4-adducts **17** with an all-carbon quaternary stereocenter in good yields and ees (Scheme 10).[16] In this process, a catalytic amount of water assisted the catalytic turnover and accelerated the reaction. This methodology provided an efficient route to synthesize β-aryl-β′-methylcyclopentanones with an *ortho*-substituted aryl group, precursors of biologically active molecules.

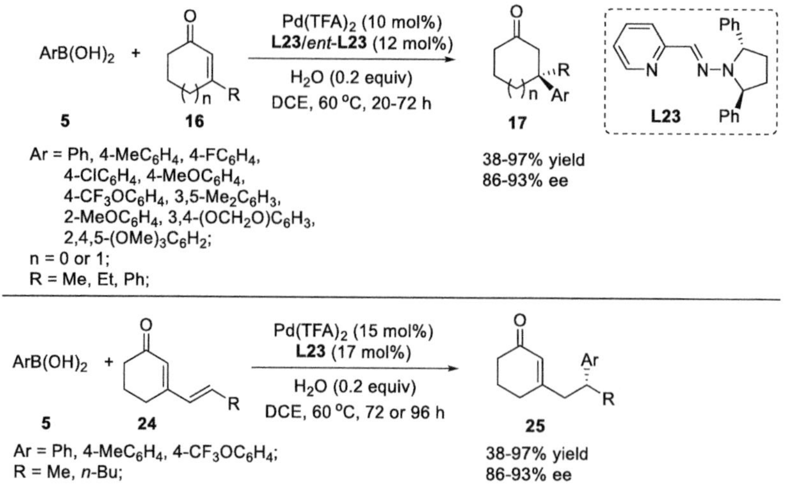

Scheme 10 Pd-catalyzed asymmetric addition of arylboronic acids to cyclic enones/dienones.

Furthermore, cyclic dienones **24** were used in the addition of arylboronic acids under the same catalytic system, which gave the 1,6-adducts **25** in moderate yields and enantioselectivities.

Various mono- and bispalladacycle complexes have been employed as efficient catalysts in the asymmetric addition of α-cyanoacetates **26** to cyclic enones **27**, furnishing the corresponding addition products **28** with adjacent quaternary and tertiary stereocenters (Scheme 11).[17] In the presence of bispalladacycle catalyst, different diastereomers of the corresponding adducts were observed in good yields and high enantioselectivities with (R,R)-diastereomers as the major product. In the presence of mono-palladacycle catalyst, the selectivity can be switched to (S,R)-diastereomers as the major product. In both cases, a Brønsted acid (HOAc) was used as a cocatalyst to avoid the undesired β-hydride elimination.

An interesting chemo-selectivity has been reported in the C,P-palladacycle catalyzed asymmetric addition of diarylphosphines **29** to 2,5-diphenylbenzoquinone **30** (Scheme 12).[18] The reaction in the presence of triethylamine in chloroform at −45 °C, followed by the treatment of sulfur, produced enantiomerically enriched 2,5-diphenyl-4-hydroxyphenyl phosphinites **33** in excellent yields. It is noteworthy that only diarylphosphines where one of the aryl groups is a sterically demanding aryl group are efficient for this reaction. When an *ortho*-monosubstituted substrate was used, much lower enantioselectivities were obtained.

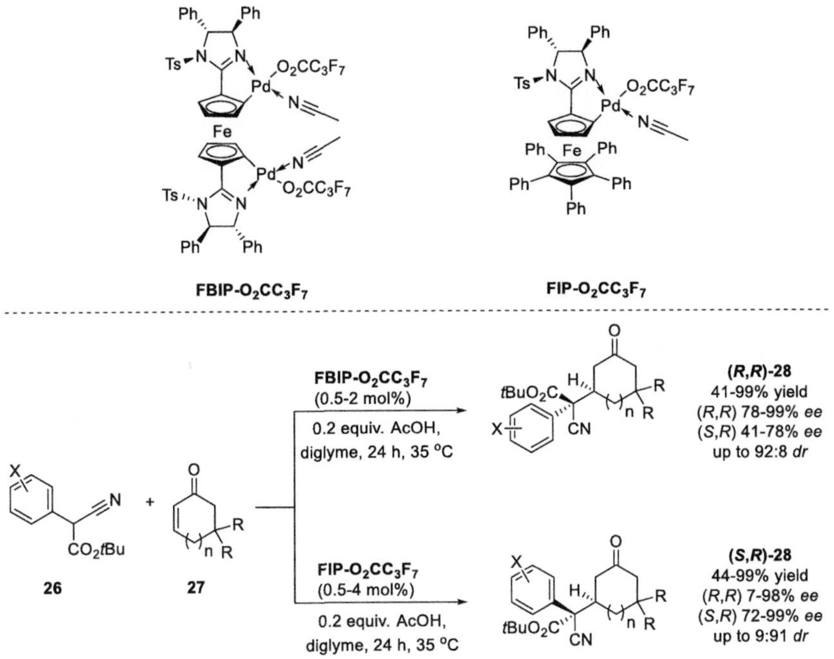

Scheme 11 Pd-catalyzed asymmetric addition of α-cyanoacetates and cyclic enones.

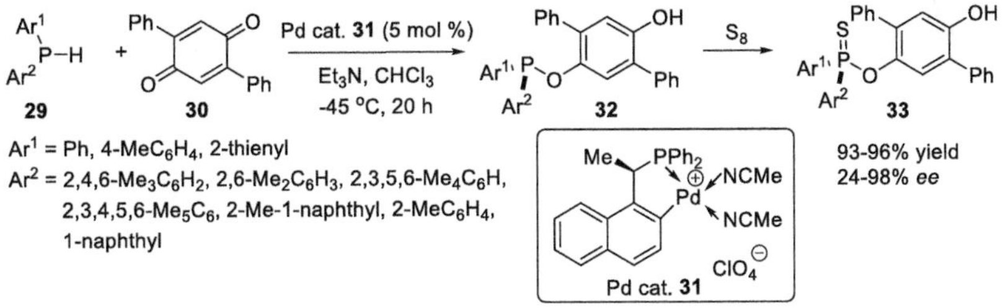

Scheme 12 C,P-palladacycle catalyzed addition of diarylphosphines to benzoquinones.

2.2 Palladium-catalyzed asymmetric conjugate addition to linear electrophiles

Compared to cyclic enones or other cyclic conjugate acceptors, the asymmetric conjugate addition to linear electrophiles is more challenging due to their flexible conformation. Enantioselective conjugate addition is an efficient method for differentiation of the two carbons of the double bonds and building the stereogenic carbon center. Although most of research focused on the Rh(I)-catalysis due to their high activity for the addition reaction, several Pd catalytic systems have emerged to enable the enantioselective additions to linear electrophiles. This section highlights the development of the asymmetric conjugate addition reactions of linear electrophiles with different nucleophiles, that is, the palladium-catalyzed asymmetric conjugate addition of arylboronic acids, diarylphosphines, and other nucleophiles.

2.2.1 Palladium-catalyzed asymmetric conjugate addition of arylboronic acids

Over the past years, significant efforts have been made to realize enantioselective conjugate addition of arylboronic acids to acyclic α,β-unsaturated ketones using palladium catalysts.[19] Ohta et al. revealed in 2009 that Pd(dba)$_2$ and (S,Rp)-[1-(2-bromoferrocenyl)ethyl]diphenylphosphine (*S,Rp*)-35 could act as a chiral catalyst for the asymmetric conjugate addition to enones 34 with arylboronic acids 5 (Scheme 13).[20] In the presence of 5 mol% of palladium complex and 1.0 equiv. of K$_2$CO$_3$ at room temperature for 3 h, the corresponding products were obtained in 45–99% yields with 4–79% ee.

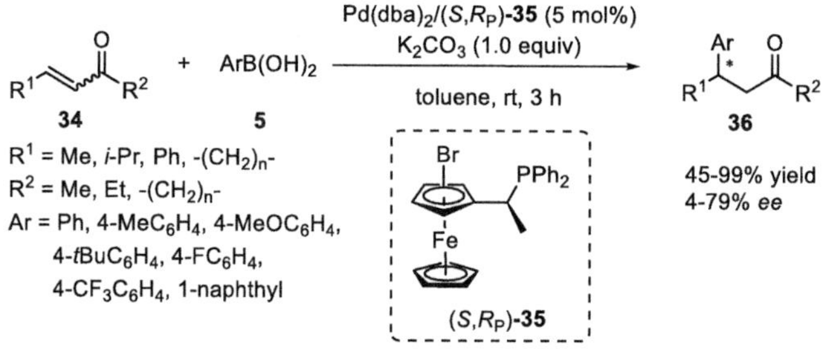

Scheme 13 Pd-(*S,Rp*)-35 catalyzed 1,4-addition of arylboronic acids to α,β-unsaturated ketones.

Moreover, an efficient catalytic system based on α-naphthalenyl ethanol derived chiral phosphapalladacycle catalyst 12 was disclosed by Pullarkat et al. for the enantioselective 1,4-addition of arylboronic acids to acyclic enones (Scheme 14).[6] A series of acyclic enones and boronic acids were tolerated and the addition adducts 36 were obtained in good yields with high enantioselectivites. It was found that electron-deficient boronic acids gave the corresponding products in lower yields and poorer *ees* under the same conditions.

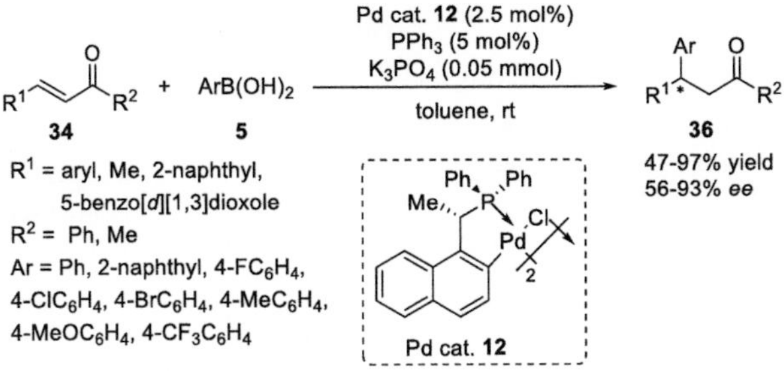

Scheme 14 Phosphapalladacycle-catalyzed asymmetric addition of arylboronic acids to enones.

Nitroalkenes as "synthetic chameleons" have gradually attracted increasing attention, which are successfully used in transition-metal-catalyzed asymmetric addition reactions. In 2014, Gutnov and co-workers discovered the first asymmetric Pd-catalyzed conjugate addition of arylboronic acids to 2-nitroacrylate derivatives **37** (Scheme 15).[21] The cationic Pd(II)/Chiraphos complex exhibits excellent activity at low catalyst loadings (0.05–0.25 mol%), providing 2-aryl-3-nitropropionamides **38** in good yields and enantioselectivities (73–89% ee). The adducts were transformed into β²-homophenylglycines in ~50% yield with >98% ee by reduction of the nitro group and subsequent amide hydrolysis.

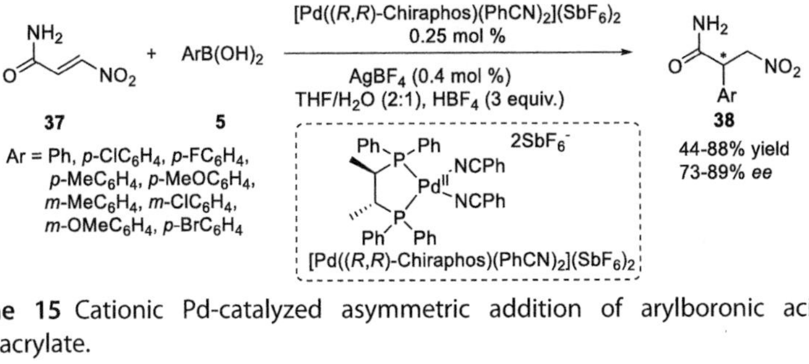

Scheme 15 Cationic Pd-catalyzed asymmetric addition of arylboronic acids to 2-nitroacrylate.

In 2015, Zhang and co-workers attempted the asymmetric conjugate addition of arylboronic acids to nitrostyrenes **39** catalyzed by Pd(TFA)$_2$/iPr-IsoQuinox in methanol under an air atmosphere (Scheme 16).[22] A series of nitrostyrenes and arylboronic acids were tolerated, giving the corresponding adducts **40** in good yields and high ees. In addition, electron-rich arylboronic acids yielded higher ees in comparison with electron-deficient arylboronic acids.

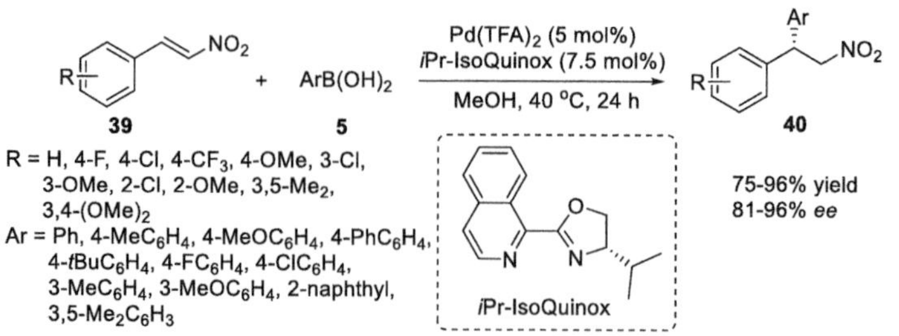

Scheme 16 Pd(TFA)$_2$/iPr-IsoQuinox-catalyzed asymmetric addition of arylboronic acids to nitrostyrenes.

Later, following Zhang's procedure, the addition of arylboronic acids to β-nitrostyrenes was explored by Chen and co-workers (Scheme 17).[23] The corresponding adducts **43** were obtained in 77–85% yield and 95–98% ee under the Pd(II)-catalytic system. In addition, this method served as the key step in the enantioselective total synthesis of a wide variety of isopavine alkaloids.

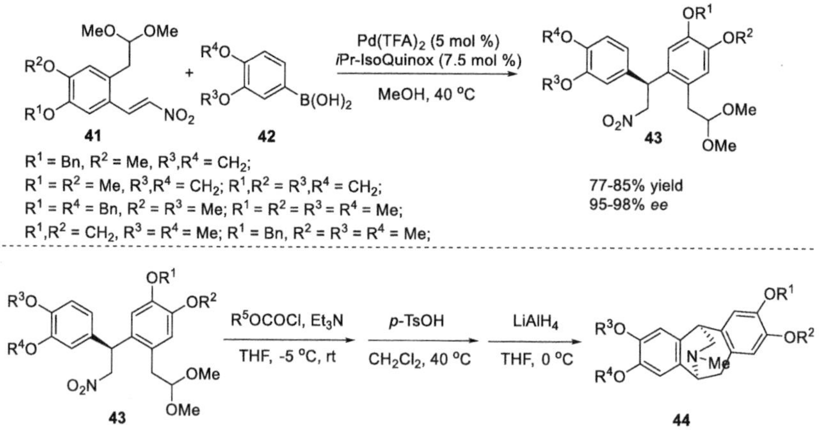

Scheme 17 Total synthesis of isopavine alkaloids.

The 1,6-addition of arylboronic acids to electron-deficient dienes has been a challenge, owing to the difficulty of controlling both regioselectivity and enantioselectivity of the reaction. Zhang and co-workers realized the first efficient route to achieve the asymmetric 1,6-addition of Meldrum's acid-derived dienes **45** using Pd(TFA)$_2$/In-Pyrox as a suitable catalyst (Scheme 18).[24] The corresponding adducts **46** were obtained in up to

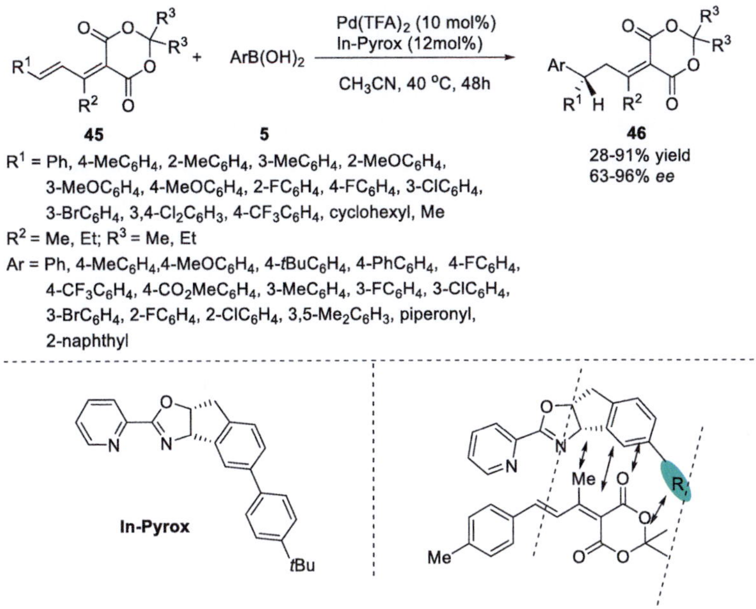

Scheme 18 Pd(II)/In-Pyrox-catalyzed asymmetric 1,6-addition of arylboronic acids to dienes.

91% yield and 96% *ee* under the optimized conditions. However, lower yields were gained for electron-deficient and *ortho*-substituents arylboronic acids albeit with high *ees*. The dienes **45** with an alkyl substituent as R^1 group were also compatible but resulted in lower *ees*. In addition, the dienes **45** with an ethyl as R^2 group afforded the adduct in 61% yield and 90% *ee*. A ligand design principle was proposed by the authors, indicating that the size of the substituent of the ligands plays a significant role in achieving high enantioselectivity due to the remote steric interactions between the substrate and ligand.

2.2.2 Palladium-catalyzed asymmetric conjugate addition of diarylphosphines

The asymmetric conjugate addition of diarylphosphines to electron-deficient olefins has attracted much attention due to the efficient synthesis of chiral P—C bonded phosphines. The research groups of Song, Duan, Leung and Zhang have developed various types of enantioselective hydrophosphination reactions using chiral pincer-Pd or phosphapalladacycle as the efficient catalyst. In 2010, the Duan group described the first stereoselective addition of diarylphosphines to α,β-unsaturated enones using pincer-Pd-OAc as catalyst (Scheme 19A).[25] The corresponding phosphination products **49** were obtained in 69–93% yield with 90–99% ee. In their recent work, they developed a bis(phosphine) (PCP) pincer-Pd catalyzed asymmetric addition of phosphorus nucleophiles to α,β-unsaturated enones **34** bearing a 2-pyridyl moiety at the β-position for the synthesis of chiral N,P-compounds (Scheme 19B).[26] In this system, not only diarylphosphines but also the secondary phosphine with an isopropyl and a phenyl substituent gave the corresponding product in 99% ee with >20:1 dr.

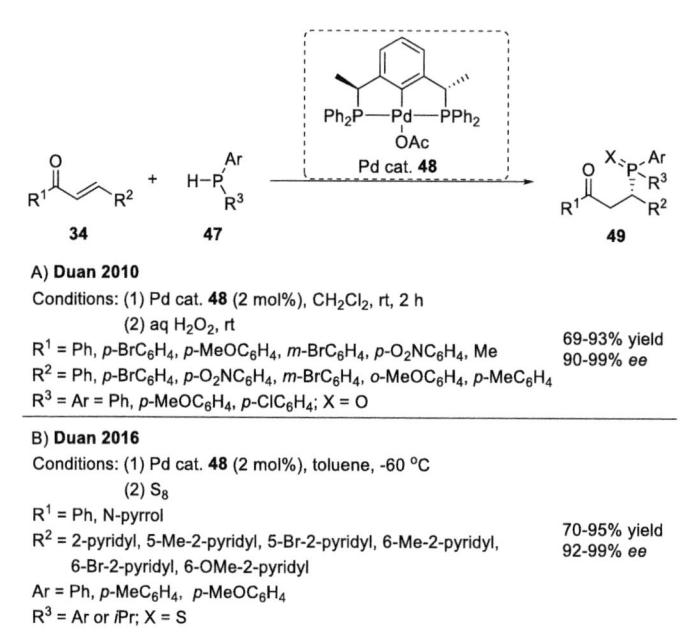

A) Duan 2010
Conditions: (1) Pd cat. **48** (2 mol%), CH$_2$Cl$_2$, rt, 2 h
(2) aq H$_2$O$_2$, rt
R^1 = Ph, p-BrC$_6$H$_4$, p-MeOC$_6$H$_4$, m-BrC$_6$H$_4$, p-O$_2$NC$_6$H$_4$, Me
R^2 = Ph, p-BrC$_6$H$_4$, p-O$_2$NC$_6$H$_4$, m-BrC$_6$H$_4$, o-MeOC$_6$H$_4$, p-MeC$_6$H$_4$
R^3 = Ar = Ph, p-MeOC$_6$H$_4$, p-ClC$_6$H$_4$; X = O

69–93% yield
90–99% ee

B) Duan 2016
Conditions: (1) Pd cat. **48** (2 mol%), toluene, -60 °C
(2) S$_8$
R^1 = Ph, N-pyrrol
R^2 = 2-pyridyl, 5-Me-2-pyridyl, 5-Br-2-pyridyl, 6-Me-2-pyridyl, 6-Br-2-pyridyl, 6-OMe-2-pyridyl
Ar = Ph, p-MeC$_6$H$_4$, p-MeOC$_6$H$_4$
R^3 = Ar or iPr; X = S

70–95% yield
92–99% ee

Scheme 19 Chiral PCP-pincer Pd-catalyzed addition of secondary phosphines to enones.

According to the proposed mechanism, the pincer-Pd catalyst reacted with diarylphosphine through a transphosphination sequence to generate a palladium phosphido complex (Scheme 20). A nucleophilic attack of palladium phosphido complex to enones produces an oxa-π-allylpalladium intermediate, which is converted into the product via a protonolysis process and regenerate the Pd catalyst.

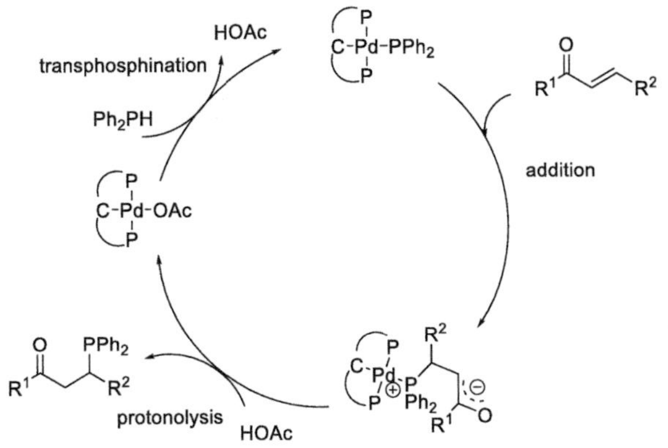

Scheme 20 Proposed mechanism for Duan's catalyst system.

The same group further elaborated the asymmetric conjugate addition to α,β-unsaturated N-acylpyrroles,[27] sulfonic esters,[28] carboxylic esters,[29] aldehydes[30] and nitroalkenes[31] using diarylphosphines as nucleophiles catalyzed by PCP-pincer Pd complex (Scheme 21). A set of chiral phosphine derivatives **51–55** were produced in good to high yields with excellent enantioselectivities. Only a few cases such as α,β-unsaturated carboxylic esters containing a cyclohexyl afford the 1,4-adducts with low enantioselectivity. More importantly, the hydrophosphination reaction of α,β-unsaturated aldehydes proceeds well with only 0.05 mol % of the pincer-Pd catalyst, delivering the adducts **54** in 85% yield and 95% ee.

In 2013, Zhang and co-workers designed and synthesized a novel P-stereogenic pincer-Pd complex **56** prepared from optically pure tert-butylmethylphosphineborane.[32] The employment of P-stereogenic pincer-Pd as the catalyst in the asymmetric 1,4-addition of diarylphosphines to nitroalkenes **39** and β,γ-unsaturated α-keto esters **58** can be used to afford the corresponding hydrophosphination products **57** or **59** in moderate to high yields with moderate to good enantioselectivities (Scheme 22). The reaction tolerates substrates **58** with alkyl R groups, providing chiral organophosphorus derivatives in 33–69% ee.

Song and co-workers developed a series of chiral PCN and NCN pincer Pd complexes **60–62**, which were also successfully applied to enantioselective conjugate addition of diarylphosphine to various enones

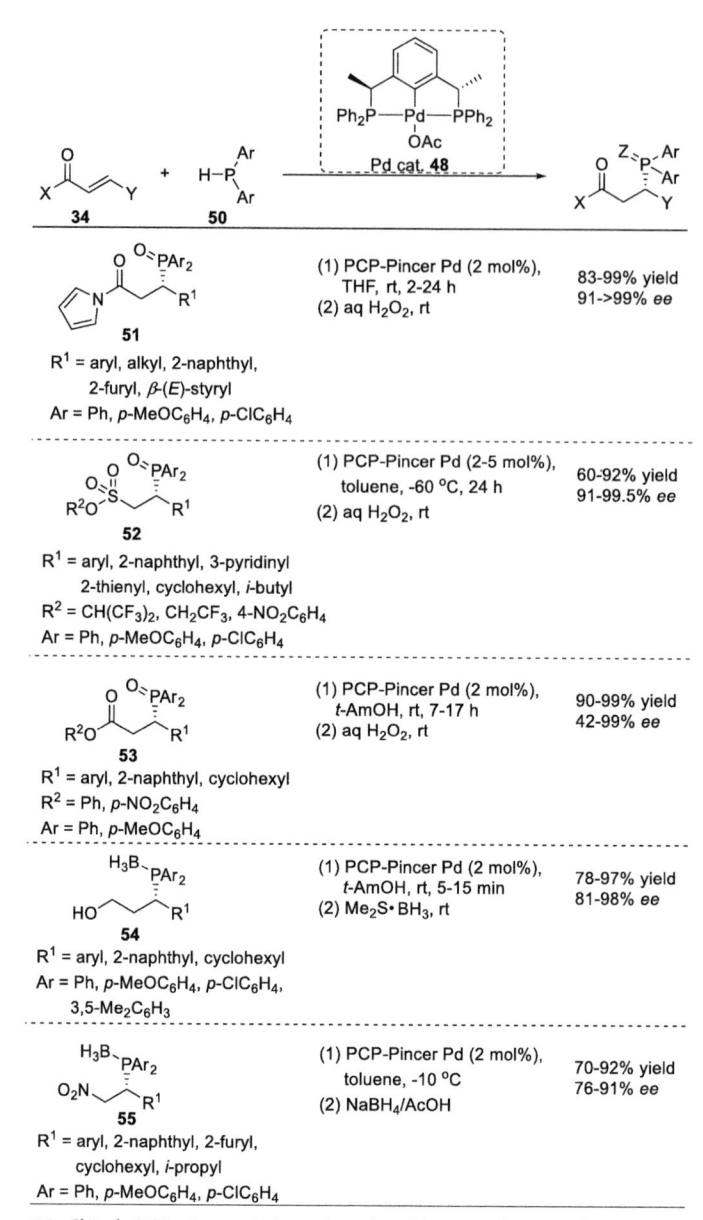

Scheme 21 Chiral PCP-pincer Pd-catalyzed addition of secondary phosphines to electron-deficient alkenes.

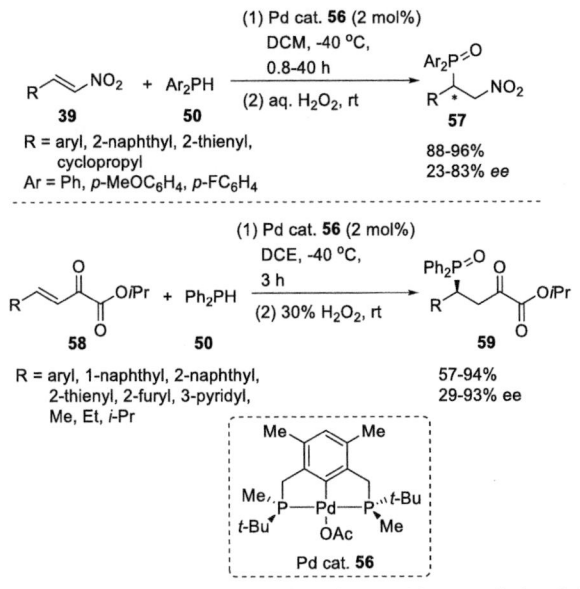

Scheme 22 P-stereogenic pincer-Pd catalyzed 1,4-addition of diarylphosphines to electron-deficient alkenes.

(Scheme 23).[33] Generally, substrates bearing electron-donating aromatic rings (R^2) gave a lower *ee* value. Notably, 2-alkenoyl pyridine *N*-oxides bearing various aromatic substituents, heteroaromatic substituents or 1-naphthyl substituent readily underwent hydrophosphination to afford the corresponding 1,4-adducts **65** in the presence of NCN pincer Pd catalyst **62b** (59–90% yield, 2–89% *ee*). Duan adopted a similar methodology for asymmetric 1,4-addition of secondary alkylphenylphosphines to enones catalyzed by NCN pincer Pd complex **62c**, which provided a series of chiral phosphorus compounds **66** bearing P- and C-stereogenic centers with both diastereomers in good to excellent *ee*.[34]

In addition, it was demonstrated that this type of transformation could be catalyzed by N,C- or P,C-palladacycle complexes. Since 2010, Leung and co-workers developed the asymmetric hydrophosphination reaction of aromatic enones catalyzed by C,P-palladacycle complex **31** (Scheme 24).[35] This transformation proceeds with 0.5 equiv. of Et$_3$N at −80 °C, affording chiral phosphine compounds **67** in good yields with excellent enantioselectivities. Specially, the employment of secondary alkylphenylphosphines and asymmetric diarylphosphines as the hydrophosphinating agent affords chiral phosphines **68** with both phosphorus and carbon chiral centers in 71–79% *ee* with moderate diastereoselectives.[35b]

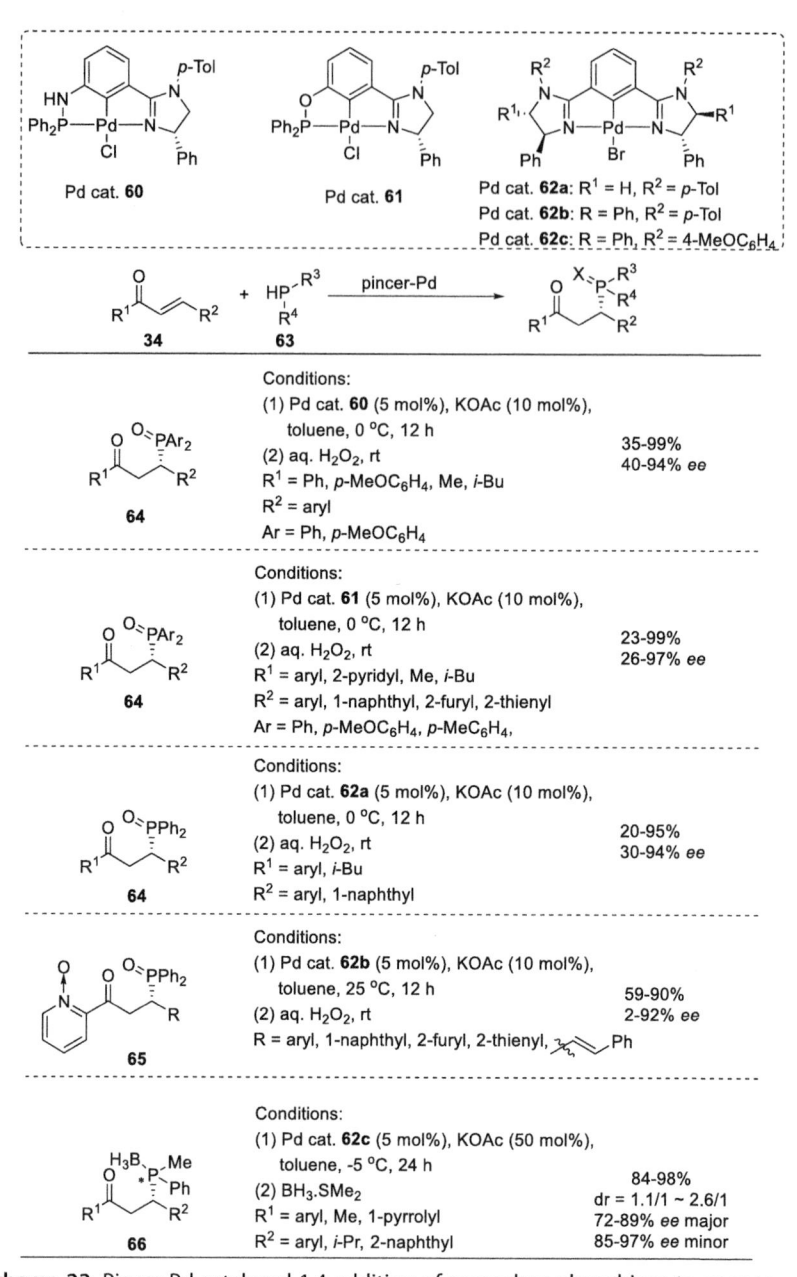

Scheme 23 Pincer-Pd catalyzed 1,4-addition of secondary phosphines to enones.

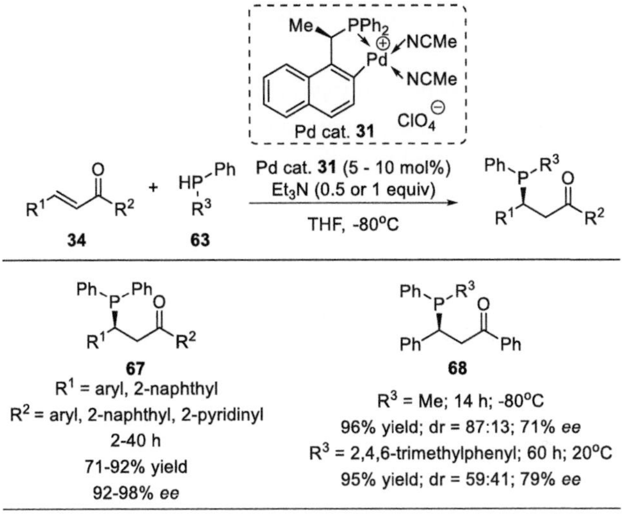

Scheme 24 C,P-palladacycle catalyzed 1,4-addition of secondary phosphines to enones.

Furthermore, this approach was applied to the synthesis of chiral diphosphines via (*R*)-C,P-palladacycle **31** catalyzed hydrophosphination of dienones (Scheme 25).[36a] The reaction tolerates a variety of aryl or heteroaryl substituted dienones **69** and gives chiral PCP-pincer ligands **70** almost in >99% *ee*. Later, they reported a diastereo- and enantioselective (*S*)-C,P-palladacycle **71** catalyzed stepwise double hydrophosphination of

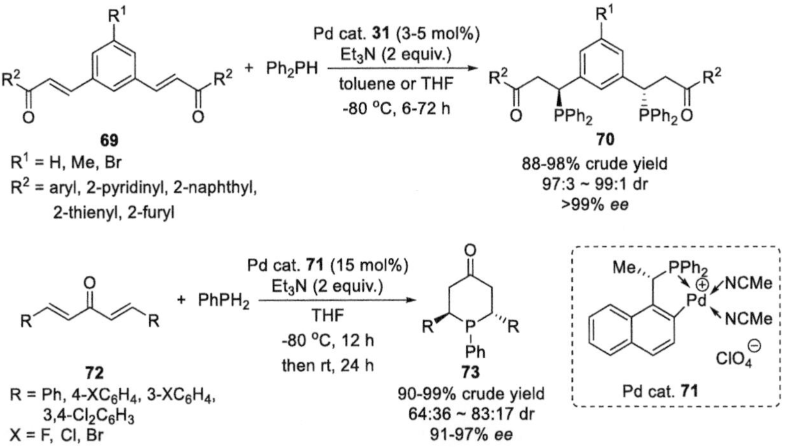

Scheme 25 C,P-palladacycle catalyzed 1,4-addition of PhPH$_2$ to dienones or bis(enones).

bis(enones) **72** to give chiral P-heterocycles **73** with excellent enantioselectivities (Scheme 25).[36b] However, this transformation is limited to bis(enones) with electron-withdrawing aryl substituents.

The asymmetric hydrophosphination reactions can be extended to synthesize C-chiral monophosphines from α,β-unsaturated ketimines **75** or (E)-3-methyl-4-nitro-5-alkenylisoxazoles **77** under the C,P-palladacycle catalyst **71** (Scheme 26(1) and (2)).[37a,b] Interestingly, N,C-palladacycle complex **74** is able to efficiently catalyze the enantioselective hydrophosphination of ferrocenyl enones **79** with Ph$_2$PH (Scheme 26 (3)).[37c] The ferrocenyl phosphine **80** has been used as substrates for the synthesis of ferrocenyl phosphapalladacycle via the diastereospecific *ortho*palladation.

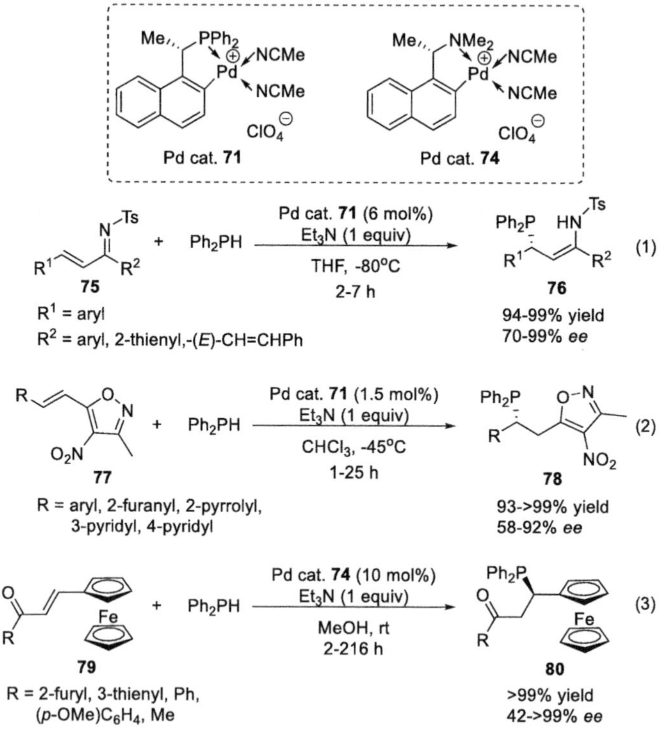

Scheme 26 Palladacycle catalyzed 1,4-addition of Ph$_2$PH to electron-deficient alkenes.

Moreover, enantioselective hydrophosphinations of activated electron poor alkenes including α,β-unsaturated substituted esters, amides, and N-enoylbenzotriazoles have been described (Scheme 27).[38]

The corresponding adducts **81–83** were obtained in high yield and excellent *ee*s. The benzotriazole substituted adducts **83** allows further functionalization and coordinated to gold, which exhibits valuable tumor suppression activity.

It is known that asymmetric 1,6-addition of conjugated dienes is more challenging compared to 1,4-addition, due to the difficulty in controlling the regioselectivity of the addition reaction.

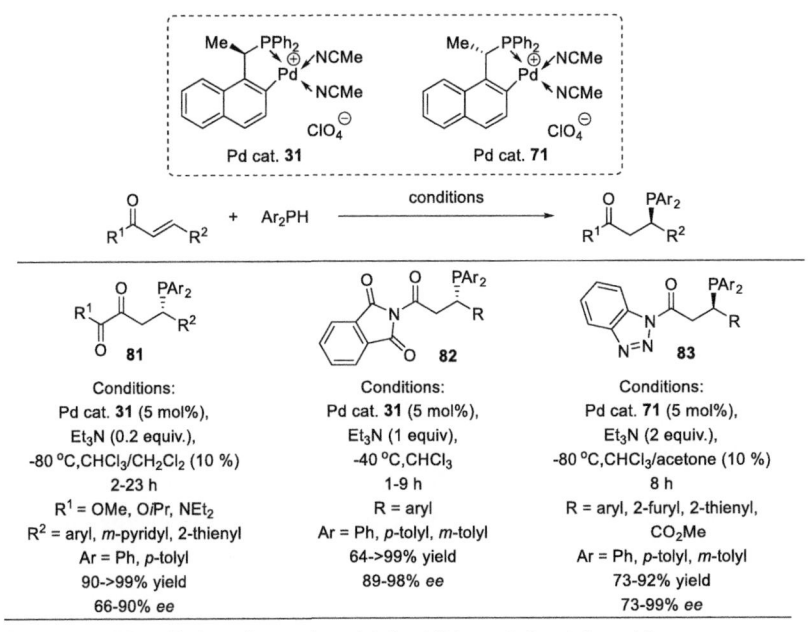

Scheme 27 C,P-palladacycle catalyzed 1,4-addition of diarylphosphines to enones.

The highly regioselective construction of C—P bonds via an asymmetric conjugate addition to dienes becomes inherently difficult. It was shown by Duan and co-workers that diarylphosphines reacts with $\alpha,\beta,\gamma,\delta$-unsaturated bisphosphate esters or sulfonic esters to produce chiral allylic phosphine derivatives **85** and **86** in the presence of pincer-PdOAc catalyst **48** at −60 °C (Scheme 28).[39] It was found that the reaction favored 1,6-selectivity when the EWG of the dienes is bulky enough. For $\alpha,\beta,\gamma,\delta$-unsaturated bisphosphate esters bearing a substituent at the *ortho*-position of aryl (R), the 1,6-adducts **85** could be obtained with good yields (78–84% yield) and *ee* values (86–91% *ee*).

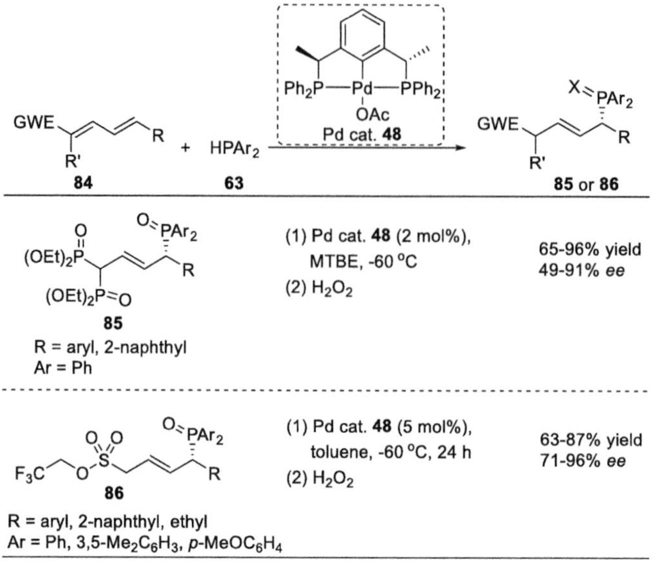

Scheme 28 Chiral PCP-pincer Pd-catalyzed 1,6-addition of diarylphosphines to electron-deficient dienes.

A similar PCP-pincer Pd catalyst **88** was used for the regioselective addition of diphenylphosphine to α,β,γ,δ-unsaturated malonate ester **87** by Leung and co-workers, providing 1,6-adduct **89** albeit in a low *ee* value (Scheme 29).[40] In their later work, the enantioselectivity of the 1,6-adduct **89** increased to >99% ee with 81% yield in the presence of a modified chiral PCP-pincer Pd catalyst **92**.[41a] Interestingly, they showed that the regioselectivity of the addition reaction was switched to provide 1,4-adducts **90** as the sole product with >99% *ee* by using a phosphapalladacycle catalyst **71**. The results indicated that the regioselectivity as well as stereoselectivity of these transformations could be controlled by the choice of the ligand associated with the Pd catalyst.

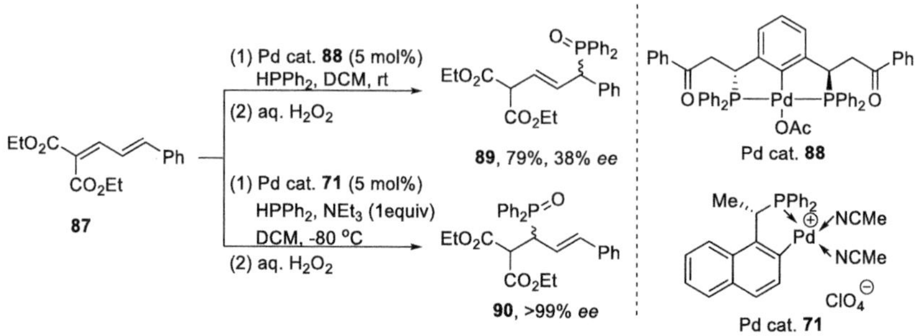

Scheme 29 Regioselective hydrophosphinations of electron-deficient dienes.

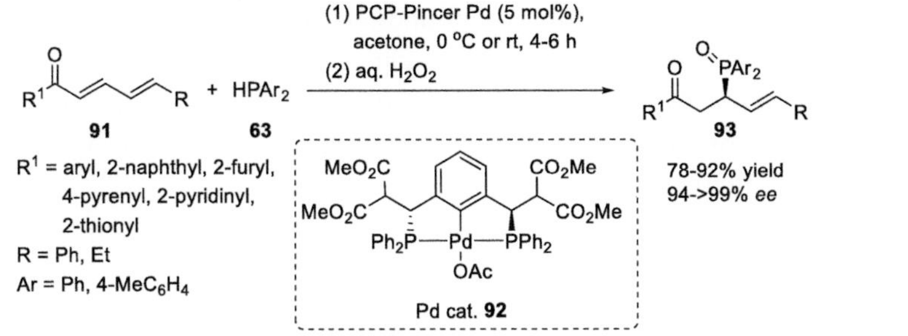

Advances in Pd-catalyzed asymmetric addition reactions 349

Scheme 30 Chiral PCP-pincer Pd-catalyzed 1,4-addition of diarylphosphines to dienones.

Furthermore, they demonstrated an analogous transformation leading to 1,4-adducts as the major product from α,β,γ,δ-unsaturated dienones **91** and diarylphosphines **63**, which was catalyzed by ester-functionalized PCP-pincer Pd **92** (Scheme 30).[41] It is worth mentioning that the asymmetric hydrophosphination reaction tolerates a wide range of substituted dienones **91**, affording chiral allylic phosphines **93** in good yields and excellent *ee*.

2.2.3 Palladium-catalyzed asymmetric conjugate addition of other substrates

A great interest for the development of 1,4-addition of aryltriethylsiloxanes,[42] amines[43] and other substrates as nucleophiles has been extensively described. Kim and co-workers reported the enantioselective 1,4-addition of aromatic amines **95** to fumarates **94** using a chiral cationic BINAP-palladium complex **96** (Scheme 31).[44] The reaction provides a variety of aspartic acid derivatives **97** in 80–95% yields with 83–96% *ee*s. In the same year, Sodeoka and co-workers described the bifunctional chiral Pd complex-catalyzed reaction of aromatic amine salts and *N*-benzyloxycarbonyl-protected acrylamides **98**

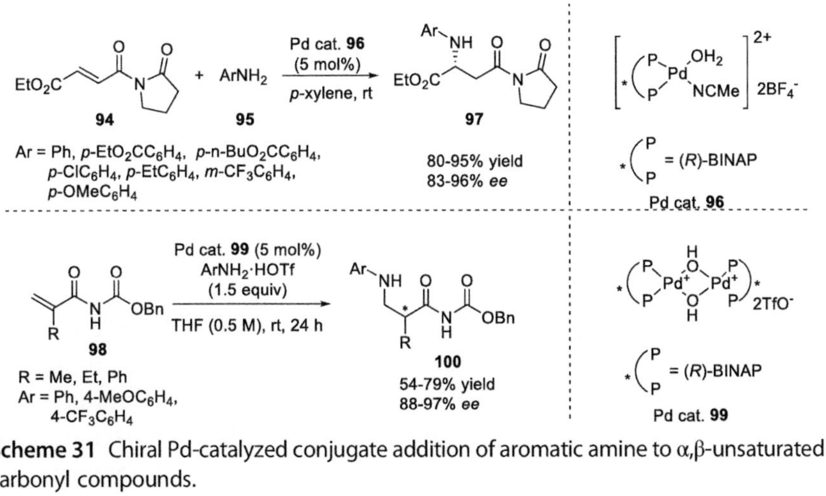

Scheme 31 Chiral Pd-catalyzed conjugate addition of aromatic amine to α,β-unsaturated carbonyl compounds.

to produce β-amino carbonyl compounds **100** with a C-stereogenic center at the α-position under mild reaction conditions.[45] The formed chiral β-amino acid derivatives are obtained in good yields and excellent *ee*s. It is noteworthy that the reaction of electron-deficient amines gave better yields by adding the corresponding free amine.

The conjugate addition of carbon nucleophiles to electron-deficient alkenes has been a powerful tool to form carbon-carbon bond. Sodeoka and co-workers developed an enantioselective addition reaction of α-substituted β-ketoesters **101** to methyl vinyl ketone **102** proceeding in the presence of an air and moisture-stable chiral Pd aqua complex **103** (Scheme 32).[46] This reaction provided the adducts **104** with a chiral quaternary carbon center in high yields and enantioselectivities. Mechanistically, the Pd aqua complex plays a double role as an acid-base catalyst, activating the enones as a protic acid and releasing the chiral Pd enolate. The scope of the reaction can be extended to α,β-unsaturated aldehydes, affording the desired acetals with good to excellent *ee* (87–99%) and moderate diastereoselectivity (dr = 3.8:1).

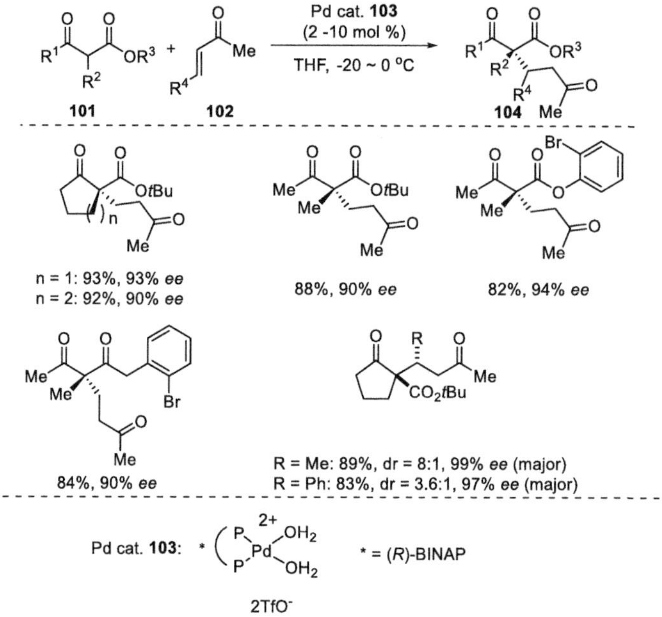

Scheme 32 Chiral Pd aqua complex-catalyzed conjugate addition of β-ketoesters to enones.

Azlactones appeared to be suitable substrates as well for Pd-catalyzed asymmetric 1,4-additions. Peter and co-workers showed that the dimeric precatalyst [FBIP-Cl]$_2$, activated by using silver salt promoted the asymmetric conjugate additions of azlactones **105** and enones **34**, affording the desired adducts **106** with an adjacent quaternary and tertiary stereocenter in moderate to good yields with up to 99% *ee* and >98:2 dr (Scheme 33,

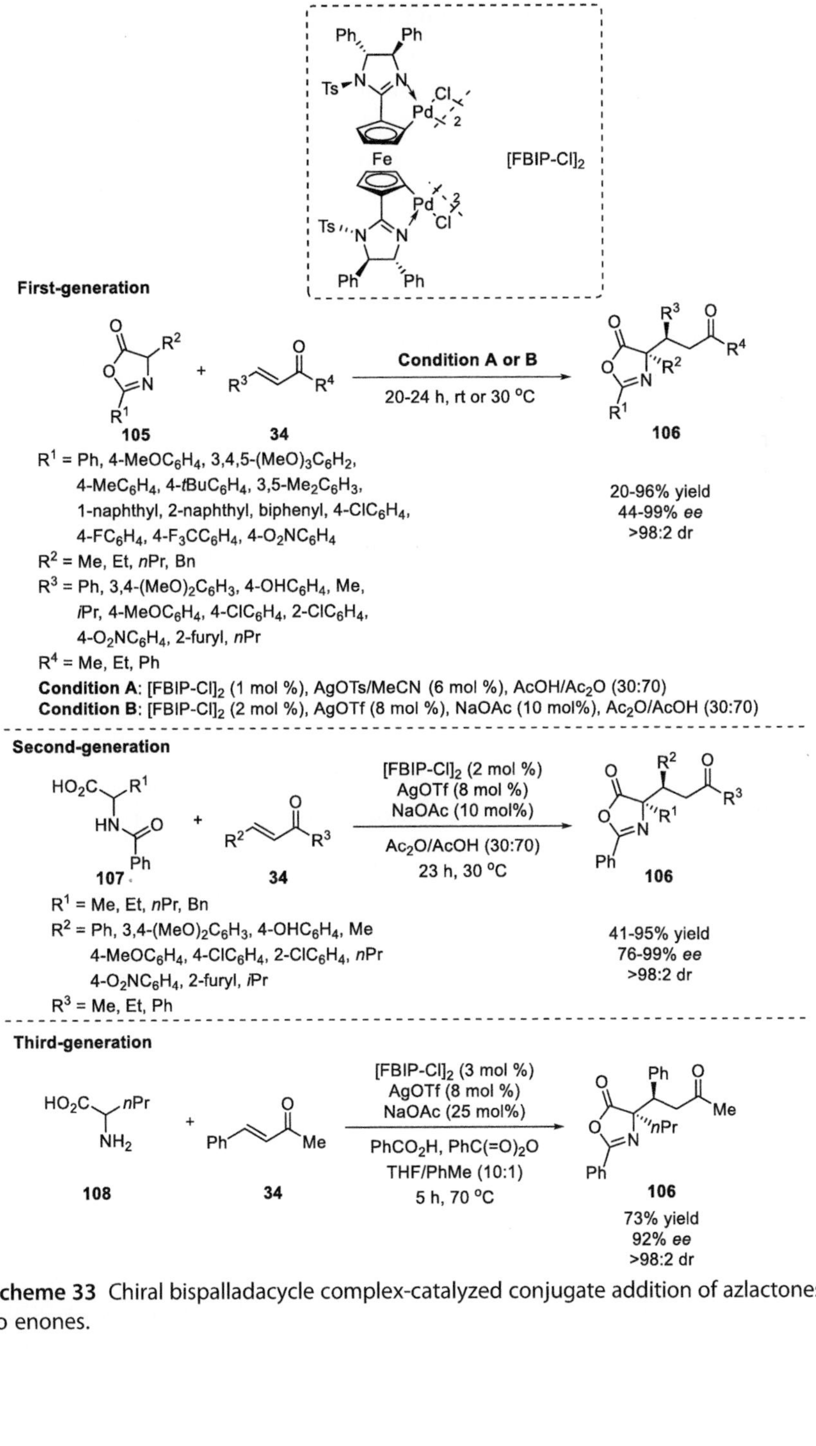

Advances in Pd-catalyzed asymmetric addition reactions   351

First-generation

[FBIP-Cl]$_2$

105 + **34** → **106**

Condition A or B, 20-24 h, rt or 30 °C

R^1 = Ph, 4-MeOC$_6$H$_4$, 3,4,5-(MeO)$_3$C$_6$H$_2$,
4-MeC$_6$H$_4$, 4-tBuC$_6$H$_4$, 3,5-Me$_2$C$_6$H$_3$,
1-naphthyl, 2-naphthyl, biphenyl, 4-ClC$_6$H$_4$,
4-FC$_6$H$_4$, 4-F$_3$CC$_6$H$_4$, 4-O$_2$NC$_6$H$_4$

R^2 = Me, Et, nPr, Bn

R^3 = Ph, 3,4-(MeO)$_2$C$_6$H$_3$, 4-OHC$_6$H$_4$, Me,
iPr, 4-MeOC$_6$H$_4$, 4-ClC$_6$H$_4$, 2-ClC$_6$H$_4$,
4-O$_2$NC$_6$H$_4$, 2-furyl, nPr

R^4 = Me, Et, Ph

20-96% yield
44-99% ee
>98:2 dr

Condition A: [FBIP-Cl]$_2$ (1 mol %), AgOTs/MeCN (6 mol %), AcOH/Ac$_2$O (30:70)
Condition B: [FBIP-Cl]$_2$ (2 mol %), AgOTf (8 mol %), NaOAc (10 mol%), Ac$_2$O/AcOH (30:70)

Second-generation

107 + **34** → **106**

[FBIP-Cl]$_2$ (2 mol %)
AgOTf (8 mol %)
NaOAc (10 mol%)
Ac$_2$O/AcOH (30:70)
23 h, 30 °C

R^1 = Me, Et, nPr, Bn
R^2 = Ph, 3,4-(MeO)$_2$C$_6$H$_3$, 4-OHC$_6$H$_4$, Me
4-MeOC$_6$H$_4$, 4-ClC$_6$H$_4$, 2-ClC$_6$H$_4$, nPr
4-O$_2$NC$_6$H$_4$, 2-furyl, iPr
R^3 = Me, Et, Ph

41-95% yield
76-99% ee
>98:2 dr

Third-generation

108 + **34** → **106**

[FBIP-Cl]$_2$ (3 mol %)
AgOTf (8 mol %)
NaOAc (25 mol%)
PhCO$_2$H, PhC(=O)$_2$O
THF/PhMe (10:1)
5 h, 70 °C

73% yield
92% ee
>98:2 dr

Scheme 33 Chiral bispalladacycle complex-catalyzed conjugate addition of azlactones to enones.

first-generation).[47] Later, they found that azlactones formed in situ from racemic N-benzoylated tertiary amino acids **107** exhibited excellent activities and enantioselectivities in the conjugate addition to enones (Scheme 33, second-generation). Furthermore, the reaction with unprotected racemic α-amino acids **108** also succeeded via a tandem N- and O-amino acid acylation/cyclization/asymmetric addition process, giving the corresponding product **106** in 73% yield with 92% ee and >98:2 dr (Scheme 33, third-generation).[48] In these transformations, the cooperative activation by a chiral ferrocene bispalladacycle catalyst, a Brønsted acid and a Brønsted base is essential to achieve high catalytic efficiency and excellent asymmetric induction. The above approaches provide a rapid method for the preparation of various unnatural quaternary amino acid derivatives.

Later, the same group developed a tandem double Michael addition of divinylketones **110** and an in situ generated azlactone from N-benzoyl glycine **109** in the presence of planar chiral ferrocene bispalladacycle catalyst [FBIP-Cl]$_2$ (Scheme 34).[49] The spirocyclic azlactones **111** with three contiguous stereocenters were produced with moderate diastereoselectivities and good enantioselectivities.

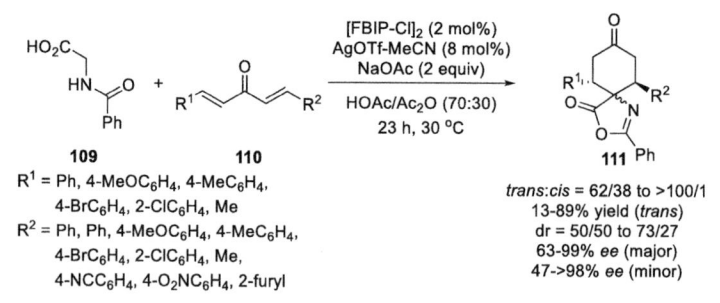

Scheme 34 Tandem double conjugate addition to form the spirocyclic azlactones.

Moreover, planar chiral ferrocene bispalladacycle [FBI-Cl]$_2$ was applied to the cascade reaction of N-benzoyl α-amino acids **112** and nitroolefins **39** (Scheme 35).[50] This reaction produces the corresponding quaternary amino succinimides **113** through a 1,4-addition/Nef-type process. It is noteworthy that the aid of Mn(OAc)$_2$ is crucial for high stereoselectivity to avoid epimerization.

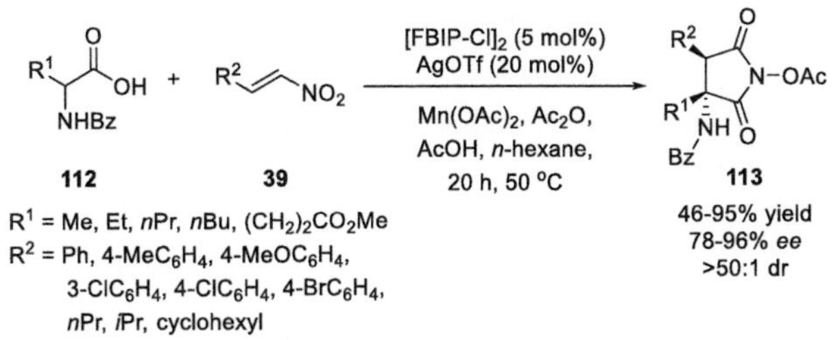

Scheme 35 Forming the quaternary amino succinimides via cascade 1,4-addition/Nef-type reaction.

In 2011, a chiral dicationic Pd-catalyzed Friedel-Crafts (F—C) alkylation of indole/pyrrole was reported by Mikami and co-workers (Scheme 36).[51] It was shown that the reaction of indoles and β,γ-unsaturated α-ketoesters **115** in the presence of (R)-DTBM-SEGPHOS-Pd proceeded smoothly to give the corresponding products **116** in 16–94% yields with 12–89% ees. Similarly, N-methyl pyrrole can also undergo a F—C alkylation into the corresponding 1,4-addition products **119** in 60–97% yields with 84–99% ees. It is important to point out that the 1,4-metal chelated species

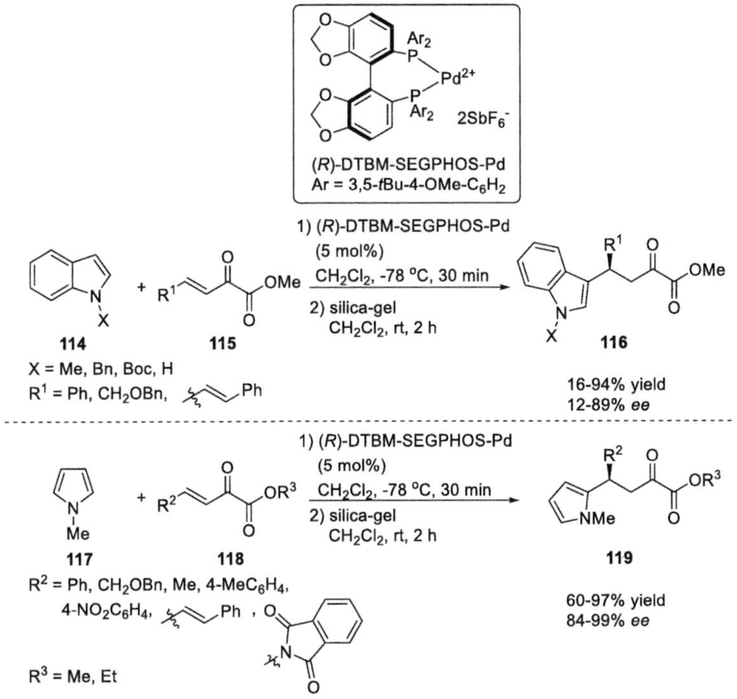

Scheme 36 Chiral Pd-catalyzed asymmetric Friedel-Crafts alkylation.

formed from α-ketoesters and chiral dicationic Pd catalyst is crucial to form the products with a good asymmetric induction. These adducts could be further transformed into functionalized α-hydroxyl esters via a catalytic carbonyl-ene reaction.

A very mild chiral Pd-catalyzed version for the asymmetric Friedel-Crafts alkylation of indoles and α,β-unsaturated carbonyl compounds was developed by Kim and co-workers (Scheme 37).[52] It was found that

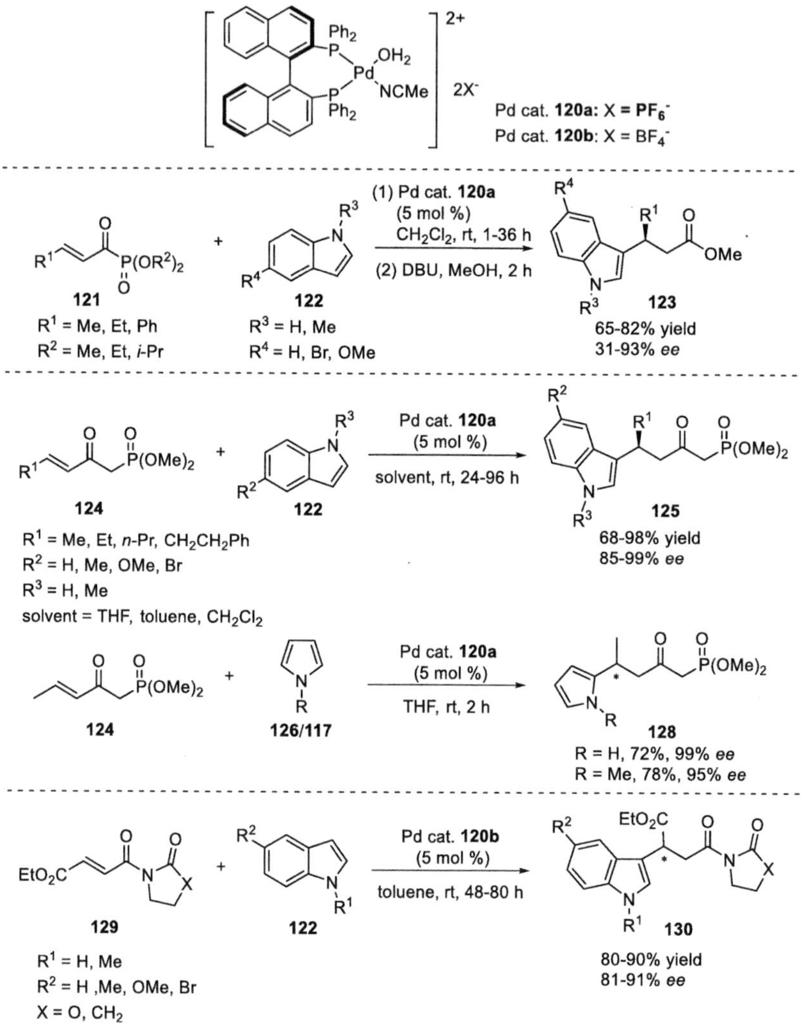

Scheme 37 Chiral Pd-catalyzed Friedel-Crafts alkylation of various α,β-unsaturated carbonyl compounds.

β,γ-unsaturated α-keto phosphonates **121** have been proven to be efficient substrates for asymmetric Friedel-Crafts alkylation reactions. The reactions lead to the intermediate γ-indolyl-α-keto phosphonates that are subsequently converted into the corresponding methyl esters **123** in 65–82% yields with 31–93% *ee*s via direct addition of methanol and DBU.

Furthermore, the same group extended this chemistry to chiral dicationic palladium **120a** mediated the F—C alkylation of indoles with γ,δ-unsaturated β-keto phosphonates **124**, which provided the adducts **125** in excellent yields and *ee*s. However, in this transformation dimethyl 2-oxo-4-phenyl-but-3-enylphosphonate (R^1 = Ph) failed to produce the alkylation product. In addition to indoles, pyrroles can also be used, and in this case the corresponding adducts **128** were obtained in good yields and excellent *ee*s. Later, the catalytic asymmetric Friedel-Crafts alkylation reaction of indoles with fumarate derivatives **129** was tested utilizing air- and moisture-stable chiral dicationic palladium **120b** as the catalyst at room temperature. The alkylated products **130** were generated in 80–90% yields with 81–91% *ee*s.

Kobayashi and co-workers have made progress by performing Pd(II)/2,2′-bipyridine-catalyzed asymmetric F—C alkylation reactions of indoles **131** with α,β-unsaturated ketones **34** by adding an anionic surfactant SDS and an additive PhNMe$_2$ in water for the synthesis of the adducts with up to 91% *ee* (Scheme 38).[53] In these transformations, α,β-unsaturated ketones **34** with an alkyl group substituent on the ketone moiety or a phenyl group substituent on the alkene moiety (R^1 = *t*Bu or R^2 = Ph) afforded the corresponding products **133** in very low enantioselectivities. Indoles **131** with the methyl group at the C2- or C7-position provided racemic products.

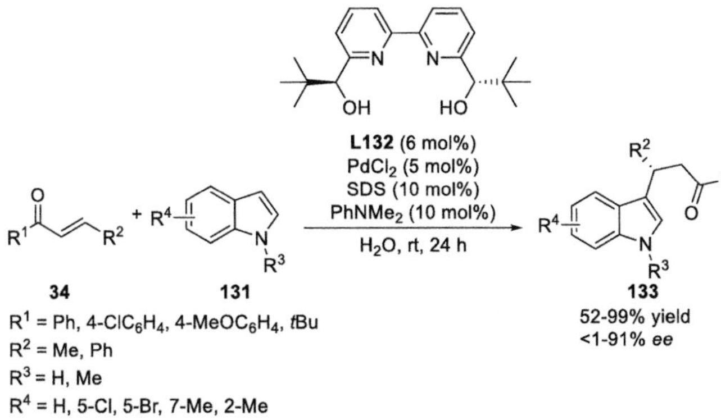

Scheme 38 Pd-catalyzed asymmetric Friedel-Crafts alkylation of indoles and α,β-unsaturated ketones.

3. Palladium-catalyzed asymmetric 1,2-addition

The 1,2-addition reactions of carbon and heteroatom nucleophiles to C=X (X=O, N, C) have emerged as one of the most important methods to from carbon-carbon or carbon-heteroatom bond. Significant efforts have been made for developing palladium-catalyzed asymmetric 1,2-addition to aldehyde, ketones, imines, unactivated olefins and allenes.[54] Compared with aldehydes, ketones and imines, unactivated olefins are more challenging in the asymmetric 1,2-addition reactions due to their lower reactivity. To solve the reactivity and stereoselectivity problems, several novel catalytic systems and new reaction modes were developed.

3.1 Palladium-catalyzed asymmetric 1,2-addition to aldehydes/ketones

The catalytic asymmetric 1,2-addition to aldehydes or ketones is an efficient C—C bond-forming process, which provided enantiomerically enriched secondary or tertiary alcohols. Although some successful asymmetric additions of nucleophiles and isatins have been established,[55] the Pd-catalyzed version remains limited. In 2009, Qin and co-workers reported the synthesis of 3-aryl-3-hydroxyoxindoles **136** via the Pd(OAc)$_2$/(Ra,S)-phosphinoimine ligand **135** catalyzed asymmetric 1,2-addition of arylboronic acids to N-benzylisatin **134** in the presence of 4 equiv. of BF$_3$·Et$_2$O (Scheme 39).[56] The products **136** were obtained in 51–78% yields with 38–73% *ees*. Subsequently, Shi and co-workers employed a chiral cationic N-heterocyclic carbene Pd^{2+} diaqua complex as catalyst to promote the asymmetric arylation of isatins **137**, leading to the corresponding adducts **138** in 79–94% yields with 50–94% *ees*.[57] In this protocol, the in situ generated LiOAr from aromatic alcohols and lithium hydroxide acted as an additive to benefit the catalytic cycles through coordination of LiOAr to the carbonyl groups of isatins. Further improvement of Pd-catalyzed asymmetric 1,2-addition of arylboronic acids to isatins could be based on the development of a novel catalytic system using chiral phosphine-oxazolines **140** as ligands.[58] In addition, the enantiopurity of 3-aryl-3-hydroxyoxindoles can be increased up to 98% via a simple recrystallization.

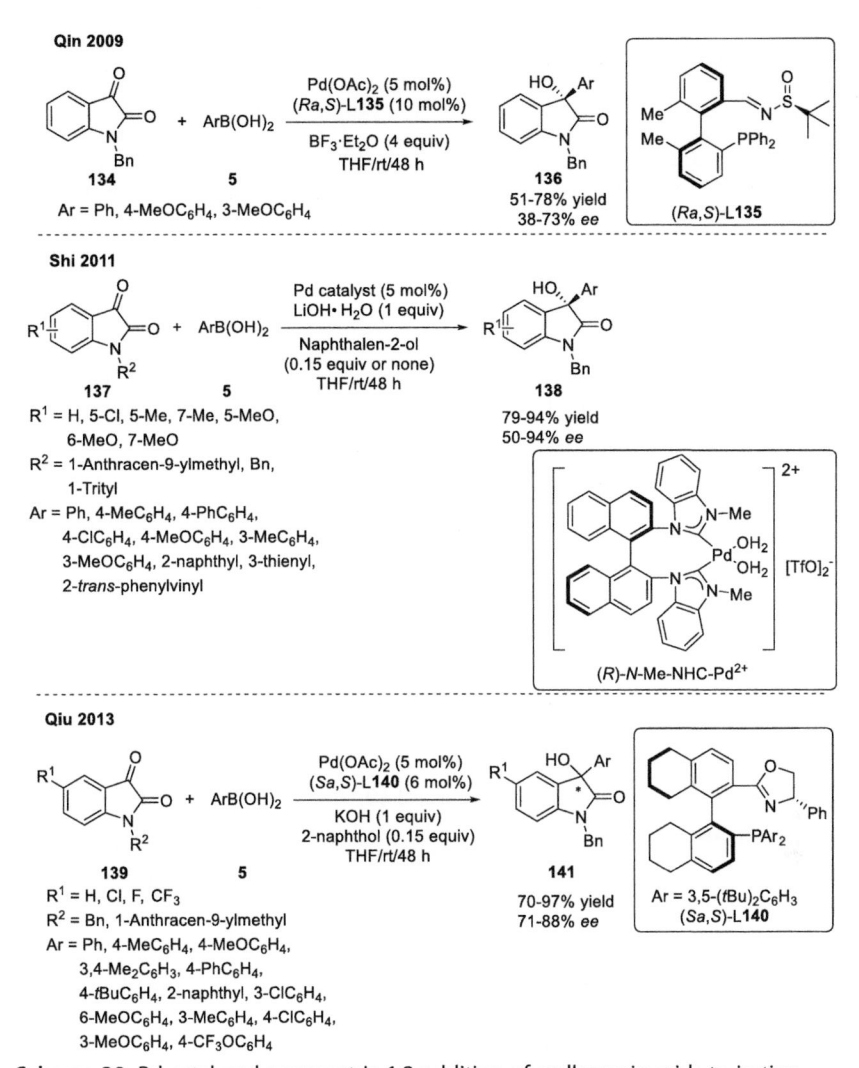

Scheme 39 Pd-catalyzed asymmetric 1,2-addition of arylboronic acids to isatins.

Chiral *N*-heterocyclic carbene Pd(II) complexes **143a** and **143b** synthesized from (*R*)-BINAM or (*R*)-H$_8$-BINAM were found to be fairly efficient catalysts to achieve the asymmetric 1,2-addition of arylboronic acids and aldehydes under mild conditions (Scheme 40).[59] In this way the corresponding chiral secondary alcohols **144** can be synthesized in good yields with moderate enantioselectivities. However, electron-donating arylaldehydes **142** led to the corresponding products in 0–1% *ees*.

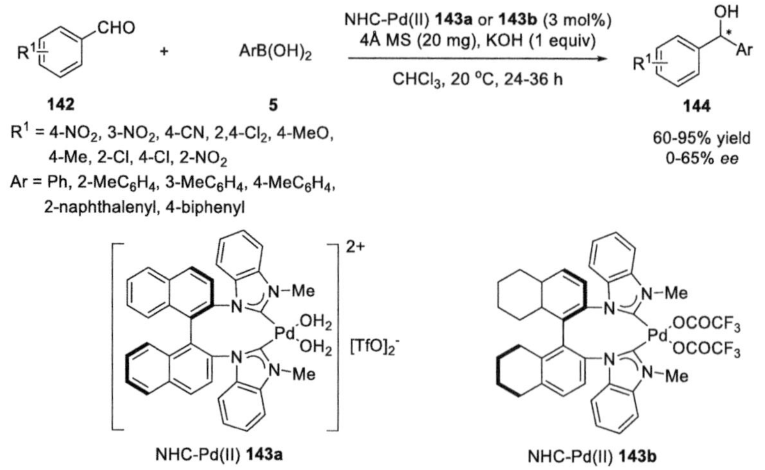

Scheme 40 Pd-catalyzed asymmetric 1,2-addition of arylboronic acids to arylaldehydes.

Mikami and co-workers developed a series of dicationic Pd-catalyzed asymmetric addition of enol silyl ethers to ketones.[60] The reaction of acetone silyl enol ether **147** and *N*-benzyl-protected isatins **146** proceeded smoothly using (*R*)-SEGPHOS as the ligand, furnishing the corresponding ene products **148** in high yields with excellent *ee*s (Scheme 41). Moreover, for the trimethylsilyl ketene thioacetal **149**, the aldol products **150** were obtained in moderate to excellent yields and good *ee*s after desilylation by treatment with 1 N HCl/THF. It should be noted that isatins with electron-withdrawing substituents in the *ortho* or *para* positions of the amide as an electrophile were completely unreactive under the optimized conditions. With these catalytic systems, keto esters and diketone derivatives **151** were found to be suitable electrophiles, providing the desired aldol products **153** in moderate to excellent diastereoselectivities and enantioselectivities in the presence of dicationic (*R*)-BINAP-Pd complex **145b**. Generally, bidentate coordination of diketones to the palladium center induced a high enantioselectivity and increased the reactivity in all cases.

The same authors also described direct addition of vinylsilane to ethyl glyoxylate **155** in the presence of dicationic (*S*)-BINAP-Pd complex (Scheme 42).[61] Optically active allylic alcohols **156** were obtained in 55–91% yield with 91–99% *ee*. Interestingly, the reaction could be extended to dienylsilanes and trienylsilanes, with excellent enantioselectivity outcomes.

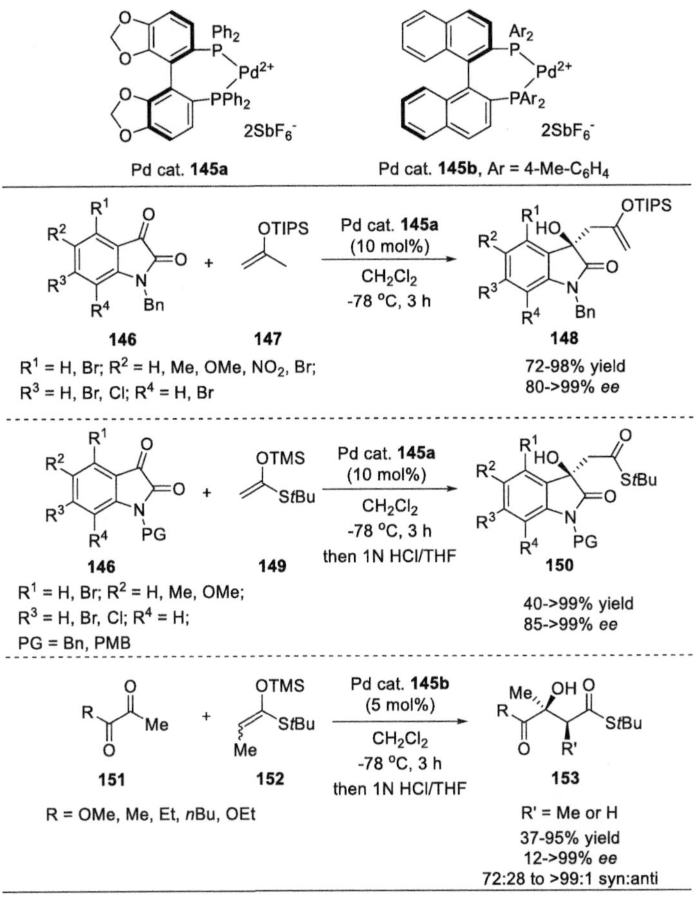

Scheme 41 Pd-catalyzed asymmetric ene reactions and Aldol reactions.

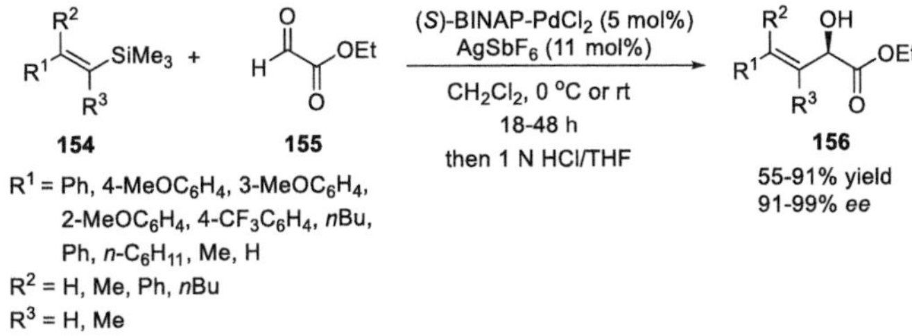

Scheme 42 Pd-catalyzed asymmetric alkenylation of ethyl glyoxylates.

Subsequent studies showed that dicationic (S)-BINAP-Pd complex is also efficient for the asymmetric addition of alkynylsilanes **157** and trifluoropyruvate **158** (Scheme 43).[62] Various alkynylsilanes with a TMS substituent gave α-trifluoromethyl-substituted tertiary alcohols with excellent *ee*s. Notably, the alkynyl products **159** can be efficiently converted into a chiral allene with a little loss in enantioselectivity. This method was further applied to polyynes, delivering the corresponding adducts with 96- >99% *ee*.

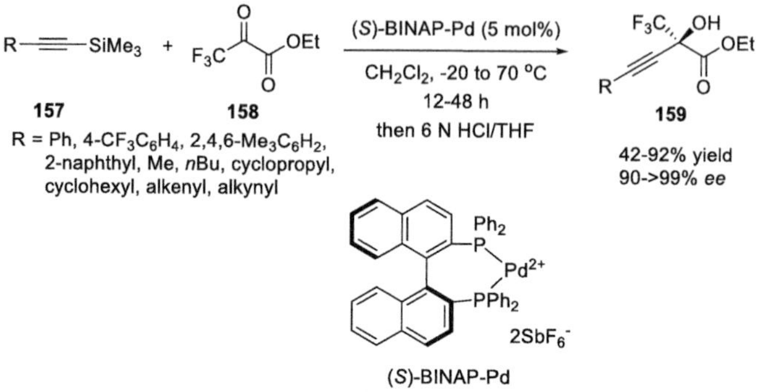

Scheme 43 Pd-catalyzed asymmetric alkynylation of trifluoropyruvate.

The palladium-catalyzed asymmetric Friedel-Crafts type reaction is an efficient approach to the formation of C—C bonds. Notably, furan, thiophene, and pyrrole derivatives as nucleophiles are able to react with trifluoropyruvate **158** to provide the Friedel-Crafts products **161** using in situ prepared dicationic Pd catalyst (Scheme 44).[63] A mixture of products with subtituents at the 2- and 3-position were observed when the *N*-Boc protected pyrrole was used. Very interestingly, in the presence of ethyl glyoxylate **155**, the reaction led to another regioselectivity with 2-trimethylsilyl furan, thiophene, and pyrrole derivatives allowing the formation of heteroarylated products **163** in good yields with high *ee*s (Scheme 45).

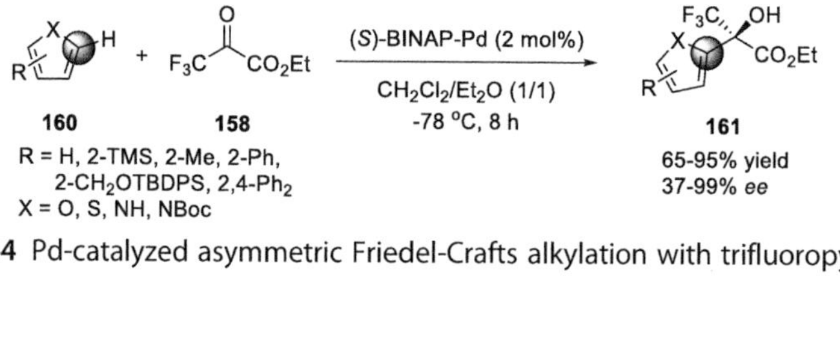

Scheme 44 Pd-catalyzed asymmetric Friedel-Crafts alkylation with trifluoropyruvate.

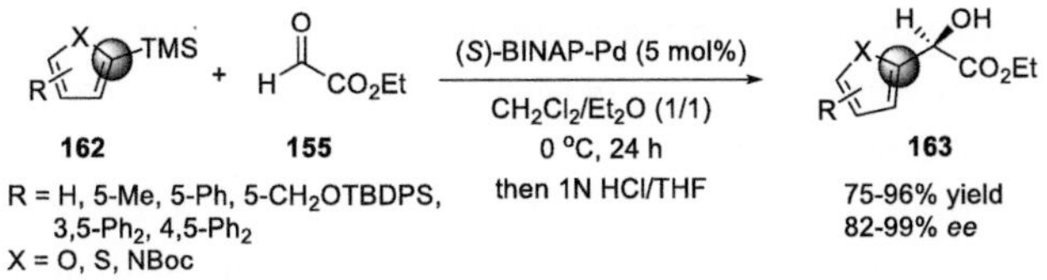

Scheme 45 Pd-catalyzed asymmetric Friedel-Crafts type reaction with ethyl glyoxylate.

Many efforts have been devoted to the Pd-catalyzed asymmetric additions of allyl to aldehydes and ketones via umpolung reactions of π-allyl palladium complexes with Et$_2$Zn, Et$_3$B and other activating reagents.[64] The group of Shi described an asymmetric umpolung Tsuji-Trost reaction of cyclohex-2-enyl acetate **164** with aldehydes **142** in the presence of 10 mol% of chiral bis(NHC)-Pd(II) complex **165** and 3.5 equiv. of Et$_2$Zn, giving the corresponding alcohols **166** in good yields (58–96%) with moderate enantioselectivities (54–66% ee) and good *syn* diastereoselectivities (84:16 to >99:1) (Scheme 46, top).[65] In this case, acyclic cinnamyl acetate afforded the corresponding alcohols in 72% yield and 55:45 *syn*:*anti* with 1.8% ee (*anti*) and 0.5% ee (*syn*). Furthermore, the authors explored the use of chiral phosphine-schiff base type ligands **167** in the umpolung allylation of aldehydes **142** with cyclohex-2-enyl acetate **164**

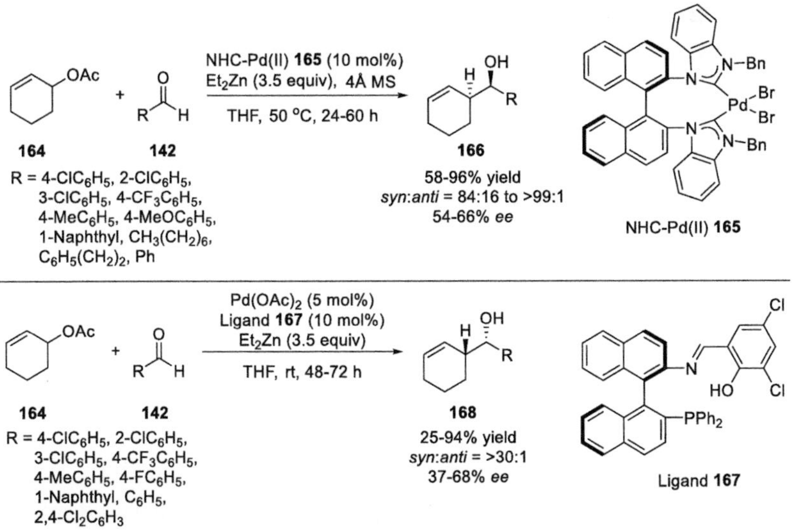

Scheme 46 Pd-catalyzed Et$_2$Zn-mediated umpolung allylation of aldehydes with 2-cyclohexenyl acetate.

and homoallylic alcohols **168** were obtained in good yields (25–94%), with moderate enantioselectivities (37–68%) and excellent *syn* diastereoselectivities (>30:1) (Scheme 46, bottom).[66]

Later, Zhou and co-workers investigated the Pd-catalyzed enantioselective umpolung allylations of aldehydes **142** with acyclic allylic alcohols and their derivatives **169** in the presence of Et$_3$B or Et$_2$Zn (Scheme 47).[67] By using bulky monodentate spiro phosphite ligands **170**, a series of homoallylic alcohols **171** were obtained with good to excellent enantioselectivities. Different aldehydes **142**, including aromatic, aliphatic, and α,β-unsaturated aldehydes, can be reacted to give the corresponding chiral alcohols **171** in good yields. A proposed mechanism includes oxidative addition of allylic compound and Pd(0) leading to electrophilic η^3-allylpalladium species. Then an ethyl group from Et$_3$B or Et$_2$Zn is transferred to the palladium specie to form a η^1-allylpalladium complex, which is nucleophilic. The addition of η^1-allylpalladium complex to aldehydes gave the homoallylic palladium alcoholate via a six-member ring transition state. Subsequent transmetallation of the alcoholate and β-elimination regenerated the Pd(0) species and the homoallylic alcohols.

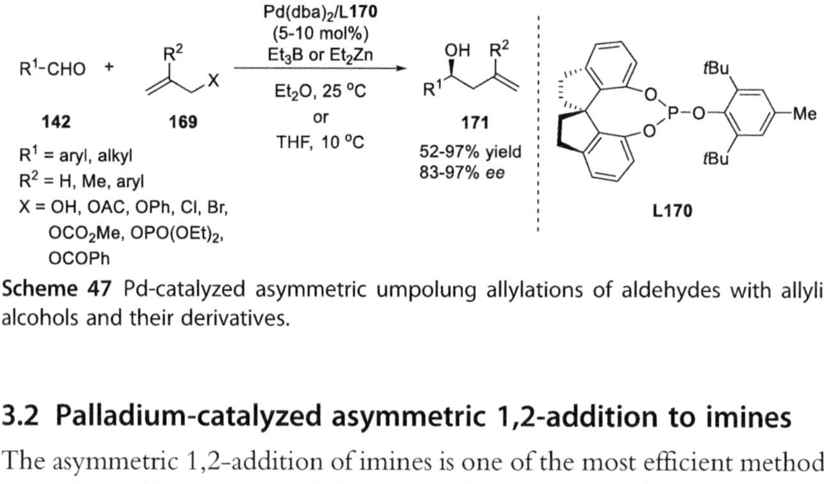

Scheme 47 Pd-catalyzed asymmetric umpolung allylations of aldehydes with allylic alcohols and their derivatives.

3.2 Palladium-catalyzed asymmetric 1,2-addition to imines

The asymmetric 1,2-addition of imines is one of the most efficient methods to produce chiral amines and these strategies are extensively applied in natural products synthesis and drug discovery. Although there are some difficulties involving the poor electrophilicity of azomethine carbon and the tendency of enolizable imines to undergo deprotonation, many efforts have been devoted to transition metal-catalyzed addition reactions of imines over the past years. In the pioneering work by Lu, the palladium-catalyzed addition of arylboronic acids to the C—N double bonds reveals a new way to afford the optically active amines.[68] For example, Lu reported a Pd(II)-catalyzed enantioselective addition of arylboronic acids to *N*-tosylaldimines **169** using (*S*)-pymox as ligand, providing diarylmethylamines **170** in moderate to good *ee* (Scheme 48).[68c] Since then, many studies have focused on the enantioselective addition of arylboronic acids to imines using various chiral ligands and palladium catalysts. In general, aldimines is more reactive than ketimines and some of electron-withdrawing substituted ketimines are as reactive as aldimines. Shi and co-workers developed a similar catalytic addition of arylboronic acids to *N*-tosylaldimines **169** using a NHC-Pd^{2+} diaqua catalyst **171** in the presence of 1 equiv. of K$_3$PO$_4$·3H$_2$O and powdered 4 Å MS in THF at 4 °C, which gives the adduct **170** in 64–99% yields and 60–94% *ees* (Scheme 48).[69] It should be noted that the reactions tolerate a wide substrate scope which enables the use of aromatic, heterocyclic or aliphatic *N*-tosyl imines and electron-donating or -withdrawing substituted arylboronic acids. This catalytic system was further applied in the arylation of *N*-Boc imines generated in situ from α-carbamoyl sulfones with arylboronic acids, giving the desired products in moderate to good yield and *ee*.[70]

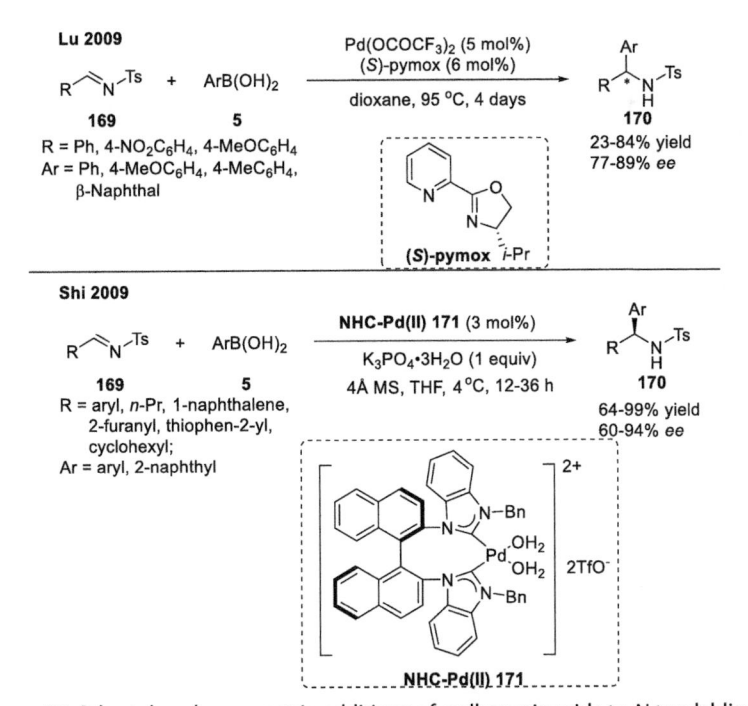

Scheme 48 Pd-catalyzed asymmetric additions of arylboronic acids to *N*-tosylaldimines.

Shi also disclosed that using indoles **122** as the nucleophiles in the asymmetric 1,2-addition to *N*-tosylaldimines **169** led to the desired products **173** in 71–89% yields with 24–66% *ee*s under a similar catalytic system (Scheme 49).[71] It was found that the addition of 4-nitrobenzoic acid provided the adduct in a higher *ee* and a small amount of achiral double-alkylation product.

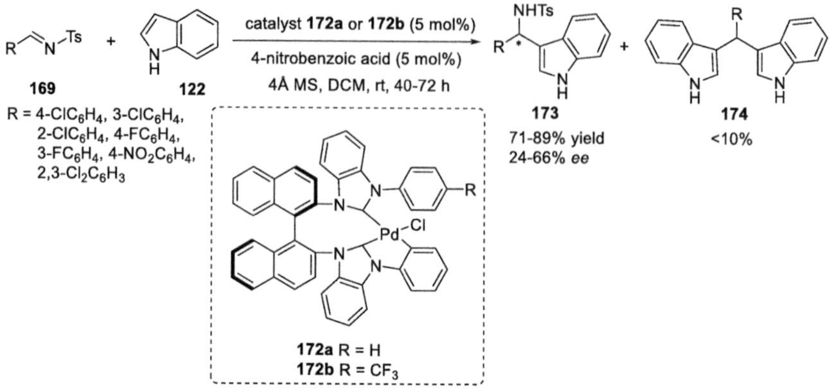

Scheme 49 Pd-catalyzed asymmetric additions of indoles to *N*-tosylaldimines.

In 2015, Peter and co-workers described a planar chiral ferrocene-derived palladacycle-catalyzed asymmetric arylation of imines **175** with arylboroxines **176** in the presence of 2 mol% AgOAc in chlorobenzene at 65 °C (Scheme 50).[72] Gratifyingly, a wide range of imines and arylboroxines with various substituents underwent the highly enantioselective 1,2-addition reactions, giving the corresponding benzylic amines **177** with excellent yields and enantioselectivities, except imine with a Ph(CH$_2$)$_2$ group on R^1, which produced the adduct in lower yield. It should be noted

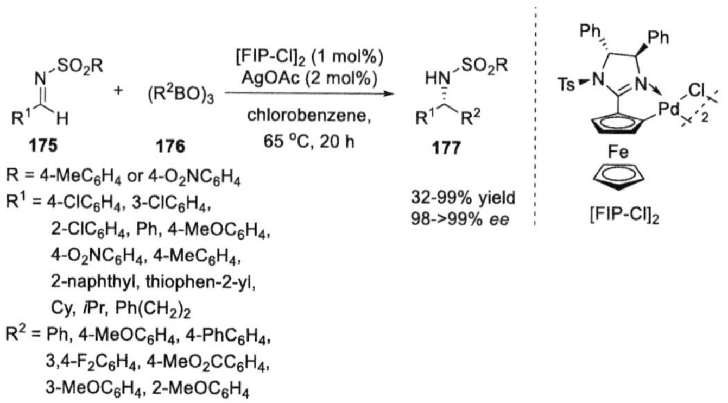

Scheme 50 Palladacycle-Catalyzed asymmetric 1,2-addition of imines and arylboroxines.

that the anionic ligand AcO⁻ increased the reactivity and avoided the imine hydrolysis and the addition of exogenous base to active the boroxines.

In 2012, Zeng developed a Pd(II)-catalyzed asymmetric addition of N-aryl-C-acylimines **178** and arylboronic acids **5** (Scheme 51).[73] After screening a series of ligands, chiral BOX ligand showed the best enantioselective induction ability. This method provides a valuable pathway for the synthesis of the arylglycine derivatives **180** in high yield with excellent ee, which can be further transformed to chiral amino acids and chiral amino alcohols.

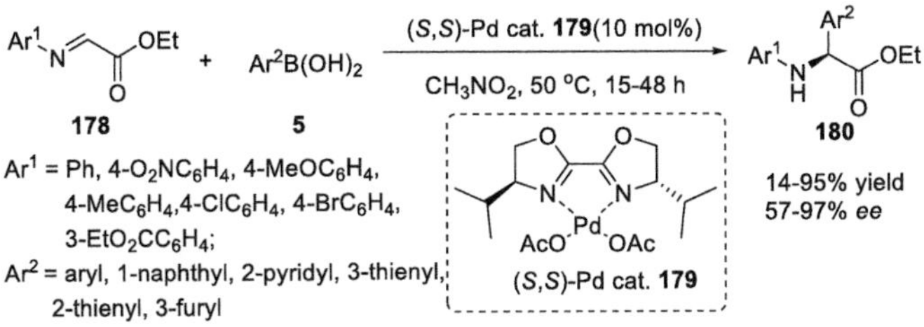

Scheme 51 Pd-catalyzed asymmetric additions of arylboronic acids to C-Acylimines.

Cyclic N-sulfonyl imines as electrophiles have attracted great attention in the asymmetric 1,2-addition reactions due to their high reactivity. Zhang and co-workers reported their elegant works on the Pd(II)-catalyzed enantioselective arylation of cyclic N-sulfonyl imines **181**.[74] The reaction of arylboronic acids and cyclic ketimines were carried out in unpurified TFE (trifluoroethanol) at 40–80 °C under air atmosphere using Pd(TFA)₂ as catalyst and tBu-Nicox as ligand (Scheme 52). The reactions are

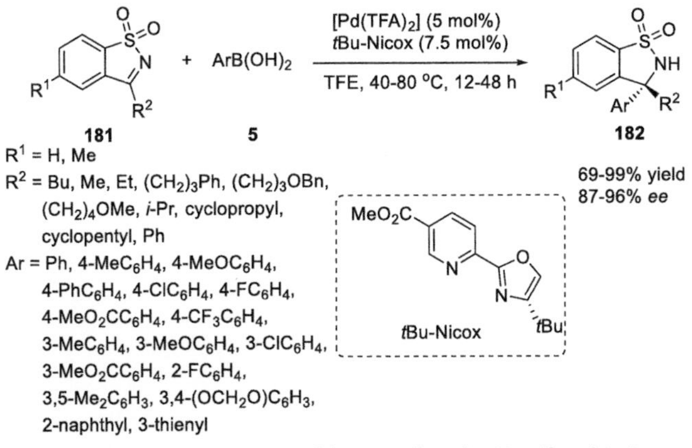

Scheme 52 Pd-catalyzed asymmetric additions of cyclic N-sulfonyl imines.

operationally simple and tolerate a wide range of arylboronic acids and cyclic ketimines, which produce adducts **182** with high enantioselectivities. Additionally, the chiral pyrrolidine compound with a α-tertiary amine motif can be synthesized in three steps with this method.

Subsequently, they also reported a Pd(II)/Nicox-catalyzed addition of six-membered cyclic *N*-sulfonyl ketimine esters **183** and arylboronic acids **5**, giving chiral amino esters **184** in excellent yields and enantioselectivities (Scheme 53).[75] On the basis of DFT calculations, it was found that the imine insertion step is the rate and stereoselectivity-determining step.

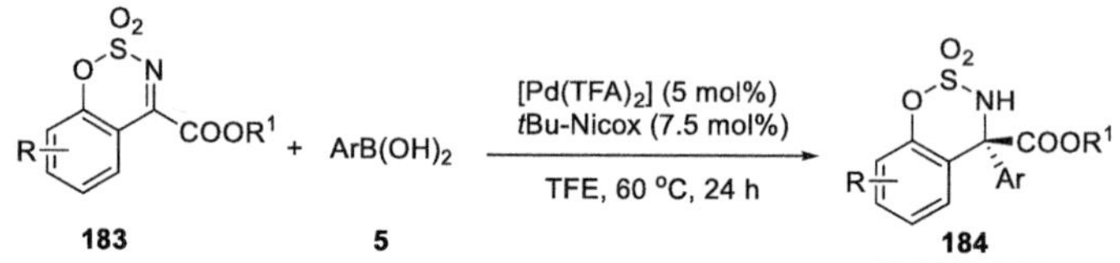

R = H, 2-Me, 3-Me, 4-Me, 4-Et, 3-OMe, 4-OMe, 3-Cl;
R^1 = Me, Et, *n*Pr;
Ar = Ph,3-MeC_6H_4, 4-MeC_6H_4, 4-$MeOC_6H_4$, 4-PhC_6H_4, 4-FC_6H_4, 4-ClC_6H_4, 4-BrC_6H_4, 4-$MeO_2CC_6H_4$, 2-Naphthyl, 3-Thienyl

Scheme 53 Pd-catalyzed asymmetric additions of cyclic *N*-sulfonyl ketimine esters.

Later, they developed the first Pd-catalyzed asymmetric addition of arylboronic acids to isatin derived imines **185**, which is similar to the previously mentioned Pd-catalyzed asymmetric addition reactions (Scheme 54).[76] Chiral 3-aryl-3-amino-2-oxindoles **186** can be obtained using Pd/In-Pyrox as catalyst in TFE at 70 °C under air atmosphere in good yield with excellent enantioselectivities. This method tolerates a wide range of 3-ketimino oxindoles **185** and arylboronic acids **5**.

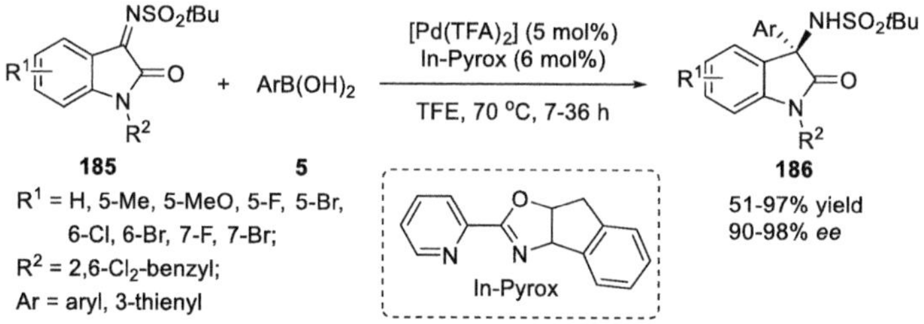

R^1 = H, 5-Me, 5-MeO, 5-F, 5-Br, 6-Cl, 6-Br, 7-F, 7-Br;
R^2 = 2,6-Cl_2-benzyl;
Ar = aryl, 3-thienyl

Scheme 54 Pd-catalyzed asymmetric additions of isatin derived imines.

Following Zhang's works on the construction of chiral α-polysubstituted amines by Pd-catalyzed asymmetric arylation of imines, several groups further developed the Pd-catalyzed 1,2-addition reaction of cyclic imines using phosphinooxazoline,[77] phosphine-imine,[78] or pyridinohydrazone[79] as ligands. In Lu/Hayashi's systems, ligand screening showed that ligand phosphinooxazoline (S)-iPr-phox gave the best results in the palladium-catalyzed asymmetric arylation of cyclic imines.[77a,b] The enantioselective addition of arylboronic acids to six-membered cyclic N-sulfonyl ketimines **187** in the presence of a cationic palladium complex PdCl$_2$[(S)-iPr-phox] and additives produced the chiral cyclic sulfamidates **188** bearing a tetrasubstituted stereogenic center in 51–99% yields with 96–99.9% *ee*s (Scheme 55). Subsequently, a new class of cyclic imines **189** with fluoroalkyl groups were used as suitable electrophiles for the Pd-catalyzed asymmetric arylation with a similar catalytic system, which afforded chiral trifluoromethylated and perfluoroalkylated 2-quinazolinones **190** in moderate to excellent yields with more than 99% *ee*s (Scheme 56). No reaction was observed when the substrate

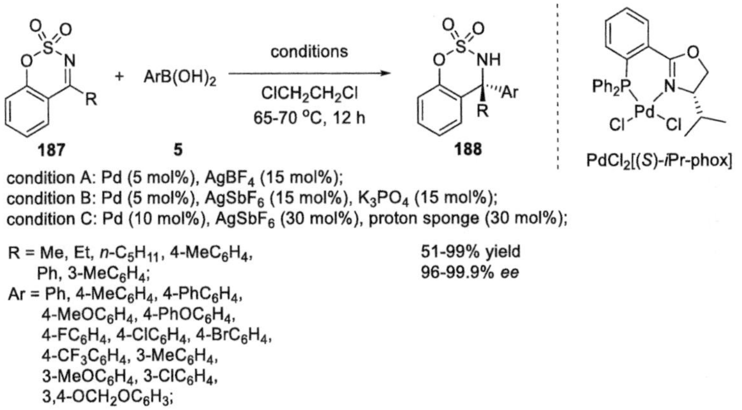

Scheme 55 Pd-catalyzed asymmetric additions of arylboronic acids to six-membered cyclic N-sulfonyl ketimines.

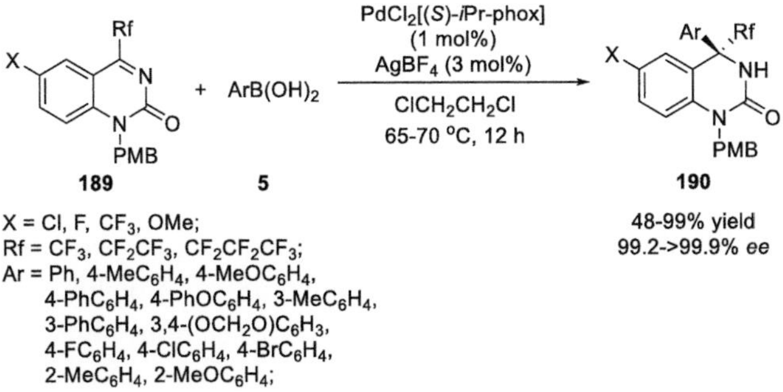

Scheme 56 Pd-catalyzed asymmetric additions of arylboronic acids to trifluoromethylated/perfluoroalkylated 2-quinazolinones.

imines without the perfluoroalkyl group at the imine carbon were used. This result strongly suggested that the electron-withdrawing perfluoroalkyl group in imines **189** plays a critical role in the activation of the imino group by the addition of an aryl-palladium intermediate.

Zhou and co-workers described the Pd(OCOCF$_3$)$_2$/(S)-tBu-phox-catalyzed asymmetric arylation of five- and six-membered cyclic α-ketiminophosphonates **191/194** with arylboronic acids, and it does not require the use of an external silver salt (Scheme 57).[77c] Various functional groups substituted cyclic α-ketiminophosphonates and arylboronic acids were tolerated and the corresponding quaternary α-aminophosphonates **193/195** were obtained in high yields with excellent enantioselectivities. Moreover, the six-membered addition product **193** can be easily prepared on gram scale in the presence of 1 mol% Pd catalyst without loss of efficiency and enantioselectivity.

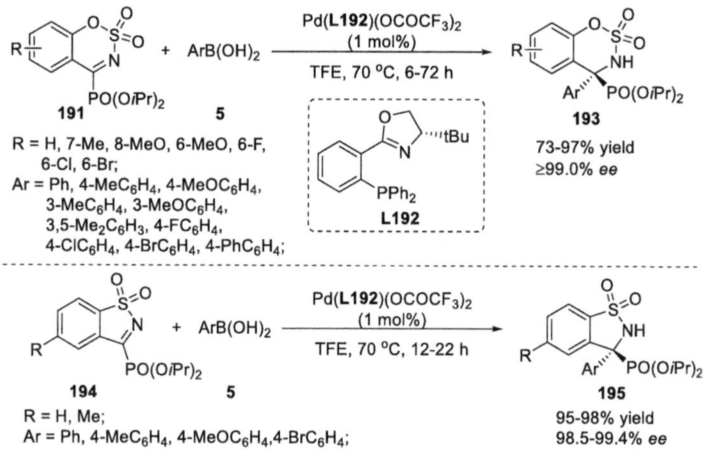

Scheme 57 Pd-catalyzed asymmetric additions of arylboronic acids to cyclic α-ketiminophosphonates.

By using a new chiral phosphine-imine ligand featuring a bulky aromatic group on the imine moiety and a sterically demanding group at the stereogenic carbon center position, the asymmetric arylation of cyclic N-sulfonyl aldimines and ketimines occurred to give enantioenriched amines in excellent yields and ees (Scheme 58).[78] Notably, only *ortho*-substituted arylboronic acids gave a low enantioselctivity, but with a high

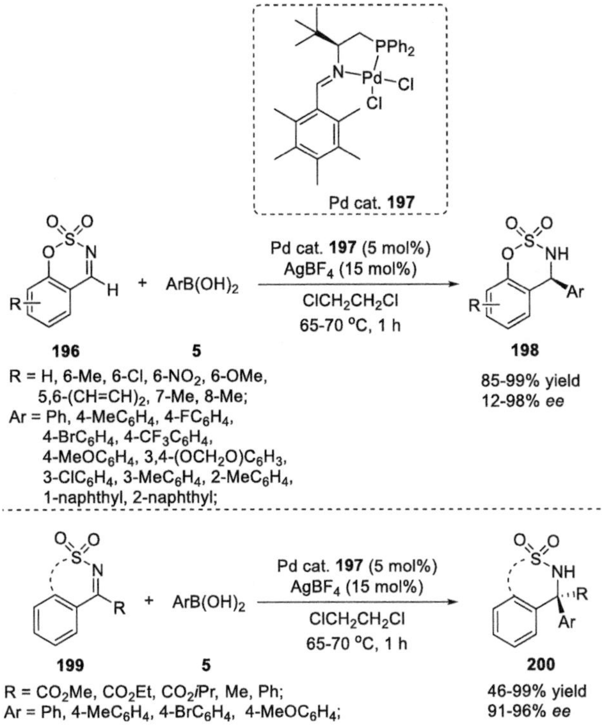

Scheme 58 Pd-catalyzed asymmetric additions of arylboronic acids to cyclic *N*-sulfonyl imines.

yield. Moreover, the more challenging cyclic *N*-sulfonyl ketiminines bearing ester, methyl and phenyl were tolerated under the same catalytic system.

Lassaletta and co-workers designed and synthesized a series of pyridine-hydrazone ligands, which could efficiently promote the asymmetric addition reaction of arylboronic acids to cyclic sulfonyl ketimines (Scheme 59).[79] The reaction between saccharin-derived cyclic ketimines **181**, bearing an aryl, alkyl or ethoxycarbonyl group on the R substituent, and arylboronic acids **5** with 5 mol% Pd(TFA)$_2$ as a palladium source, 7.5 mol% of pyridine-hydrazone **201** as ligand, in trifluoroethanol (TFE) at 40 °C for 12–48 h, led to the corresponding adducts **202** in 75–99% yields with 86–98% *ee*s. Specifically, the high regioselectivities along with high yields and enantioselectivities were achieved when the more challenging substrates 3,4-disubstituted 1,2,5-thiadiazole 1,1-dioxides **203** were used.

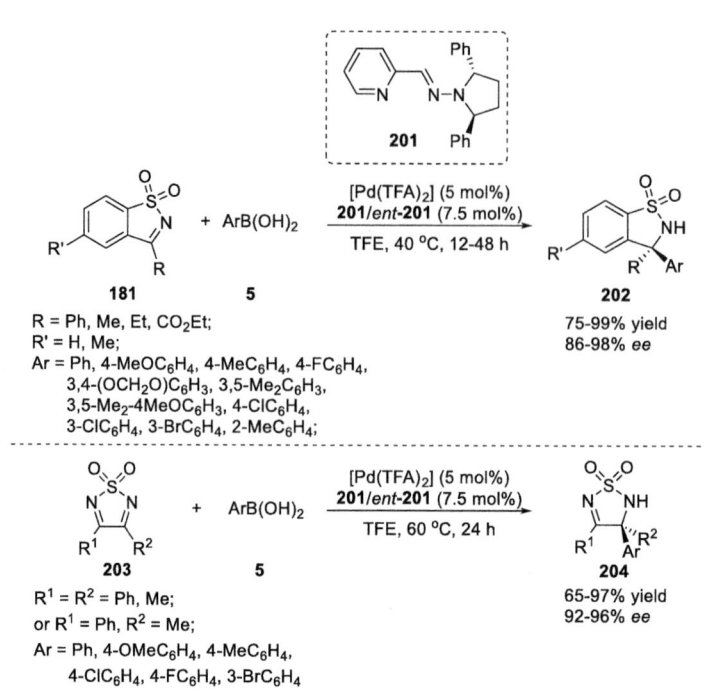

Scheme 59 Pd/N,N'-ligand-catalyzed asymmetric addition reaction of arylboronic acids to cyclic sulfonylketimines.

Catalytic asymmetric Mannich-type reactions of β-ketoesters provided a powerful route to synthesize many natural and unnatural compounds. Sodeoka and co-workers described a Mannich-type reaction of β-ketoesters with N-*tert*-butoxycarbonyl(Boc)-protected imines **205** employing Pd^{2+} diaqua complexes **103**, which provided the corresponding Mannich adducts **206** containing vicinal tertiary and quaternary carbon centers in good yields with excellent enantioselectivities and moderate to good diastereoselectivities (Scheme 60).[46] Moreover, a one-pot procedure starting directly from ethyl 2-oxoacetate and 3-chloroaniline proceeded smoothly to form the corresponding Mannich adduct **210** in 85% yield with 98% *ee* (major).

Shi and co-workers also developed an enantioselective Mannich-type reaction of cyclic β-ketoesters **211** with N-Boc imines **205** using a chiral cationic Pd^{2+} NHC diaqua complex **143a** as catalyst (Scheme 61).[80] Generally, this reaction tolerates a wide range of aromatic and aliphatic N-Boc imines and affords the corresponding products **212** in moderate to good yields and *ee*s under mild conditions.

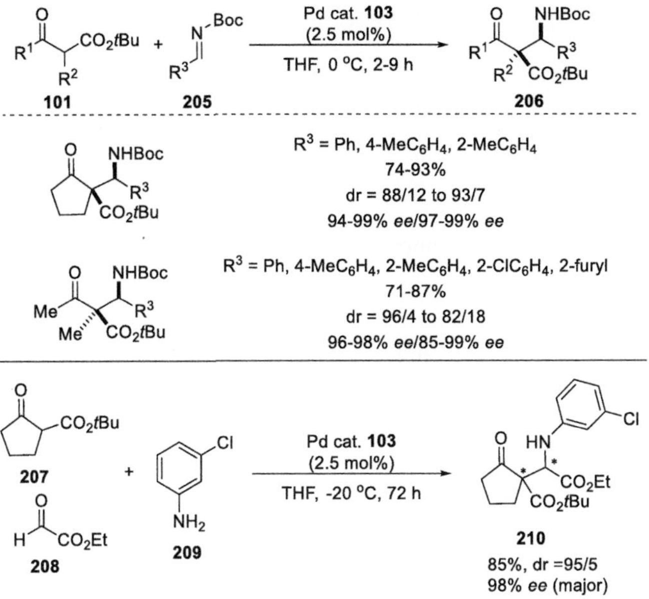

Scheme 60 Pd-catalyzed asymmetric Mannich-type reaction with N-Boc imines.

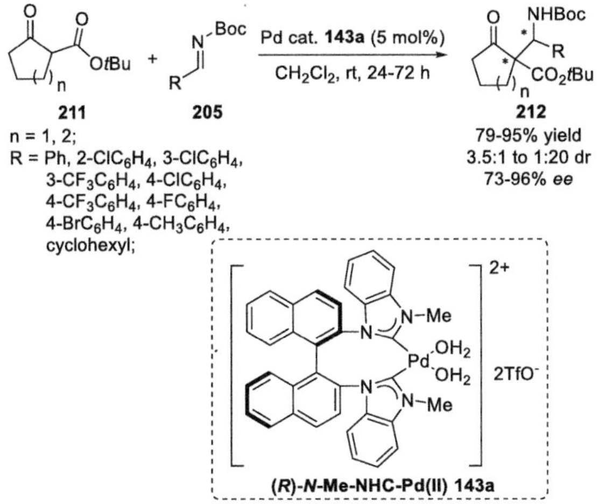

Scheme 61 Pd-catalyzed asymmetric Mannich-type reaction of cyclic β-ketoesters with N-Boc imines.

Shibata and Nakamura invested the enantioselective reactions of benzyl nitriles **214** with N-tosylimines **213** using chiral bis(imidazoline)-palladium catalysts (Scheme 62).[81] The best enantioselectivity was observed when the reaction was carried out at −20 °C in THF using 1,3-bis(imidazoline-2-yl)

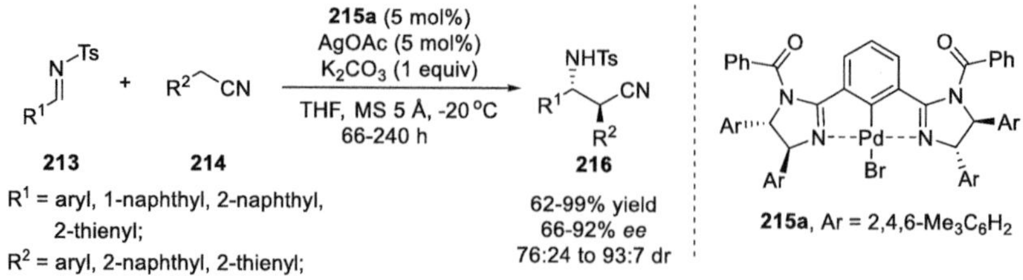

Scheme 62 Pd-catalyzed asymmetric reaction of benzyl nitriles with N-tosylimines.

benzene-palladium(II) complex **215a** in combination with AgOAc and K$_2$CO$_3$ in the presence of molecular sieves 5 Å. A wide range of functionalized β-amino nitriles **216** were prepared in good yields with moderate diastereoselectivities and good enantioselectivities. Furthermore, they applied the bis(imidazoline)/palladium system to the enantioselective aza-Morita–Baylis–Hillman reaction of acrylonitrile **217** with imines **213** (Scheme 63).[82] This reaction provides a simple and efficient method for the synthesis of α-methylene-β-aminonitrile derivatives **218**.

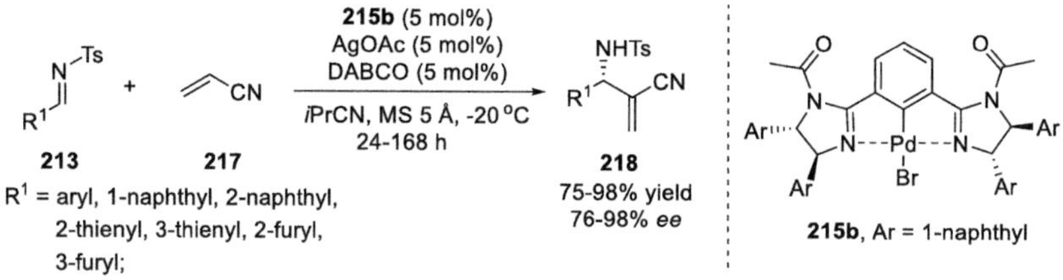

Scheme 63 Pd-catalyzed asymmetric aza-MBH reaction of acrylonitrile with N-tosylimines.

Lam and co-workers described a novel catalytic asymmetric addition of 2-alkylazaarenes **219** and N-Boc imines **205**. The reactions were treated with a chiral Pd(II)-bis(oxazoline) complex at 50 °C or rt. in CHCl$_3$ (used without any drying treatment) under air atmosphere, giving products **221** with high levels of diastereo- and enantioselectivities (Scheme 64).[83] Importantly, incorporation of an electron-withdrawing group such as nitro, cyano or ester group into the substrate azaarene **219** enhanced the acidity of α-protons and promoted the reaction occurred. Similar enantioselective additions of 2-alkylazaarenes to nitroalkenes proceeded smoothly in the same catalytic system.

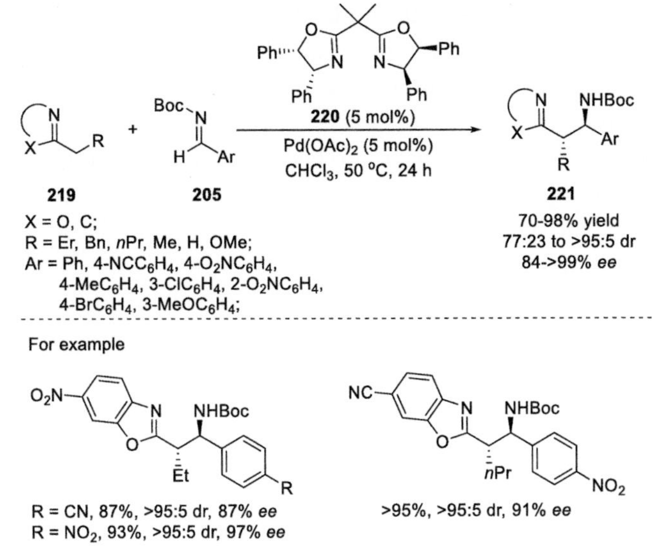

Advances in Pd-catalyzed asymmetric addition reactions

Scheme 64 Pd-catalyzed asymmetric addition of 2-alkylazaarenes and imines.

Analogously to Pd-catalyzed asymmetric umpolung allylations of aldehydes, Zhou and co-workers developed a similar asymmetric umpolung allylations of imines **222** with allylic alcohols **223** (Scheme 65).[84] This process, carried out with Pd(OAc)$_2$ as the catalyst and chiral spiro phosphoramidite **224** as ligand in the presence of Et$_3$B, afforded homoallylic

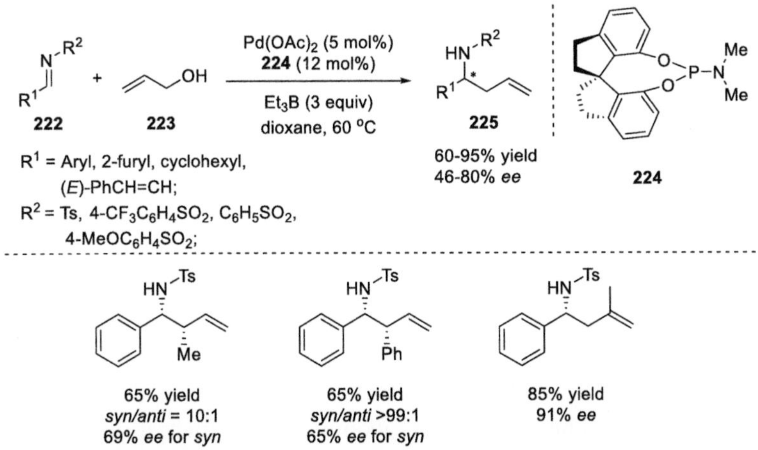

Scheme 65 Pd-catalyzed asymmetric allylation of *N*-tosylimines with various allylic alcohols.

amines **225** in good yields with moderate *ee*s. Different substituted allylic alcohols were suitable substrates to produce the corresponding homoallylic amines with excellent diastereoselectivities and moderate to good enantioselectivities.

Heteroatom nucleophiles have also been used to the Pd-catalyzed asymmetric additions of imines. Duan and co-workers developed a PCP-pincer palladium catalytic system, which was used for the enantioselective addition of diarylphosphines **50** to *N*-tosylimines **213** (Scheme 66).[85] Chiral phosphine sulfides **226** were synthesized in excellent yields and *ee*s under mild conditions.

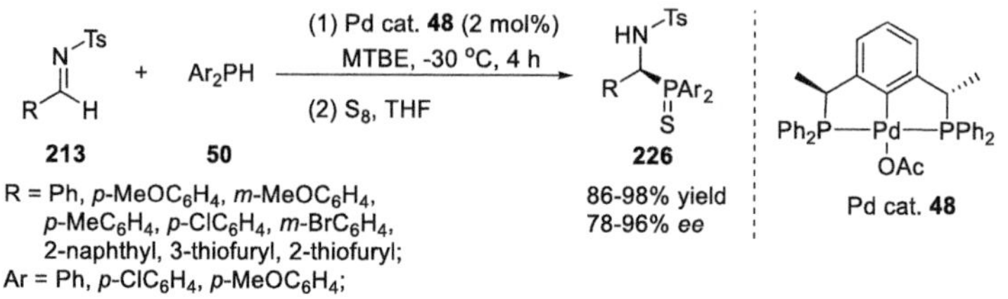

Scheme 66 Pd-catalyzed asymmetric addition of diarylphosphines to *N*-tosylimines.

3.3 Palladium-catalyzed asymmetric 1,2-addition to olefins/allenes/cumulenes

Functionalization of olefins is a significant transformation in organic synthesis and considerable efforts have been made toward their enantioselective transformations. However, Pd(II)-catalyzed asymmetric nucleophilic addition to unactivated olefins remains problematic and challenging.[86] Sigman and co-workers developed intramolecular Wacker-type cyclization reactions of a specific alkene class that incorporate the addition of two different nucleophiles.[87–89] In Sigman's studies, alkenes **227** bearing a pendant alcohol nucleophile were chosen as substrates, which reacted with exogenous alcohols as another nucleophile resulting in the formation of oxygen containing heterocycles **228** with contiguous chiral centers (Scheme 67A).[87] When using Pd(MeCN)$_2$Cl$_2$ as catalyst and (S)-*i*PrQuinox as the chiral ligand in the presence of CuCl, base and O$_2$, the desired products **228** were obtained in moderate yields and with highly enantioselectivities and diastereoselectivities. The asymmetric alkene difunctionalization reaction is proposed by intramolecular oxypalladation followed by the addition of alcohols to an *ortho*-quinone methide. A beneficial effect of copper cocatalyst

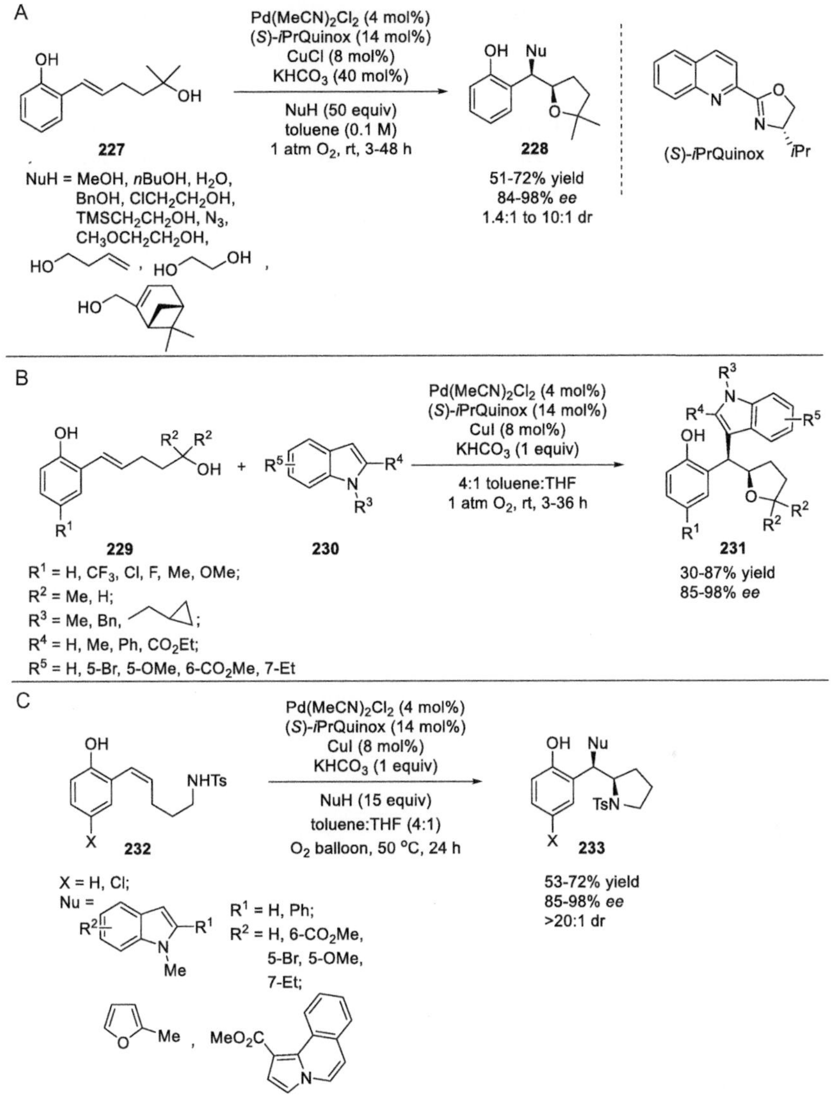

Scheme 67 Pd-catalyzed asymmetric difunctionalization of o-vinyl phenols.

resulted in a higher reactivity without loss of enantioselectivity when the appropriate chiral ligand was used to coordinate to both palladium and copper.

Subsequently, the successfully use of indole derivatives **230** as nucleophile to trap the quinone methide intermediate, afforded 3-substituted indoles **231** as a single diastereomer in good yields with excellent *ee*s

(Scheme 67B).[88] Several new indole products exhibits modest biological activity in MCF-7 tumor cells in comparison with MCF-10A normal breast cells.

Furthermore, this catalytic system was successfully applied to the asymmetric difunctionalization of alkenes **232** with a tethered protected amine with a second nucleophile (Scheme 67C).[89] A wide range of the second nucleophile including *N*-methyl indoles, furans and indolizines were tolerated and the resulting products **233** could be isolated in 53–72% yields and 85–98% *ee*s with >20:1 dr.

After the pioneer work of Sigman, a number of catalytic enantioselective oxidative reactions of alkenes with a tethered protected amine were developed. The key point is an alkyl-palladium(II) intermediate generated by aminopalladation of alkenes, undergoing subsequent various transformations to construct nitrogen-containing heterocycles (Scheme 68). In 2009, Yang and co-workers described a Pd-catalyzed enantioselective oxidative cascade cyclization of *o*-allyl anilides **234** to synthesize chiral indolines **235** as a single diastereomer (Scheme 69).[90] High enantioselectivities could be achieved with a catalytic amount of quinoline-oxazoline ligand ((*S*)-*t*BuQuinox) and Pd(OAc)$_2$ in the presence of 2,6-lutidine and HNTf$_2$ over activated 3 Å molecular sieves under an oxygen atmosphere. This reaction is proposed occurring via *cis*-aminopalladation of the internal alkene followed by

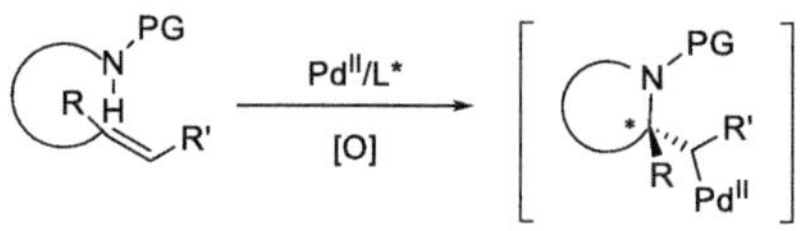

Scheme 68 A key alkylpalladium(II) intermediate generated by aminopalladation.

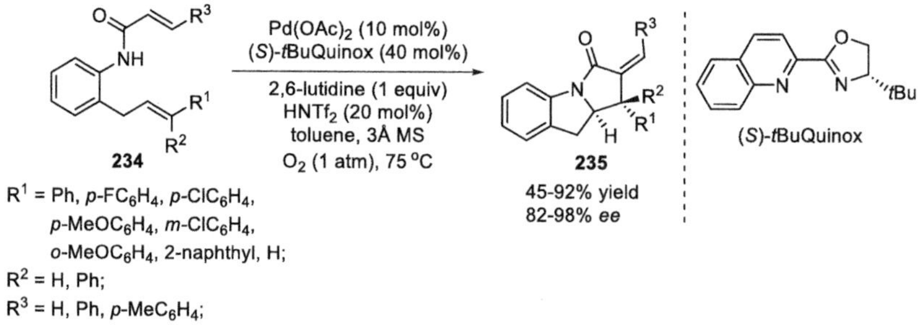

Scheme 69 Pd-catalyzed asymmetric oxidative cascade cyclization of disubstituted olefins.

Advances in Pd-catalyzed asymmetric addition reactions 377

cis-carbopalladation of the terminal alkene. It is noteworthy that the relative stereochemistry of the product was dependent of the olefin geometry of the substrates. Recently, they explored the challenging asymmetric oxidative tandem cyclization of alkene-tethered aliphatic acrylamides **236** (Scheme 70).[91] By using Pd(TFA)$_2$/(S,S)-diPh-pyrox catalyst, a variety of aliphatic alkenyl acrylamides **236** gave pyrrolizidine derivatives **237** in up to 99% yield with up to 93% *ee* under mild aerobic conditions.

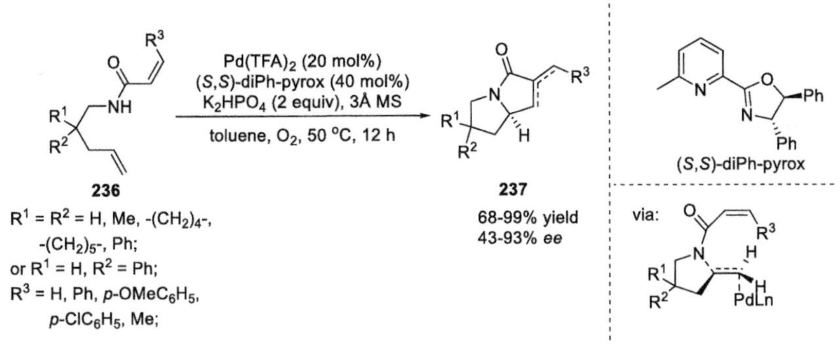

Scheme 70 Pd-catalyzed asymmetric oxidative cascade cyclization of aliphatic alkenyl acrylamides.

An enantioselective Wacker-type oxidative cyclization of γ-alkenyl sulfonamides **238** via aminopalladation of alkenes and subsequent β-hydride elimination was showcased by Stahl and co-workers (Scheme 71).[92] Good yields and excellent enantioselectivities were observed for the reaction, when a Pd(TFA)$_2$/pyrox catalyst was used. It was believed that electronic asymmetry of the pyrox ligand in combination with steric asymmetry controlled the stereochemical course of aminopalladation.

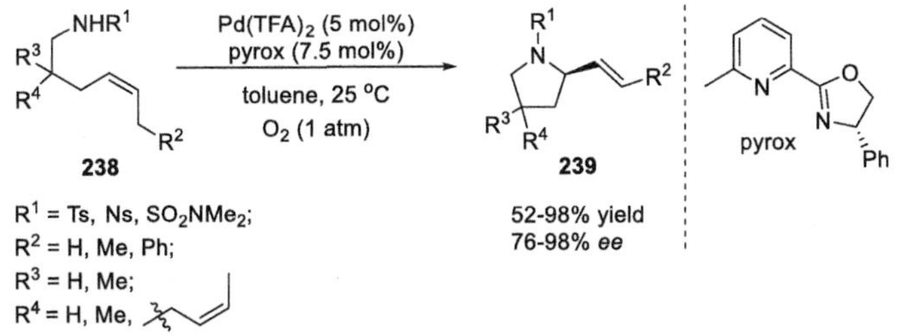

Scheme 71 Pd-catalyzed asymmetric cyclization of γ-alkenyl sulfonamides.

Zhang also described a catalytic enantioselective aza-Wacker-type cyclization reaction of trisubstituted olefins **240** using *t*Bu-Pyrox as the chiral ligand and Pd(TFA)$_2$ as the palladium source without an additive (Scheme 72).[93] In this case, the N-protecting group of the substrates has a strong influence on the enantioselectivity. Under the optimal conditions, a series of isoindolinones **241** with a chiral quaternary carbon adjacent to the nitrogen atom were obtained in excellent yields and enantioselectivities.

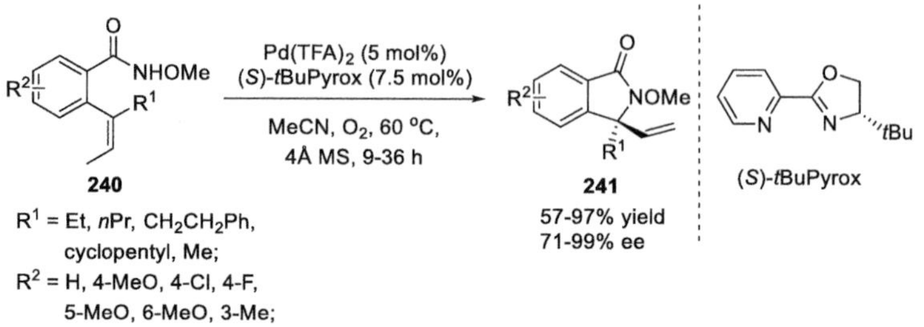

Scheme 72 Pd-catalyzed asymmetric aza-Wacker-type cyclization reaction.

Michael and co-workers surveyed a variety of pybox and quinox ligands in Pd-catalyzed asymmetric diamination of alkenes **242**, and the phenyl-substituted quinoline-oxazoline ligand resulted in good yields and enantioselectivities (Scheme 73).[94] This reaction produced protected cyclic 1,2-diamines **244** using N-fluorobenzenesulfonimide (NFBS) as both an oxidant and a source of nitrogen through a Pd^{II}/Pd^{IV} catalytic cycle. The alkylpalladium intermediate complex that resulted from aminopalladation of substrate alkenes was characterized by X-ray crystallography. Mechanistic experiments established that aminopalladation is the enantiodetermining step of this transformation.

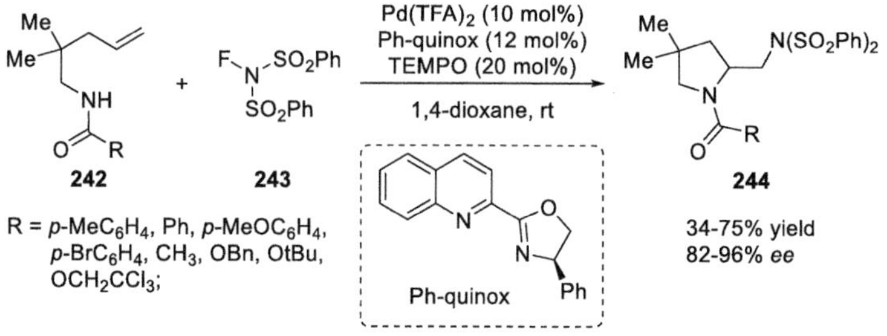

Scheme 73 Pd-catalyzed asymmetric diamination of alkenes.

In Liu's studies, a Pd-catalyzed enantioselective oxidative aminoarylation of alkenes **245** was proved to be an efficient method to construct optically pure polycyclic heterocycles **246** bearing a quaternary carbon center (Scheme 74).[95] The reaction catalyzed by Pd(OAc)$_2$ and Qox (quinoline-oxazoline) using Ag$_2$CO$_3$ as the oxidant is initiated by aminopalladation and generates a alkyl-palladium(II) species, which then undergoes C—H activation in a cascade cyclization. The addition of phenyl-glyoxylic acid shows a favorable effect on the enantioselectivity.

Scheme 74 Pd-catalyzed asymmetric aminoarylation of alkenes.

There are a few reports focused on enantioselective Pd-catalyzed 6-*endo* oxidative cyclization of alkenes. More recently, Liu and co-workers showed the first Pd-catalyzed asymmetric 6-*endo* oxidative aminoacetoxylation of unactivated alkenes **247** to provide structurally diverse 3-acetoxy piperidines **249** (Scheme 75).[96] Good regio- and enantioselectivities could be achieved using a novel pyridinyl-oxazoline ligand. Mechanistic studies showed that a stereoselective *trans*-aminopalladation and subsequent reductive elimination with stereoretention determined the enantioselectivity of the products.

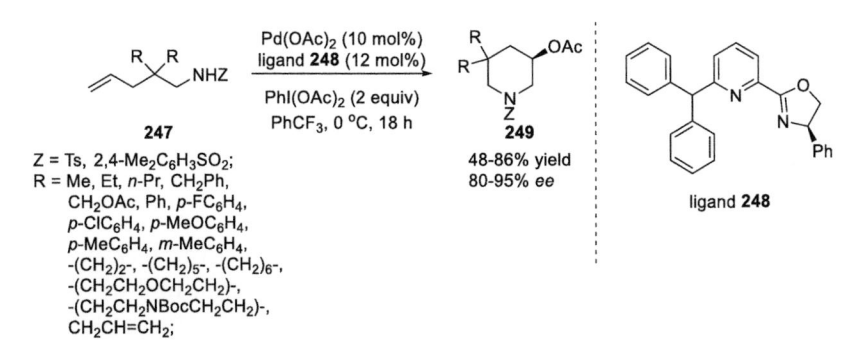

Scheme 75 Pd-catalyzed asymmetric aminoacetoxylation of unactivated alkenes.

Wolfe and co-workers developed another class of asymmetric Pd-catalyzed alkene difunctionalization reactions, in which N-Boc-pent-4-enylamines **250** and aryl or alkenyl bromides **251** underwent carboamination to produce enantioselective 2-(arylmethyl)- and 2-(alkenylmethyl)pyrrolidine derivatives (Scheme 76A).[97] With the best ligand (R)-Siphos-PE, the reaction proceeded with 2.5 mol% Pd$_2$(dba)$_3$ and 1–2 equiv. of NaOtBu in toluene at 90 °C to the product **252** in 61–80% yields and 72–94% ees. This method could be applied in the synthesis of (−) tylophorine, which is found in bioactive molecules such as enantiomerically enriched phenanthroindolizidine alkaloids.

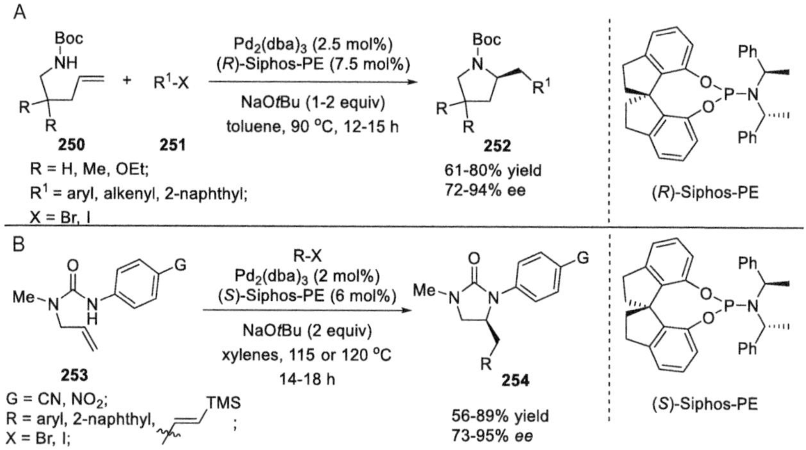

Scheme 76 Pd-catalyzed asymmetric carboamination of unactivated alkenes.

Additionally, they developed an asymmetric carboamination of alkenes between N-allyl urea derivatives **253** and aryl or alkenyl halides **251** with the [Pd$_2$(dba)$_3$]/(S)-Siphos-PE catalytic system (Scheme 76B).[98] The enantioenriched 4-(arylmethyl)-imidazolidin-2-ones **254** were obtained in good yields and ees. Importantly, the electronic property of N-aryl substituents and anionic additive have great impact on the product enantioselectivity, indicating that C—C bond-forming reductive elimination may be the enantiodetermining step in this transformation.

In recent years, palladium-catalyzed asymmetric reductive Heck reactions or cascade cyclization reactions involving a reductive Heck process has also attracted much attention and become a powerful tool to realize functionalization of alkenes (Scheme 77).[2a] In 2007, Buchwald first described the Pd-catalyzed asymmetric intramolecular reductive Heck

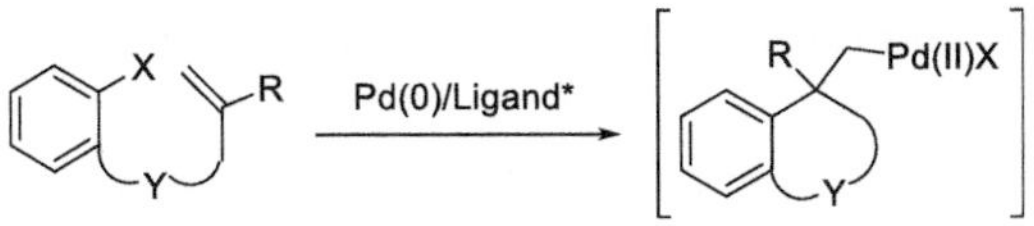

Scheme 77 A key palladium(II) intermediate generated by Heck cyclization.

reaction of 2′-perfluoroalkyl-sulfonated aryl α,β-unsaturated ketones, which afforded the optically active 3-substituted indanones up to 94% *ee*.[99] Since then, only limited developments based on the carbopalladation-initiated enantioselective transformations of activated and unactivated olefins were reported.[100] In 2011, Lautens and co-workers reported one example of Pd-catalyzed asymmetric carboiodination reaction of alkene-tethered aryl bromide **255**.[101] Cyclized iodide **256** was obtained in 14% yield with 94% *ee* using Josiphos as a chiral ligand (Scheme 78).

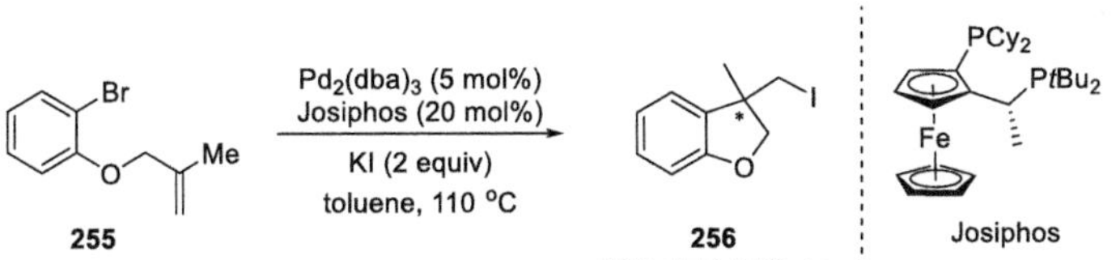

255 **256** Josiphos
14% yield, 94% *ee*

Scheme 78 Pd-catalyzed asymmetric carboiodination reaction of alkene-tethered aryl bromide.

Tong and co-workers described a catalytic asymmetric vinylborylation reaction of (*Z*)-1-iodo-dienes **257** and B_2pin_2 **258** using Pd(0)/(*S*)-*p*-CF_3-BnPHOX as the catalyst in the presence of two equivalents of Ag_2CO_3 in toluene at 25 °C (Scheme 79).[102] A series of 3,3-disubstituted tetrahydropyridines **259** were efficiently constructed in 44–95% yields with 79–97% *ees*. Using an alkene with a lengthening tether as the starting

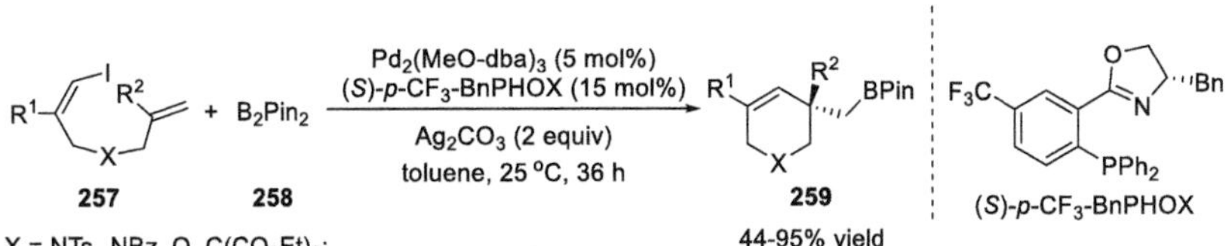

257 **258** **259** (*S*)-*p*-CF_3-BnPHOX

44-95% yield
79-97% *ee*

X = NTs, NBz, O, $C(CO_2Et)_2$;
R^1 = Ph, 4-MeC_6H_4, 4-$MeOC_6H_4$, Et, H;
R^2 = Me, *n*Bu, *i*Bu, Bn, Cy, Ph, CH_2OTBS, $(CH_2)_3OTBS$, CH_2OAr;

Scheme 79 Pd-catalyzed asymmetric vinylborylation reaction of (*Z*)-1-iodo-dienes and B_2pin_2.

material, the corresponding 7-membered cyclic product was obtained in 52% yield with only 33% ee. Control experiments highlighted that transmetallation with B_2pin_2 is prior to the insertion of the pendant alkene. The authors proposed this transformation initiated by oxidative addition, followed by transmetallation, alkene insertion and reductive elimination.

Zhang and co-workers achieved a significant breakthrough with the development of a Pd^0-catalyzed highly enantioselective reductive Heck reaction of allyl aryl ethers **260** (Scheme 80).[103] Using a novel *N*-Me-Xu-Phos ligand that they developed, the intramolecular hydroarylation reaction of allyl aryl ethers **260** in the presence of $Pd_2(dba)_3 \cdot CHCl_3$ with a mixed toluene/methanol solvent system gave the corresponding chiral 2,3-dihydrobenzofurans **261** with a quaternary stereocenter in high yields (up to 97% yield) and excellent enantioselectivities (up to 95% *ee*). To witness this method's efficiency, a variety of CB2 receptor agonists could be synthesized in 70–91% yields with 91–95% *ee*s under the standard reaction conditions. Deuterium labeling experiments revealed that the hydride donor comes from DCO_2Na.

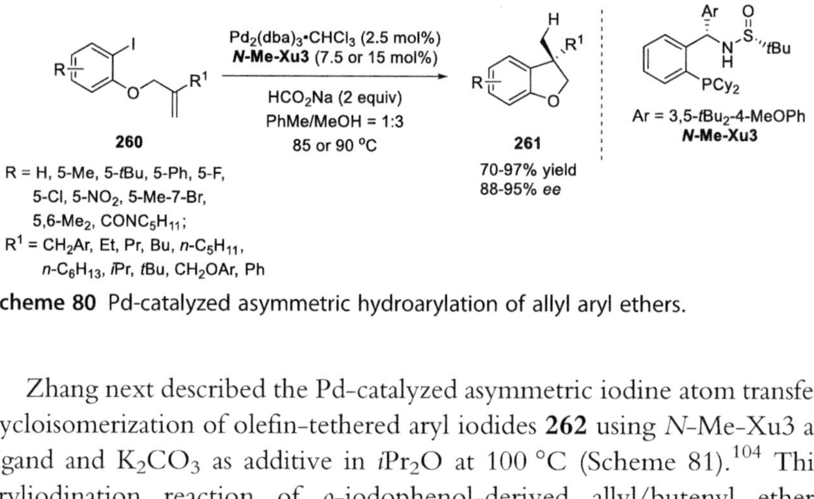

Scheme 80 Pd-catalyzed asymmetric hydroarylation of allyl aryl ethers.

Zhang next described the Pd-catalyzed asymmetric iodine atom transfer cycloisomerization of olefin-tethered aryl iodides **262** using *N*-Me-Xu3 as ligand and K_2CO_3 as additive in *i*Pr$_2$O at 100 °C (Scheme 81).[104] This aryliodination reaction of *o*-iodophenol-derived allyl/butenyl ethers provided a powerful approach to construct highly enantioselective 2,3-dihydrobenzofuran and chromanes by a tandem sequence involving the cascade oxidative addition, alkene insertion and C—I reductive elimination. In this case, direct reduction reaction and reductive Heck reaction were inhibited totally. Similarly, the aryliodination reaction of *o*-iodoaniline-derived allyl amines afforded the desired indolines with an

Scheme 81 Pd-catalyzed asymmetric aryliodination of olefin-tethered aryl iodides.

alkyl iodide substituent when using toluene as the solvent. Theoretical studies performed by the DFT calculations disclosed that the alkene insertion was the rate-determining step which led to the high reactivity and enantioselectivity.

Very recently, they demonstrated the first asymmetric dicarbofunctionalization of olefin-tethered aryl halides **264** with boronic acids **5** by a Pd-catalyzed tandem Heck/Suzuki-coupling reaction (Scheme 82).[105] The process afforded a variety of 3,3-disubstituted

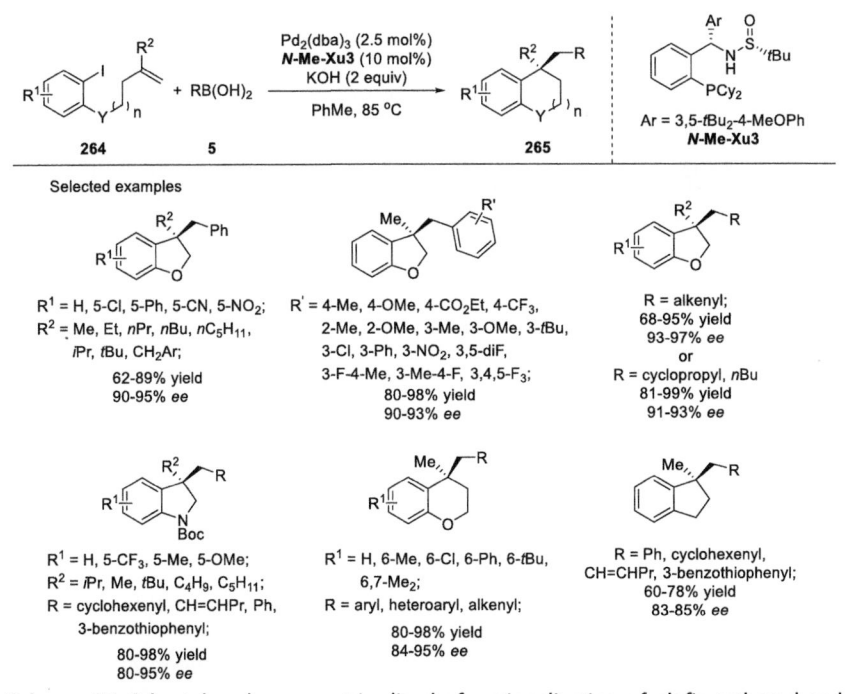

Scheme 82 Pd-catalyzed asymmetric dicarbofunctionalization of olefin-tethered aryl halides.

dihydrobenzofurans with a quaternary stereocenter in excellent enantioselectivity when employing Xu-Phos ligands that they developed. The reaction scope shows aryl-, alkyl- and alkenylboronic acids were competent coupling partners in this tandem intramolecular cyclization/cross-coupling reaction. Furthermore, a series of enantioselective indolines, chromanes and indanes were successfully obtained using the same strategy. Notably, a pair of stereoisomeric products are accessible through simple inversion the coupling sequence.

Intermolecular asymmetric addition of carbon nucleophiles to unactivated alkenes with excellent regio- and enantioselectivities is challenging. A breakthrough appeared when Sigman and co-workers reported their elegant works on the Pd-catalyzed enantioselective intermolecular Heck arylation reactions of acyclic alkenyl alcohols employing chiral pyridine oxazoline ligands (PyrOx).[106–109] They successfully established an asymmetric redox-relay addition reaction of electron-deficient aryldiazonium salts **267** to racemic allylic alcohols **266** catalyzed by Pd$_2$dba$_3$, which is processed by the transposition of the palladium, initial migratory insertion and selective β-H elimination coupled with oxidation of the alcohol (Scheme 83A).[106] The adducts **268** with a chiral center far away from the resultant carbonyl could be constructed in 52–84% yields with 82–94% ees. Notably, alkenes of both (E)- and (Z)-configuration were compatible in these transformations, providing the opposite enantiomers of the products. Importantly, this redox-relay strategy enables to synthesize γ-aryl ketones or aldehydes from secondary or primary homoallylic alcohols. Additionally, the reaction of hept-5-en-2-ol or hex-4-enal with aryldiazonium salt under the same conditions gave the δ-substituted products in good enantioselectivities along with 20% of the γ-substituted products.

A variety of aryl boronic acids **5** proved to be suitable aryl source in the Pd-catalyzed enantioselective redox-relay oxidative Heck reactions and the resulting remotely functionalized carbonyl products were obtained in excellent ees and high site selectivities (Scheme 83B).[107] Theoretical studies revealed that has the electronic effects of the alkenes has a remarkable influence on the site selectivity. Furthermore, trisubstituted alkenyl alcohols **271** were reacted with arylboronic acids **5** with a similar catalytic system, including a cationic Pd salt, a catalytic amount of Cu(OTf)$_2$ and chiral pyridine oxazoline ligands under an O$_2$ atmosphere, to access a series of functionalized carbonyl products **272** bearing a quaternary stereocenter in good to excellent enantioselectivity (Scheme 83C).[108] Notably, the alkenyl alcohols with different chain lengths provided the corresponding β,γ,δ,ε-aryl carbonyl compounds with the high site selectivity on the more hindered position.

A

PyrOx /tBu (F$_3$C-pyridine-oxazoline)

(±) **266** + R^2-Ar-N$_2$PF$_6$ **267** → **268**

Pd$_2$dba$_3$ (3 mol%)
PyrOx (7 mol%)
DMF, rt, 3-24 h

R = Me, C$_8$H$_{17}$, Bn, (CH$_3$)$_2$CHCH$_2$, C$_5$H$_{11}$, iPr;
R^1 = Me, (CH$_2$)$_4$OH, C$_4$H$_9$, C$_3$H$_7$, (CH$_2$)$_3$CO$_2$Me;
R^2 = 4-CO$_2$Me, 4-COMe, 4-NO$_2$, 3-I;

52-84% yield
82-94% ee

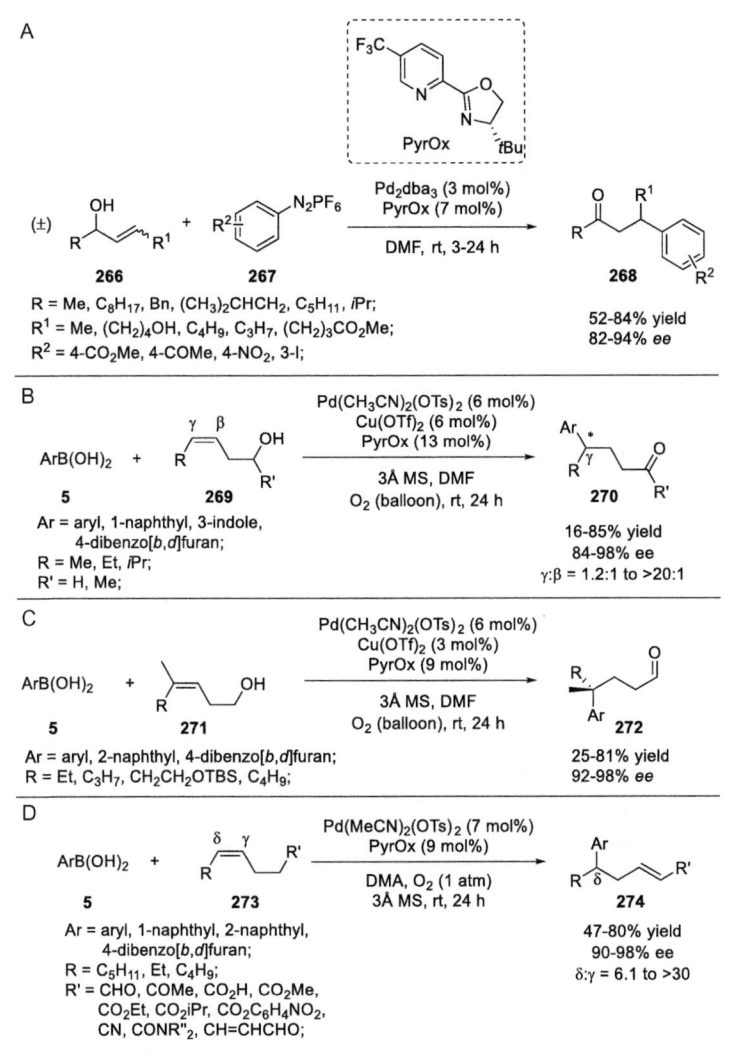

B

ArB(OH)$_2$ **5** + **269** (γ,β alkenyl alcohol) → **270**

Pd(CH$_3$CN)$_2$(OTs)$_2$ (6 mol%)
Cu(OTf)$_2$ (6 mol%)
PyrOx (13 mol%)
3Å MS, DMF
O$_2$ (balloon), rt, 24 h

Ar = aryl, 1-naphthyl, 3-indole, 4-dibenzo[b,d]furan;
R = Me, Et, iPr;
R' = H, Me;

16-85% yield
84-98% ee
γ:β = 1.2:1 to >20:1

C

ArB(OH)$_2$ **5** + **271** → **272**

Pd(CH$_3$CN)$_2$(OTs)$_2$ (6 mol%)
Cu(OTf)$_2$ (3 mol%)
PyrOx (9 mol%)
3Å MS, DMF
O$_2$ (balloon), rt, 24 h

Ar = aryl, 2-naphthyl, 4-dibenzo[b,d]furan;
R = Et, C$_3$H$_7$, CH$_2$CH$_2$OTBS, C$_4$H$_9$;

25-81% yield
92-98% ee

D

ArB(OH)$_2$ **5** + **273** (δ,γ) → **274**

Pd(MeCN)$_2$(OTs)$_2$ (7 mol%)
PyrOx (9 mol%)
DMA, O$_2$ (1 atm)
3Å MS, rt, 24 h

Ar = aryl, 1-naphthyl, 2-naphthyl, 4-dibenzo[b,d]furan;
R = C$_5$H$_{11}$, Et, C$_4$H$_9$;
R' = CHO, COMe, CO$_2$H, CO$_2$Me, CO$_2$Et, CO$_2$iPr, CO$_2$C$_6$H$_4$NO$_2$, CN, CONR"$_2$, CH=CHCHO;

47-80% yield
90-98% ee
δ:γ = 6.1 to >30

Scheme 83 Pd-catalyzed asymmetric redox-relay arylations of alkenyl alcohols.

The Pd-catalyzed asymmetric relay Heck reaction is applicable to alkenes **273** with a carbonyl linker, affording a range of α,β-unsaturated carbonyl compounds **274** with a δ-substituted aryl group in high site selectivity and enantioselectivity (Scheme 83D).[109] This reaction has a broad substrate scope of arylboronic acids and carbonyl derivatives **273**, including aldehydes, ketones, carboxylic acids, esters, nitriles, amide, and α,β-unsaturated aldehydes. Geometry minimizations and IR frequency calculations revealed that polarization of the alkenes during migratory insertion had a remote effect on the site selectivity of arylation.

Recently, He, Peng and Chen developed an asymmetric version of hydrocarbofunctionalization of unactivated alkenes **275** with an amide-linked aminoquinoline directing group using a novel monodentate oxazoline ligand (Scheme 84).[110] The reactions were carried out with Pd(OAc)$_2$/MOXin in the presence 1 equiv. of *ortho*-phenyl benzoic acid (*o*PBA) in MeOH at 60 °C, which provided γ-substituted products **277** in good yields and enantioselectivities. A wide range of nucleophiles **276** including indoles, *N*-methylindoles, cyclic 1,3-diones or phenols have been successfully applied in these transformations. The same enantioenriched products were observed when *cis* and *trans* alkenes were used. It is worth noting that the AQ directing group controls the regioselectivity of addition resulting in the suppression of a competing β-H elimination process.

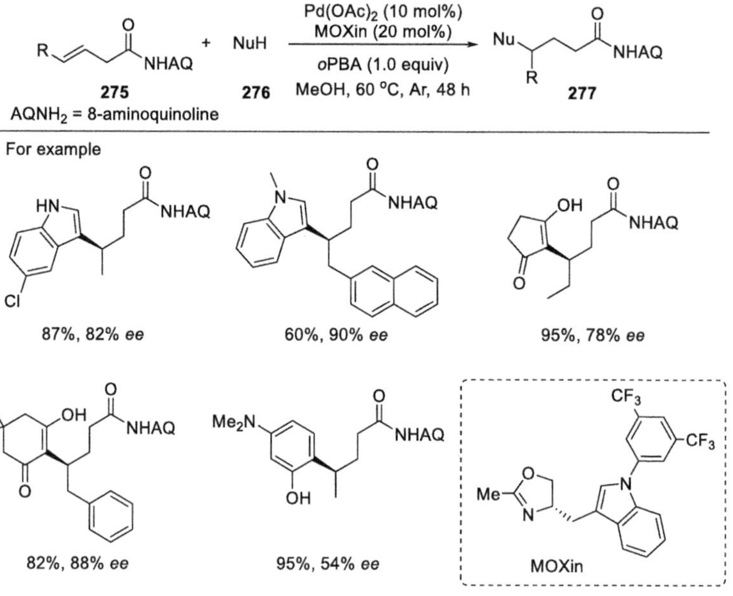

Scheme 84 Pd-catalyzed asymmetric hydrocarbofunctionalization of unactivated alkenes.

Gong and co-workers reported the enantioselective addition of cyclic ketones **279** to alkenes **278** with an amide substituent catalyzed by Pd(CH$_3$CN)$_2$Cl$_2$ and chiral amine organocatalyst (Scheme 85).[111] The palladium catalyst coordination with alkene and 8-aminoquinoline formed a Pd

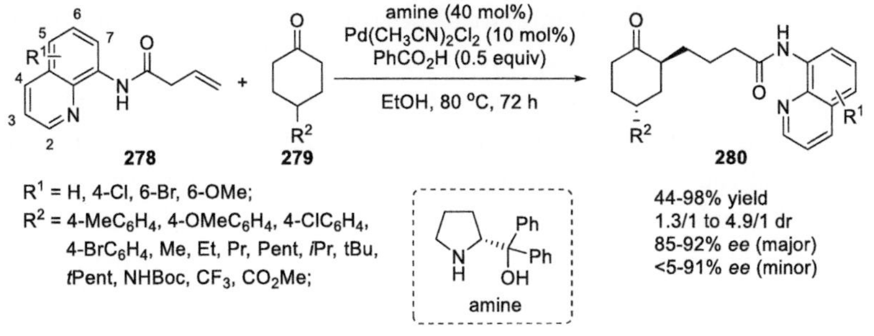

Scheme 85 Pd-catalyzed asymmetric addition of C-nucleophile to unactivated alkenes.

complex, which underwent a Wacker-type nucleopalladation with enamine generated from the condensation of cyclic ketone and chiral amine organocatalyst. The γ-addition products **280** were obtained in moderate to excellent yields with high enantioselectivities for the major diastereomers and lower enantioselectivities for the minor diastereomers. The internal alkenes were also shown to promote the asymmetric addition reactions of unactivated alkenes, giving the desired products in moderate yields, diastereoselectivities and enantioselectivities.

Methylenecyclopropanes are important building blocks to construct various organic scaffolds through C—C bond cleavage. A Pd-catalyzed kinetic resolution of racemic methylenecyclopropanes **282** with MePh$_2$Si-B(pin) **281** was described by Suginome and co-workers (Scheme 86).[112] Chiral phosphoramidite ligand **283** exhibits good product selectivity, giving the corresponding products **284** and **285** in a ratio of 67:33 to 86:14 with high *ee*s for major products. Substrates with siloxyalkyl, acetoxypropyl, chloro-substituted alkyl or phthalimido-substituted alkyl substituent were tolerated in this reaction.

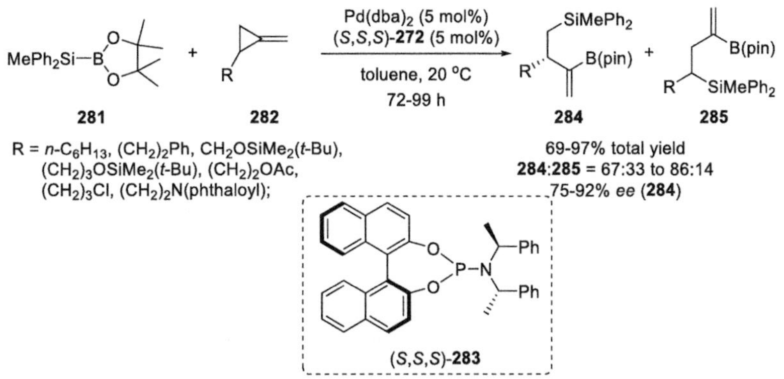

Scheme 86 Pd-catalyzed kinetic resolution of racemic methylenecyclopropanes.

Asymmetric catalytic addition to allenes is an efficient method for the synthesis of functionalized allylic compounds. For example, Trost and co-workers developed a Pd-catalyzed asymmetric addition to allenes **287** using 3-aryloxindoles **286** as carbon-based nucleophiles, and synthesized formal asymmetric allylic alkylation reaction products **289** with two adjacent chiral centers in high yields, diastereoselectivity, enantioselectivity and regioselectivity (with up to 99% yield, 18:1 dr, 93% ee, branched product) (Scheme 87).[113] The reaction took place in the presence of $Pd_2(dba)_3$, Trost ligand (*R,R*)-**288** and an acid cocatalyst in THF at room temperature for 2–24 h. Notably, the electronic nature of the acid cocatalyst has an impact in the stereochemical outcomes during the formation of transition state. This method was successfully applied to the synthesis of pyrrolidinoinoline core of the gliocladins.

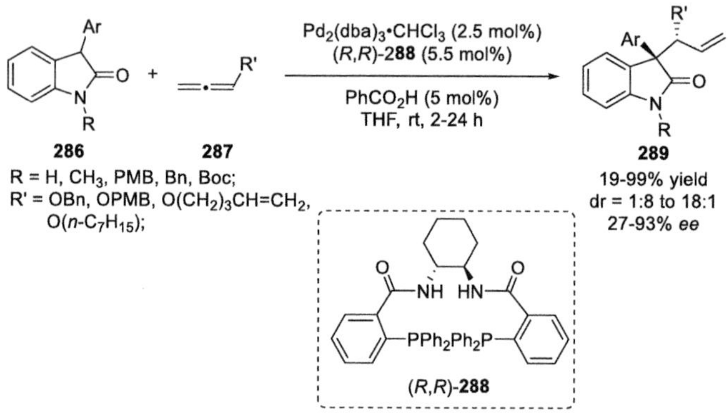

Scheme 87 Pd-catalyzed asymmetric addition of oxindoles to allenes.

In 2012, Dixon and co-workers combined a chiral amine with $Pd(OAc)_2$ in a catalytic asymmetric carbocyclization of aldehyde-linked allenes **290**, affording vinyl-substituted cyclopentane and pyrrolidine products **291** in 48–72% yield and with 13:1 to 20:1 dr and 51–82% *ee* (Scheme 88).[114]

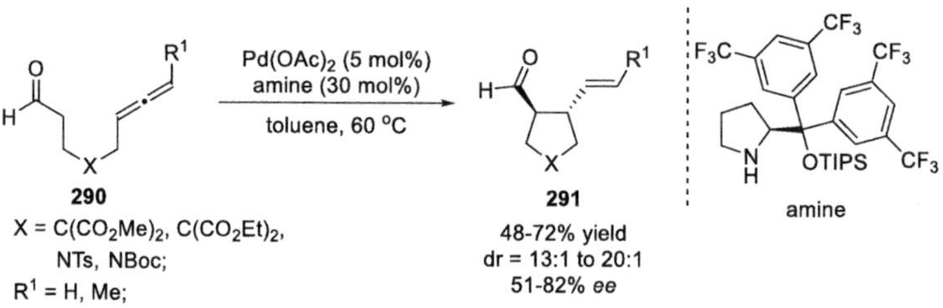

Scheme 88 Pd-catalyzed asymmetric carbocyclization of aldehyde-linked allenes.

The process achieved the nucleophilic addition of allenes **290** via an enamine and Pd-activated intermediate.

The formation of two new C—C and C-X (X=C, O) bonds using Pd-catalyzed cascade reactions between allenyl derivatives and organic halides is one useful method to synthesize bisubstituted alkenes. In 2007, Ma and co-workers reported the asymmetric cyclization reaction of optically active 3,4-allenylic hydrazines **292** with organic halides **293** catalyzed by Pd(OAc)$_2$/(R,R)-Bn-Box (Scheme 89, A).[115] In general, the reaction carried out in THF at 80 °C in the presence of Ag$_3$PO$_4$ gave the corresponding pyrazolidines **294** with two stereocenters in good yields and high diastereoselectivites with excellent *ee*s (up to 99%). Next, a novel chiral spiro-bisoxazoline ligand (*Ra,S,S*)-**296** was developed by Ma group, which was applied to the synthesis of optically activated pyrazolidine derivatives **297** via the Pd(0)-catalyzed enantioselective cyclization reaction of 3,4-allenyl hydrazines **295** with aromatic iodides **293** (Scheme 89B).[116]

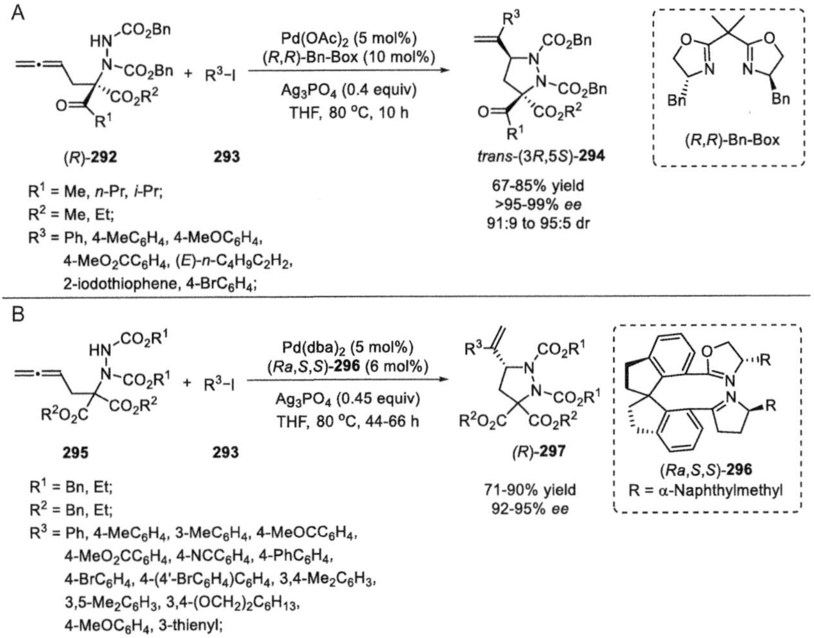

Scheme 89 Pd-catalyzed asymmetric cyclization of 3,4-allenylic hydrazines and organic iodides.

Using this strategy, the nucleophile was further expanded to oxygen for the construction of chiral O-heterocycles. The cascade coupling/cyclization reaction of terminal γ-allenols **298** with aryl iodides using Pd(dba)$_2$/ bisphosphine ligand (R,R)-**288** as the catalyst and K$_3$PO$_4$ as the base in a mixed CH$_3$CN/(CH$_3$)$_3$CCN solvent at 90 °C afforded the desired tetrahydrofurans **299** bearing a styrenyl substituent in 60–86% yields with 85–92% ees (Scheme 90).[117] Notably, aryl iodides with an electron-donating group need lower concentration and less reaction time compared to the ones with an electron-withdrawing group in this transformation.

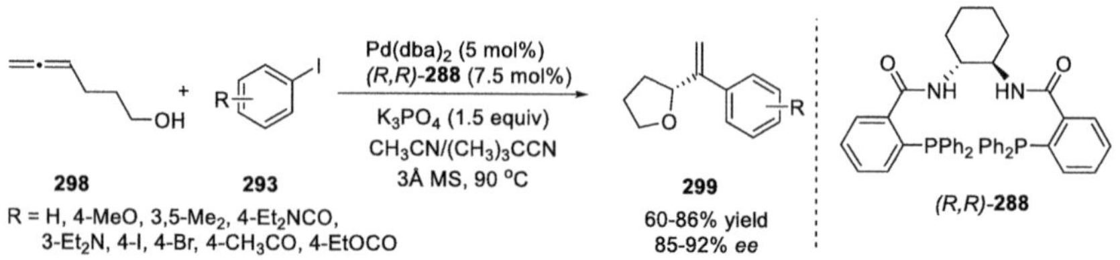

Scheme 90 Pd-catalyzed asymmetric coupling cyclization reaction of terminal γ-allenols with aryl iodides.

A similar cascade process involving a coupling of allenes **300** with organic iodides **293** and nucleophilic attack of indanone ketoamides carried out with catalytic amounts of Pd(OAc)$_2$ and chiral bis(oxazoline) ligand **301** in the presence of silver phosphate, afforded aryl- or vinyl-substituted spirolactam products **302** (Scheme 91).[118] A wide range of aryl or vinyl iodides and substituted allene-linked indanone ketoamide underwent the highly diastereo- and enantioselective cascade cyclization reaction to give the corresponding products **302** as a single diastereomer in

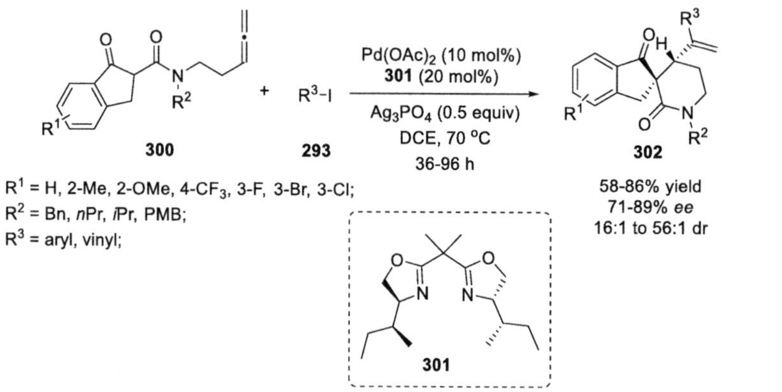

Scheme 91 Pd-catalyzed asymmetric carbocyclization of indanone ketoamide-linked allenes and aryl or vinyl iodides.

58–86% yield and 71–89% *ee*. This reaction provided a novel method for complex molecule synthesis.

In 2014, Ree and co-workers reported a two-step protocol involving a Pd-catalyzed intermolecular asymmetric addition of unactivated aliphatic alcohols **303** to alkoxyallenes **304** and a subsequent ring-closing-metathesis (RCM) reaction, which led to five to seven-membered cyclic acetals **306** in good yield and high *ee* with >10:1 ratio (Scheme 92).[119] The use of Trost ligand **288** as chiral ligand resulted in a high level of enantioselectivity in the hydroalkoxylation step. This strategy offers an efficient approach to synthesize divergent mono- and oligosaccharides.

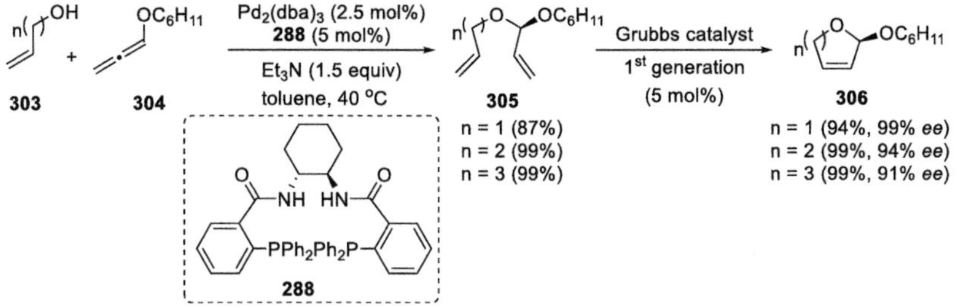

Scheme 92 Pd-catalyzed asymmetric hydroalkoxylation of alkoxyallenes.

Later, Wang and Ding described the preparation of enantioselective α-methylene-β-arylamino acid esters **308** starting from terminal allenes, CO, methanol and arylamines through a cascade alkoxycarbonylation-amination process (Scheme 93).[120] The Pd(OAc)$_2$/aromatic spiroketal-based diphosphine was the most efficient catalytic system, where the copper salt was used as an oxidant. The reaction tolerates a series of terminal allenes **307** bearing an aromatic or aliphatic substituent, affording the corresponding products **308** in good yields and excellent *ee*s with high branched regioselectivity.

In 2017, Luo and co-workers developed a new type of Pd-catalyzed asymmetric addition of α-branched β-ketocarbonyls **310** and allenes **311** under chiral primary amine/achiral palladium catalysis via enamine intermediates (Scheme 94).[121] The reaction gave an excellent linear selectivity with a wide range of allenes **311** with aryl, aliphatic and 1,1-disubstituents, providing allylic products **313** in good yields and high enantioselectivities. Additionally, α-branched aldehydes **314** were also applied in the asymmetric addition reaction, leading to the corresponding adducts **315** bearing an all-carbon sterocenter in 70–82% yields with 83–91% *ee*s.

Scheme 93 Pd-catalyzed asymmetric cascade alkoxycarbonylation-amination of terminal allenes.

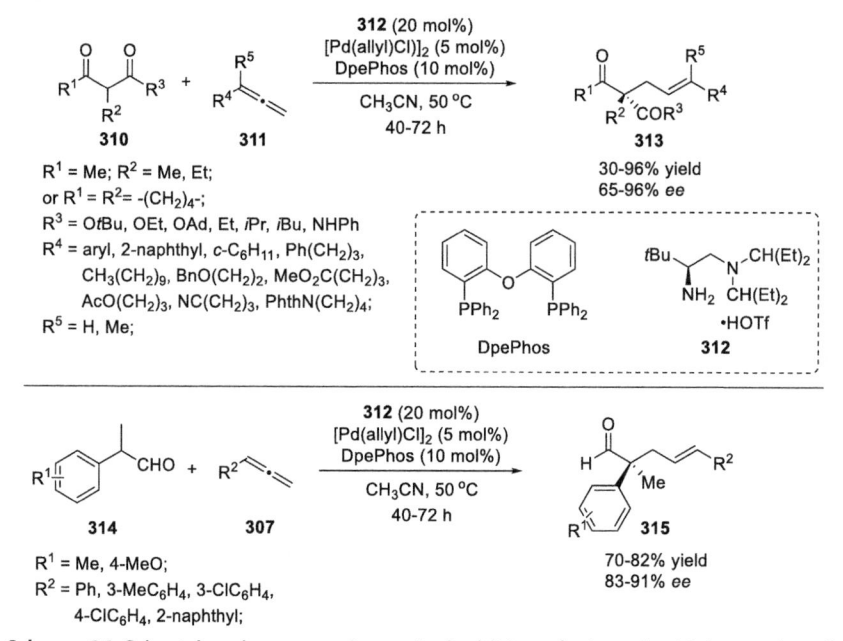

Scheme 94 Pd-catalyzed asymmetric terminal addition of α-branched β-ketocarbonyls to allenes.

Shi and co-workers described an enantioselective hydroarylation of cumulene derivatives **316** with organoboronic reagents **5** catalyzed by (R)-[(N-Me-NHC)]Pd^{2+}-diaqua complex **143a** to provide chiral allenic products **317** in 51–96% yields with 50–94% ees (Scheme 95).[122] The asymmetric reaction of diverse substituted phenyl, naphthyl, phenylvinyl or thienylboronic acids **5** to cumulenes **316** with two aryl substituents (R^1=aryl) gave the corresponding products **317** with excellent regioselectivity, which could be transformed into synthetically useful compounds.

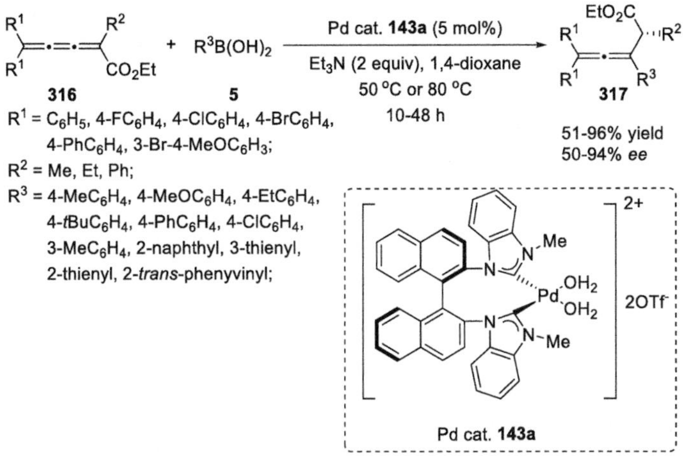

Scheme 95 Pd-catalyzed asymmetric addition of cumulene derivatives and arylboronic acids.

4. Conclusion and outlook

Palladium-catalyzed asymmetric addition reactions have been developed quite widely in the past decades, constituting one of the most powerful methodologies in organic synthesis for the formation of chiral compounds. The asymmetric conjugate addition of nucleophiles to electron-deficient carbon–carbon double bonds was extensively used in the formation of the carbon–carbon and carbon–heteroatom bonds. The asymmetric 1,2-addition of nucleophiles to aldehydes, ketones, aldimines, ketimines, olefins, allenes and cumulenes provided a powerful approach to construct chiral alcohols, amines and allylic compounds. A variety of chiral palladium complexes based on chiral ligands have been discussed in the Pd-catalyzed asymmetric addition reactions, including chiral bidentated imine ligands, chiral oxazoline ligands, chiral bisoxazoline ligands, chiral NHC ligands, Trost ligands and many more. In some cases, a dual catalytic system involving a chiral secondary amine and a palladium salt could lead to the stereoselective addition of aldehyde or ketone to alkenes or allenes. Moreover, the cascade processes involving Pd-catalyzed asymmetric addition were discovered, which offered an efficient approach toward the chiral skeletons not easily obtained through other methods. These strategies were also used as a key method for the construction of natural products and medicine molecules. Despite the excellent recent progress in Pd-catalyzed asymmetric additions, more efficient catalytic systems are still highly needed to expand the reaction substrates to more challenging compounds under mild conditions and to induce high control in the reaction regioselectivity and enantioselectivity.

Acknowledgments

We are grateful to the National Natural Science Foundation of China (21672067, 21702063), Shanghai Sailing Program (16YF1402800), 973 Programs (2015CB856600), and Changjiang Scholars and Innovative Research Team in University (PCSIRT) for financial support.

References

1. (a) Miyaura N. Celebrating 20 years of *SYNLETT*—special account on palladium(II)-catalyzed additions of arylboronic acids to electron-deficient alkenes, aldehydes, imines, and nitriles. *Synlett*. 2009;13:2039–2050. (b) Sun YW, Zhu PL, Xu Q, Shi M. Development of Pd catalyzed asymmetric additions in the last five years. *RSC Adv*. 2013;3:3153–3168. (c) Shockley SE, Holder JC, Stoltz BM. Palladium-catalyzed asymmetric conjugate addition of arylboronic acids to α,β-unsaturated cyclic electrophiles. *Org Process Res Dev*. 2015;19:974–981. (d) Wu L, Shen J, Yang G, Zhang W. Recent advances in the Pd(II)-catalyzed asymmetric addition of arylboronic acids to electron-deficient olefins. *Tetrahedron Lett*. 2018;59:4055–4062.
2. (a) Ping Y, Li Y, Zhu J, Kong W. Construction of quaternary stereocenters by palladium-catalyzed carbopalladation-initiated cascade reactions. *Angew Chem Int Ed*. 2019;58:1562–1573. (b) Wu X, Gong LZ. Palladium(0)-catalyzed difunctionalization of 1,3-dienes: from racemic to enantioselective. *Synthesis*. 2019;51:122–134.
3. (a) Christoffers J, Koripelly G, Rosiak A, Rössle M. Recent advances in metal-catalyzed asymmetric conjugate additions. *Synthesis*. 2007;9:1279–1300. (b) Zheng K, Liu X, Feng X. Recent advances in metal-catalyzed asymmetric 1,4-conjugate addition (ACA) of nonorganometallic nucleophiles. *Chem Rev*. 2018;118:7586–7656.
4. (a) Zhang T, Shi M. Chiral bidentate bis(N-heterocyclic carbene)-based palladium complexes bearing carboxylate ligands: highly effective catalysts for the enantioselective conjugate addition of arylboronic acids to cyclic enones. *Chem Eur J*. 2008;14:3759–3764. (b) Xu Q, Zhang R, Zhang T, Shi M. Asymmetric 1,4-addition of arylboronic acids to 2,3-dihydro-4-pyridones catalyzed by axially chiral NHC-Pd(II) complexes. *J Org Chem*. 2010;75:3935–3937.

5. Morisaki Y, Imoto H, Hirano K, Hayashi T, Chujo Y. Synthesis of enantiomerically pure P-stereogenic diphosphacrowns and their palladium complexes. *J Org Chem.* 2011;76:1795–1803.
6. Wong J, Gan K, Chen HJ, Pullarkat SA. Evaluation of palladacycles as a non-rhodium based alternative for the asymmetric conjugate 1,4-addition of arylboronic acids to α,β-unsaturated enones. *Adv Synth Catal.* 2014;356:3391–3400.
7. Tamura M, Ogata H, Ishida Y, Takahashi Y. Design and synthesis of chiral 1,10-phenanthroline ligand, and application in palladium catalyzed asymmetric 1,4-addition reactions. *Tetrahedron Lett.* 2017;58:3808–3813.
8. Lin S, Lu X. Cationic Pd(II)/bipyridine-catalyzed conjugate addition of arylboronic acids to β,β-disubstituted enones: construction of quaternary carbon centers. *Org Lett.* 2010;12:2536–2539.
9. Kikushima K, Holder JC, Gatti M, Stoltz BM. Palladium-catalyzed asymmetric conjugate addition of arylboronic acids to five-, six-, and seven-membered β-substituted cyclic enones: enantioselective construction of all-carbon quaternary stereocenters. *J Am Chem Soc.* 2011;133:6902–6905.
10. (a) Holder JC, Zou L, Marziale AN, et al. Mechanism and enantioselectivity in palladium-catalyzed conjugate addition of arylboronic acids to β-substituted cyclic enones: insights from computation and experiment. *J Am Chem Soc.* 2013;135:14996–15007. (b) Holder JC, Goodman ED, Kikushima K, Gatti M, Marziale AN, Stoltz BM. Synthesis of diverse β-quaternary ketones via palladium-catalyzed asymmetric conjugate addition of arylboronic acids to cyclic enones. *Tetrahedron.* 2015;71:5781–5792.
11. Holder JC, Marziale NM, Mao GB, Stoltz BM. Palladium-catalyzed asymmetric conjugate addition of arylboronic acids to heterocyclic acceptors. *Chem Eur J.* 2013;19:74–77.
12. Shockley SE, Holder JC, Stoltz BM. A catalytic, enantioselective formal synthesis of (+)-dichroanone and (+)-taiwaniaquinone H. *Org Lett.* 2014;16:6362–6365.
13. Kadam AA, Ellern A, Stanley LM. Enantioselective, palladium-catalyzed conjugate additions of arylboronic acids to form bis-benzylic quaternary stereocenters. *Org Lett.* 2017;19:4062–4065.
14. Gottumukkala AL, Matcha K, Lutz M, de Vries JG, Minnaard AJ. Palladium-catalyzed asymmetric quaternary stereocenter formation. *Chem Eur J.* 2012;18:6907–6914.
15. Buter J, Moezelaar R, Minnaard AJ. Enantioselective palladium catalyzed conjugate additions of ortho-substituted arylboronic acids to β,β-disubstituted cyclic enones: total synthesis of herbertenediol, enokipodin A and enokipodin B. *Org Biomol Chem.* 2014;12:5883–5890.
16. Retamosa MG, Álvarez-Casao Y, Matador E, et al. Pyridine-hydrazone ligands in asymmetric palladium-catalyzed 1,4- and 1,6-additions of arylboronic acids to cyclic (Di)enones. *Adv Synth Catal.* 2019;361:176–184.
17. Eitel SH, Jautze S, Frey W, Peters R. Asymmetric michael additions of α-cyanoacetates by soft Lewis acid/hard Brønsted acid catalysis: stereodivergency with bi- vs. monometallic catalysts. *Chem Sci.* 2013;4:2218–2233.
18. Huang Y, Li Y, Leung PH, Hayashi T. Asymmetric synthesis of P-stereogenic diarylphosphinites by palladium-catalyzed enantioselective addition of diarylphosphines to benzoquinones. *J Am Chem Soc.* 2014;136:4865–4868.
19. (a) Nishikata T, Kiyomura S, Yamamoto Y, Miyaura N. Asymmetric 1,4-addition of arylboronic acids to α,β-unsaturated esters catalyzed by dicationic palladium(II)-chiraphos complex for short-step synthesis of SmithKline Beecham's endothelin receptor antagonist. *Synlett.* 2008;16:2487–2490. (b) Nishikata T, Yamamotob Y, Miyaurab N. Palladium(II)-catalyzed 1,4-addition of arylboronic acids to β-arylenals for enantioselective syntheses of 3,3-diarylalkanals: a short synthesis of (+)-(R)-CDP

840. *Tetrahedron Lett.* 2007;48:4007–4010. c Nishikata T, Yamamoto Y, Miyaurab N. Palladium(II)-catalyzed 1,4-addition of arylboronic acids to β-arylenones for enantioselective synthesis of 4-aryl-4H-chromenes. *Adv Synth Catal.* 2007;349: 1759–1764. d Nishikata T, Kobayashi Y, Kobayashi K, Yamamoto Y, Miyaura N. Tandem conjugate addition-aldol cyclization to give optically active 1-aryl-1H-indenes via asymmetric Pd^{2+}-catalyzed 1,4-addition of aryl-boronic acids. *Synlett.* 2007;15:3055–3057.
20. Suzuma Y, Hayashi S, Yamamoto T, Oe Y, Ohta T, Ito Y. Asymmetric 1,4-addition of organoboronic acids to α,β-unsaturated ketones and 1,2-addition to aldehydes catalyzed by a palladium complex with a ferrocene-based phosphine ligand. *Tetrahedron: Asymmetry.* 2009;20:2751–2758.
21. Petri A, Seidelmann O, Eilitz U, et al. Pd-catalyzed asymmetric conjugate addition of arylboronic acids to 2-nitroacrylates: a facile route to β^2-homophenylglycines. *Tetrahedron Lett.* 2014;55:267–270.
22. He Q, Xie F, Fu G, et al. Palladium-catalyzed asymmetric addition of arylboronic acids to nitrostyrenes. *Org Lett.* 2015;17:2250–2253.
23. Sun L, Li D, Zhou X, et al. General and catalytic enantioselective approach to isopavine alkaloids. *J Org Chem.* 2017;82:12899–12907.
24. Chen S, Wu L, Shao G, Yang G, Zhang W. Pd(II)-catalyzed asymmetric 1,6-conjugate addition of arylboronic acids to Meldrum's acid-derived dienes. *Chem Commun.* 2018;54:2522–2525.
25. Feng JJ, Chen XF, Shi M, Duan WL. Palladium-catalyzed asymmetric addition of diarylphosphines to enones toward the synthesis of chiral phosphines. *J Am Chem Soc.* 2010;132:5562–5563.
26. Song YC, Dai GF, Xiao F, Duan WL. Palladium-catalyzed enantioselective hydrophosphination of enones for the synthesis of chiral P,N-compounds. *Tetrahedron Lett.* 2016;57:2990–2993.
27. Du D, Duan WL. Palladium-catalyzed 1,4-addition of diarylphosphines to α,β-unsaturated N-acylpyrroles. *Chem Commun.* 2011;47:11101–11103.
28. Lu JZ, Ye JX, Duan WL. Palladium-catalyzed asymmetric addition of diarylphosphines to α,β-unsaturated sulfonic esters for the synthesis of chiral phosphine sulfonate compounds. *Org Lett.* 2013;15:5016–5019.
29. Du D, Lin ZQ, Lu JZ, Li C, Duan WL. Palladium-catalyzed asymmetric 1,4-addition of diarylphosphines to α,β-unsaturated carboxylic esters. *Asian J Org Chem.* 2013;2:392–394.
30. Chen YR, Duan WL. Palladium-catalyzed 1,4-addition of diarylphosphines to α,β-unsaturated aldehydes. *Org Lett.* 2011;13:5824–5826.
31. Feng JJ, Huang M, Lin ZQ, Duan WL. Palladium-catalyzed asymmetric 1,4-addition of diarylphosphines to nitroalkenes for the synthesis of chiral P,N-compounds. *Adv Synth Catal.* 2012;354:3122–3126.
32. (a) Ding B, Zhang Z, Xu Y, et al. P-Stereogenic PCP pincer-Pd complexes: synthesis and application in asymmetric addition of diarylphosphines to nitroalkenes. *Org Lett.* 2013;15:5476–5479. (b) Xu Y, Yang Z, Ding B, et al. Asymmetric michael addition of diphenylphosphine to β,γ-unsaturated α-keto esters catalyzed by a P-stereogenic pincer-Pd complex. *Tetrahedron.* 2015;71:6832–6839.
33. (a) Yang MJ, Liu YJ, Gong JF, Song MP. Unsymmetrical chiral PCN pincer palladium(II) and nickel(II) complexes with aryl-based aminophosphine-imidazoline ligands: synthesis via aryl C-H activation and asymmetric addition of diarylphosphines to enones. *Organometallics.* 2011;30:3793–3803. (b) Hao XQ, Huang JJ, Wang T, Lv J, Gong JF, Song MP. PCN pincer palladium(II) complex catalyzed enantioselective hydrophosphination of enones: synthesis of pyridine-functionalized chiral phosphine oxides as NC$_{sp3}$O pincer preligands. *J Org Chem.* 2014;79:9512–9530. (c) Hao XQ,

Zhao YW, Yang JJ, Niu JL, Gong JF, Song MP. Enantioselective hydrophosphination of enones with diphenylphosphine catalyzed by bis(imidazoline) NCN pincer palladium(II) complexes. *Organometallics.* 2014;33:1801–1811.
34. Li C, Bian QL, Xu S, Duan WL. Palladium-catalyzed 1,4-addition of secondary alkylphenylphosphines to α,β-unsaturated carbonyl compounds for the synthesis of phosphorus- and carbon-stereogenic compounds. *Org Chem Front.* 2014;1: 541–545.
35. (a) Huang Y, Pullarkat SA, Li Y, Leung PH. Palladium(II)-catalyzed asymmetric hydrophosphination of enones: efficient access to chiral tertiary phosphines. *Chem Commun.* 2010;46:6950–6952. (b) Huang Y, Pullarkat SA, Li Y, Leung PH. Palladacycle-catalyzed asymmetric hydrophosphination of enones for synthesis of C*- and P*-chiral tertiary phosphines. *Inorg Chem.* 2012;51:2533–2540. (c) Yang XY, Jia YX, Tay WS, Li Y, Pullarkat SA, Leung PH. Mechanistic insights into the role of PC- and PCP-type palladium catalysts in asymmetric hydrophosphination of activated alkenes incorporating potential coordinating heteroatoms. *Dalton Trans.* 2016;45:13449–13455.
36. (a) Huang YH, Chew RJ, Li YX, Pullarkat SA, Leung PH. Direct synthesis of chiral tertiary diphosphines via Pd(II)-catalyzed asymmetric hydrophosphination of dienones. *Org Lett.* 2011;13:5862–5865. (b) Huang Y, Pullarkat SA, Teong S, Chew RJ, Li Y, Leung PH. Palladacycle-catalyzed asymmetric intermolecular construction of chiral tertiary P-heterocycles by stepwise addition of H-P-H bonds to bis(enones). *Organometallics.* 2012;31:4871–4875.
37. (a) Chew RJ, Huang Y, Li Y, Pullarkat SA, Leung PH. Enantioselective addition of diphenylphosphine to 3-methyl-4-nitro-5-alkenylisoxazoles. *Adv Synth Catal.* 2013;355:1403–1408. (b) Huang Y, Chew RJ, Pullarkat SA, Li Y, Leung PH. Asymmetric synthesis of enaminophosphines via palladacycle-catalyzed addition of Ph$_2$PH to α,β-unsaturated imines. *J Org Chem.* 2012;77:6849–6854. (c) Gan K, Sadeer A, Xu C, Li Y, Pullarkat SA. Asymmetric construction of a ferrocenyl phosphapalladacycle from achiral enones and a demonstration of its catalytic potential. *Organometallics.* 2014;33:5074–5076.
38. (a) Chew RJ, Lu Y, Jia YX, et al. Palladacycle catalyzed asymmetric P-H addition of diarylphosphines to N-enoyl phthalimides. *Chem Eur J.* 2014;20:14514–14517. (b) Chew RJ, Teo KY, Huang Y, et al. Enantioselective phospha-Michael addition of diarylphosphines to β,γ-unsaturated α-ketoesters and amides. *Chem Commun.* 2014;50:8768–8770. (c) Chew RJ, Sepp K, Li BB, et al. An approach to the efficient syntheses of chiral phosphino-carboxylic acid esters. *Adv Synth Catal.* 2015;357:3297–3302.
39. (a) Huang J, Zhao MX, Duan WL. Palladium-catalyzed asymmetric 1,6-addition of diphenylphosphine to (4-aryl-1,3-butadienylidene)bis(phosphonates) for the synthesis of chiral phosphines. *Tetrahedron Lett.* 2014;55:629–631. (b) Lu J, Ye J, Duan WL. Palladium-catalyzed asymmetric 1,6-addition of diarylphosphines to $\alpha,\beta,\gamma,\delta$-unsaturated sulfonic esters: controlling regioselectivity by rational selection of electron-withdrawing groups. *Chem Commun.* 2014;50:698–700.
40. Yang XY, Gan JH, Li Y, Pullarkat SA, Leung PH. Palladium catalyzed asymmetric hydrophosphination of α,β- and $\alpha,\beta,\gamma,\delta$-unsaturated malonate esters-efficient control of reactivity, stereo- and regio-selectivity. *Dalton Trans.* 2015;44:1258–1263.
41. (a) Yang XY, Tay WS, Li Y, Pullarkat SA, Leung PH. Asymmetric 1,4-conjugate addition of diarylphosphines to $\alpha,\beta,\gamma,\delta$-unsaturated ketones catalyzed by transition-metal pincer complexes. *Organometallics.* 2015;34:5196–5201. (b) Yang XY, Tay WS, Li Y, Pullarkat SA, Leung PH. The synthesis and efficient one-pot catalytic "self-breeding" of asymmetrical NC(sp^3)E-hybridised pincer complexes. *Chem Commun.* 2016;52:4211–4214.

42. Gini F, Hessen B, Feringa BL, Minnaard AJ. Enantioselective palladium-catalyzed conjugate addition of arylsiloxanes. *Chem Commun.* 2007;710–712.
43. (a) Li K, Hii KK. Dicationic [(BINAP)Pd-(solvent)$_2$]$^{2+}$[TfO$^-$]$_2$: enantioselective hydroamination catalyst for alkenoyl-*N*-oxazolidinones. *Chem Commun.* 2003; 1132–1133. (b) Li K, Cheng X, Hii KK. Asymmetric synthesis of β-amino acid and amide derivatives by catalytic conjugate addition of aromatic amines to *N*-alkenoylcarbamates. *Eur J Org Chem.* 2004;2004:959–964. (c) Hamashima Y, Somei H, Shimura Y, Tamura T, Sodeoka M. Amine-salt-controlled, catalytic asymmetric conjugate addition of various amines and asymmetric protonation. *Org Lett.* 2004;6:1861–1864. (d) Phua PH, de Vries JG, Hii KK. Palladium-catalysed enantioselective conjugate addition of aromatic amines to α,β-unsaturated *N*-imides. *Adv Synth Catal.* 2005;347:1775–1780. (e) Phua PH, White AJP, de Vries JG, Hii KK. Enabling ligand screening for palladium-catalysed enantioselective Aza-Michael addition reactions. *Adv Synth Catal.* 2006;348:587–592. (f) Phua PH, Mathew SP, White AJP, de Vries JG, Blackmond DG, Hii KK. Elucidating the mechanism of the asymmetric Aza-Michael reaction. *Chem Eur J.* 2007;13:4602–4613. (g) Sheshenev AE, Smith AM, Hii KK. Preparation of dicationic palladium catalysts for asymmetric catalytic reactions. *Nat Protoc.* 2012;7:1765–1773.
44. Kang SH, Kang YK, Kim DY. Catalytic enantioselective conjugate addition of aromatic amines to fumarate derivatives: asymmetric synthesis of aspartic acid derivatives. *Tetrahedron.* 2009;65:5676–5679.
45. Hamashima Y, Tamura T, Suzuki S, Sodeoka M. Enantioselective protonation in the Aza-Michael reaction using a combination of chiral Pd-μ-hydroxo complex with an amine salt. *Synlett.* 2009;10:1631–1634.
46. Sodeoka M, Hamashima Y. Chiral Pd aqua complex-catalyzed asymmetric C-C bond-forming reactions: a Brønsted acid-base cooperative system. *Chem Commun.* 2009;5787–5798.
47. Weber M, Jautze S, Frey W, Peters R. Bispalladacycle-catalyzed Michael addition of in situ formed azlactones to enones. *Chem Eur J.* 2012;18:14792–14804.
48. Weber M, Frey W, Peters R. Catalytic asymmetric synthesis of functionalized α,α-disubstituted α-amino acid derivatives from racemic unprotected α-amino acids via in-situ generated azlactones. *Adv Synth Catal.* 2012;354:1443–1449.
49. Weber M, Frey W, Peters R. Catalytic asymmetric synthesis of spirocyclic azlactones by a double Michael-addition approach. *Chem Eur J.* 2013;19:8342–8351.
50. Weber M, Frey W, Peters R. Asymmetric palladium(II)-catalyzed cascade reaction giving quaternary amino succinimides by 1,4-addition and a Nef-type reaction. *Angew Chem Int Ed.* 2013;52:13223–13227.
51. Aikawa K, Honda K, Mimura S, Mikami K. Highly enantioselective Friedel-Crafts alkylation of indole and pyrrole with β,γ-unsaturated α-ketoester catalyzed by chiral dicationic palladium complex. *Tetrahedron Lett.* 2011;52:6682–6686.
52. (a) Kang YK, Suh KH, Kim DY. Catalytic enantioselective Friedel-Crafts alkylation of indoles with β,γ-unsaturated α-keto phosphonates in the presence of chiral palladium complexes. *Synlett.* 2011;8:1125–1128. (b) Kang YK, Kwon BK, Mang JY, Kim DY. Chiral Pd-catalyzed enantioselective Friedel–Crafts reaction of indoles with γ,δ-unsaturated β-keto phosphonates. *Tetrahedron Lett.* 2011;52:3247–3249. (c) Lee HJ, Kim DY. Catalytic enantioselective Friedel-Crafts alkylation of indoles with fumarate derivatives in the presence of chiral palladium complexes. *Synlett.* 2012;23: 1629–1632.
53. Kitanosono T, Miyo M, Kobayashi S. Surfactant-aided chiral palladium(II) catalysis exerted exclusively in water for the C-H functionalization of indoles. *ACS Sustainable Chem Eng.* 2016;4:6101–6106.

54. (a) Quan M, Wu L, Yang G, Zhang W. Pd(II), Ni(II) and Co(II)-catalyzed enantioselective additions of organoboron reagents to ketimines. *Chem Commun.* 2018;54:10394–10404.
55. (a) Shintani R, Inoue M, Hayashi T. Rhodium-catalyzed asymmetric addition of aryl- and alkenylboronic acids to isatins. *Angew Chem Int Ed.* 2006;45:3353–3356. (b) Toullec PY, Jagt RBC, de Vries JG, Feringa BL, Minnaard AJ. Rhodium-catalyzed addition of arylboronic acids to isatins: an entry to diversity in 3-aryl-3-hydroxyoxindoles. *Org Lett.* 2006;8:2715–2718. (c) Shintani R, Takatsu K, Hayashi T. Copper-catalyzed asymmetric addition of arylboronates to isatins: a catalytic cycle involving alkoxocopper intermediates. *Chem Commun.* 2010;46:6822–6824. (d) Gui J, Chen G, Cao P, Liao J. Rh(I)-catalyzed asymmetric addition of arylboronic acids to NH isatins. *Tetrahedron: Asymmetry.* 2012;23:554–563.
56. Lai H, Huang ZH, Wu Q, Qin Y. Synthesis of novel enantiopure biphenyl *P,N*-ligands and application in palladium-catalyzed asymmetric addition of arylboronic acids to *N*-benzylisatin. *J Org Chem.* 2009;74:283–288.
57. Liu Z, Gu P, Shi M, McDowell P, Li G. Catalytic asymmetric addition of arylboronic acids to isatins using C_2-symmetric cationic *N*-heterocyclic carbenes (NHCs) Pd^{2+} diaqua complexes as catalysts. *Org Lett.* 2011;13:2314–2317.
58. Li Q, Wan P, Wang S, et al. Synthesis of a class of new phosphine-oxazoline ligands and their applications in palladium-catalyzed asymmetric addition of arylboronic acids to isatins. *Appl Catal, A.* 2013;458:210–216.
59. Zhang R, Xu Q, Zhang X, Zhang T, Shi M. Axially chiral C_2-symmetric *N*-heterocyclic carbene (NHC) palladium complexes-catalyzed asymmetric arylation of aldehydes with arylboronic acids. *Tetrahedron: Asymmetry.* 2010;21:1928–1935.
60. (a) Mikami K, Kawakami Y, Akiyama K, Aikawa K. Enantioselective catalysis of ketoester-ene reaction of silyl enol ether to construct quaternary carbons by chiral dicationic palladium(II) complexes. *J Am Chem Soc.* 2007;129:12950–12951. (b) Aikawa K, Mimura S, Numata Y, Mikami K. Palladium-catalyzed enantioselective ene and aldol reactions with isatins, keto esters, and diketones: reliable approach to chiral tertiary alcohols. *Eur J Org Chem.* 2011;2011:62–65.
61. Aikawa K, Hioki Y, Mikami K. Highly enantioselective alkenylation of glyoxylate with vinylsilane catalyzed by chiral dicationic palladium(II) complexes. *J Am Chem Soc.* 2009;131:13922–13923.
62. Aikawa K, Hioki Y, Mikami K. Highly enantioselective alkynylation of trifluoropyruvate with alkynylsilanes catalyzed by the BINAP-Pd complex: access to α-trifluoromethyl-substituted tertiary alcohols. *Org Lett.* 2010;12:5716–5719.
63. Aikawa K, Yuya A, Hioki Y, Mikami K. Catalytic and highly enantioselective Friedel-Crafts type reactions of heteroaromatic compounds with trifluoropyruvate and glyoxylate by a dicationic palladium complex. *Tetrahedron: Asymmetry.* 2014;25:1104–1115.
64. (a) Zanoni G, Gladiali S, Marchetti A, Piccinini P, Tredici I, Vidari G. Enantioselective catalytic allylation of carbonyl groups by umpolung of π-allyl palladium complexes. *Angew Chem Int Ed.* 2004;43:846–849. (b) Zhu SF, Yang Y, Wang LX, Liu B, Zhou QL. Synthesis and application of chiral spiro phospholane ligand in Pd-catalyzed asymmetric allylation of aldehydes with allylic alcohols. *Org Lett.* 2005;7:2333–2335. (c) Howell GP, Minnaard AJ, Feringa BL. Asymmetric allylation of aryl aldehydes: studies on the scope and mechanism of the palladium catalysed diethylzinc mediated umpolung using phosphoramidite ligands. *Org Biomol Chem.* 2006;4:1278–1283.
65. Wang W, Zhang T, Shi M. Chiral bis(NHC)-palladium(II) complex catalyzed and diethylzinc-mediated enantioselective umpolung allylation of aldehydes. *Organometallics.* 2009;28:2640–2642.

66. Jiang JJ, Wang D, Wang WF, et al. Pd(II)-catalyzed and diethylzinc-mediated asymmetric umpolung allylation of aldehydes in the presence of chiral phosphine-Schiff base type ligands. *Tetrahedron: Asymmetry*. 2010;21:2050–2054.
67. Zhu SF, Qiao XC, Zhang YZ, Wang LX, Zhou QL. Highly enantioselective palladium-catalyzed umpolung allylation of aldehydes. *Chem Sci*. 2011;2: 1135–1140.
68. (a) Dai H, Lu X. Diastereoselective synthesis of arylglycine derivatives by cationic palladium(II)-catalyzed addition of arylboronic acids to *N-tert*-butanesulfinyl imino esters. *Org Lett*. 2007;9:3077–3080. (b) Dai H, Yang M, Lu X. Palladium(II)-catalyzed one-pot enantioselective synthesis of arylglycine derivatives from ethyl glyoxylate, *p*-toluenesulfonyl isocyanate and arylboronic acids. *Adv Synth Catal*. 2008;350:249–253. (c) Dai H, Lu X. Palladium(II)/2,2-bipyridine-catalyzed addition of arylboronic acids to *N*-tosyl-arylaldimines. *Tetrahedron Lett*. 2009;50:3478–3481.
69. Ma GN, Zhang T, Shi M. Catalytic enantioselective arylation of *N*-tosylarylimines with arylboronic acids using C_2-symmetric cationic N-heterocyclic carbene Pd^{2+} diaquo complexes. *Org Lett*. 2009;11:875–878.
70. Liu Z, Shi M. Catalytic asymmetric addition of arylboronic acids to *N*-Boc imines generated in situ using C_2-symmetric cationic *N*-heterocyclic carbenes (NHCs) Pd^{2+} diaquo complexes. *Tetrahedron*. 2010;66:2619–2623.
71. Liu Z, Shi M. Efficient chirality switching in the asymmetric addition of indole to *N*-tosylarylimines in the presence of axially chiral cyclometalated bidentate N-heterocyclic carbene palladium(II) complexes. *Tetrahedron: Asymmetry*. 2009;20: 119–123.
72. Schrapel C, Peters R. Exogenous-base-free palladacycle-catalyzed highly enantioselective arylation of imines with arylboroxines. *Angew Chem Int Ed*. 2015;54: 10289–10293.
73. Chen J, Lu X, Lou W, Ye Y, Jiang H, Zeng W. Palladium(II)-catalyzed enantioselective arylation of α-imino esters. *J Org Chem*. 2012;77:8541–8548.
74. (a) Yang G, Zhang W. A palladium-catalyzed enantioselective addition of arylboronic acids to cyclic ketimines. *Angew Chem Int Ed*. 2013;52:7540–7544.
75. Quan M, Yang G, Xie F, Gridnev ID, Zhang W. Pd(II)-catalyzed asymmetric addition of arylboronic acids to cyclic N-sulfonyl ketimine esters and a DFT study of its mechanism. *Org Chem Front*. 2015;2:398–402.
76. He Q, Wu L, Kou X, Butt N, Yang G, Zhang W. Pd(II)-catalyzed asymmetric addition of arylboronic acids to isatin-derived ketimines. *Org Lett*. 2016;18:288–291.
77. (a) Jiang C, Lu Y, Hayashi T. High performance of a palladium phosphinooxazoline catalyst in the asymmetric arylation of cyclic *N*-sulfonyl ketimines. *Angew Chem Int Ed*. 2014;53:9936–9938. (b) Zhou B, Jiang C, Gandi VR, Lu Y, Hayashi T. Palladium-catalyzed asymmetric arylation of trifluoromethylated/perfluoroalkylated 2-quinazolinones with high enantioselective. *Chem Eur J*. 2016;22:13068–13071. (c) Yan Z, Wu B, Gao X, Zhou YG. Enantioselective synthesis of quaternary α-aminophosphonates by Pd-catalyzed arylation of cyclic α-ketiminophosphonates with arylboronic acids. *Chem Commun*. 2016;52:10882–10885.
78. Zhou B, Li K, Jiang C, Lu Y, Hayashi T. Modified amino acid-derived phosphine-imine ligands for palladium-catalyzed asymmetric arylation of cyclic *N*-sulfonyl imines. *Adv Synth Catal*. 2017;359:1969–1975.
79. Álvarez-Casao Y, Monge D, Álvarez E, Fernández R, Lassaletta JM. Pyridine-hydrazones as *N,N'*-ligands in asymmetric catalysis: Pd(II)-catalyzed addition of boronic acids to cyclic sulfonylketimines. *Org Lett*. 2015;17:5104–5107.
80. Liu Z, Shi M. Catalytic enantioselective addition of cyclic β-keto esters with activated olefins and N-Boc imines using chiral C_2-summetric cationic Pd^{2+} N-heterocyclic carbene (NHC) diaqua complexes. *Organometallics*. 2010;29:2831–2834.

81. Hyodo K, Nakamura S, Tsuji K, Ogawa T, Funahashi Y, Shibata N. Enantioselective reaction of imines and benzyl nitriles using palladium pincer complexes with C_2-symmetric chiral bis(imidazoline)s. *Adv Synth Catal*. 2011;353:3385–3390.
82. Hyodo K, Nakamura S, Shibata N. Enantioselective Aza-Morita-Baylis-Hillman reactions of acrylonitrile catalyzed by palladium(II) pincer complexes having C_2-symmetric chiral bis(imidazoline) ligands. *Angew Chem Int Ed*. 2012;51:10337–10341.
83. Best D, Kujawa S, Lam HW. Diastereo- and enantioselective Pd(II)-catalyzed additions of 2-alkylazaarenes to N-Boc imines and nitroalkenes. *J Am Chem Soc*. 2012;134: 18193–18196.
84. Qiao XC, Zhu SF, Chen WQ, Zhou QL. Palladium-catalyzed asymmetric umpolung allylation of imines with allylic alcohols. *Tetrahedron: Asymmetry*. 2010;21:1216–1220.
85. Huang M, Li C, Huang J, Duan WL, Xu S. Palladium-catalyzed asymmetric addition of diarylphosphines to N-tosylimines. *Chem Commun*. 2012;48:11148–11150.
86. (a) McDonald RI, Liu G, Stahl SS. Palladium(II)-catalyzed alkene functionalization via nucleopalladation: stereochemical pathways and enantioselective catalytic applications. *Chem Rev*. 2011;111:2981–3019. (b) Wang D, Weinstein AB, White PB, Stahl SS. Ligand-promoted palladium-catalyzed aerobic oxidation reactions. *Chem Rev*. 2018;118:2636–2679.
87. Jensen KH, Pathak TP, Zhang Y, Sigman MS. Palladium-catalyzed enantioselective addition of two distinct nucleophiles across alkenes capable of quinone methide formation. *J Am Chem Soc*. 2009;131:17074–17075.
88. Pathak TP, Gligorich KM, Welm BE, Sigman MS. Synthesis and preliminary biological studies of 3-substituted indoles accessed by a palladium-catalyzed enantioselective alkene difunctionalization reaction. *J Am Chem Soc*. 2010;132:7870–7871.
89. Jana R, Pathak TP, Jensen KH, Sigman MS. Palladium(II)-catalyzed enantio- and diastereoselective synthesis of pyrrolidine derivatives. *Org Lett*. 2012;14:4074–4077.
90. He W, Yip KT, Zhu NY, Yang D. Pd(II)/*t*Bu-quinolineoxazoline: an air-stable and modular chiral catalyst system for enantioselective oxidative cascade cyclization. *Org Lett*. 2009;11:5626–5628.
91. Du W, Gu Q, Li Y, Lin Z, Yang D. Enantioselective palladium-catalyzed oxidative cascade cyclization of aliphatic alkenyl amides. *Org Lett*. 2017;19:316–319.
92. (a) McDonald RI, White PB, Weinstein AB, Tam CP, Stahl SS. Enantioselective Pd(II)-catalyzed aerobic oxidative amidation of alkenes and insights into the role of electronic asymmetry in pyridine-oxazoline ligands. *Org Lett*. 2011;13:2830–2833. (b) Weinstein AB, Stahl SS. Reconciling the stereochemical course of nucleopalladation with the development of enantioselective Wacker-type cyclizations. *Angew Chem Int Ed*. 2012;51:11505–11509.
93. Yang G, Shen C, Zhang W. An asymmetric aerobic Aza-Wacker-type cyclization: synthesis of isoindolinones bearing tetrasubstituted carbon stereocenters. *Angew Chem Int Ed*. 2012;51:9141–9145.
94. Ingalls EL, Sibbald PA, Kaminsky W, Michael FE. Enantioselective palladium-catalyzed diamination of alkenes using N-fluorobenzenesulfonimide. *J Am Chem Soc*. 2013;135:8854–8856.
95. Zhang W, Chen P, Liu G. Enantioselective palladium(II)-catalyzed intramolecular aminoarylation of alkenes by dual N-H and aryl C-H bond cleavage. *Angew Chem Int Ed*. 2017;56:5336–5340.
96. Qi X, Chen C, Hou C, Fu L, Chen P, Liu G. Enantioselective Pd(II)-catalyzed intramolecular oxidative 6-*endo* aminoacetoxylation of unactivated alkenes. *J Am Chem Soc*. 2018;140:7415–7419.
97. Mai DN, Wolfe JP. Asymmetric palladium-catalyzed carboamination reactions for the synthesis of enantiomerically enriched 2-(arylmethyl)- and 2-(alkenylmethyl) pyrrolidines. *J Am Chem Soc*. 2010;132:12157–12159.

98. Hopkins BA, Wolfe JP. Synthesis of enantiomerically enriched imidazolidin-2-ones through asymmetric palladium-catalyzed alkene carboamination reactions. *Angew Chem Int Ed.* 2012;51:9886–9890.
99. Minatti M, Zheng X, Buchwald SL. Synthesis of chiral 3-substituted indanones via an enantioselective reductive-Heck reaction. *J Org Chem.* 2007;72:9253–9258.
100. (a) Yue G, Lei K, Hirao H, Zhou J. Palladium-catalyzed asymmetric reductive Heck reaction of aryl halides. *Angew Chem Int Ed.* 2015;54:6531–6535. (b) Kong W, Wang Q, Zhu J. Synthesis of diversely functionalized oxindoles enabled by migratory insertion of isocyanide to a transient σ-alkylpalladium(II) complex. *Angew Chem Int Ed.* 2016;55:9714–9718. (c) Mannathan S, Raoufmoghaddam S, Reek JNH, de Vries JG, Minnaard AJ. Enantioselective intramolecular reductive Heck reaction with a palladium/monodentate phosphoramidite catalyst. *ChemCatChem.* 2017;9:551–554. (d) Kong W, Wang Q, Zhu J. Water as a hydride source in palladium-catalyzed enantioselective reductive Heck reactions. *Angew Chem Int Ed.* 2017;56:3987–3991. (e) Bao X, Wang Q, Zhu J. Palladium-catalyzed enantioselective Narasaka-Heck reaction/direct C-H alkylation of arenes: iminoarylation of alkenes. *Angew Chem Int Ed.* 2017;56:9577–9581.
101. Newman SG, Howell JK, Nicolaus N, Lautens M. Palladium-catalyzed carbohalogenation: bromide to iodide exchange and domino processes. *J Am Chem Soc.* 2011;133:14916–14919.
102. Jiang Z, Hou L, Ni C, Chen J, Wang D, Tong X. Enantioselective construction of quaternary tetrahydropyridines by palladium-catalyzed vinylborylation of alkenes. *Chem Commun.* 2017;53:4270–4273.
103. Zhang Z, Xu B, Qian Y, et al. Palladium-catalyzed enantioselective reductive heck reactions: convenient access to 3,3-disubstituted 2,3-dihydrobenzofuran. *Angew Chem Int Ed.* 2018;57:10373–10377.
104. Zhang Z, Xu B, Wu L, et al. Palladium/XuPhos-catalyzed enantioselective carboiodination of olefin-tethered aryl iodides. *J Am Chem Soc.* 2019;141:8110–8115.
105. Zhang Z, Xu B, Wu L, et al. Enantioselective dicarbofunctionalization of unactivated alkenes by Pd-catalyzed tandem Heck/Suzuki-coupling reaction. *Angew Chem Int Ed.* 2019;58:14653–14659. https://doi.org/10.1002/anie.201907840.
106. Werner EW, Mei TS, Burckle AJ, Sigman MS. Enantioselective Heck arylations of acyclic alkenyl alcohols using a redox-relay strategy. *Science.* 2012;338:1455–1458.
107. Mei TS, Werner EW, Burckle AJ, Sigman MS. Enantioselective redox-relay oxidative Heck arylations of acyclic alkenyl alcohols using boronic acids. *J Am Chem Soc.* 2013;135:6830–6833.
108. Mei TS, Patel HH, Sigman MS. Enantioselective construction of remote quaternary stereocentres. *Nature.* 2014;508:340–344.
109. Zhang C, Santiago CB, Kou L, Sigman MS. Alkenyl carbonyl derivatives in enantioselective redox relay Heck reactions: Accessing α,β-unsaturated systems. *J Am Chem Soc.* 2015;137:7290–7293.
110. Wang H, Bai Z, Jiao T, et al. Palladium-catalyzed amide-directed enantioselective hydrocarbofunctionalization of unactivated alkenes using a chiral monodentate oxazoline ligand. *J Am Chem Soc.* 2018;140:3542–3546.
111. Shen HC, Zhang L, Chen SS, et al. Enantioselective addition of cyclic ketones to unactivated alkenes enabled by amine/Pd(II) cooperative catalysis. *ACS Catal.* 2019;9:791–797.
112. Ohmura T, Taniguchi H, Suginome M. Kinetic resolution of racemic 1-alkyl-2-methylenecyclopropanes via palladium-catalyzed silaborative C-C cleavage. *Org Lett.* 2009;11:2880–2883.
113. Trost BM, Xie J, Sieber JD. The palladium catalyzed asymmetric addition of oxindoles and allenes: an atom-economical versatile method for the construction of chiral indole alkaloids. *J Am Chem Soc.* 2011;133:20611–20622.

114. Li M, Datta S, Barber DM, Dixon DJ. Dual amine and palladium catalysis in diastereo- and enantioselective allene carbocyclization reactions. *Org Lett.* 2012;14:6350–6353.
115. Yang Q, Jiang X, Ma S. Highly diastereoselective palladium-catalyzed cyclizations of 3,4-allenylic hydrazines and organic halides-highly stereoselective synthesis of optically active pyrazolidine derivatives and the prediction of the stereoselectivity. *Chem Eur J.* 2007;13:9310–9316.
116. (a) Shu W, Yang Q, Jia G, Ma S. Studies on palladium-catalyzed enantioselective cyclization of 3,4-allenylic hydrazines with organic halides. *Tetrahedron.* 2008;64:11159–11166. (b) Shu W, Ma S. Synthesis of a new spiro-BOX ligand and its application in enantioselective allylic cyclization based on carbopalladation of allenyl hydrazines. *Chem Commun.* 2009;6198–6200.
117. Xie X, Ma S. Palladium-catalyzed asymmetric coupling cyclization of terminal γ-allenols with aryl iodides. *Chem Commun.* 2013;49:5693–5695.
118. Li M, Hawkins A, Barber DM, Bultinck P, Herrebout W, Dixon DJ. Enantio- and diastereoselective palladium catalysed arylative and vinylative allene carbocyclisation cascades. *Chem Commun.* 2013;49:5265–5267.
119. Lim W, Kim J, Rhee YH. Pd-catalyzed asymmetric intermolecular hydroalkoxylation of allene: an entry to cyclic acetals with activating group-free and flexible anomeric control. *J Am Chem Soc.* 2014;136:13618–13621.
120. Liu J, Han Z, Wang X, Wang Z, Ding K. Highly regio- and enantioselective alkoxycarbonylative amination of terminal allenes catalyzed by a spiroketal-based diphosphine/Pd(II) complex. *J Am Chem Soc.* 2015;137:15346–15349.
121. Zhou H, Wang Y, Zhang L, Cai M, Luo S. Enantioselective terminal addition to allenes by dual chiral primary amine/palladium catalysis. *J Am Chem Soc.* 2017;139:3631–3634.
122. Liu Z, Gu P, Shi M. Asymmetric addition of arylboronic acids to cumulene derivatives catalyzed by axially chiral N-heterocyclic carbene–Pd^{2+} complexes. *Chem Eur J.* 2011;17:5796–5799.

CHAPTER SEVEN

Titanium catalyzed synthesis of amines and N-heterocycles

Laurel L. Schafer*, Manfred Manßen, Peter M. Edwards, Erica K.J. Lui, Samuel E. Griffin, Christine R. Dunbar

Department of Chemistry, University of British Columbia, Vancouver, B.C, Canada
*Corresponding author: e-mail address: schaferl@mail.ubc.ca

Contents

1. Introduction	405
2. Hydroamination	407
2.1 Overview	407
2.2 Mechanism	409
2.3 Alkyne hydroamination	411
2.4 Allene hydroamination	414
2.5 Alkene hydroamination	416
2.6 Hydroamination featured in tandem sequential synthetic methods	418
2.7 Summary and outlook	426
3. Hydroaminoalkylation	427
3.1 Overview	427
3.2 Mechanism	428
3.3 Intramolecular hydroaminoalkylation	431
3.4 Intermolecular hydroaminoalkylation	433
3.5 N-heterocycle synthesis by hydroaminoalkylation	439
3.6 Summary and outlook	440
4. Titanium redox chemistry for catalytic amine synthesis	441
4.1 Introduction	441
4.2 Ti(II) catalysis for the synthesis of amines and N-heterocycles	444
4.3 Ti(III) catalysis for the synthesis of amines and N-heterocycles	453
4.4 Summary and outlook	457
References	458

1. Introduction

Titanium is earth abundant and inexpensive, as it is the second most abundant transition metal and the ninth most over all elements in the earth's crust. It is used as a catalyst for olefin polymerization, one of the key industrial processes. Furthermore, its low toxicity and biocompatibility make it attractive for application in fine chemical synthesis. While titanium continues to be used in powerful reagents that mediate selective bond formations, a significant drawback of such approaches remains the stoichiometric amount of TiO_2 by-product that ultimately remains upon workup. However, such stoichiometric transformations have been an inspiration for the development of catalytic variants of a growing list of useful reactions. Typically, titanium is used in the Ti(IV) oxidation state and is oxophilic, making it moisture sensitive, but owing to its fully oxidized state, it is not oxygen sensitive. Another attractive opportunity in catalysis is the ligand lability that can be achieved with this Lewis acidic metal center, which affords ready access to the reactive metal center. Titanium has also shown promise for its catalytic use in redox reactivity, where oxidative additions and reductive eliminations can result in the formation of cyclic products.[1]

Titanium has been shown to be a versatile catalyst for small molecule synthesis in a broad range of reactions including hydrogenation reactions,[2] cyclization reactions,[3-5] polymerization reactions,[6,7] aldol and allylic additions to ketones and aldehydes,[8,9] epoxidation of alkenes,[10] and carbonate formation with epoxides and CO_2.[11,12] Relevant to this chapter is the use of titanium for the catalytic synthesis of amines including its use in two related hydrofunctionalization reactions, hydroamination and hydroaminoalkylation. These transformations can be used to assemble selectively substituted amine small molecules and N-heterocycles. Amines and N-heterocycles are critical components of naturally occurring and synthetic biologically active compounds of relevance to the pharmaceutical and agrochemical industries.[13] Furthermore, amine containing materials have a broad range of applications ranging from anti-microbial coatings to CO_2 capture materials. Indeed, atom economical N-centered chemistry has been identified as one of the "10 most sought after chemical transformations"[14] sparking intense interest in developing new routes to achieve this transformation. This demand has resulted in recent developments in the synthesis of N-heterocycles using titanium catalyzed redox transformations and other select reactions. This chapter includes a historical perspective on the use of titanium for this chemistry and the focus of the reviewed literature is on titanium catalyzed amine and N-heterocycle synthesis.

2. Hydroamination
2.1 Overview

Catalytic hydroamination (Scheme 1) is the metal-catalyzed addition of an N—H bond across a C—C multiple bond. This is an attractive transformation as the reaction is 100% atom economic. This reaction was initially explored in the early and mid-20th century with heterogeneous metal oxides and salts under forcing conditions for the amination of olefins. Over the next few decades, main group and late transition metal catalysts improved functional group tolerance with less harsh conditions, and managed to achieve some regioselectivity favoring the Markovnikov addition products.[15]

Scheme 1 Representation of titanium catalyzed hydroamination.

The first example of intermolecular alkyne hydroamination catalyzed by titanium was published by Doye in 1999, using dimethyltitanocene (2).[16] The catalytically active species in this work has been proposed to be a titanium imido complex. These advances prompted further exploration of non-Cp ligand sets and now titanium is known to be useful for the intermolecular hydroamination of alkynes and allenes, although titanium catalyzed alkene hydroamination is limited to intramolecular cyclohydroamination reactions using specialized aminoalkene substrates. Furthermore, due to mechanistic details (vide infra) titanium catalyzed hydroamination is limited to the use of primary amine substrates. This results in the efficient synthesis of secondary amines, which can undergo further reactivity. Furthermore, the use of primary amines is complementary to typical reactivity trends realized with late transition metal hydroamination catalysts, which often require secondary or protected amine substrates.

Due to advances in the use of titanium catalysts for this transformation, there are several commercially available titanium (pre-)catalysts that can be used with catalyst loadings between 2 and 10 mol% (Scheme 2).[17,18] However, such complexes and similar variants can also be readily prepared from commercially available $Ti(NMe_2)_4$ (**1**) and suitably protic ligands using a simple protonolysis reaction (Scheme 3). This facile assembly using simple syringe techniques has also resulted in the use of in situ prepared catalysts with comparable outcomes.

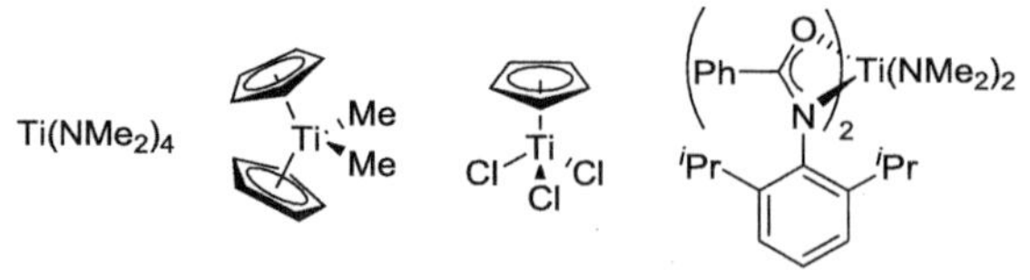

1 **2** **3a** **4a**

Scheme 2 Selected examples of commercially available titanium catalysts.

$$2\,LH + Ti(NMe_2)_4 \longrightarrow L_2Ti(NMe_2)_2 + 2\,HNMe_2$$

or

$$LH_2 + Ti(NMe_2)_4 \longrightarrow LTi(NMe_2)_2 + 2\,HNMe_2$$

Scheme 3 Protonolysis routes for facile assembly of reactive titanium complexes.

Hydroamination is a challenging reaction as it is not kinetically favored, due to the difficulties associated with reacting an electron-rich amine with an electron-rich alkene. This kinetic challenge can be addressed using a transition metal catalyst, and in the case of titanium catalysts, the reaction proceeds via amine activation (vide infra). In addition to challenging reaction kinetics, there are also thermodynamic limitations associated with this transformation. First and foremost, desirable intermolecular reactions with simple amine and hydrocarbon substrates are entropically disfavored. Furthermore, hydroamination reactions of alkenes have a minimal thermodynamic driving force,[17–19] which accounts for the fact that titanium catalysts that display outstanding intermolecular reactivity with alkynes and allenes are ineffective with alkene substrates. To date, intermolecular alkene hydroamination with titanium catalysts is an unmet goal.

This chapter will summarize previously reviewed achievements in titanium hydroamination catalysis and review new advances in the development of improved titanium catalysts. Finally, this section concludes with a review of applications of hydroamination catalysis in the synthesis of more complex target molecules, such as *N*-heterocycles.

2.2 Mechanism

Mechanistic details for group 4 catalyzed hydroamination were established by Bergman and coworkers using well-established Cp_2Zr complexes.[20,21] Careful kinetic, stoichiometric and computational mechanistic investigations showed that hydroamination proceeds via an intermediate group 4 metal imido complex (**A**, Scheme 4) formed upon N—H activation of the requisite primary amine substrates. Thus, a common experimental test for the intermediacy of metal imido intermediates is the attempted reaction with secondary amine substrates. If the reaction is not successful, that is preliminary evidence to support the proposal of imido intermediates. Indeed, such experimental tests have consistently supported the intermediate role of titanium imido complexes, even though computational investigations have suggested that the direct insertion of C=C bonds into Ti—N sigma bonds is energetically accessible.[22]

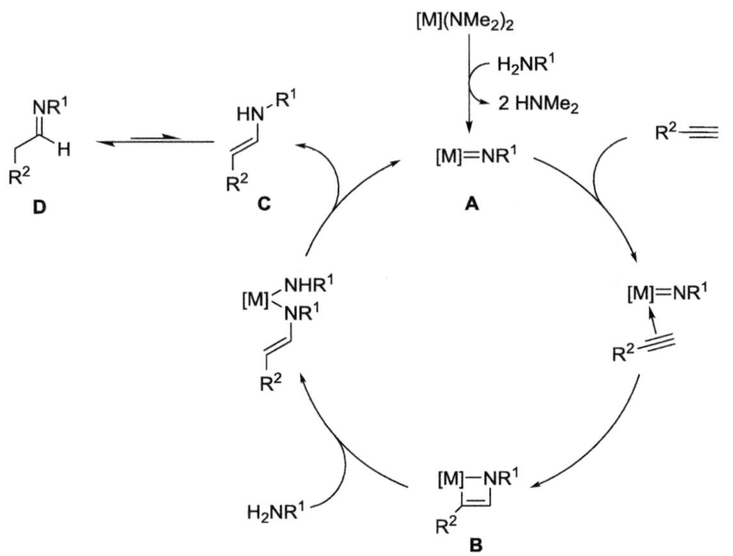

Scheme 4 Catalytic cycle for the intermolecular, anti-Markovnikov hydroamination of terminal alkynes.

Since the first experiments demonstrating titanium catalyzed hydroamination using Cp_2TiMe_2 precatalysts (**2**), research groups around the world have shown that titanium complexes bearing a variety of ligands catalyze the reaction through such imido mediated reactivity.[23] Further evidence of such intermediates includes the formation of discrete titanium imido complexes that are catalytically competent (Scheme 5).[17,24–30] Reaction monitoring of some precatalysts has resulted in the observation of an induction period, which is proposed to correspond to the formation of the requisite imido species in situ.

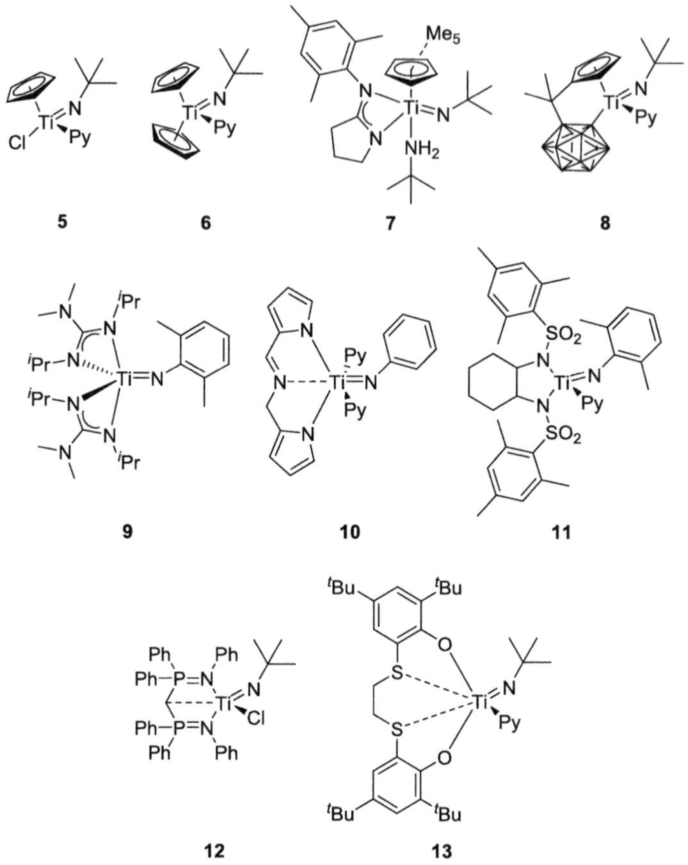

Scheme 5 Selected titanium imido complexes for hydroamination catalysis.

Upon formation of the titanium imido complexes, a concerted [2+2] cycloaddition step results in the formation of the four-membered metallacycle **B** (Scheme 4). Likewise, isolated examples of such [2+2] cycloaddition products have been prepared by adding an equivalent of alkyne to preformed imido species.[23,31,32] Subsequent protonation events from an incoming primary amine, regenerates the imido catalyst **A** and liberates the enamine product **C**. Typically, under the Lewis acidic conditions of titanium catalyzed hydroamination such enamine products undergo tautomerization and are most commonly observed as imine products **D**. Notably, the enamine/imine tautomerization can result in an equilibrium mixture of products, where the substituents of the amine and alkyne starting materials define the more favored tautomer.

Mechanistic evaluation of this reaction has been extensive and thorough. At this point the mechanistic details are well established, although clear mechanistic rationale for varied regioselectivities between complexes and the consistent lack of intermolecular reactivity with alkenes has not been presented for titanium: this challenge has been addressed through the use of the alternative group 4 metal zirconium. Titanium has shown particular promise in applications toward organic synthesis. Of all early transition metals, titanium is the most commonly used for the organic synthetic community. As will be shown in Section 2.6, both enamine and imine products from hydroamination are well suited for the development of tandem sequential reactions to assemble more complex amine and *N*-heterocyclic products.

2.3 Alkyne hydroamination

In the case of titanium catalyzed alkyne hydroamination, both terminal and internal alkynes are suitable substrates and a broad range of catalysts have been explored for these transformations; this topic has been reviewed multiple times.[18,19] Ligand modifications have been shown to impact both catalyst activity and regioselectivity. Ligands that promote more electrophilic character at the metal center have been reported to be more catalytically active, while sterically accessible metal centers promote enhanced reactivity. The design of sterically accessible titanium complexes with enhanced reactivity is challenging, as complexes with too many available coordination sites at the metal center can result in the formation of catalytically inactive bridging titanium imido complexes. To address this challenge hemilabile ligands for example have been exploited.

Control of regioselectivity is an important ligand design challenge. Regioselectivity can be modified by the incorporation of steric bulk into the ligand set, although regioselectivity is known to be very substrate dependent and is impacted by electronic effects. One area of particular interest is the development of catalysts suitable for anti-Markovnikov hydroamination of terminal alkynes. This is the product that cannot be readily achieved using traditional acid-catalyzed routes and is the typical product of late transition metal-catalyzed hydroamination. To address this problem a bis(amidate)titanium catalyst **4b** (which can be used interchangeably with **4a**) has been reported to complete this regioselective transformation with a broad range of terminal alkynes with both alkyl and most aryl amines (Scheme 6).[33,34] The excellent regioselectivity of this catalyst has been attributed to the hemi-lability of the amidate ligand, which can promote the formation of a more sterically accessible metal center upon modification of the ligand binding motif from the experimentally observed κ^2-chelate to the proposed reactive κ^1-species. This increased steric accessibility allows for the incorporation of the bulky terminal alkyne substituent away from the N-substituent, resulting in the formation of the anti-Markovnikov product. Interestingly, other titanium catalysts yield the more commonly accessed Markovnikov selective product. Such catalysts have been developed empirically and to date there is a lack of predictive mechanistic tools to design systems with predictable regioselective outcomes.

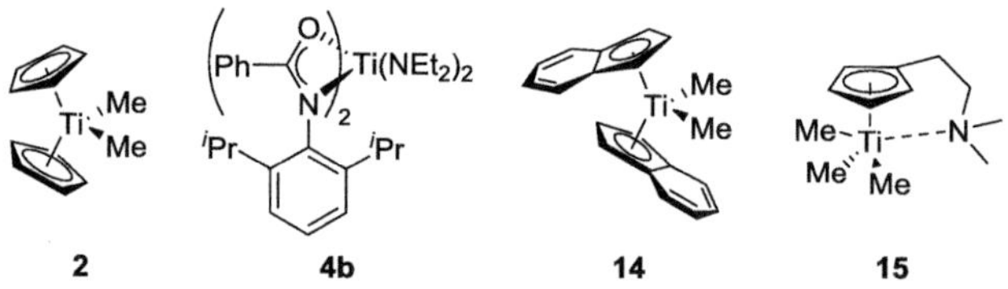

Scheme 6 Selected titanium complexes reported for anti-Markovnikov hydroamination.

Titanium complexes are also well established for anti-Markovnikov hydroamination[16,33–37] with aryl-alkyl-substituted internal alkynes. These complexes can also be used with terminal alkynes as substrates, however the scope of reactivity and reliable regioselectivity is improved with aryl-alkyl-substituted internal alkynes.

One of the perceived shortcomings of early transition metal catalysis generally and titanium more specifically is the functional group intolerance of these oxophilic metal centers. Recent hydroamination advances with **4** have shown that in fact ligand design, specifically the incorporation of hard donors into the ligand framework, can help to increase functional group tolerance (Scheme 7). With catalyst **4** substrates including ethers and protected alcohols, esters and halides can be used and full conversion is observed in all cases.[38] The incorporation of such functional groups allows for further reactivity of the resultant amine products to make more complex small molecules.

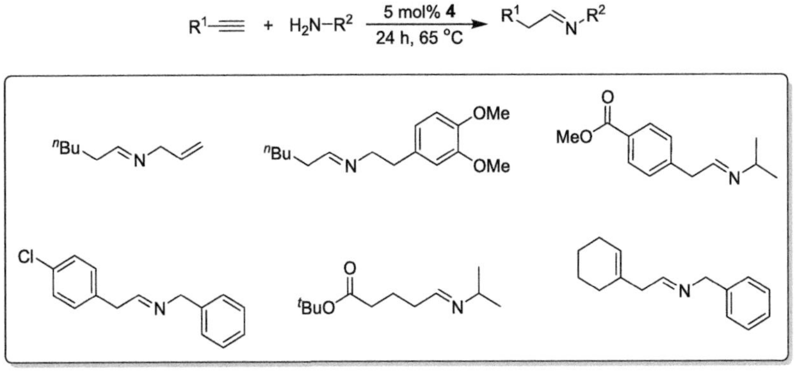

Titanium catalyzed synthesis of amines and N-heterocycles 413

Scheme 7 Selected products of intermolecular hydroamination of terminal alkynes, demonstrating functional group tolerance.

While homogeneous titanium catalysts have been extensively investigated for this transformation, the development of heterogeneous variants offers an attractive opportunity for synthetic application. To this end, Odom and coworkers recently disclosed silica supported titanium amido complexes for alkyne hydroamination.[39] The facile synthesis of the Ti(NMe$_2$)$_2$/SiO$_2^{200}$ results in a thermally robust catalyst system that favors the formation of the Markovnikov product between alkyl alkynes and aniline (Scheme 8).

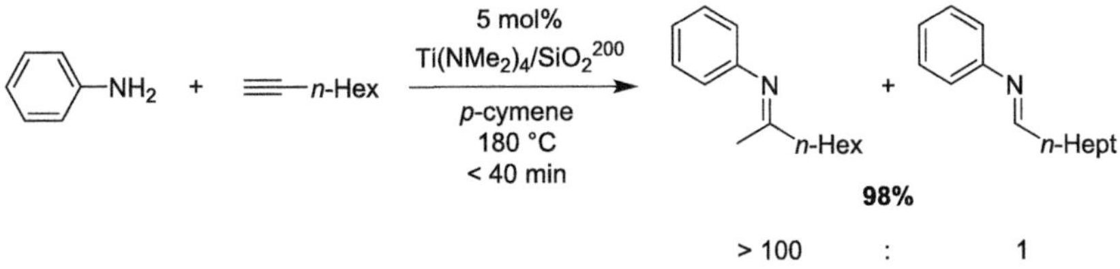

Scheme 8 Ti(NMe$_2$)$_2$/SiO$_2^{200}$ as a solid supported catalyst system for alkyne hydroamination.

Another frontier in hydroamination catalysis is the use of ammonia as a substrate to yield primary amine products. To date there has been one report of using ammonia as a substrate in combination with a gold catalyst.[40] While, there are no examples of titanium for this desirable transformation, a recent contribution (Scheme 9) showed that silylamines can be used as masked ammonia equivalents in combination with bis(amidate)titanium catalyst **4**.[41] This reaction yields silylated enamine products that are resistant to

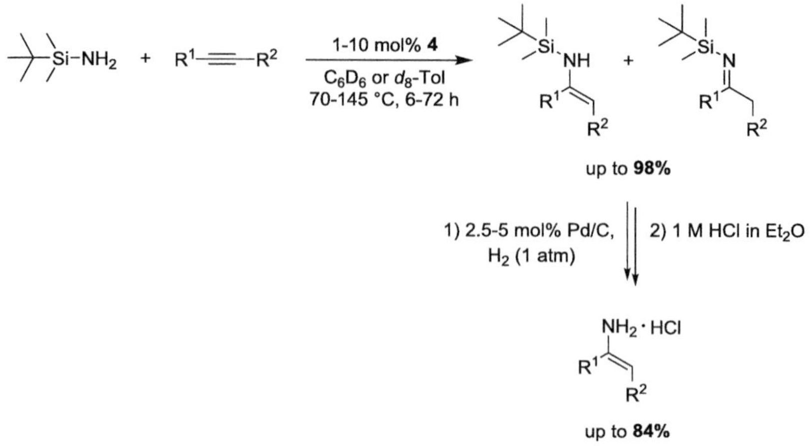

Scheme 9 Synthesis of primary amines by hydroamination of silylamines.

tautomerization, yet susceptible to subsequent hydrogenation with Pd/C and H$_2$ to give primary amine products.

Alkynes are preferred and well-established substrates for titanium catalyzed hydroamination. Although no examples of intramolecular alkyne hydroamination are discussed here, it should be noted that any catalyst that performs intermolecular alkyne hydroamination is also capable of intramolecular transformations. A word of caution regarding intramolecular alkyne hydroamination is warranted; this transformation is very easily catalyzed and virtually any Lewis acidic metal, including molecular sieves, can mediate five- and six-membered ring formation by cyclohydroamination. For catalyst development purposes more significant reactivity challenges remain in allene and alkene hydroamination.

2.4 Allene hydroamination

Allene hydroamination has been less commonly reported as few substrates are commercially available (Selected titanium catalysts are shown in Scheme 10). Allene hydroamination also presents a more significant challenge for developing regioselective reactivity. In the case of terminal allenes there are three possible products that can result (Scheme 11). For these reasons few investigations of intermolecular allene hydroamination have been reported, in spite of the fact that this route has many interesting opportunities for the synthesis of useful allylamine substrates.

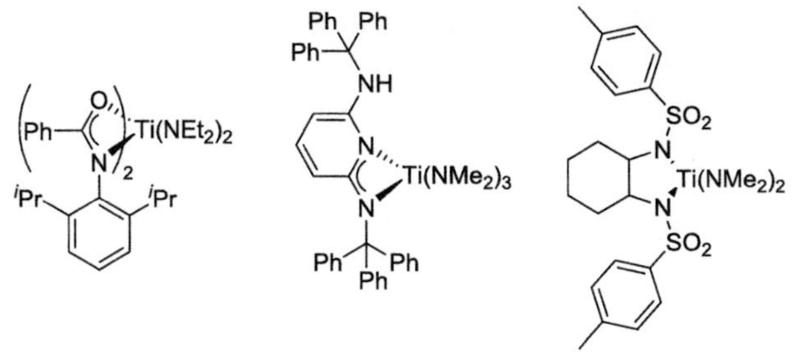

Scheme 10 Selected titanium complexes reported for allene hydroamination.

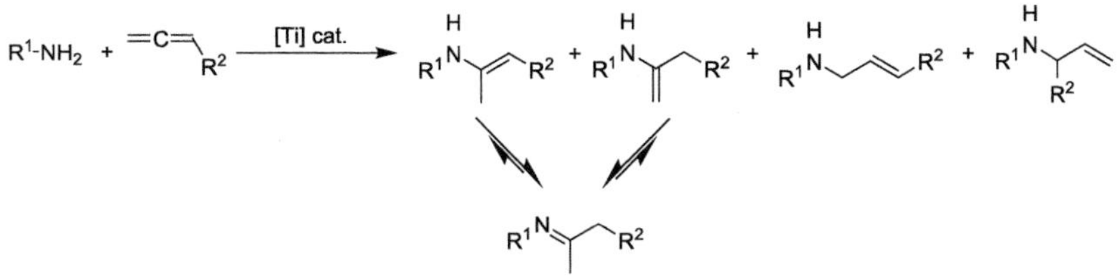

Scheme 11 Possible regioisomeric products from the intermolecular hydroamination of allenes.

Early examples showed that complex **4** is an active catalyst for the intermolecular hydroamination of allenes (Scheme 12).[42] Using **4**, a number of aryl and alkyl-substituted amines are tolerated to selectively give the Markovnikov branched enamine product, which can be isolated as the isomerized imine or reduced to yield the secondary amine. The functional group tolerance of **4** was extended to include reactivity with selective oxygen-substituted allenes.[43] More recently, similar results were achieved with a bulky aminopyridinate complex **16**.[44]

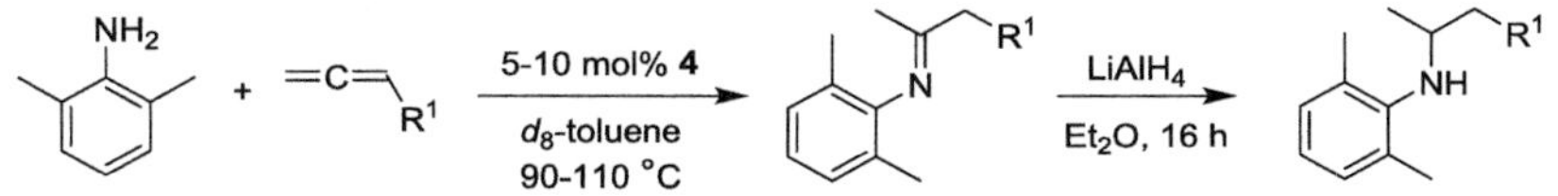

Scheme 12 Markovnikov product of intermolecular allene hydroamination.

Allenes have also been used in intramolecular hydroamination with titanium. The reactivity for intramolecular allene hydroamination was explored using a titanium complex bearing a diamide chelating ligand with N-tosyl groups, which can adopt an N,O-bonding motif.[29] For the intramolecular hydroamination of allenes, this catalyst is effective at room temperature. This contrasts with Ti(NMe$_2$)$_4$ (**1**) which requires slightly elevated temperatures (75 °C), and dimethyltitanocene (**2**) which requires even more heating (135 °C) to efficiently catalyze the same reaction (Scheme 13).

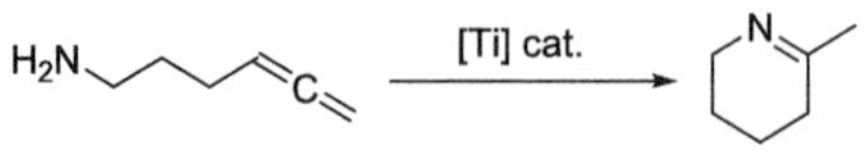

[Ti] cat.	T (°C)	t (h)	yield
Cp$_2$TiMe$_2$ **2**	135	3	74%
Ti(NMe$_2$)$_4$ **1**	75	3	quant.
NSO Ti **17**	25	5	quant.

Scheme 13 Selected results for titanium catalyzed intramolecular hydroamination of aminoallenes.

An interesting opportunity in cyclohydroamination of aminoallene substrates is the possibility of realizing enantioselective versions of this reaction. Regioselective allene hydroamination to form the allylamine product affords a stereocenter. Thus, success in this transformation demands shifting both regioselectivity and stereoselectivity. Chelating N,O-ligands derived from chiral aminoalcohols have been used in combination with titanium to generate enantioenriched product with modest yields and ee's (Scheme 14).[45] It should be noted that although late transition metals can be used for enantioselective allene hydroamination,[46] late transition metals are not compatible with the primary amine substrates preferred here.

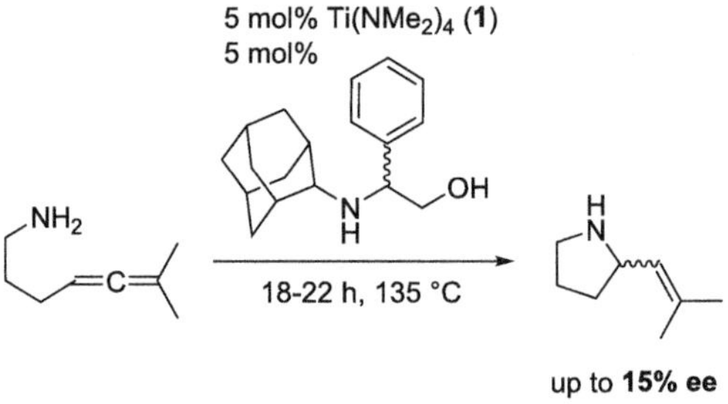

Scheme 14 Enantioselective intramolecular aminoallene hydroamination.

2.5 Alkene hydroamination

A more challenging transformation is the titanium catalyzed intermolecular alkene hydroamination. To date, there is no report of a titanium catalyst that can mediate this reaction. Thus, attention has turned toward exploring titanium catalysts for the intramolecular variant for this reaction. In 2005, Schafer reported that the simple and commercially available Ti(NMe$_2$)$_4$ (**1**) can be used as a catalyst for the intramolecular hydroamination of aminoalkenes.[47] This result drew renewed attention to group 4 metals for this challenging transformation, although quickly it was determined that zirconium is generally more reactive for this more sterically demanding transformation.[48] This has been attributed to the formation of sp^3-hybridized carbon centers in the key metallacyclic intermediate (analogue of **B**, Scheme 4). However, with the use of elevated temperatures, early examples could be realized using titanium. Thus, intramolecular alkene hydroamination became a useful probe for developing hydroamination catalysts with enhanced reactivity. Due to the negative entropy term associated with intermolecular hydroamination, it is critical to develop catalytic systems that can mediate this transformation at lower temperatures. These efforts resulted in the identification of various titanium catalysts that could realize cyclohydroamination reactions of aminoalkenes at a reasonable rate (e.g., less than 24 h and at ambient temperatures) (Scheme 15).[44,49–54]

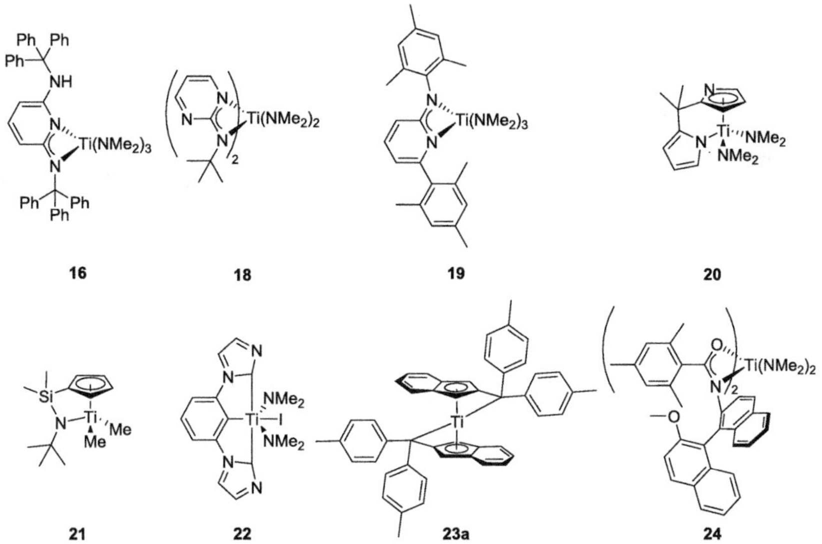

Scheme 15 Selected titanium catalysts for the intramolecular hydroamination of aminoalkenes.

Efforts to develop more active catalysts focus on generating more electrophilic catalysts, with sterically accessible metal centers. The more sterically accessible, mono-ligated *N,N*-chelating system of aminopyridinates

(**16**, **18** and **19**), when installed on titanium, have shown great promise for delivering room temperature reactivity for intramolecular aminoalkene hydroamination.[44,49] Designing ligand sets that yield mono-ligated complexes demands the introduction of significant steric bulk. For example, pincer complexes can also afford titanium complexes that mediate this challenging reaction (**22**).[53] Alternatively, steric accessibility can be revealed at the metal center by selecting ligands that have variable binding modes, such as tethered bipyrrole ligands in which one of the pyrroles can bind either through a κ^5-binding mode or a κ^1-N-coordination (**20**).[54] This flexibility in bonding can be responsive to the coordination sphere of the metal center throughout the catalytic cycle. Another approach sees the use of pentafulvene as a reactive ligand that can reveal a coordination site upon reactivity with the amine substrate (**23a**).[52] Control and design of steric bulk is important for supporting the reactive terminal imido intermediates required for the desired reactivity.

An important area of investigation in cyclohydroamination is the investigation of enantioselective transformations. Titanium amidate complexes (**24**) can be used for the conversion of aminoalkenes to chiral pyrrolidine products (Scheme 16).[55] Unfortunately, the chiral bis(amidate) titanium complexes are not as reactive as their zirconium analogues and enantioselectivities are also lower than state-of-the-art examples in zirconium asymmetric cyclohydroamination catalysis.[56,57]

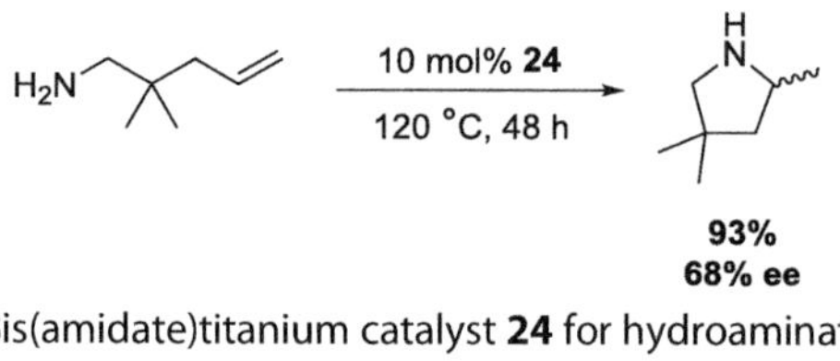

Scheme 16 Chiral Bis(amidate)titanium catalyst **24** for hydroamination of 1-amino-2,2-dimethylpent-1-ene.

2.6 Hydroamination featured in tandem sequential synthetic methods

Titanium catalyzed intermolecular hydroamination yields reactive small molecules that can be incorporated as building blocks in the assembly of more complex amines and N-heterocycles. Various catalysts have been used to prepare selectively substituted five-membered rings, such as pyrroles, indoles and pyrazoles as well as six-membered rings, such as piperidines, piperazines and tetrahydroquinoline derivatives.

2.6.1 Amine containing small molecules and oligomers

Regioselective hydroamination results in the high yielding synthesis of aldimine and/or ketimine intermediates. These reactive intermediates can undergo subsequent reactivity. Most commonly such imine derivatives verify stoichiometric reduction using hydride reducing agents (e.g., NaCNBH$_3$, LiAlH$_4$) to prepare and isolate the corresponding amine products. However, it was shown that tandem catalytic reduction methods can result in the efficient synthesis of amine products from alkyne starting materials. For example, PhSiH$_3$ can be used in combination with the titanium hydroamination catalyst to realize catalytic imine hydrosilylation.[58,59] Alternatively, the fully atom economic synthesis of amines can be carried out by using tandem hydrogenation of imine intermediates.[34] Such imines are also susceptible to nucleophilic attack, such that tandem sequential addition of metal alkyl reagents can result in the efficient assembly of quaternary centers adjacent to the amine nitrogen.[60,61] An alternative nucleophile that can be useful for introducing further functionality is the addition of TMSCN to give α-cyanoamines that can then be further reacted to access amino acid derivatives (Scheme 17).[62,63] This sequence represents a modified Strecker amino acid synthesis and offers access to tautomerizable α-alkylated cyanoamine derivatives which are challenging to prepare using traditional protocols.

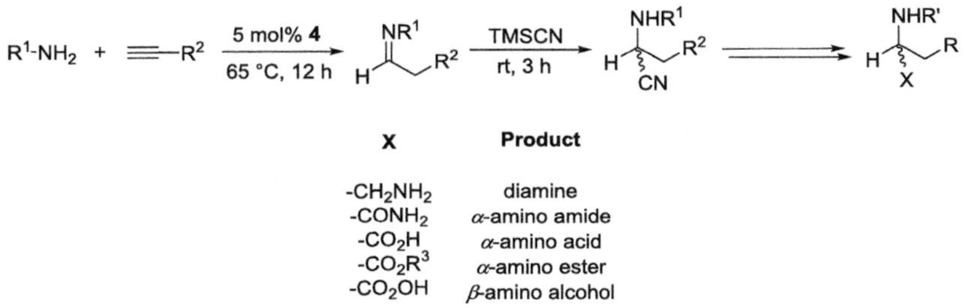

Scheme 17 Modified strecker amino acid synthesis via titanium catalyzed hydroamination and addition of TMSCN.

Hydroamination has been extensively exploited in the synthesis of small molecules. However the advent of highly reactive catalysts allows for the synthesis of functional amine containing materials.[35] Regioselective hydroamination can be exploited in the synthesis of conjugated amines and notably, due to the extended conjugation of these products, there is no observed tautomerization of the secondary enamine products to the imine products with reduced conjugation (Scheme 18).[35] This work also showed that titanium catalysts can be compatible with Lewis acidic heteroatom substituents, such as boranes, to generate highly luminescent products that display solvatochromism.

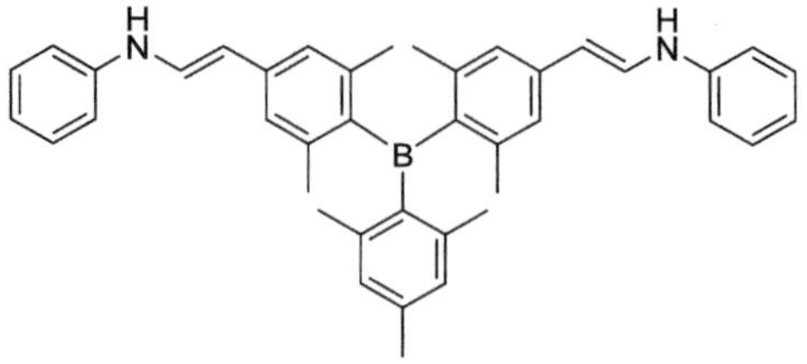

Scheme 18 Boron containing donor-acceptor conjugated dienamine.

2.6.2 Five-membered ring heterocycles

Pyrrolidines are typically prepared using cyclohydroamination of aminoalkene derivatives. An alternative approach can be enabled by intermolecular hydroamination of cyclopropyl alkynes with Ind$_2$TiMe$_2$ (**14**, Scheme 19), followed by cyclopropylimine rearrangement and reduction to yield 2-arylmethylpyrrolidine products (Scheme 20).[64]

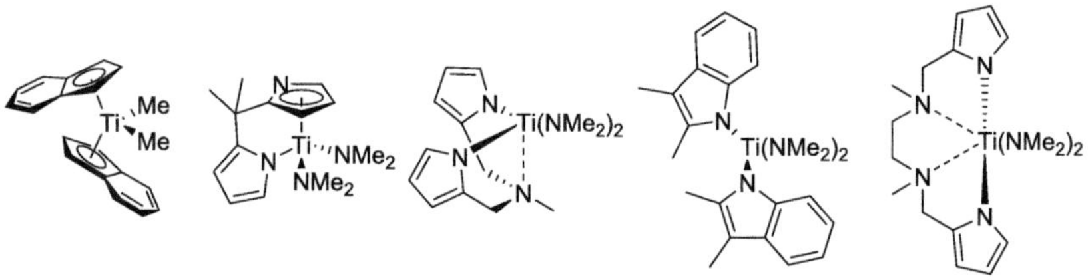

Scheme 19 Selected titanium catalysts for the synthesis of heterocycles via hydroamination.

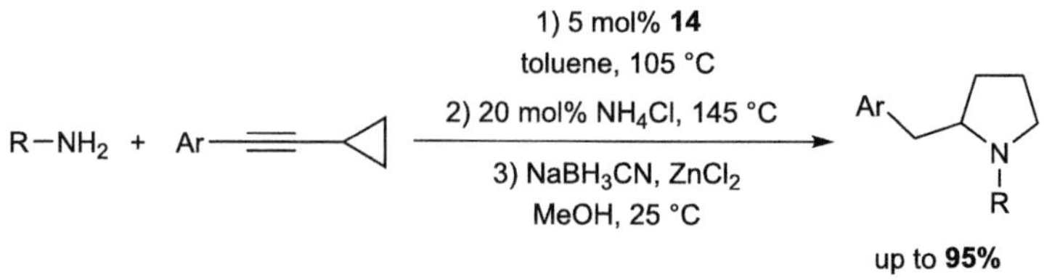

Scheme 20 Intermolecular hydroamination of cyclopropyl alkynes to pyrrolidines.

Monosubstituted pyrroles can be assembled using TiCl$_4$ and chloroenynes (Scheme 21).[65] This reaction is best characterized as an example of Lewis acid-catalyzed hydroamination.

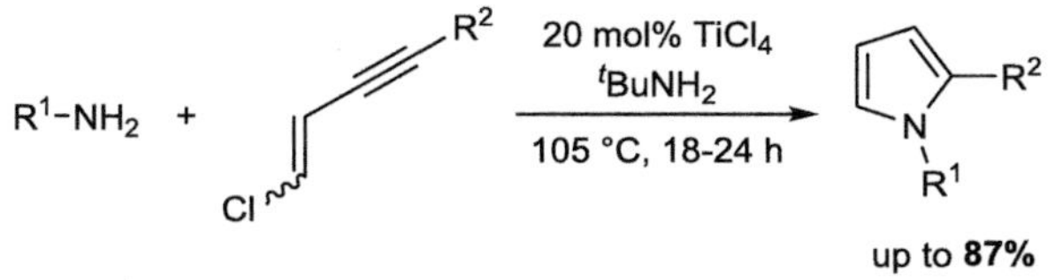

Scheme 21 Titanium catalyzed reaction sequences for syntheses of pyrroles from (E/Z)-chloroenynes.

Disubstituted pyrroles can be assembled from diynes and primary amines (Scheme 22).[66] Strategic selection of the catalyst is required such that if terminal alkynes are incorporated into the substrate, the tridentate titanium complex is preferred, while all internal alkyne containing substrates feature the use of the bidentate titanium complex, presumably to accommodate the more sterically demanding substrates.

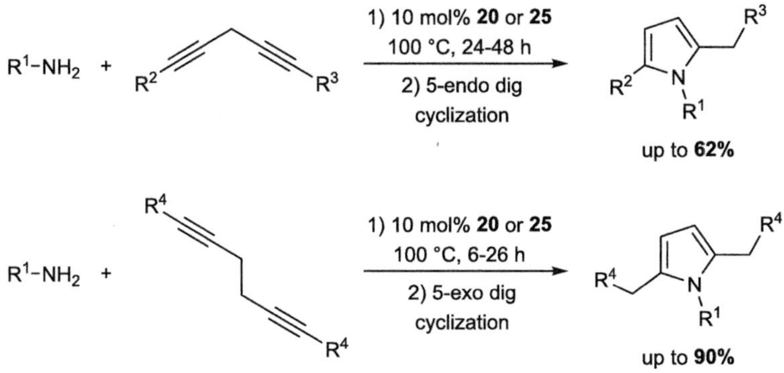

Scheme 22 Sequential double titanium catalyzed hydroamination reactions toward the synthesis of pyrroles.

Tetra-substituted pyrroles can be prepared using a multicomponent coupling strategy developed by the Odom group.[67] This approach takes advantage of the reactive four-membered metallacycle, which forms upon [2+2] cycloaddition, to undergo isonitrile insertion giving five-membered metallacycle ring formation. Further heteroatom containing substrates can be incorporated into this strategy to yield a variety of five-membered ring products. For example, using of this approach the synthesis of highly functionalized 2,3-diaminopyrroles can be efficiently assembled (Scheme 23).

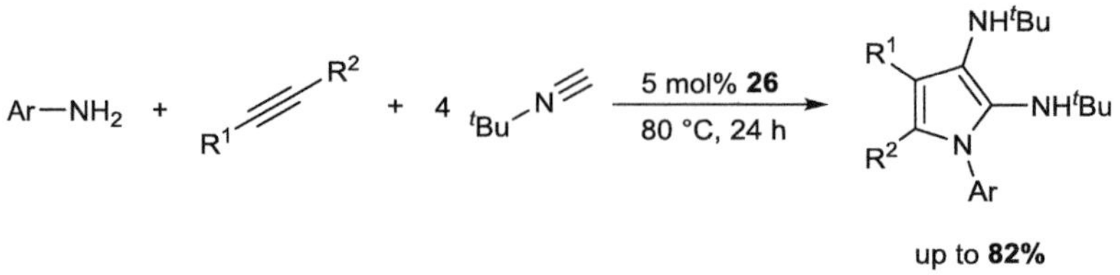

Scheme 23 Sequential titanium catalyzed hydroamination-iminoamination-cyclization reactions toward the synthesis of pyrroles.

Alternatively, the titanium catalyzed coupling of 1 equiv. of isonitrile, 1 equiv. of amine and 1 equiv. of alkyne, gave 1,3-diimines which undergo in situ cyclization with hydrazine to give pyrazoles (Scheme 24).[68]

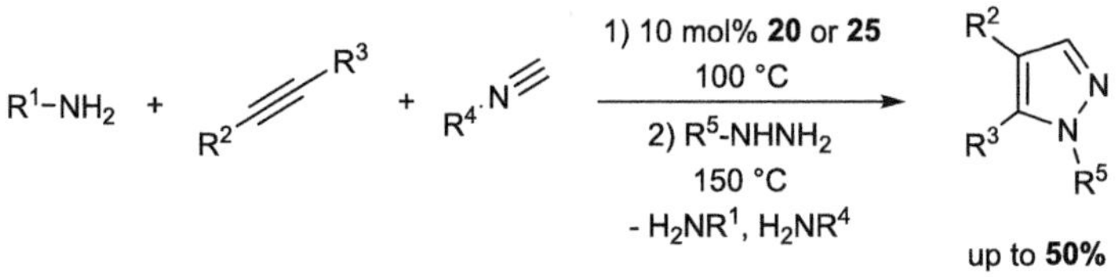

Scheme 24 Sequential titanium catalyzed hydroamination-iminoamination-hydrazine addition reactions toward the synthesis of pyrazoles.

A related approach can be used to synthesize 4- or 3,4-substituted isoxazoles from a multicomponent coupling of alkyne, primary amine and isonitrile followed by the addition of hydroxylamine (Scheme 25).[69]

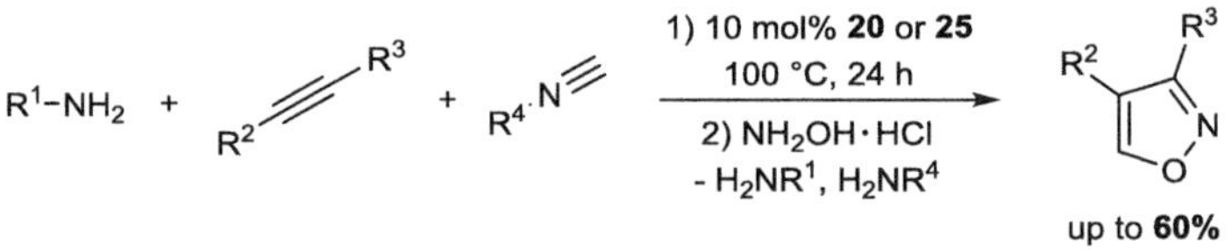

Scheme 25 Sequential titanium catalyzed hydroamination-iminoamination-hydroxylamine addition reactions toward the synthesis of mono- and disubstituted isoxazoles.

Hydroamination can also be applied to access indole heterocycles. Metals from across the periodic table have been used in hydroaminative approaches to access this important heterocycle motif.[70] Hydrohydrazination followed by Fischer–indole synthesis can yield variably substituted indoles (Scheme 26).[71]

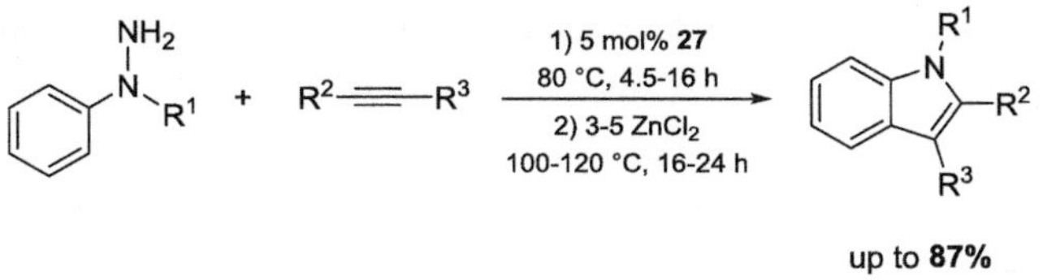

Titanium catalyzed synthesis of amines and N-heterocycles   423

Scheme 26 Titanium catalyzed hydrohydrazination to access selectively substituted indole heterocycles.

2.6.3 Six-membered N-heterocycles

Substituted piperidines are commonly accessed via cyclohydroamination of the appropriate aminoalkene substrates.[44,49,50,53] Multicomponent strategies from the Odom group can also be applied toward the synthesis of six-membered rings. For example 1,3-diimines accessed through intermolecular hydroamination of alkynes and isonitrile insertion, followed by treatment with malononitrile in the presence of base, yields 2-amino-3-cyanopyridines in a one-pot procedure (Scheme 27).[72] This procedure affords a single isomer of the final product with ^{t}butylamine as a side product.

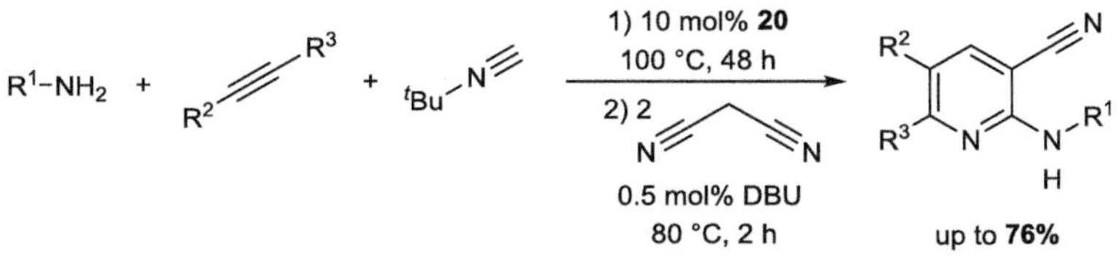

Scheme 27 Sequential titanium catalyzed hydroamination-iminoamination-dimroth rearrangement reactions toward the synthesis of 2-amino-3-cyanopyridines.

An alternative approach toward the synthesis of selectively substituted pyridines has been realized using readily accessed silylenamine intermediates from regioselective hydroamination followed by treatment with α,β-unsaturated carbonyl derivatives (Scheme 28). Subsequent oxidation results

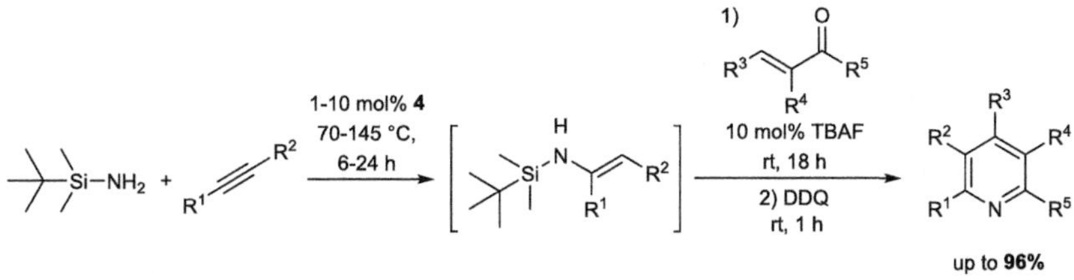

Scheme 28 N-silylenamines as reactive intermediates for the synthesis substituted pyridines.

in the efficient assembly of selectively substituted pyridine products with impressive regioselectivity for up to five-different substituents.[73]

A three-component strategy could be applied to synthesize quinolines from aniline, alkyne and ^{t}butylisonitrile (Scheme 29).[74] Once again, the tridentate pyrrolyl titanium complex **25** was chosen as the catalyst when terminal alkynes were used, while the bidentate pyrrolyl titanium complex **20** was chosen for internal alkynes.

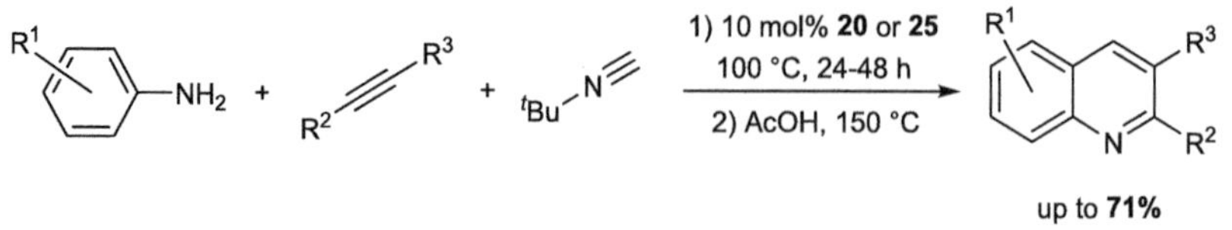

Scheme 29 Multicomponent coupling titanium catalyzed hydroamination-iminoamination-cyclization reactions toward the synthesis of quinolines.

Reduced quinolone derivatives can also be prepared using titanium catalyzed hydroamination. For example using catalyst **4**, intermolecular hydroamination followed by an acid-catalyzed Pictet-Spengler reaction yields tetrahydroisoquinolines (Scheme 30).[38]

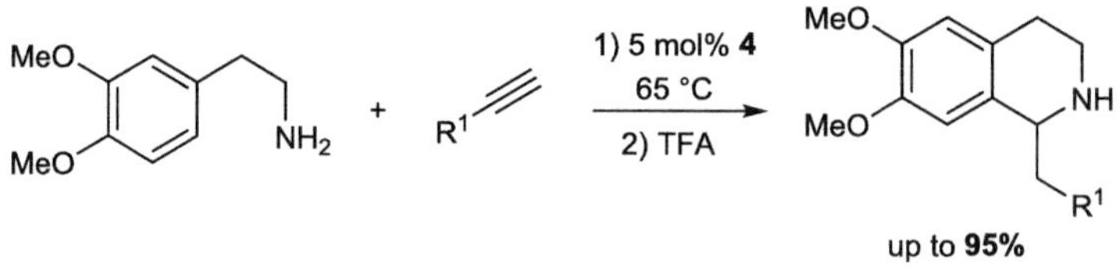

Scheme 30 Anti-Markovnikov selective catalytic hydroamination enables the synthesis of tetrahydroisoquinolines.

The synthesis of heterocycles that contain multiple heteroatoms is important for readily accessing a variety of commonly occurring structural motifs. To this end, simple TiCl$_4$ can be used to realize the Lewis acid-catalyzed hydroamination of N-ethoxycarbonyl-2-alkynylindole, to access an intermediate that undergoes intramolecular nucleophilic attack to access pyrimido[1,6-α]indolones (Scheme 31).[75]

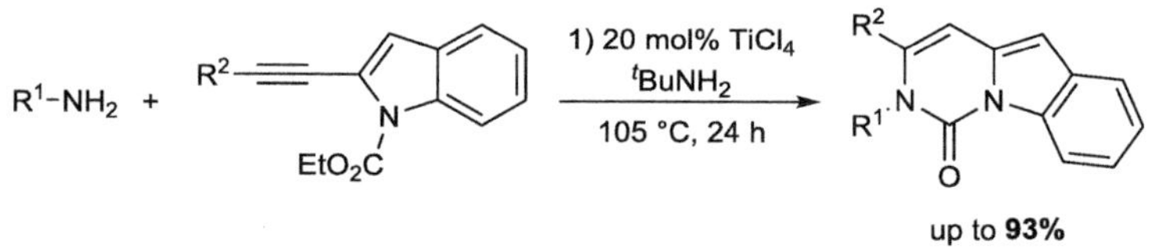

Scheme 31 Sequential titanium catalyzed hydroamination-cyclization reactions toward the synthesis of pyrimido[1,6-α]indolones.

Alternatively, using a multicomponent coupling strategy substituted pyrimidines can be accessed via a one-pot procedure through the aforementioned formation of 1,3-diimides followed by the addition of amidines (Scheme 32).[76] This transformation proceeds in a regioselective fashion, whereby the regioselectivity is determined by the titanium catalyst employed.

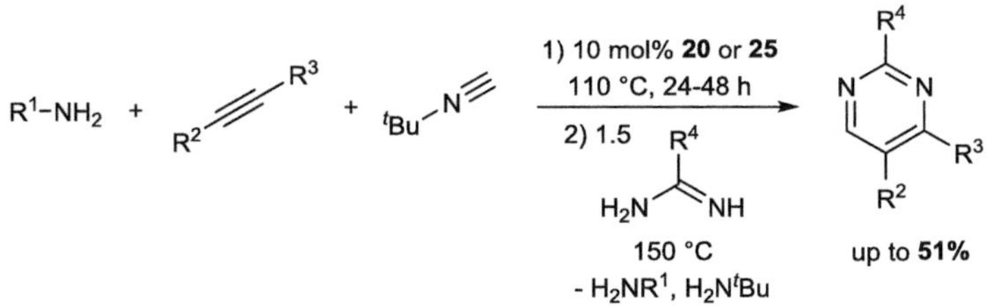

Scheme 32 Sequential titanium catalyzed hydroamination-iminoamination-amidine addition reactions toward the synthesis of pyrimidines.

In addition to selectively substituted planar N-heterocycles there is immense opportunity in the synthesis selectively substituted saturated heterocycles. For example, the modular synthesis of a series of ether containing aminoalkynes enabled the use of intramolecular hydroamination to give cyclic imine products in excellent yield. These products do not need to be isolated, but rather can be used directly in an asymmetric transfer-hydrogenation reaction with Noyori-Ikariya catalyst to give enantioenriched morpholines in excellent yield and enantiopurities (>99% ee) (Scheme 33).[77] This same strategy can expanded to prepare enantioenriched piperazines.[78]

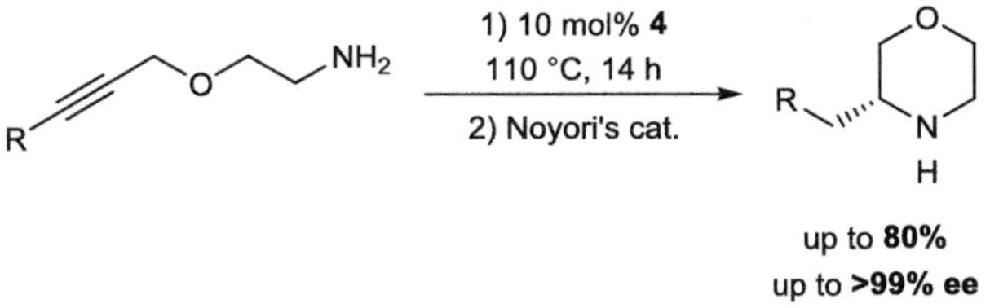

Scheme 33 Enantioselective synthesis of chiral morpholines via intramolecular hydroamination and asymmetric transfer-hydrogenation.

Various hydroamination catalyst technologies and transformations developed in the Schafer group can be assembled to realize the diastereoselective synthesis of 2,5-disubstituted piperazines. A combination of regioselective terminal alkyne hydroamination with allyl amine, followed by TMSCN addition and diastereoselective intramolecular alkene hydroamination using a zirconium ureate catalyst yields a variety of the targeted N-heterocycles with excellent diastereoselectivity (Scheme 34).[77–79]

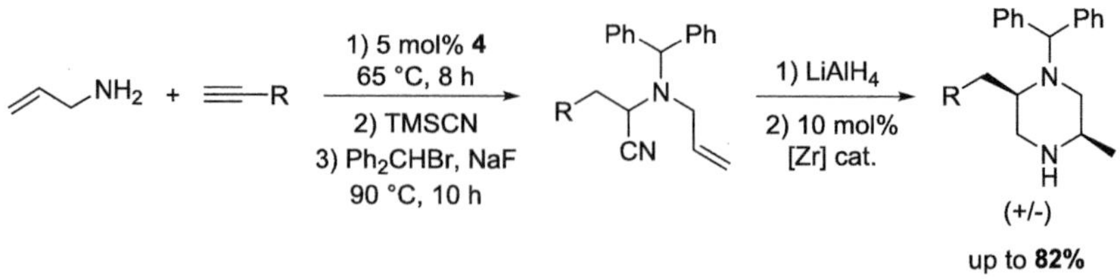

Scheme 34 Tandem and sequential reactions with alkyne hydroamination using titanium catalyst **4**.

2.7 Summary and outlook

Titanium catalyzed alkyne hydroamination is well established and mechanistically well understood. Interestingly, computational approaches have suggested that hydroamination is energetically feasible via insertion of C—C unsaturation into Ti—N σ-bonded amido ligands, although no experimental examples have been reported to date. While the empirical development of various non-Cp catalyst systems has resulted in the preparation of highly reactive catalysts, some with excellent regioselectivity, there is minimal predictive power in the development of ligand sets. Notably, recent advances in the establishment of ligand design parameters for early transition metals[80] can be applied toward the development of enhanced reactivity in hydroamination, by promoting the formation of more electrophilic metal centers. These trends may be used to realize the challenging intermolecular alkene hydroamination reaction, although such advances will likely also need to be coupled with advances in reaction protocol/reactor design to address the fact that there is a poor thermodynamic driving force for this transformation. To date there are catalysts that mediate the preferred formation of the anti-Markovnikov product, but there are no titanium catalysts that can rival late transition metal catalysts for the Markovnikov selective variant of this reaction. Thus, mechanistic rationale to predict regioselectivity outcomes would be helpful. Finally, the latter section of this chapter demonstrates that titanium catalyzed hydroamination is an enabling tool in the synthesis of selectively substituted amine small molecules, materials and N-heterocycles. Notably the development of titanium catalyzed multicomponent reactions have resulted in a flexible approach toward the synthesis of a range of N-heterocycles.

3. Hydroaminoalkylation
3.1 Overview

A complementary strategy for the addition of amines to alkenes is hydroaminoalkylation (Scheme 35). This catalytic reaction generates C_{sp}^3—C_{sp}^3 bonds by hydrofunctionalization of an alkene with a C—H bond α to nitrogen. Hydroaminoalkylation has the benefit of being atom economic, while avoiding pre-functionalized substrates and features readily available alkene feedstocks. Hydroaminoalkylation of alkenes benefits from the enhanced thermodynamic driving force of creating a new C—C bond rather than the C—N bond of hydroamination. This offers approximately 10 kcal/mol improved thermodynamic driving force, but the reaction is still challenged by significant thermodynamic and kinetic barriers, including the challenging activation of C_{sp}^3 hybridized C—H bonds. This results in reactions that demand high temperatures, and thus one challenge in catalyst design is the synthesis of complexes that are thermally robust. Furthermore, alkene hydroamination presents challenges in reaction regioselectivity (Scheme 35), with observed regioselectivity being largely determined by the metal center used for the transformation. Notably, titanium catalysts reported to date offer mixtures of linear **A** and branched **B** products, with significant substrate control being exploited to realize the formation of linear products.[81,82] The linear product **A** is attractive since this product offers an alternative approach for accessing the desired anti-Markovnikov amine products that have proven challenging to access by hydroamination. On-going efforts focus on developing catalyst-controlled regioselectivity for this transformation.[83]

Scheme 35 Amine directed C—C bond formation by hydroaminoalkylation.

From the 1980s to the early 2000s stoichiometric reactions of group 4 metallaaziridines with alkynes, alkenes or allenes were reported for the synthesis of selectively substituted amines.[84,85] Such reactive metallacycles can be prepared via ligand exchange reactions, rearrangement of aminoacyl compounds, thermolysis of metal amides or reaction of metallocene alkyl complexes.[86–91] These stoichiometric synthetic methods have been elaborated into impressive stereochemically controlled reactions to deliver selectively substituted amines. However, such stoichiometric reactivity resulted in the formation of significant amounts of waste by-products. As catalytic early transition metal chemistry has shown across a variety of reaction manifolds, non-Cp supported complexes with enhanced electrophilicity and improved steric accessibility to the metal center can often result in catalytic variants of such stoichiometric transformations.[92]

The catalytic hydroaminoalkylation reaction was first reported for homoleptic dimethylamido metal complexes of various early transition metals in 1980.[93] These early reactions exhibited low yields (10–38%) with few substrates and required long times and harsh conditions. Since then, late transition metals have also been reported for this transformation.[94] One notable difference when using late transition metals is the prevalent use of directing group-functionalized amine substrates.[95] While several reports before 1990 were disclosed for this transformation, it largely remained unexplored in the literature for over two decades, until it was revisited by Herzon and Hartwig in 2007 using early transition metals.[96] That work demonstrated the importance of using aryl amine substrates for enhanced reactivity and the effectiveness of tantalum for this transformation. It also reported modest or poor results using other early transition metal centers. It should be noted the Ti(IV) and Ta(V) octahedral complexes have comparable ionic radii,[97] suggesting that with the appropriate ligand environment, d^0 Ti(IV) may be effective for this attractive transformation. Indeed, in 2008 Doye reported the unanticipated intramolecular hydroaminoalkylation reaction while working to develop improved catalysts for aminoalkene cyclohydroamination.[98] Hydroaminoalkylation has evolved to become a fruitful area of investigation that has been recently reviewed.[92] Here we summarize a range of titanium catalysts that can be used to realize catalytic hydroaminoalkylation. Mechanistic insights are presented and applications toward organic synthesis are also highlighted.

3.2 Mechanism

Hydroaminoalkylation results from C—H activation adjacent to an amine functionality, and subsequent addition of that C—H bond across an alkene (Scheme 36).[99] In the case of early transition metal-catalyzed variants of this reaction, such as titanium, this reaction first proceeds via the N—H

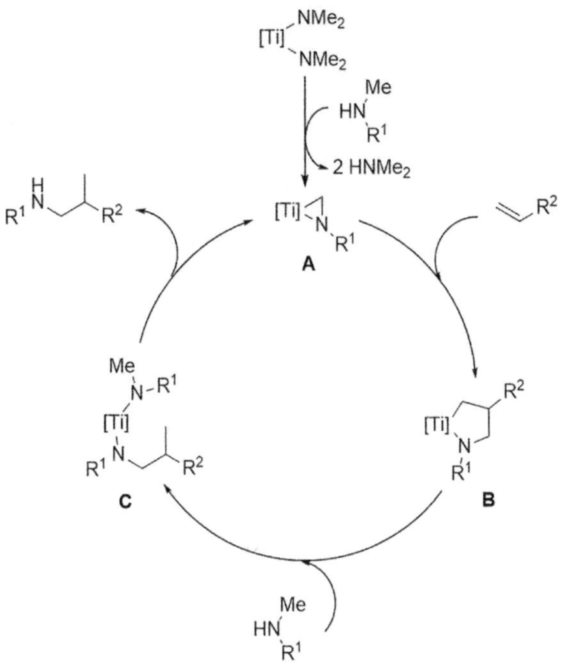

Scheme 36 General mechanism of the hydroaminoalkylation of alkenes by titanium catalysts.

activation of either primary or secondary amines. It should be noted that the intramolecular variant of this reaction can proceed with primary amines while the intermolecular version of this reaction demands secondary amines at this time. After the initial Ti—N bond formation places the substrate in proximity to the electrophilic metal center, there is a subsequent C—H bond activation event to yield a metallaaziridine catalytically active species **A**.

Such titanium aziridines and intermediates have been isolated, and characterized (Scheme 37).[100] Furthermore, their relevance to hydroaminoalkylation has also been demonstrated in stoichiometric investigations.

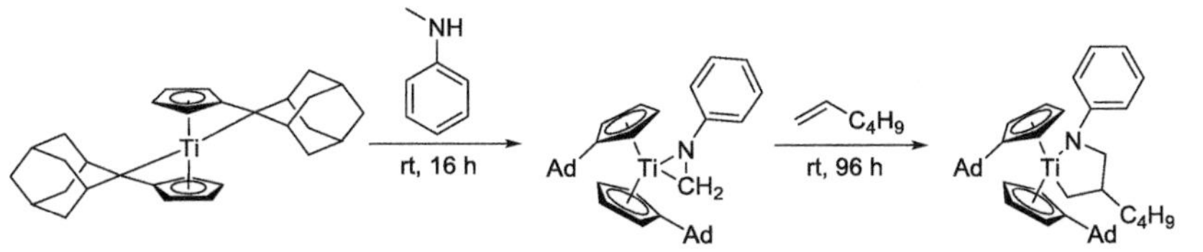

Scheme 37 First isolation of all intermediates of the titanium catalyzed hydroaminoalkylation of alkenes.

The strained catalytically active titanium aziridine can undergo alkene insertion to yield metallacyclic intermediate **B**. In the case of catalytic reactivity, an incoming equivalent of amine substrate is sufficient for protonolysis of the metal–carbon bond of the metallacycle, resulting in the formation of a new bis(amido) intermediate **C** that can then undergo subsequent C—H activation to regenerate the requisite metallaaziridine catalytically active species. This catalytic mechanistic proposal builds from the literature of related stoichiometric transformations.[100] However, the mechanism as illustrated is a significantly simplified picture, as there are many equilibria in this reaction, some of which generate secondary amine products from the dimethyl amine resulting from activation of the bis(amido) titanium precatalyst to prepare **A** in situ. Such amine exchange equilibria can complicate the generation of efficient catalytic systems.[101]

To date several titanium catalysts have been developed for alkene hydroaminoalkylation (Scheme 38).[81,83,102–105] In general, these catalysts typically require high reaction temperatures and long reaction times, in comparison to other state-of-the-art hydroaminoalkylation catalysts.[92] However, latest results indicate that by applying specific conditions, reaction times can be reduced from days to minutes (see Section 3.4).[106] No particular titanium

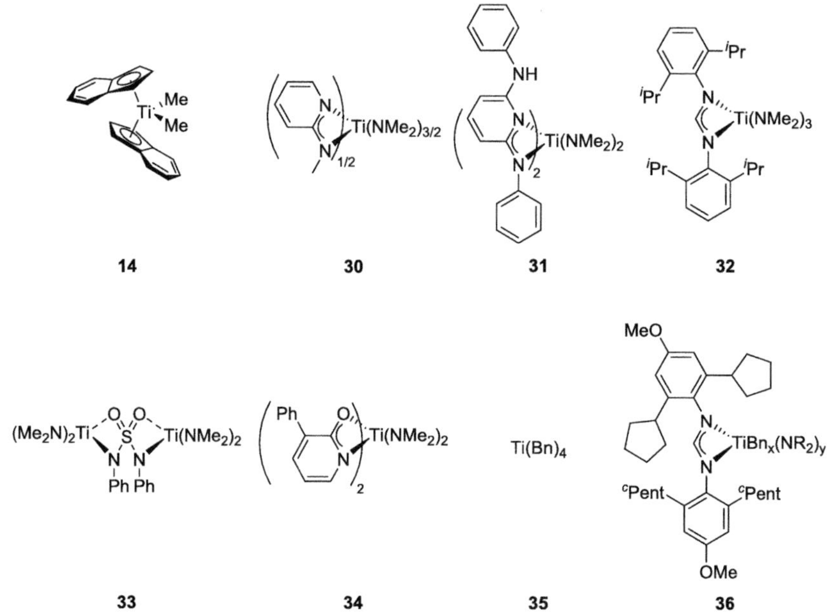

Scheme 38 Selected titanium catalysts for hydroaminoalkylation.

catalyst addresses all the regioselectivity, diastereoselectivity and substrate scope challenges of this transformation, but titanium does offer some unique reactivity trends that can be exploited to access products that cannot be prepared using other approaches. The distinguishing features of titanium catalyzed hydroaminoalkylation catalysis are highlighted below.

3.3 Intramolecular hydroaminoalkylation

The first example of titanium catalyzed hydroaminoalkylation was reported as a by-product observed during attempted cyclohydroamination of aminoalkene substrates.[82,107] Most notably, it was during attempts to form the challenging seven-membered azepane products by cyclohydroamination that the six-membered amine substituted cyclohexane products were observed (Scheme 39). This reaction, under substrate control, is chemoselective for hydroaminoalkylation over hydroamination. A diastereomeric mixture of products results, where the modest diastereoselectivities are also under substrate control.

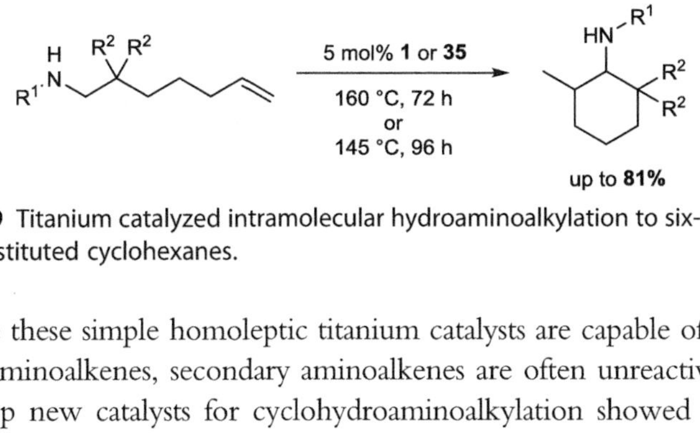

Scheme 39 Titanium catalyzed intramolecular hydroaminoalkylation to six-membered amine substituted cyclohexanes.

While these simple homoleptic titanium catalysts are capable of cyclizing primary aminoalkenes, secondary aminoalkenes are often unreactive. Efforts to develop new catalysts for cyclohydroaminoalkylation showed that aminopyridinate titanium catalyst **30** is capable of using secondary aminoalkenes to access the desired products (Scheme 40), albeit with poor diastereoselectivity.[108] Due to the fact that intermolecular hydroaminoalkylation actually demands the use of secondary amines (vide infra), it is surprising that

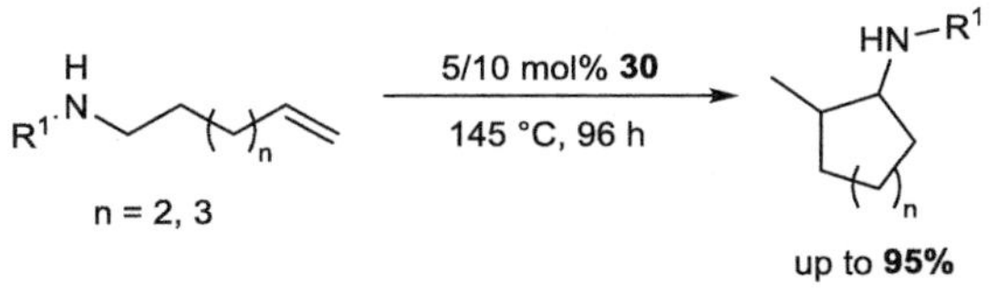

Scheme 40 Intramolecular hydroaminoalkylation with titanium aminopyridonate catalyst **30**.

intramolecular hydroaminoalkylation is such a challenge. Furthermore, it is remarkable that while titanium catalyzed intramolecular hydroaminoalkylation is known for both primary and secondary aminoalkene substrates, other early transition metals reported to date do not promote intramolecular variants of this reaction.

For intramolecular hydroaminoalkylation, a major limitation is the selective formation of the desired hydroaminoalkylation product over the hydroamination product. However, it is notable that varying the steric bulk on the ancillary ligands of the titanium complex results in changes in the chemoselectivity of cyclohydroaminoalkylation vs. cyclohydroamination. Specifically, more sterically bulky ligand sets favor cyclohydroamination over cyclohydroaminoalkylation.[104] Furthermore, during catalytic amination investigations with benzyl amine as substrate, a bridged titanaziridine product could be isolated, which would result from the C—H activation of bridged titanium imido species that are known to form in cases where there is insufficient steric bulk (Scheme 41).[109]

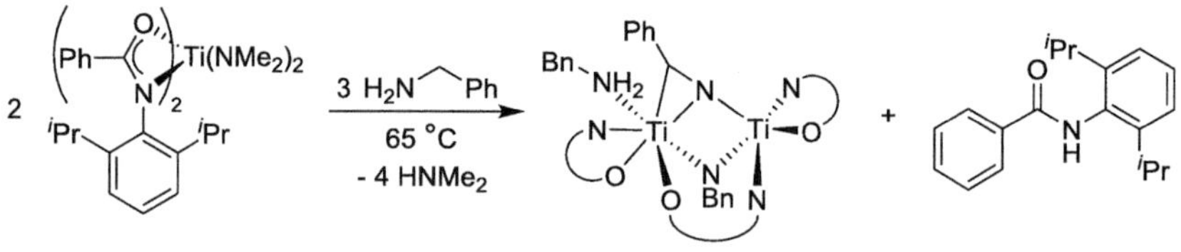

Scheme 41 Synthesis of a bridging titanaaziridine complex (ligands simplified for clarity).

This isolated reactive intermediate is competent for hydroaminoalkylation and was the basis for a hypothesis that such bridged bimetallic complexes may be involved in primary amine cyclohydroamination, as illustrated in Scheme 42.[104] This proposal would be consistent with the observation that ligands with significant steric bulk that favor the formation of terminal imido species, would preferentially undergo cyclohydroamination and reduce the amount of amine substituted cyclohexane product obtained.

This hypothesis was further tested by preparing a series of cyclohydroaminoalkylation precatalysts using pyridonate ligands of varied steric bulk (Scheme 43).[104] The more challenging chemoselective reaction of promoting cyclopentane formation (hydroaminoalkylation) over

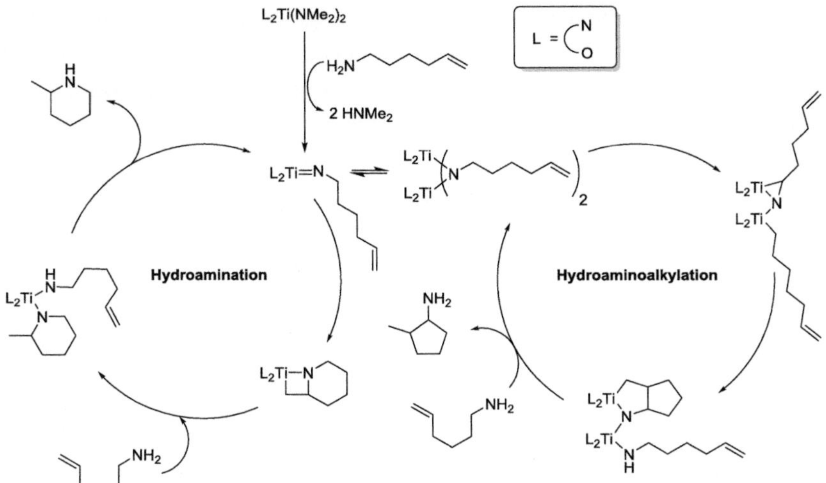

Scheme 42 Competing titanium-mediated hydroamination and hydroaminoalkylation catalytic cycles with primary aminoalkene substrates.

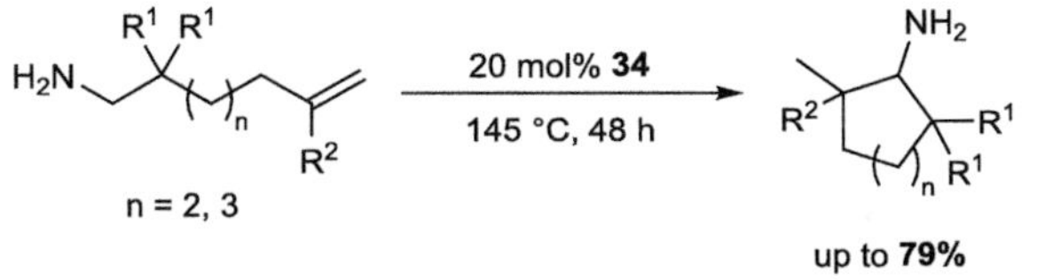

Scheme 43 Intramolecular hydroaminoalkylation with titanium pyridonate catalyst **34**.

piperidine formation (hydroamination) was used to identify preferred catalysts. In this work, the bulkier, but yet readily accessed 3-phenyl substituted pyridonate is the only reported system for catalyst-controlled chemoselectivity of hydroaminoalkylation over hydroamination. This catalyst also offered substrate dependent diastereoselectivities that could be excellent.

3.4 Intermolecular hydroaminoalkylation

The titanium catalyzed addition of amines to alkenes by C—N bond formation remains unknown. However, an alternative and complementary transformation that promotes the addition of amines to alkenes is through the C—C bond formation of hydroaminoalkylation. This reaction has an appreciably improved thermodynamic driving force (approx. 10 kcal, depending upon substrate combination) that results in high yielding syntheses of the desired products. In order to achieve products equivalent to those achieved using anti-Markovnikov hydroamination, regioselective intermolecular hydroaminoalkylation to prepare linear amines must be realized. To date, only hydroaminoalkylation as mediated by late transition metals can reliably complete this transformation. Unfortunately, late transition metal hydroaminoalkylation typically requires the installation of protecting/directing groups on the amine and to date the use of simple amine substrates is not broadly accessible.[95] Notably, titanium shows promise for this transformation and while complete regioselectivity remains elusive, of all early transition metals that can be used for this reaction, titanium does offer catalytic hydroaminoalkylation with preference for the linear product with select substrates.[81]

In experiments using simple homoleptic Ti(NMe$_2$)$_4$ (**1**) and Ti(Bn)$_4$ (**35**) complexes, titanium provided some access to linear products in comparison to previous reports that reported only branched product formation (Scheme 44).[82]

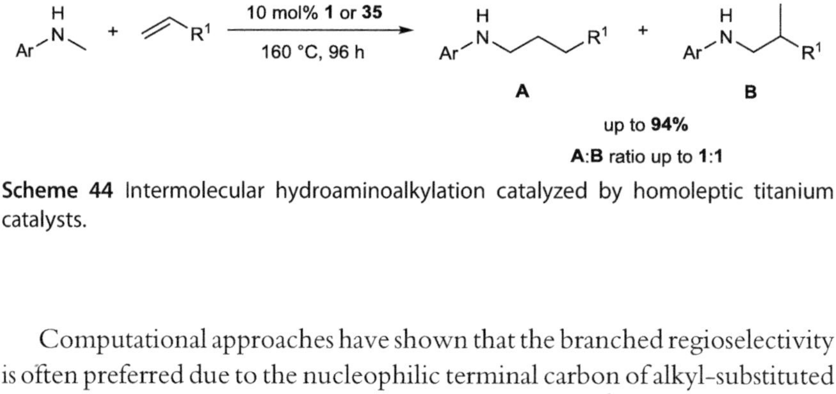

Scheme 44 Intermolecular hydroaminoalkylation catalyzed by homoleptic titanium catalysts.

Computational approaches have shown that the branched regioselectivity is often preferred due to the nucleophilic terminal carbon of alkyl-substituted terminal alkenes interacting with the electrophilic d^0 metal of the metallaaziridine during alkene insertion.[110] These calculations show that the energy of the alkene insertion transition states are significantly affected by electronic properties. Thus, substrates, such as styrene derivatives, dienes and vinylsilanes, with different electronic profiles, can undergo substrate-controlled insertion to yield the preferred linear product. In addition to substrate control, it has been noted that catalyst design can influence insertion regioselectivity, where bulky *N,N*-chelating ligands in combination with titanium promote enhanced formation of the linear product. In this context, the latest results from the Doye group demonstrated excellent selectivity for the formation of the linear product when using the titanium catalyst systems

30 and 31.[81,103] The selectivity for both of these catalysts are primarily restricted to styrenes as alkene substrate, however both systems offer the opportunity to change the amine substrate to something other than methyl substituted amines. In this case α-substituted amines can be assembled in a single catalytic, atom economic step (Scheme 45). Additionally, it has to be noted that only titanium catalysts have been shown to give these products reliably among early transition metals. Even Ti(NMe$_2$)$_4$ (1) can realize the synthesis of such desirable products, albeit with no regioselectivity.[82]

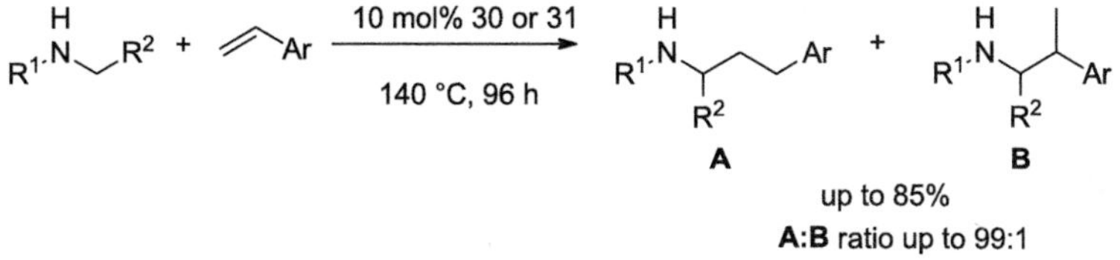

Scheme 45 Titanium catalyzed intermolecular hydroaminoalkylation for the synthesis of linear products.

Furthermore, the N,N-chelated titanium catalyst 30 is the only catalyst from across the periodic table that can successfully α-alkylate unprotected pyrrolidine with styrene to give the linear hydroaminoalkylation product preferentially (Scheme 46),[81] whereas late transition metal catalysts can only achieve the same transformation by using directing groups.[95]

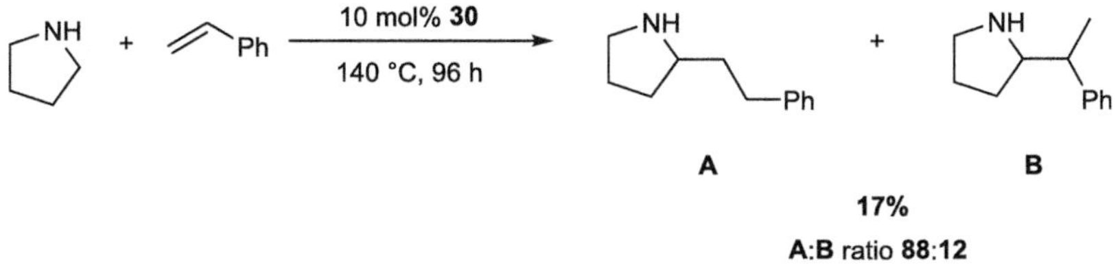

Scheme 46 Titanium catalyzed intermolecular hydroaminoalkylation for the synthesis of substituted pyrrolidine.

Catalyst 32 showed good selectivities for the formation of the linear product with diene substrates when using N-benzylmethylamine (Scheme 47).[83] Nevertheless, the substrate dependence of this system is illustrated by the fact that the use of the common N-methylaniline substrate results in the preferred formation of the branched product with the same diene. These results show that more research is needed to develop reliable and catalyst-controlled synthesis of the linear regioisomer.

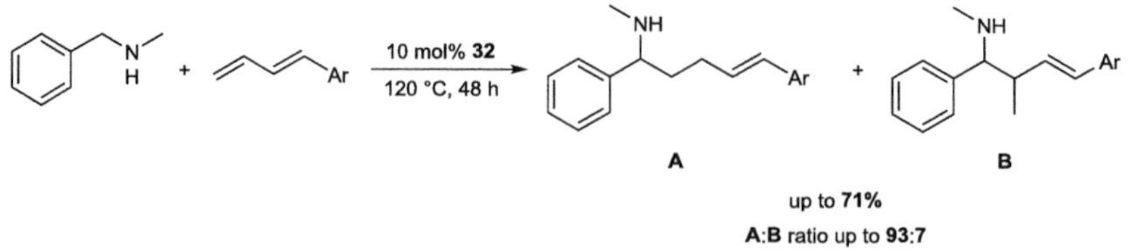

Scheme 47 Titanium catalyzed intermolecular hydroaminoalkylation of dienes with *n*-benzylmethylamine for the synthesis of linear products.

In addition to *N,N*-chelating ligands, *N,O*-chelating ligands have been evaluated in intermolecular hydroaminoalkylation. Sulfonamidate complexes are unique in that two perpendicular *N,O*-chelates are accessible, allowing for the formation of dinuclear complexes with one ligand bridging the metal centers.[105] For the preparation of this catalytically active species both 1 and 2 equiv. of sulfonamidate ligand were evaluated for hydroaminoalkylation and only the complex with 1 equiv. of ligand per two metal centers was a viable catalyst (Scheme 48). Notably, this system that offers a more electrophilic metal center, with less steric crowding, favors the formation of the branched amine product. Additionally, in intramolecular reactivity, this complex forms the hydroamination product preferentially.

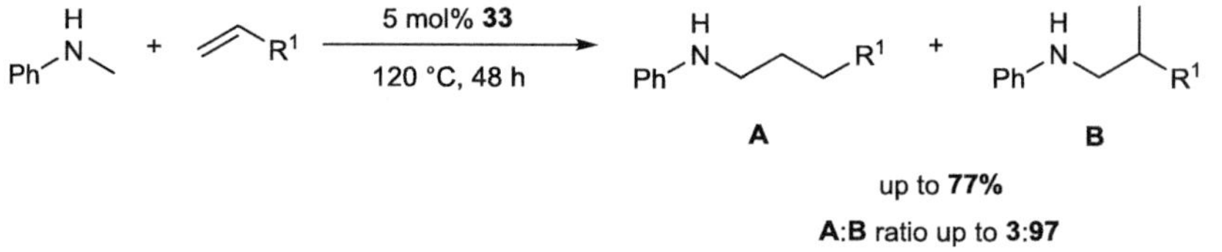

Scheme 48 Dinuclear titanium catalyzed intermolecular hydroaminoalkylation for the synthesis of branched products.

As mentioned previously, the branched product is the most commonly accessed for all early transition metal-catalyzed hydroaminoalkylation, and this result is complementary to late transition metal catalysts that give the linear product exclusively. Homoleptic complexes furnished a mixture of regioisomeric products.[82] The first titanium catalyst reported for the preferential synthesis of β-methylated amines, the branched product, was the known Ind$_2$TiMe$_2$ (**14**) (Scheme 49).[102] Although a mixture of products can result, using this titanium organometallic precatalyst the branched product can be accessed preferentially, even with substrates that are known to promote formation of the linear isomer.

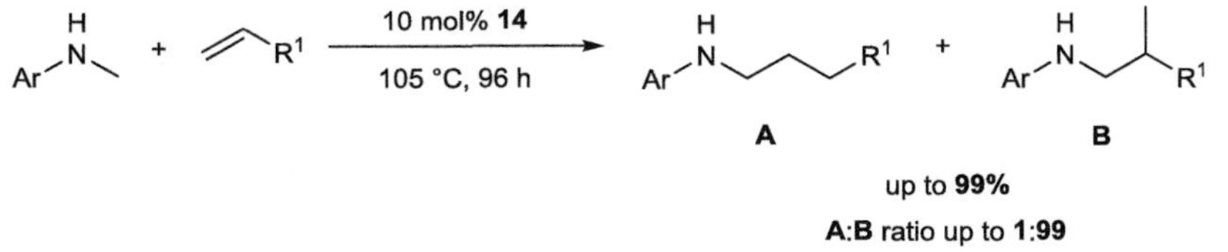

Scheme 49 Titanium catalyzed intermolecular hydroaminoalkylation for the synthesis of branched products.

While substrate control has been used to access linear products with N,N-chelated titanium complexes, similar catalysts can be used to prepare significant quantities of branched products with unactivatedalkyl-substituted alkenes. This has been used to access highly substituted amine products by using dimethylamine as a substrate with catalyst **32** (Scheme 50).[111] This synthetic approach features the use of the feedstock chemicals HNMe$_2$ and α-olefins to prepare sterically bulky amines, and either the mono- or dialkylated product can be preferentially accessed using controlled amounts of alkene substrate.

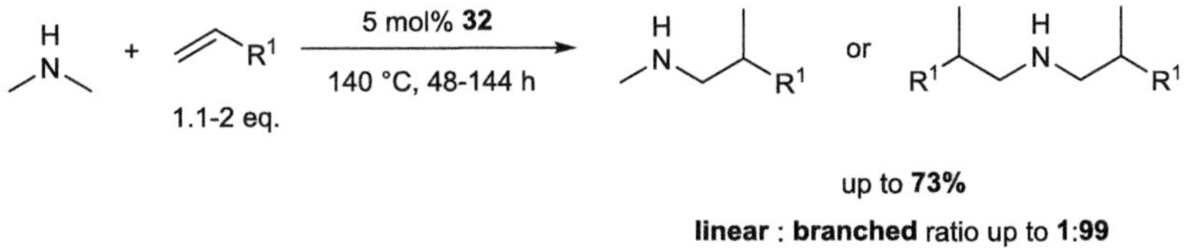

Scheme 50 Titanium catalyzed intermolecular hydroaminoalkylation for the synthesis of branched products.

Recent investigations of the Doye group using the same catalyst **32** gave good activity with N-methylanilines using reactive and gaseous ethylene (Scheme 51).[112] This catalyst also furnished up to 20% of the dialkylated product, which could not be prevented by lowering the reaction time. Additionally, industrial relevant gaseous dimethylamine was reacted with ethylene to form dipropylamine as major product.

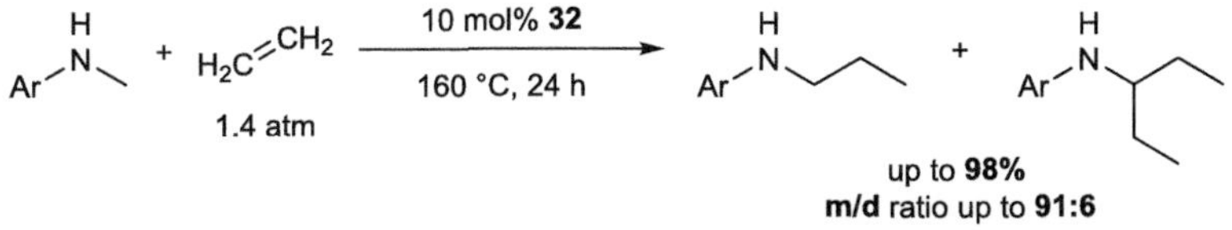

Scheme 51 Titanium catalyzed intermolecular hydroaminoalkylation of ethylene with N-methylanilines for the formation of mono- (**m**) and dialkylated (**d**) amines.

By modifying this ligand to the even more sterically enhanced N,N-chelated system **36** with in situ use of highly reactive Ti(Bn)$_4$ (**35**) as titanium source and solvent-free conditions, reaction times of the titanium catalyzed hydroaminoalkylation can be ultimately reduced from days to minutes (Scheme 52). However, the increased reactivity compared to other reported formamidinate titanium complex **32** stems largely from the titanium precursor and reaction conditions; the use of the specific ligand only doubles the catalytic activity. Additionally, TMS- and TBDMS-protected secondary amines can be reacted, which can be simply deprotected with water to yield the corresponding primary amines in good overall yields.

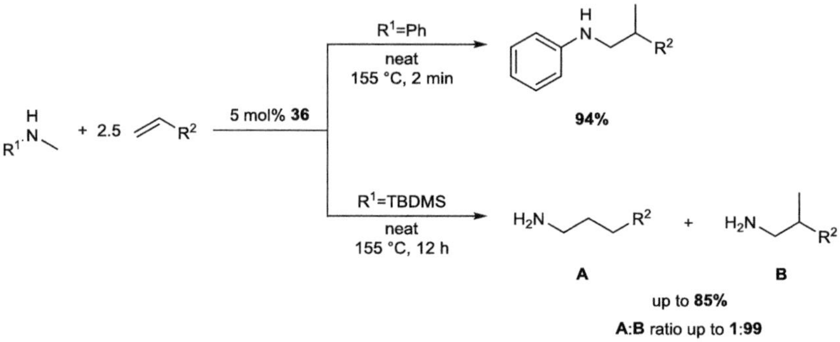

Scheme 52 Fast titanium catalyzed intermolecular hydroaminoalkylation of alkenes utilizing in situ generated titanium complex **36** and solvent-free conditions; synthesis of primary amines via silyl-protected amines.

By increasing the steric bulk of the trityl-substituted N,N-chelate a mono-ligated catalyst **16** could be accessed. This catalytic system, which was also used for the hydroamination of allenes and the intramolecular hydroamination of aminoalkenes (see Sections 2.4 and 2.5), promoted the formation of the branched product.[44,113] Even vinyl silanes, which are known to afford linear products,[96] furnish the β-methylated product here (Scheme 53).

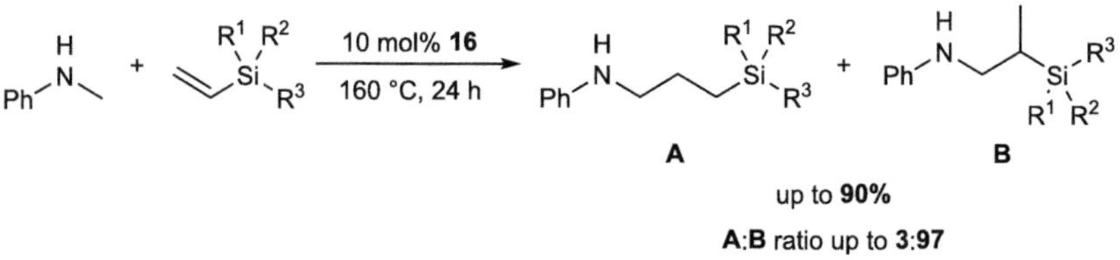

Scheme 53 Titanium catalyzed intermolecular hydroaminoalkylation of vinyl silanes.

The synthesis of α,β-disubstituted amines via hydroaminoalkylation is a challenging transformation. Formation of such products proceed via sterically demanding transition states upon alkene insertion. Although rare, early transition metals, including titanium, can promote this desirable transformation (Scheme 54).[82,107]

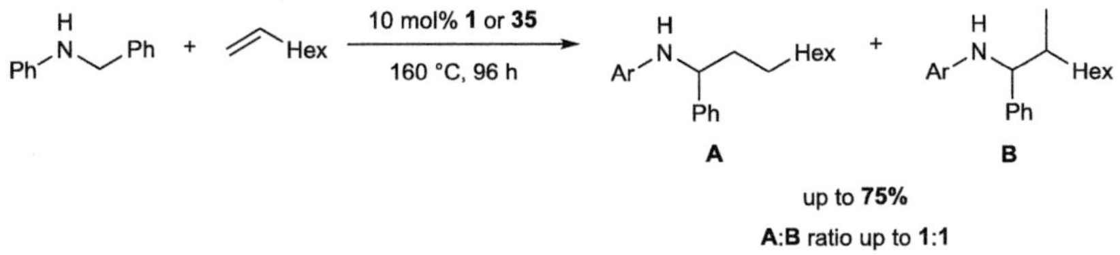

Scheme 54 Titanium catalyzed intermolecular hydroaminoalkylation of N-benzylaniline with 1-octene.

Titanium complexes are growing to become powerful catalysts for hydroaminoalkylation. This early transition metal offers reactivity trends and regioselectivies that are unique from other metal catalysts across the periodic table. For these reasons, developing new titanium catalysts for this transformation holds much promise. One of the significant challenges facing titanium catalyst development is the harsh reaction conditions required to access desired reactivity. On-going catalyst development efforts would benefit from further contributions from organometallic chemists that explore controlled coordination environments, including rigorous characterization of precatalysts, reactive intermediates and reaction kinetics. At present, a clear understanding of why titanium accesses complementary reactivity to other early transition metals has not been presented and this offers an exciting opportunity for further mechanistic investigation.

3.5 N-heterocycle synthesis by hydroaminoalkylation

With new approaches for assembling C—C bonds in an atom economic fashion, hydroaminoalkylation is ideal for assembling selectively substituted frameworks that can then undergo further reactivity. Furthermore, because titanium is a d^0 metal in this transformation, and thus not capable of undergoing traditional oxidative addition reactions with aryl halides, the catalysts developed here can incorporate halogenated aryl groups into the alkene substrate. After hydroaminoalkylation, the secondary amine generated can undergo an intramolecular Buchwald-Hartwig coupling reaction to make new, selectively substituted heterocycles. Using this strategy multiple reports using catalysts **16** and **32** have been used to synthesize a variety of six- and seven-membered *N*-heterocycles such as 1,4-benzoazasilines, 1,5-benzoazasilepines, 1,5-benzodiazepines, 1,5-benzoxazepines and 1,5-benzothiazepines in moderate to good yields (Scheme 55).[113–116]

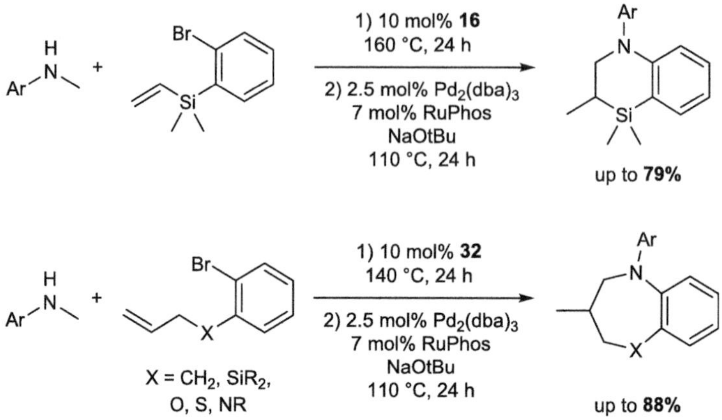

Scheme 55 Titanium catalyzed intermolecular hydroaminoalkylation and Buchwald-Hartwig amination for the synthesis of *N*-heterocycles.

3.6 Summary and outlook

Hydroaminoalkylation is quickly evolving into a reliable protocol for the direct, catalytic addition of amines to alkenes. Titanium catalyzed hydroaminoalkylation has undergone significant development in the past 10 years and now provides a facile and direct route to a broad range of selectively alkylated amines and heterocycles. The strategic selection of substrates in combination with titanium catalysts can offer access to linear regioisomers that typically can only be prepared using late transition metal catalysts. One significant advantage with early transition metals is their ability to mediate these reactions using simple amines with no need of installing auxiliary protecting/directing groups. This demonstrates the power of this reaction to access a structurally diverse set of amines from simple commercially available starting materials, making it of potential utility for industrial application.

Although there has been significant progress over the past 10 years, there are still many opportunities for further catalyst and methodology development. To date titanium is not efficient for use with internal alkenes. The development of catalyst-controlled regioselectivity for the linear products is particularly exciting, and the development of improved catalysts that can utilize a broader range of amines and alkenes is an interesting opportunity.

All of the reported catalysts are based on air- and moisture-sensitive metal alkyl and metal dimethylamido complexes, and thus require inert atmosphere for use. Most reports deal with this by performing all of the reactions in gloveboxes, or by using specialized Schlenk glassware. The sensitivity of the catalysts has also led to the use of higher catalyst loadings (10–20 mol%), as trace impurities can deactivate the catalysts over the course of the reaction. In addition, catalysts often require temperatures of 110 °C or higher, with most reactions being run in superheated solvent. Advanced catalyst development efforts could exploit titanium as an inexpensive and more broadly useful early transition element. Furthermore, advances in catalyst delivery protocols that make air-sensitive catalyst systems more easily handled could be applied toward titanium systems.[117] The requirement for high temperatures due to the high energy transition states involved are not surprising, and such temperatures are often required for C—H alkylation reactions using metals from across the periodic table.

Ligand design for titanium catalyzed hydroaminoalkylation has few guiding principles to date. The use of bidentate, hemilabile ligands has been shown to facilitate the development of active catalysts. There are few clear trends in promoting reactivity beyond preliminary observations regarding the use of ligands that promote the formation of electrophilic metal centers and the impact of steric bulk on promoting the desired regioselective reactivity.[118–120]

The past 10 years have seen the development of hydroaminoalkylation from an obscure reaction, largely forgotten in the literature, into an active area of research that shows immense promise for the synthesis of selectively alkylated and/or arylated amines. Most importantly, these advances have realized one of the few catalytic amination reactions of simple alkenes. Within the field of organometallic and catalytic chemistry there are opportunities for developing enhanced mechanistic understanding and subsequent improved reaction outcome whereas within the field of organic synthetic methodologies, new disconnection strategies can be explored for atom- and step-economic assembly of selectively substituted amines and *N*-heterocycles.

4. Titanium redox chemistry for catalytic amine synthesis

4.1 Introduction

Both of the aforementioned hydrofunctionalization reactions for the synthesis of amines proceed without a change in oxidation state. Hydrofunctionalization catalyst development efforts have focused on the design of ligand sets that support robust, Lewis acidic metal centers in the +IV oxidation state. While titanium is readily oxidized and most easily isolated in the d⁰ electron configuration, it does have readily accessible oxidation states of Ti(III) and Ti(II). Such metal centers can then be used in alternative transformations that proceed by either radical or 2e⁻ redox cycles. Indeed, extensive examples of stoichiometric chemistry using reduced titanium reagents were reported in the 1980s and 1990s[84,121–123] and more recently these reagents have been exploited in advanced synthetic protocols to realize efficient syntheses of complex natural products.[124–126] A well-known example for this chemistry is the readily accessible Nugent-RajanBabu reagent.[127] This material is prepared via reduction of Cp_2TiCl_2 (**37**) with reducing metal, such as Mg, to give a green Ti(III) solution of the bridged chloride dimer **38**, which can then undergo further reactivity in the presence of acetonitrile to make the blue acetonitrile coordinated titanocene complex.[128] This complex is so air and moisture sensitive that it is a preferred solution to be used as a glovebox indicator. However, this sensitivity makes it less attractive for use in synthetic protocols.

Alternatively, several approaches have been adopted for the generation of reduced titanium species as well-defined reagents that can be stored and used as such (Scheme 56).[129–136] Such complexes are typically prepared using initial Ti(IV) precursors that are then reduced using harsh metal

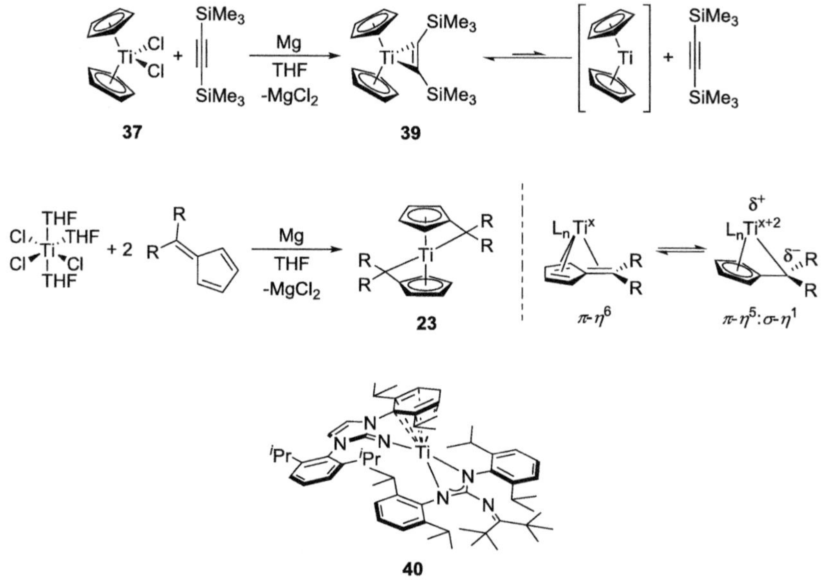

Scheme 56 Selected examples of "Masked" Ti(II) complexes.

reductants, and trapped with neutral donor ligands that are strong π-acceptor ligands (see **39**). This π-acceptor character results in the formation of "masked Ti(II)" complexes that could formally be considered d⁰ metals but upon dissociation of the neutral donor ligands, a reactive Ti(II) species is revealed. Instead of harsh reducing metal conditions, organosilane reductants have been developed (Scheme 57).[137,138] These reagents are attractive for practical use as they are crystalline, soluble solids that have sufficient reduction potentials to offer alternative routes into known reduced titanium complexes with well-established reactivity profiles.

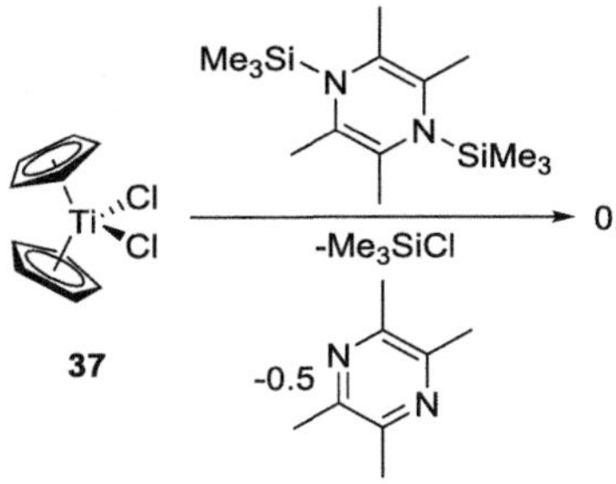

Scheme 57 Organosilane as a stoichiometric reductant for the synthesis of a Ti(III) complex.

Masked Ti(II) reagents are susceptible to unwanted decomposition if not properly stored under inert atmosphere and away from moisture. However, such reagents have been readily exploited in synthetic applications in organometallic and organic chemistry and are powerful reagents for a broad range of reductive bond coupling transformations,[4,84,121,139–147] including the stoichiometric formation of titanaaziridines from Ti(II) and imine substrates, for the stoichiometric variant of the hydroaminoalkylation reaction.[86,87]

Alternatively, it is attractive to begin with simple Ti(IV) complexes and generate such reactive species in situ with readily available reagents. For example, Ti(O^{i}Pr)$_4$ (**41**) can undergo reduction in the presence of 2 equiv. of an alkyl Grignard reagent to access an ill-defined Ti(II) species via β-hydrogen abstraction of the intermediate bis(alkyl) complexes.[123,148] This readily available Ti(II) species generated in solution using simple syringe techniques can then be exploited in synthetic protocols (Scheme 58).

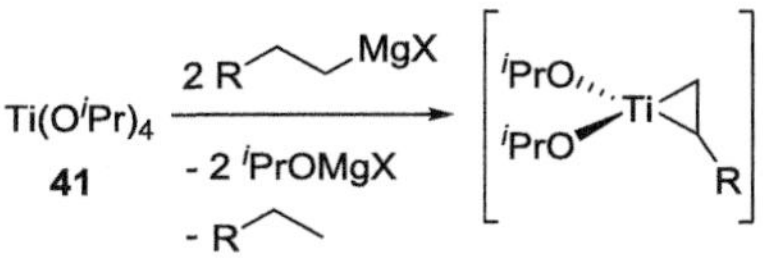

Scheme 58 In situ prepared "Masked" Ti(II) reagent for stoichiometric reduction reactions.

Reduced Ti(II) complexes can be used for the synthesis of olefin reductive coupling and the formation of substituted benzene derivatives from alkynes.[149–153] In these cases the initial Ti(II) species can either be generated in situ or prepared as a discrete complex and used in catalytic quantities, due to a β-hydride/reductive elimination step that results in regeneration of the Ti(II) reduced metal catalyst, much like late transition metal catalytic redox cycles. Notably, the application of such catalytic approaches toward the synthesis of amines or N-heterocycles is only now becoming an emerging area of research interest.

Ti(III) reagents are also very important reduced metal species for the synthesis of a broad range of organic compounds.[154–163] While reliable protocols have been developed for the preparation and isolation of Ti(III) complexes, such as Cp$_2$TiCl, these crystalline materials are rarely used in applications toward the preparation of amines, because of the challenges associated with making them. However, Ti(III) complexes can also be readily prepared in situ, simply through the treatment of Ti(IV) precursors with amines.[164]

This chapter highlights recent achievements in transitioning well-established stoichiometric protocols for these established reduced titanium complexes into catalytic variants for these reactions through strategic and selective use of ligands, co-reagents and substrates. By building from early stage results that featured the use of ill-defined reduced titanium species that could mediate catalytic transformations, new catalytic approaches, complete with mechanistic insights, are being developed for the catalytic synthesis of amines and N-heterocycles.

4.2 Ti(II) catalysis for the synthesis of amines and N-heterocycles

Early stage developments for Ti(II) applications toward organic chemistry have featured the use of Ti(O^{i}Pr)$_4$ (**41**) as an easily handled Ti(IV) precursor that can be treated with alkyl organometallic reagents, such as XMg^{i}Pr, to provide in situ access to Ti(II) reagents. This reagent has been extensively applied toward stoichiometric syntheses and has been thoroughly reviewed.[123,148] These synthetic advances generate a stoichiometric amount of TiO$_2$ upon workup which can be problematic, although the low cost and low toxicity of titanium makes these reagents very attractive for applications in organic synthesis. Notably, there are examples of catalytic Ti(II) species generated in situ, such as the Kulinkovich reaction.[123,148] This transformation was initially developed for the synthesis of cyclopropyl alcohols, but a

Titanium catalyzed synthesis of amines and N-heterocycles

variant was developed that realizes the synthesis of amine substituted cyclopropanes (Scheme 59).[123,165,166] By beginning the reaction with an organic amide that is not very electrophilic, the added Grignard reagent serves to generate the Ti(II) complex **A** in situ and also achieves catalytic turnover.

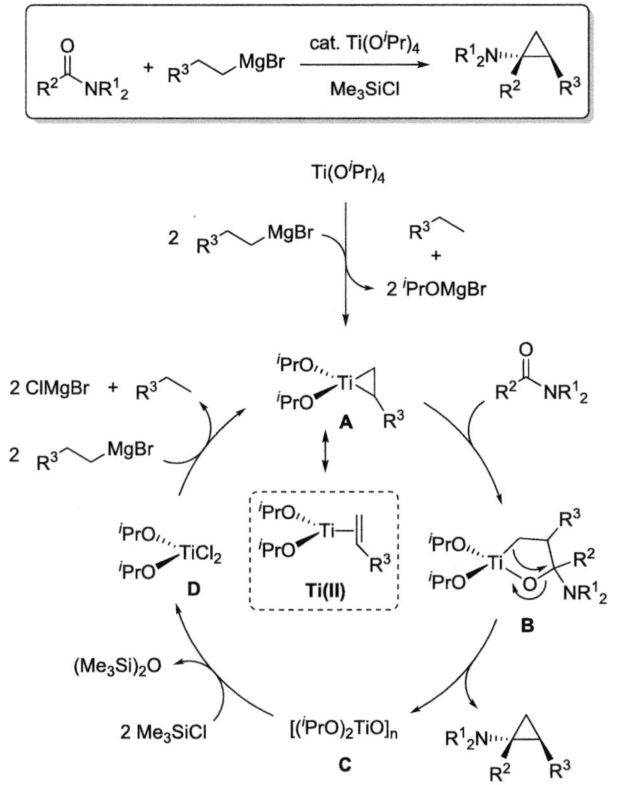

Scheme 59 A modified Kulinkovich reaction to assemble amine substituted cyclopropanes and proposed mechanism.

The generation of reduced titanium species in situ can be achieved using a variety of reaction conditions. More recently, efforts to realize intermolecular alkyne hydroamination with a biphenolate pincer complex **42** (Scheme 60) unexpectedly resulted in the observation of alkyne trimerization, in addition to the targeted hydroamination product.[167] This observation suggested that this ligand environment supported the in situ generation of Ti(II) species that could then engage in established catalytic alkyne trimerization reactions to yield substituted arenes.

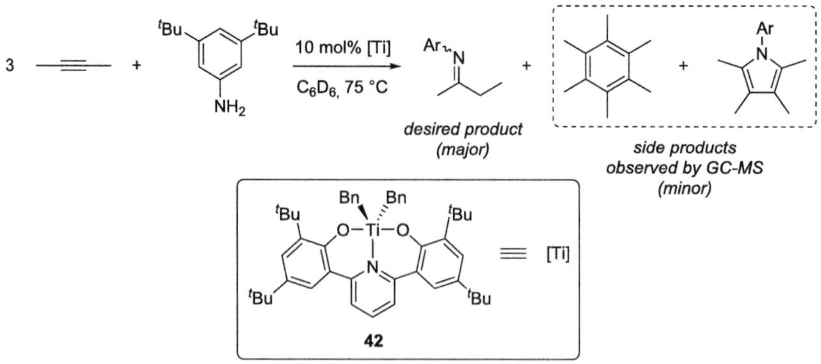

Scheme 60 Unexpected alkyne trimerization and pyrrole formation using biphenolate pincer complex **42**.

Furthermore, this example included the first report of pyrrole synthesis, presumably from imido and alkyne [2 + 2] cycloaddition followed by a subsequent alkyne insertion and finally a reductive elimination step to give the cyclic product and presumably regenerate a Ti(II) species. These initial observations inspired on-going work for the synthesis of pyrroles via titanium catalyzed nitrene transfer. Pyrroles are important N-heterocycles of biological relevance and efficient approaches for their syntheses are desirable. An optimized route that promotes the formation of selectively substituted pyrroles from simple alkyne and nitrene precursors represents an attractive opportunity for developing new titanium catalysts. By starting with a well-known and preformed imido complex **43a** in catalytic quantities, along with the excess alkyne and azobenzene as a nitrene source, the efficient synthesis of pyrroles could be realized (Scheme 61).[168] This reaction is a formal [2 + 2 + 1] cyclization reaction and is reminiscent of established Pauson-Khand chemistry for the synthesis of cyclopentenones.[169] The nitrene reaction can incorporate various substituents into the pyrrole ring and by taking advantage of regioselective alkyne insertion steps, good control over regiochemistry can be realized in many examples.

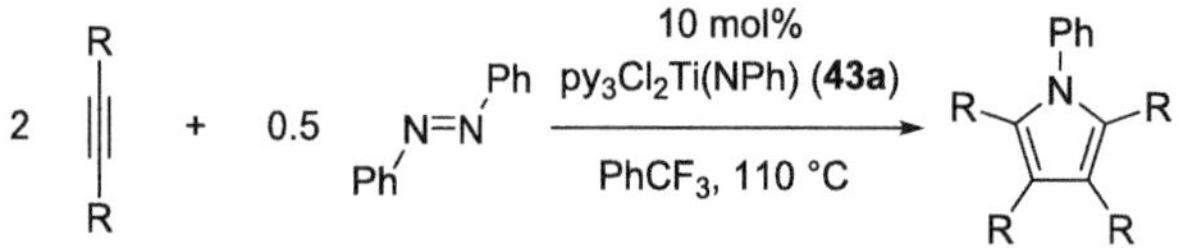

Scheme 61 Catalytic pyrrole synthesis using a titanium imido catalyst.

Importantly, catalytic turnover in this reaction demands reductive elimination as a key step in the catalytic cycle. The proposed mechanism, which has been supported by DFT investigations,[170] is presented in Scheme 62. The reductive elimination step proceeds via an electrocyclic reductive elimination, to give access to a pyrrole stabilized "masked Ti(II)" complex **D** that can then undergo ligand substitution with azobenzene to access a Ti(IV) complex **E**, which could also be considered a "masked Ti(II)" complex. Another key step to ensure catalytic turnover is the disproportionation step to regenerate the catalytically active titanium imido species **A**. This mechanism is proposed and has been calculated to proceed via a dimeric complex, although it has not yet been extensively explored experimentally.

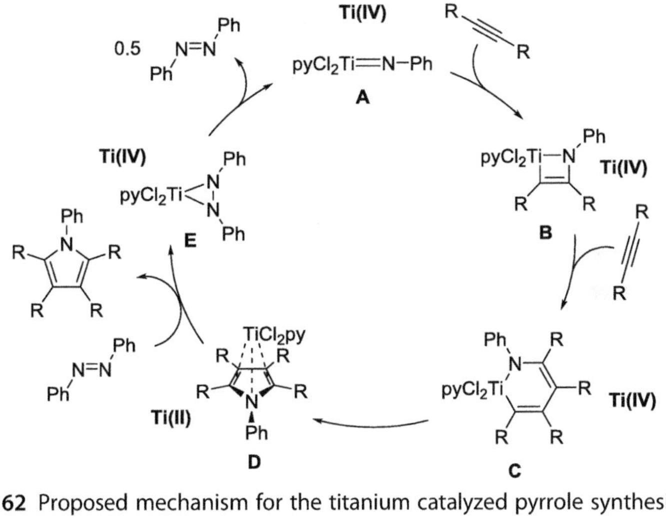

Scheme 62 Proposed mechanism for the titanium catalyzed pyrrole synthesis.

4.2.1 Reaction development

With the fundamental mechanistic understanding in place, the Tonks group has worked to extend the substrate scope that can be exploited in the synthesis of the biologically important pyrrole heterocycles. In an extension of preliminary investigations, efforts focused on the development of reactions that could assemble pyrroles with two different alkynes (Scheme 63).[171,172] This work took advantage of the different electronic properties of silyl

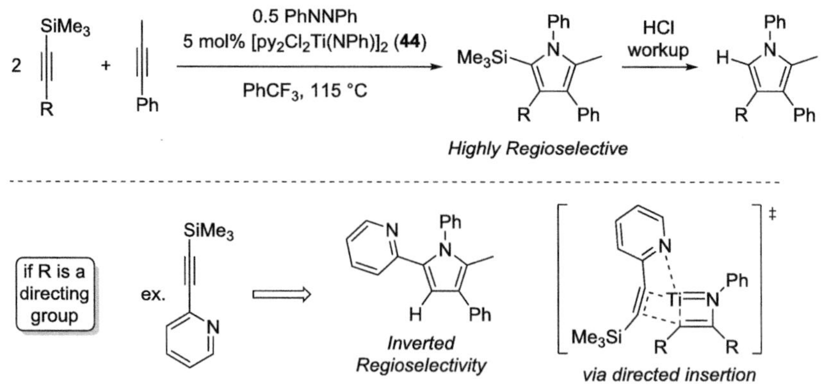

Scheme 63 Controlling regioselectivity using different electronic properties or directing groups incorporated into alkynes.

substituted alkynes from aryl and alkyl-substituted alkynes to realize the regioselective synthesis of differentially substituted pyrroles. In this case, the silyl substituted alkyne does not undergo the [2 + 2] cycloaddition with the imido intermediate, but in excess quantities it is an efficient substrate for the subsequent alkyne insertion step.[172] Additionally, due to the preference of the silyl substituent to be in the α-position in the resultant metallacycle, regioselectivity is controlled. Further control of regioselectivity is achieved using aryl-alkyl-substituted alkynes, which are known to undergo preferential C—N bond formation at the alkyl-substituted end of the alkyne. Alternatively, the incorporation of alkyne substituents that can coordinate to the metal center, such as pyridine substituents, act as a directing group to help control regioselectivity (Scheme 63).[171]

While advances in controlling substitution patterns on the carbon substituents can be realized, the incorporation of alternative *N*-substituents has been realized using an alternative imido precursor. In this case an alternative nitrogen source, a bulky alkylazide, can be used in combination with the imido precatalyst (Scheme 64) to access *N*-alkylated pyrroles.[173] Notably the effects of the auxiliary ligands are profound and optimization of the precatalyst to incorporate alternative halide counterions and the addition of more weakly bound THF donor ligands (**45**) result in optimized yields using these azide precursors.

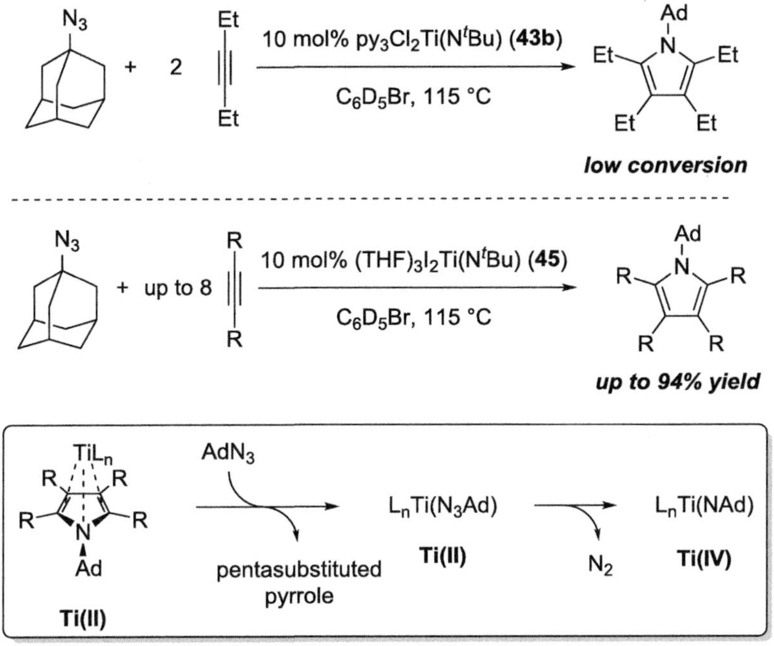

Scheme 64

Scheme 64 Modified precatalyst **45** required to use adamantylazide as an alternative nitrene precursor to access N-alkyl pyrroles.

4.2.2 Applications in synthesis

Pyrroles are important N-heterocycles in biologically active small molecules and Tonks and coworkers have shown how their regioselective approach for substituted pyrrole synthesis can be used in the formal synthesis of the cytotoxic natural product lamellarin R (Scheme 65).[172] This synthetic approach highlighted the application of silylsubstituted substrates to realize excellent regioselectivity and offered a previously reported intermediate in 24% yield after functional group interconversion.

Another synthetic advance focuses on accessing a benchtop ready protocol that uses air-stable THF adducts of titanium chloride precursors **46** in combination with the readily available reductant Zn dust to generate catalytically active species in situ (Scheme 66).[174] This synthetic contribution enables redox chemistry with titanium to be readily accessible to any chemistry laboratory. Furthermore, this contribution demonstrated that this facile synthetic approach can be used to access a range of pyrrole products.

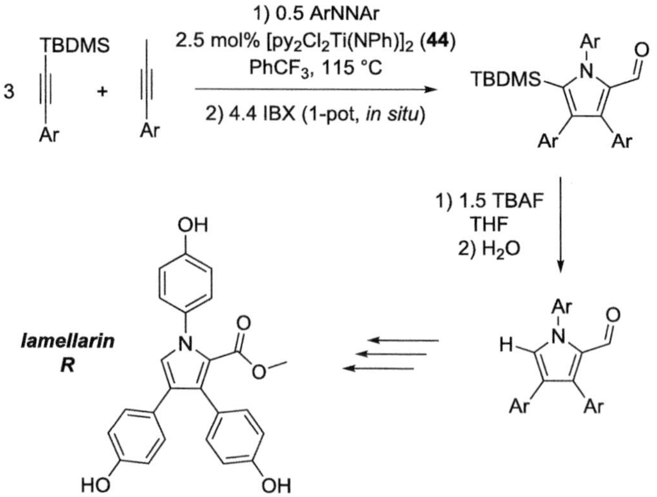

Scheme 65 Application of titanium catalyzed pyrrole synthesis in the formal synthesis of cytotoxic lamellarin R.

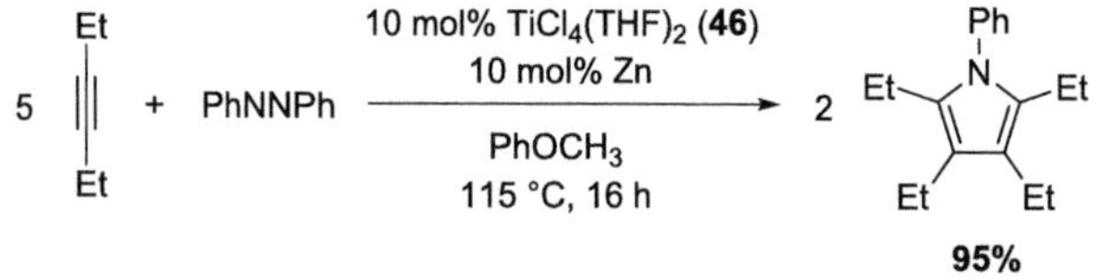

Scheme 66 General protocol for pyrrole synthesis using air-stable THF adducts of titanium chloride in situ.

In summary, new titanium redox activity to prepare pyrrole small molecules has been developed by the Tonks group. To date, established trends in substrate-controlled reactivity for [2+2] cycloadditions to titanium imido complexes and known features of alkyne insertion reactions have been leveraged to realize the regioselective synthesis of diversely substituted pyrroles. Thorough ligand modified reactivity has not been reported for this transformation, although the importance of tuning the electronic character at the metal center has been disclosed.[173,175] This suggests that there remain opportunities for enhanced reactivity, although designing ligands that can accommodate the steric bulk of key intermediates while providing sufficient steric protection of reactive terminal imido complexes presents new challenges in ligand design. The relevance of this methodology as it applies to the synthesis of more complex natural products has been demonstrated. The mechanistic strategies explored here also point toward potential opportunities for realizing the synthesis of other N-containing heterocycles by redox chemistry with titanium.

4.2.3 Carboamination

Multicomponent coupling strategies featuring redox active Ti(II) species allows for the catalytic synthesis of more complex N-containing products.[176] Alkene substrates could be used in excess to favor the insertion of alkenes into the Ti—C bond that forms upon [2+2] cycloaddition of the titanium imido with an alkyne. Notably, alkenes do not undergo productive [2+2] cycloaddition with titanium imido complexes.[177] Using this approach two products, the linear iminoalkene and the cyclopropanated imine, were observed. They are both proposed to result from the same cyclic intermediates (Scheme 67). The product distribution is under substrate control, although a clear understanding of substitution effects has not yet been presented. This reaction can also be performed in an intramolecular fashion and by modifying the substrate to eliminate the possibility of β-hydride elimination, cyclopropanated products are accessed. These reactive products are prone to further reactivity to give alternative N-substituted cyclic alkane products (Scheme 68).

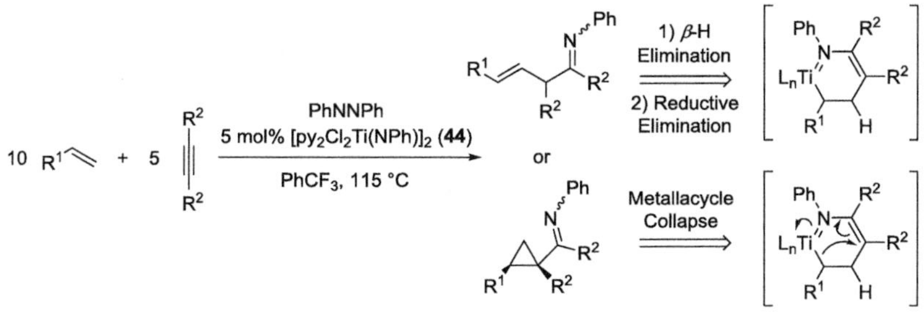

Scheme 67 Catalytic carboamination using both alkyne and alkene substrates afford more complex substituted imine products.

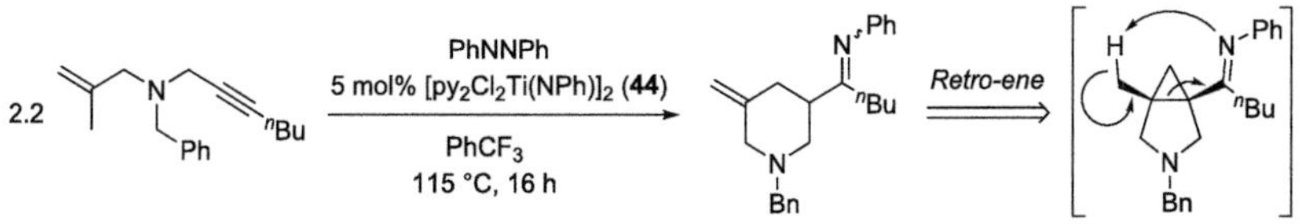

Scheme 68 Intramolecular variants generating reactive cyclopropyl intermediates.

An alternative approach to carboamination can be achieved without redox chemistry. By using titanium imido precursors that can undergo imine metathesis with aldimine substrates a reactive titanium imido complex **A** is generated that undergoes [2+2] cycloaddition chemistry with an alkyne (**B**, Scheme 69). A second equivalent of aldimine can insert into the reactive Ti—C bond to give a metallacyclic intermediate **C** that can undergo C—N bond cleavage to regenerate the catalytically active imido **A** and release the catalytically generated α,β-unsaturated imine product. The Mindiola group has generated two different sterically accessible titanium complexes that can be used as precursors suitable for imido metathesis.[178,179]

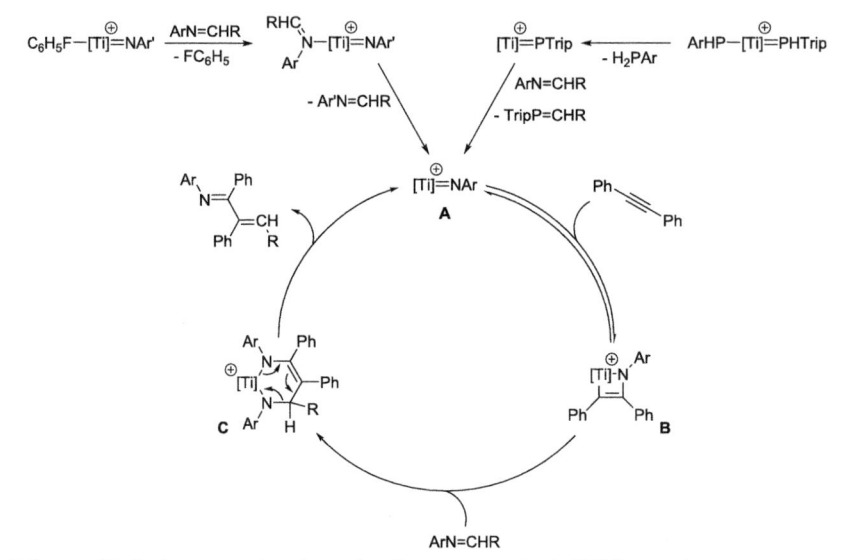

Scheme 69 Redox-neutral carboamination using cationic Ti(IV) complexes.

The incorporation of multicomponent coupling in redox generated imido complexes offers new opportunities to access selectively substituted amines. Notably, these products can also be accessed using the in situ prepared catalyst described above.[174] Gaining catalyst control over product distributions remains an unaddressed synthetic challenge. Insights into the proposed β-hydride elimination and metallacycle collapse pathways will be informative for better understanding the reductive elimination pathways with these group 4 metals in general. Substrate scope limitations in carboamination remain unexplored and thus the diversity of α,β-unsaturated imine products that may be accessible using this catalytic approach is not yet established.

4.3 Ti(III) catalysis for the synthesis of amines and N-heterocycles

Ti(III) complexes and specifically Cp$_2$TiCl can be reactive reagents generated in situ. All reactivity is consistent with the d^1 electron configuration promoting radical reactivity with a broad range of C=E multiply bonded substrates or reactive strained rings to generate selectively substituted organic products. While such stoichiometric approaches have been extensively developed, more powerful catalytic variants have also been realized. For example, Streuff and coworkers have proposed that two Ti(III) species react in concert to result in imine-nitrile cross-coupling reactions (Scheme 70).[156,158,161,162,180,181] This mechanistic proposal has been supported by both experimental and computational proposals completed for ketone-nitrile cross-coupling reactions that result in the related syntheses of α-hydroxyketones.[159] The resultant proposal for a bimetallic pathway (D), based upon online ESI-MS experiments, offers interesting opportunities for ligand design to further control such reactivity patterns.

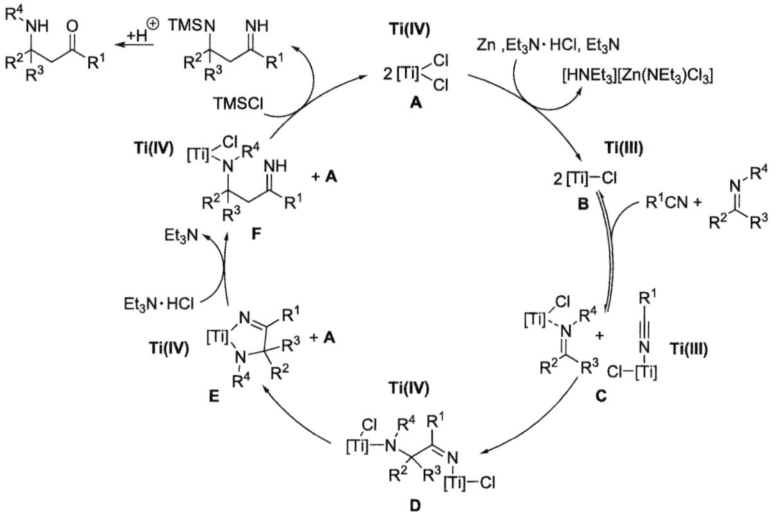

Scheme 70 Mechanism for the Ti(III) catalyzed imine-nitrile cross-coupling.

This reactivity can be extended to imine-nitrile cross-coupling to prepare α-amino ketones with a variety of substitution patterns (Scheme 71).

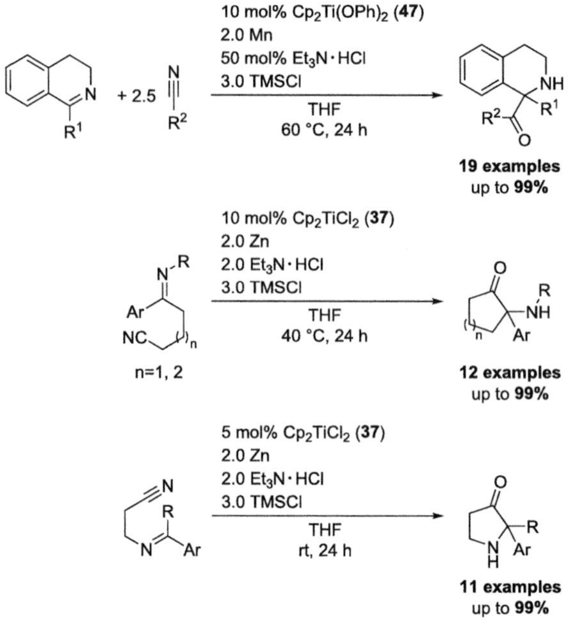

Scheme 71 Selected examples of α-amino ketone synthesis via imine-nitrile cross-coupling.

Furthermore, this strategy can be extended to the synthesis of amine substituted privileged heterocycles like indoles and pyrroles (Scheme 72).[161]

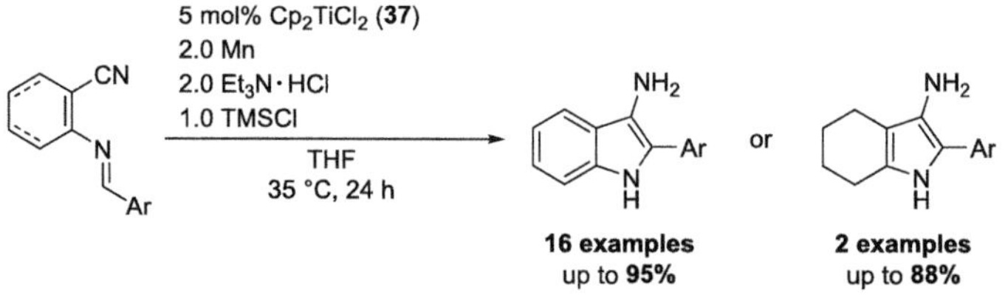

Scheme 72 Indole and pyrrole synthesis via imine-nitrile cross-coupling.

These strategies have been extended to natural product synthesis where both intramolecular and intermolecular imine-nitrile cross-coupling strategies have been explored for the racemic synthesis of demethoxyerythratidinone (Scheme 73).[162]

Single electron addition is not limited to C=X multiply bonded electron acceptors. Strained N-acylaziridines can undergo Ti(III) catalyzed addition to Michael-type acceptors to give acylated and highly substituted pyrrolidine products.[182] This reaction proceeds via radical redox-relay

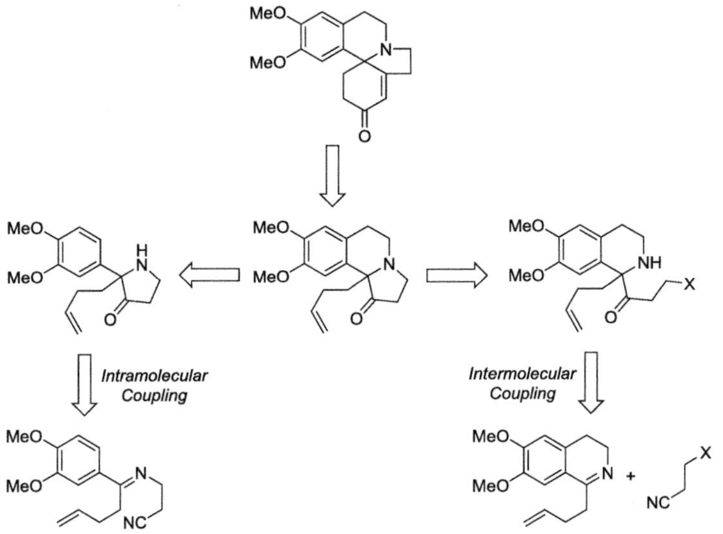

Scheme 73 Key retrosynthetic steps for the synthesis of demethoxyerythratidinone via imine-nitrile cross-coupling.

where the initial ring-opened N-acylaziridine is generated by C—N bond cleavage upon treatment with Cp*TiCl$_2$ (**37**), which is in turn generated from Cp*TiCl$_3$ (**3b**) and zinc dust (**A**, Scheme 74). The advantage of using a titanium radical catalyzed approach to aziridine ring opening is the opportunity to cleave the most highly substituted C—N bond (**B**), which is complementary to nucleophilic ring opening that would be achieved upon addition of a 2e$^-$-species. This C—N bond cleavage triggers oxidation to Ti(IV) (**C**) and a stabilized tertiary radical that can add to activated alkenes regioselectively. Upon ring-closure (**D**) the desired acylated pyrrolidine product is accessed and Ti(III) is regenerated for catalytic turnover. Evidence for the intermediacy of radical species was provided through radical trapping and characterization by EPR and competition experiments that showed enhanced reactivity with electron accepting alkenes over more nucleophilic alkenes, which may be the anticipated preferred reaction partner if a cationic species was generated upon ring opening. This approach demonstrates the potential to use Ti(III) in a catalytic fashion to generate acylated pyrrolidine products using a radical-relay reaction. Due to the radical nature of the intermediate involved, this reaction is ideally suited for employing highly substituted alkene reaction partners. This results in the ability to efficiently assemble diversely substituted pyrrolidine products.

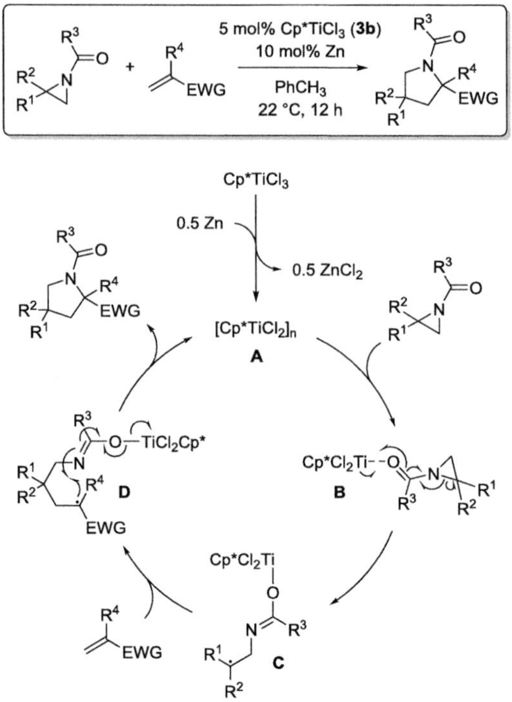

Scheme 74 Synthesis of acylated and highly substituted pyrrolidines via radical redox-relay and proposed mechanism.

Such useful Ti(III) catalysts can be generated using alternative reductive protocols. Gansäuer has shown that electrochemical approaches can be used, in the presence of select urea or thiourea additives to generate reactive Cp$_2$TiX species in solution to mediate radical arylation reactions.[183] While the focus of the reactivity was with reactive epoxide moieties, the preferred protocol demonstrated the synthesis of N-substituted indoline products (Scheme 75).

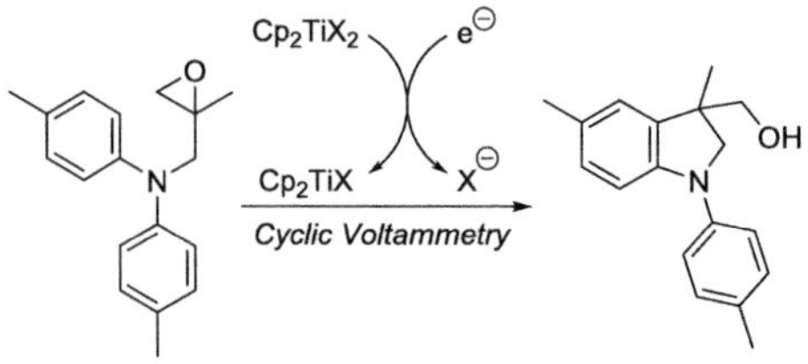

Scheme 75 Synthesis of N-substituted indolines via single electron steps by cyclic voltammetry.

Alternatively, the same group showed that photocatalysis could also be harnessed to generate reactive radical species for the same transformation.[184] By using an iridium photocatalyst **48** coupled with titanocene dichloride (**37**) optimized yields were obtained for the generation of the desired product (Scheme 76).

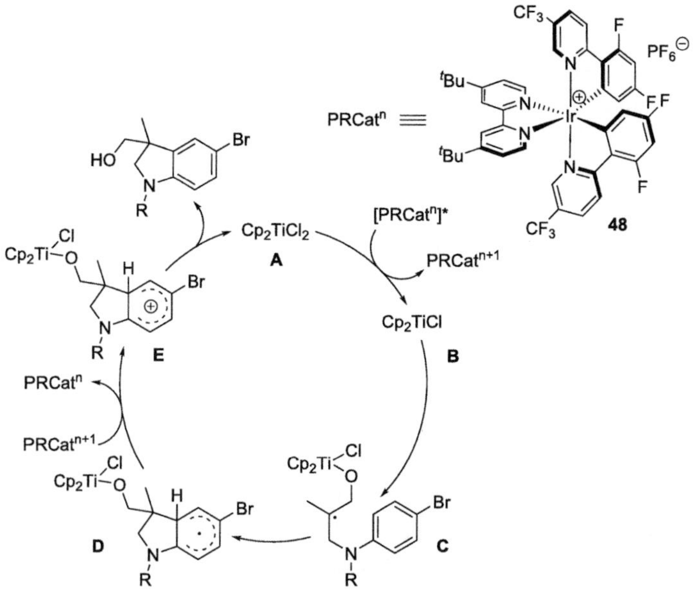

Scheme 76 Ti(III) catalysis merged with photoredox catalysis.

The attractive opportunity for both electrocatalysis and photoredox catalysis is the fact that chemical reductants are avoided, resulting in simplified product isolation and purification.

4.4 Summary and outlook

Recent advances in titanium catalyzed transformations have expanded to include redox reactions that exploit the accessible Ti(II) and Ti(III) oxidation states. In the former, two electron organometallic reaction mechanisms, including insertion reactions and cycloaddition reactions that result in reductive elimination at the metal center have been exploited in the catalytic synthesis of pyrroles. Notably, this catalytic pathway invokes the important intermediacy of "masked" Ti(II) species that react like free Ti(II) complexes, but can be isolated or calculated to exist as weak Ti(IV) adducts. Such masked Ti(II) adducts are well established in the literature,[130,131] and have been typically exploited in an array of stoichiometric transformations. Indeed, the precedent for such transformations point the way to new catalytic reactions that could employ the abundant and inexpensive titanium metal more broadly in organic synthetic chemistry. As ligand modified redox potentials are better understood, new synthetic approaches to generate reduced species in situ can be realized.

In addition to established reactivity using two electron mechanisms, Ti(III) has also been extensively used in stoichiometric radical transformations.[185,186] Recent advances use chemical, electrochemical and photochemical reductants to access reactive Ti(III) species can undergo radical reactions and subsequent catalytic turnover to generate N-heterocycles, such as pyrrolidine and indoline products. Once again, opportunities to modify titanium catalyst redox potentials and catalytic activity through ligand modification are underexplored. Also, developing systems for stereochemical control is largely underexplored. To date, single electron chemistry reactions have featured cyclopentadienyl derivatives of titanium, leaving much room for expanded reactivity.

Catalytic titanium chemistry, featuring redox reactivity, is at the cutting edge of the field of titanium catalyzed amine and N-heterocycle synthesis. The diversity of catalytic reactions established to date is limited. However, stoichiometric reactions that feature reduced titanium complexes as reagents that can be employed with imines, carbonyls, alkenes, alkynes and strained rings are much more prevalent and consequently, there are significant opportunities to expand this field.

References

1. Davis-Gilbert ZW, Tonks IA. Titanium redox catalysis: insights and applications of an earth-abundant base metal. *Dalton Trans*. 2017;46:11522–11528.
2. Chen J, Lu Z. Asymmetric hydrofunctionalization of minimally functionalized alkenes via earth abundant transition metal catalysis. *Org Chem Front*. 2018;5:260–272.
3. Kablaoui NM, Buchwald SL. Reductive cyclization of enones by a titanium catalyst. *J Am Chem Soc*. 1995;117:6785–6786.
4. Urabe H, Takeda T, Hideura D, Sato F. Ti(II)-mediated allenyne cyclization as a new tool for generation of chiral organotitanium compounds. Asymmetric generation of cyclic allyltitanium reagents with no chiral auxiliary. *J Am Chem Soc*. 1997;119: 11295–11305.
5. Morcillo SP, Miguel D, Campaña AG, Álvarez de Cienfuegos L, Justicia J, Cuerva JM. Recent applications of Cp$_2$TiCl in natural product synthesis. *Org Chem Front*. 2014;1:15–33.
6. Makio H, Terao H, Iwashita A, Fujita T. FI catalysts for olefin polymerization—a comprehensive treatment. *Chem Rev*. 2011;111:2363–2449.
7. Collins RA, Russell AF, Mountford P. Group 4 metal complexes for homogeneous olefin polymerisation: a short tutorial review. *Appl Petrochem Res*. 2015;5:153–171.

8. Solsona JG, Nebot J, Romea P, Urpí F. Highly stereoselective aldol reaction based on titanium enolates from (S)-1-benzyloxy-2-methyl-3-pentanone. *J Org Chem.* 2005;70:6533–6536.
9. Koch G, Loiseleur O, Fuentes D, Jantsch A, Altmann K-H. Diastereoselective titanium enolate aldol reaction for the total synthesis of epothilones. *Org Lett.* 2002;4:3811–3814.
10. Katsuki T, Martin V. Asymmetric epoxidation of allylic alcohols: the katsuki–sharpless epoxidation reaction. In: *Organic Reactions.* John Wiley and Sons, Inc. 2004:1–299
11. North M, Pasquale R, Young C. Synthesis of cyclic carbonates from epoxides and CO_2. *Green Chem.* 2010;12:1514–1539.
12. Bai D, Jing H, Liu Q, Zhu Q, Zhao X. Titanocene dichloride–Lewis base: an efficient catalytic system for coupling of epoxides and carbon dioxide. *Catal Commun.* 2009;11:155–157.
13. Salvatore RN, Yoon CH, Jung KW. Synthesis of secondary amines. *Tetrahedron.* 2001;57:7785–7811.
14. Constable DJC, Dunn PJ, Hayler JD, et al. Key green chemistry research areas—a perspective from pharmaceutical manufacturers. *Green Chem.* 2007;9:411–420.
15. Huang L, Arndt M, Gooßen K, Heydt H, Gooßen LJ. Late transition metal-catalyzed hydroamination and hydroamidation. *Chem Rev.* 2015;115:2596–2697.
16. Haak E, Bytschkov I, Doye S. Intermolecular hydroamination of alkynes catalyzed by dimethyltitanocene. *Angew Chem Int Ed.* 1999;38:3389–3391.
17. Bytschkov I, Doye S. Group-IV metal complexes as hydroamination catalysts. *Eur J Org Chem.* 2003;2003:935–946.
18. Severin R, Doye S. The catalytic hydroamination of alkynes. *Chem Soc Rev.* 2007;36:1407–1420.
19. Pohlki F, Doye S. The catalytic hydroamination of alkynes. *Chem Soc Rev.* 2003;32:104–114.
20. Walsh PJ, Baranger AM, Bergman RG. Stoichiometric and catalytic hydroamination of alkynes and allene by zirconium bisamides $Cp_2Zr(NHR)_2$. *J Am Chem Soc.* 1992;114:1708–1719.
21. Baranger AM, Walsh PJ, Bergman RG. Variable regiochemistry in the stoichiometric and catalytic hydroamination of alkynes by imidozirconium complexes caused by an unusual dependence of the rate law on alkyne structure and temperature. *J Am Chem Soc.* 1993;115:2753–2763.
22. Müller C, Koch R, Doye S. Mechanism of the intramolecular hydroamination of alkenes catalyzed by neutral indenyltitanium complexes: a DFT study. *Chem A Eur J.* 2008;14:10430–10436.
23. Hazari N, Mountford P. Reactions and applications of titanium imido complexes. *Acc Chem Res.* 2005;38(11):839–849.
24. Ong T-G, Yap GPA, Richeson DS. Formation of a guanidinate-supported titanium imido complex: a catalyst for alkyne hydroamination. *Organometallics.* 2002;21(14):2839–2841.
25. Normand AT, Massard A, Richard P, et al. Titanium imido complexes stabilised by bis(iminophosphoranyl)methanide ligands: the influence of N-substituents on solution dynamics and reactivity. *Dalton Trans.* 2014;43(40):15098–15110.
26. Chen Z, Li L, Chen Y, et al. Titanium amido- and imido-complexes supported by a tridentate pyrrolyl ligand: syntheses, characterisation and catalytic activities. *J Chem Res.* 2012;36(5):249–253.
27. Weitershaus K, Ward BD, Kubiak R, et al. Titanium hydroamination catalysts bearing a 2-aminopyrrolinato spectator ligand: monitoring the individual reaction steps. *Dalton Trans.* 2009;23:4586–4602.
28. Lian B, Spaniol TP, Horrillo-Martínez P, Hultzsch KC, Okuda J. Imido and amido titanium complexes that contain a [OSSO]-type Bis(phenolato) ligand: Synthesis, structures, and hydroamination catalysis. *Eur J Inorg Chem.* 2009;2009(3):429–434.

29. Ackermann L, Bergman RG, Loy RN. Use of group 4 Bis(sulfonamido) complexes in the intramolecular hydroamination of alkynes and allenes. *J Am Chem Soc.* 2003;125(39):11956–11963.
30. Wang H, Chan H-S, Xie Z. Synthesis, characterization, and reactivity of terminal titanium imido complexes incorporating constrained-geometry carboranyl ligands. *Organometallics.* 2005;24(15):3772–3779.
31. Manßen M, de Graaff S, Meyer M-F, Schmidtmann M, Beckhaus R. Direct access to titanocene imides via Bis(η5:η1-penta-fulvene)titanium complexes and primary amines. *Organometallics.* 2018;37(23):4506–4514.
32. Vujkovic N, Ward BD, Maisse-François A, Wadepohl H, Mountford P, Gade LH. Imido-alkyne coupling in titanium complexes: new insights into the alkyne hydroamination reaction. *Organometallics.* 2007;26(23):5522–5534.
33. Zhang Z, Schafer LL. Anti-Markovnikov intermolecular hydroamination: a Bis(amidate) titanium precatalyst for the preparation of reactive aldimines. *Org Lett.* 2003;5(24):4733–4736.
34. Lui EKJ, Schafer LL. Facile synthesis and isolation of secondary amines via a sequential titanium(IV)-catalyzed hydroamination and palladium-catalyzed hydrogenation. *Adv Synth Catal.* 2016;358(5):713–718.
35. Hao H, Thompson KA, Hudson ZM, Schafer LL. Ti-catalyzed hydroamination for the synthesis of amine-containing π-conjugated materials. *Chem A Eur J.* 2018;24(21):5562–5568.
36. Esteruelas MA, López AM, Mateo AC, Oñate E. New titanium complexes containing a cyclopentadienyl ligand with a pendant aminoalkyl substituent: preparation, behavior of the amino group, and catalytic hydroamination of alkynes. *Organometallics.* 2005;24(21):5084–5094.
37. Heutling A, Pohlki F, Doye S. [Ind2TiMe2]: a general catalyst for the intermolecular hydroamination of alkynes. *Chem A Eur J.* 2004;10(12):3059–3071.
38. Zhang Z, Leitch DC, Lu M, Patrick BO, Schafer LL. An easy-to-use, regioselective, and robust bis(amidate) titanium hydroamination precatalyst: mechanistic and synthetic investigations toward the preparation of tetrahydroisoquinolines and benzoquinolizine alkaloids. *Chem A Eur J.* 2007;13(7):2012–2022.
39. Aldrich KE, Odom AL. Titanium-catalyzed hydroamination and multicomponent coupling with a simple silica-supported catalyst. *Organometallics.* 2018;37(23):4341–4349.
40. Lavallo V, Frey GD, Donnadieu B, Soleilhavoup M, Bertrand G. Homogeneous catalytic hydroamination of alkynes and allenes with ammonia. *Angew Chem Int Ed.* 2008;47(28):5224–5228.
41. Lui EKJ, Brandt JW, Schafer LL. Regio- and stereoselective hydroamination of alkynes using an ammonia surrogate: synthesis of N-silylenamines as reactive synthons. *J Am Chem Soc.* 2018;140(15):4973–4976.
42. Ayinla RO, Schafer LL. Bis(amidate) titanium precatalyst for the intermolecular hydroamination of allenes. *Inorg Chim Acta.* 2006;359(9):3097–3102.
43. Ayinla RO, Schafer LL. Intermolecular hydroamination of oxygen-substituted allenes. New routes for the synthesis of N,O-chelated zirconium and titanium amido complexes. *Dalton Trans.* 2011;40(30):7769–7776.
44. Lühning LH, Brahms C, Nimoth JP, Schmidtmann M, Doye S. A new N-trityl-substituted aminopyridinato titanium catalyst for hydroamination and hydroaminoalkylation reactions—unexpected intramolecular C–H bond activation. *Z Anorg Allg Chem.* 2015;641(12–13):2071–2082.
45. Hoover JM, Petersen JR, Pikul JH, Johnson AR. Catalytic intramolecular hydroamination of substituted aminoallenes by chiral titanium amino-alcohol complexes. *Organometallics.* 2004;23(20):4614–4620.

46. LaLonde RL, Sherry BD, Kang EJ, Toste FD. Gold(I)-catalyzed enantioselective intramolecular hydroamination of allenes. *J Am Chem Soc*. 2007;129(9):2452–2453.
47. Bexrud JA, Beard JD, Leitch DC, Schafer LL. Intramolecular hydroamination of unactived olefins with Ti(NMe2)4 as a precatalyst. *Org Lett*. 2005;7(10): 1959–1962.
48. Thomson RK, Bexrud JA, Schafer LL. A pentagonal pyramidal zirconium imido complex for catalytic hydroamination of unactivated alkenes. *Organometallics*. 2006;25(17): 4069–4071.
49. Chong E, Qayyum S, Schafer LL, Kempe R. 2-Aminopyridinate titanium complexes for the catalytic hydroamination of primary aminoalkenes. *Organometallics*. 2013;32(6): 1858–1865.
50. Brahms C, Tholen P, Saak W, Doye S. An (aminopyrimidinato)titanium catalyst for the hydroamination of alkynes and alkenes. *Eur J Org Chem*. 2013;2013(33):7583–7592.
51. Gräbe K, Pohlki F, Doye S. Neutral Ti complexes as catalysts for the hydroamination of alkynes and alkenes: do the labile ligands change the catalytic activity? *Eur J Org Chem*. 2008;2008(28):4815–4823.
52. Janssen T, Severin R, Diekmann M, et al. Bis(η5:η1-pentafulvene)titanium complexes: catalysts for intramolecular alkene hydroamination and reagents for selective reactions with N–H acidic substrates. *Organometallics*. 2010;29(7):1806–1817.
53. Helgert TR, Hollis TK, Valente EJ. Synthesis of titanium CCC-NHC pincer complexes and catalytic hydroamination of unactivated alkenes. *Organometallics*. 2012;31(8): 3002–3009.
54. Majumder S, Odom AL. Group-4 dipyrrolylmethane complexes in intramolecular olefin hydroamination. *Organometallics*. 2008;27(6):1174–1177.
55. Wang Q, Song H, Zi G. Synthesis, structure, and catalytic activity of group 4 complexes with new chiral biaryl-based NO2 ligands. *J Organomet Chem*. 2010;695(10): 1583–1591.
56. Yonson N, Yim JCH, Schafer LL. Alkene hydroamination with a chiral zirconium catalyst. Connecting ligand design, precatalyst structure and reactivity trends. *Inorg Chim Acta*. 2014;422:14–20.
57. Wood MC, Leitch DC, Yeung CS, Kozak JA, Schafer LL. Chiral neutral zirconium amidate complexes for the asymmetric hydroamination of alkenes. *Angew Chem Int Ed*. 2007;46(3):354–358.
58. Gruber-Woelfler H, Khinast JG, Flock M, et al. Titanocene-catalyzed hydrosilylation of imines: experimental and computational investigations of the catalytically active species. *Organometallics*. 2009;28(8):2546–2553.
59. Heutling A, Pohlki F, Bytschkov I, Doye S. Hydroamination/hydrosilylation sequence catalyzed by titanium complexes. *Angew Chem Int Ed*. 2005;44(19):2951–2954.
60. Yamada K-I, Tomioka K. Copper-catalyzed asymmetric alkylation of imines with dialkylzinc and related reactions. *Chem Rev*. 2008;108(8):2874–2886.
61. Rong J, Collados JF, Ortiz P, Jumde RP, Otten E, Harutyunyan SR. Catalytic enantioselective addition of Grignard reagents to aromatic silyl ketimines. *Nat Commun*. 2016;7(1):13780.
62. Lee AV, Sajitz M, Schafer LL. The direct synthesis of unsymmetrical vicinal diamines from terminal alkynes: a tandem sequential approach for the synthesis of imidazolidinones. *Synthesis*. 2009;2009(01):97–104.
63. Lee AV, Schafer LL. A sequential C-N, C-C bond-forming reaction: direct synthesis of α-amino acids from terminal alkynes. *Synlett*. 2006;2006(18):2973–2976.
64. Gräbe K, Zwafelink B, Doye S. One-pot procedure for the synthesis of N-substituted 2-(arylmethyl)pyrrolidines from 1-aryl-2-cyclopropylalkynes and primary amines by a hydroamination/cyclopropylimine rearrangement/reduction sequence. *Eur J Org Chem*. 2009;2009(32):5565–5575.

65. Ackermann L, Sandmann R, Kaspar LT. Two titanium-catalyzed reaction sequences for syntheses of pyrroles from (E/Z)-chloroenynes or α-haloalkynols. *Org Lett.* 2009;11(9):2031–2034.
66. Ramanathan B, Keith AJ, Armstrong D, Odom AL. Pyrrole syntheses based on titanium-catalyzed hydroamination of diynes. *Org Lett.* 2004;6(17):2957–2960.
67. Barnea E, Majumder S, Staples RJ, Odom AL. One-step route to 2,3-diaminopyrroles using a titanium-catalyzed four-component coupling. *Organometallics.* 2009;28(13): 3876–3881.
68. Majumder S, Gipson KR, Staples RJ, Odom AL. Pyrazole synthesis using a titanium-catalyzed multicomponent coupling reaction and synthesis of withasomnine. *Adv Synth Catal.* 2009;351(11–12):2013–2023.
69. Dissanayake AA, Odom AL. Regioselective conversion of alkynes to 4-substituted and 3,4-disubstituted isoxazoles using titanium-catalyzed multicomponent coupling reactions. *Tetrahedron.* 2012;68(3):807–812.
70. Müller TE, Hultzsch KC, Yus M, Foubelo F, Tada M. Hydroamination: direct addition of amines to alkenes and alkynes. *Chem Rev.* 2008;108(9):3795–3892.
71. Banerjee S, Barnea E, Odom AL. Titanium-catalyzed hydrohydrazination with monosubstituted hydrazines: catalyst design, synthesis, and reactivity. *Organometallics.* 2008;27(5):1005–1014.
72. Dissanayake AA, Staples RJ, Odom AL. Titanium-catalyzed, one-pot Synthesis of 2-amino-3-cyano-pyridines. *Adv Synth Catal.* 2014;356(8):1811–1822.
73. Lui EKJ, Hergesell D, Schafer LL. N-Silylenamines as reactive intermediates: hydroamination for the modular synthesis of selectively substituted pyridines. *Org Lett.* 2018;20(21):6663–6667.
74. Majumder S, Gipson KR, Odom AL. A multicomponent coupling sequence for direct access to substituted quinolines. *Org Lett.* 2009;11(20):4720–4723.
75. Facoetti D, Abbiati G, d'Avolio L, Ackermann L, Rossi E. Novel domino approach to fluorescent pyrimido[1,6-a]indolones. *Synlett.* 2009;14:2273–2276.
76. Majumder S, Odom AL. Titanium catalyzed one-pot multicomponent coupling reactions for direct access to substituted pyrimidines. *Tetrahedron.* 2010;66(17):3152–3158.
77. Zhai H, Borzenko A, Lau YY, Ahn SH, Schafer LL. Catalytic asymmetric synthesis of substituted morpholines and piperazines. *Angew Chem Int Ed.* 2012;51(49): 12219–12223.
78. Lau YY, Zhai H, Schafer LL. Catalytic asymmetric synthesis of morpholines. Using mechanistic insights to realize the enantioselective synthesis of piperazines. *J Org Chem.* 2016;81(19):8696–8709.
79. Borzenko A, Pajouhesh H, Morrison J-L, Tringham E, Snutch TP, Schafer LL. Modular, efficient synthesis of asymmetrically substituted piperazine scaffolds as potent calcium channel blockers. *Bioorg Med Chem Lett.* 2013;23(11):3257–3261.
80. Billow BS, McDaniel TJ, Odom AL. Quantifying ligand effects in high-oxidation-state metal catalysis. *Nat Chem.* 2017;9(9):837–842.
81. Dörfler J, Doye S. Aminopyridinato titanium catalysts for the hydroaminoalkylation of alkenes and styrenes. *Angew Chem Int Ed.* 2013;52(6):1806–1809.
82. Prochnow I, Kubiak R, Frey ON, Beckhaus R, Doye S. Tetrabenzyltitanium: an improved catalyst for the activation of sp3 C—H bonds adjacent to nitrogen atoms. *ChemCatChem.* 2009;1(1):162–172.
83. Dörfler J, Preuß T, Brahms C, Scheuer D, Doye S. Intermolecular hydroaminoalkylation of alkenes and dienes using a titanium mono(formamidinate) catalyst. *Dalton Trans.* 2015;44(27):12149–12168.
84. Gao Y, Yoshida Y, Sato F. In situ generation of titanium-imine complexes from imines and Ti(OiPr)4/2 iPrMgX, and their reactions with alkynes, nitriles and imines. *Synlett.* 1997;12(12):1353–1354.

85. Fukuhara K, Okamoto S, Sato F. Asymmetric synthesis of allyl- and α-allenylamines from chiral imines and alkynes via (η2-imine)Ti(O-i-Pr)2 complexes. *Org Lett.* 2003;5(12):2145–2148.
86. Loose F, Schmidtmann M, Saak W, Beckhaus R. Imines in the titanium coordination sphere: highly reactive titanaaziridines and larger titanacycles formed by subsequent C–C coupling reactions. *Eur J Inorg Chem.* 2015;2015(31):5171–5187.
87. Loose F, Plettenberg I, Haase D, et al. Aromatic imines in the titanocene coordination sphere—titanaaziridine vs 1-Aza-2-titanacyclopent-4-ene structures. *Organometallics.* 2014;33(23):6785–6795.
88. Buchwald SL, Watson BT, Wannamaker MW, Dewan JC. Zirconocene complexes of imines. General synthesis, structure, reactivity, and in situ generation to prepare geometrically pure allylic amines. *J Am Chem Soc.* 1989;111(12):4486–4494.
89. Durfee LD, Hill JE, Fanwick PE, Rothwell IP. Formation and characterization of .eta.2-imine and .eta.2-azobenzene derivatives of titanium containing ancillary aryloxide ligation. *Organometallics.* 1990;9(1):75–80.
90. Li L, Kristian KE, Han A, Norton JR, Sattler W. Synthesis, structural characterization, and reactivity of Cp2- and (CpMe)2-ligated titanaaziridines and titanaoxiranes with fast enantiomer interconversion rates. *Organometallics.* 2012;31(23):8218–8224.
91. Klei E, Teuben JH. Reaction of titanocene alkyls with pyridines; a novel type of cyclometallation reaction. *J Organomet Chem.* 1981;214(1):53–64.
92. Edwards PM, Schafer LL. Early transition metal-catalyzed C–H alkylation: hydroaminoalkylation for Csp3–Csp3 bond formation in the synthesis of selectively substituted amines. *Chem Commun.* 2018;54(89):12543–12560.
93. Clerici MG, Maspero F. Catalytic C-alkylation of secondary amines with alkenes. *Synthesis.* 1980;1980(04):305–306.
94. Jun C-H. Chelation-assisted alkylation of benzylamine derivatives by Ru0 catalyst. *Chem Commun.* 1998;13:1405–1406.
95. Chatani N, Asaumi T, Yorimitsu S, Ikeda T, Kakiuchi F, Murai S. Ru3(CO)12-catalyzed coupling reaction of sp3 C−H bonds adjacent to a nitrogen atom in alkylamines with alkenes. *J Am Chem Soc.* 2001;123(44):10935–10941.
96. Herzon SB, Hartwig JF. Direct, catalytic hydroaminoalkylation of uactivated olefins with N-alkyl arylamines. *J Am Chem Soc.* 2007;129(21):6690–6691.
97. Shannon R. Revised effective ionic radii and systematic studies of interatomic distances in halides and chalcogenides. *Acta Crystallogr A.* 1976;32(5):751–767.
98. Müller C, Saak W, Doye S. Neutral group-IV metal catalysts for the intramolecular hydroamination of alkenes. *Eur J Org Chem.* 2008;2008(16):2731–2739.
99. Prochnow I, Zark P, Müller T, Doye S. The mechanism of the titanium-catalyzed hydroaminoalkylation of alkenes. *Angew Chem Int Ed.* 2011;50(28):6401–6405.
100. Manßen M, Lauterbach N, Dörfler J, et al. Efficient access to titanaaziridines by C—H activation of N-methylanilines at ambient temperature. *Angew Chem Int Ed.* 2015;54(14):4383–4387.
101. Gilmour DJ, Lauzon JMP, Clot E, Schafer LL. Ta-catalyzed hydroaminoalkylation of alkenes: insights into ligand-modified reactivity using DFT. *Organometallics.* 2018;37(23):4387–4394.
102. Kubiak R, Prochnow I, Doye S. [Ind2TiMe2]: a catalyst for the hydroaminomethylation of alkenes and styrenes. *Angew Chem Int Ed.* 2010;49(14):2626–2629.
103. Dörfler J, Preuß T, Schischko A, Schmidtmann M, Doye S. A 2,6-Bis(phenylamino)pyridinato titanium catalyst for the highly regioselective hydroaminoalkylation of styrenes and 1,3-butadienes. *Angew Chem Int Ed.* 2014;53(30):7918–7922.
104. Chong E, Schafer LL. 2-Pyridonate titanium complexes for chemoselectivity. Accessing intramolecular hydroaminoalkylation over hydroamination. *Org Lett.* 2013;15(23):6002–6005.

105. Jaspers D, Saak W, Doye S. Dinuclear titanium complexes with sulfamide ligands as precatalysts for hydroaminoalkylation and hydroamination reactions. *Synlett.* 2012;23(14):2098–2102.
106. Bielefeld J, Doye S. Fast titanium-catalyzed hydroaminomethylation of alkenes and the formal conversion of methylamine. *Angew Chem Int Ed.* 2020;59:6138–6143.
107. Kubiak R, Prochnow I, Doye S. Titanium-catalyzed hydroaminoalkylation of alkenes by C—H bond activation at sp3 centers in the α-position to a nitrogen atom. *Angew Chem Int Ed.* 2009;48(6):1153–1156.
108. Dörfler J, Bytyqi B, Hüller S, et al. An aminopyridinato titanium catalyst for the intramolecular hydroaminoalkylation of secondary aminoalkenes. *Adv Synth Catal.* 2015;357(10):2265–2276.
109. Bexrud JA, Eisenberger P, Leitch DC, Payne PR, Schafer LL. Selective C−H activation α to primary amines. Bridging metallaaziridines for catalytic, intramolecular α-alkylation. *J Am Chem Soc.* 2009;131(6):2116–2118.
110. Liu F, Luo G, Hou Z, Luo Y. Mechanistic insights into scandium-catalyzed hydroaminoalkylation of olefins with amines: origin of regioselectivity and charge-based prediction model. *Organometallics.* 2017;36(8):1557–1565.
111. Bielefeld J, Doye S. Dimethylamine as a substrate in hydroaminoalkylation reactions. *Angew Chem Int Ed.* 2017;56(47):15155–15158.
112. Rosien M, Töben I, Schmidtmann M, Beckhaus R, Doye S. Titanium-catalyzed hydroaminoalkylation of ethylene. *Chem A Eur J.* 2020;26:2138–2142.
113. Lühning LH, Rosien M, Doye S. Thieme chemistry journals awardees—where are they now? Titanium-catalyzed hydroaminoalkylation of vinylsilanes and a one-pot procedure for the synthesis of 1,4-benzoazasilines. *Synlett.* 2017;28(18):2489–2494.
114. Lühning LH, Strehl J, Schmidtmann M, Doye S. Hydroaminoalkylation of allylsilanes and a one-pot procedure for the synthesis of 1,5-benzoazasilepines. *Chem A Eur J.* 2017;23(17):4197–4202.
115. Weers M, Lühning LH, Lührs V, Brahms C, Doye S. One-pot procedure for the synthesis of 1,5-benzodiazepines from N-allyl-2-bromoanilines. *Chem A Eur J.* 2017;23(6):1237–1240.
116. Kaper T, Doye S. Hydroaminoalkylation/Buchwald-Hartwig amination sequences for the synthesis of benzo-annulated seven-membered nitrogen heterocycles. *Tetrahedron.* 2019;75(32):4343–4350.
117. Sather AC, Lee HG, Colombe JR, Zhang A, Buchwald SL. Dosage delivery of sensitive reagents enables glove-box-free synthesis. *Nature.* 2015;524(7564):208–211.
118. Herzon SB, Hartwig JF. Hydroaminoalkylation of unactivated olefins with dialkylamines. *J Am Chem Soc.* 2008;130(45):14940–14941.
119. Garcia P, Lau YY, Perry MR, Schafer LL. Phosphoramidate tantalum complexes for room-temperature C—H functionalization: hydroaminoalkylation catalysis. *Angew Chem Int Ed.* 2013;52(35):9144–9148.
120. Chong E, Brandt JW, Schafer LL. 2-Pyridonate tantalum complexes for the intermolecular hydroaminoalkylation of sterically demanding alkenes. *J Am Chem Soc.* 2014;136(31):10898–10901.
121. Gao Y, Shirai M, Sato F. Synthesis of substituted pyrroles from an alkyne, an imine and carbon monoxide via an organotitanium intermediate. *Tetrahedron Lett.* 1996;37(43):7787–7790.
122. 4.03—Complexes of titanium in oxidation states 0 to ii. Chirik PJ, Bouwkamp MW, Mingos DMP, Crabtree RH, eds. *Comprehensive Organometallic Chemistry III.* Oxford: Elsevier; 2007:243–279.
123. Kulinkovich OG, de Meijere A. 1,n-Dicarbanionic titanium intermediates from monocarbanionic organometallics and their application in organic synthesis. *Chem Rev.* 2000;100(8):2789–2834.

124. Haym I, Brimble MA. The Kulinkovich hydroxycyclopropanation reaction in natural product synthesis. *Org Biomol Chem.* 2012;10(38):7649–7665.
125. Campbell AD, Taylor RJK, Raynham TM. The total synthesis of (−)-α-kainic acid using titanium-mediated diene metallabicyclisation methodology. *Chem Commun.* 1999;3:245–246.
126. Cheng X, Micalizio GC. Synthesis of neurotrophic Seco-prezizaane Sesquiterpenes (1R,10S)-2-oxo-3,4-dehydroneomajucin, (2S)-hydroxy-3,4-dehydroneomajucin, and (−)-jiadifenin. *J Am Chem Soc.* 2016;138(4):1150–1153.
127. Rosales A, Rodríguez-García I, Muñoz-Bascón J, et al. The nugent reagent: a formidable tool in contemporary radical and organometallic chemistry. *Eur J Org Chem.* 2015;2015(21):4567–4591.
128. Burgmayer SJN. Use of a titanium metallocene as a colorimetric indicator for learning inert atmosphere techniques. *J Chem Educ.* 1998;75(4):460.
129. Beweries T, Haehnel M, Rosenthal U. Recent advances in the chemistry of heterometallacycles of group 4 metallocenes. *Cat Sci Technol.* 2013;3(1):18–28.
130. Rosenthal U. Advantages of group 4 metallocene Bis(trimethylsilyl)acetylene complexes as metallocene sources towards other synthetically used systems. *ChemistryOpen.* 2019;8(8):1036–1047.
131. Rosenthal U. Recent synthetic and catalytic applications of group 4 metallocene Bis(trimethylsilyl)acetylene complexes. *Eur J Inorg Chem.* 2019;2019(7):895–919.
132. Beckhaus R. Pentafulvene complexes of group four metals: versatile organometallic building blocks. *Coord Chem Rev.* 2018;376:467–477.
133. Snead TE. Dicarbonylbis(cyclopentadienyl)titanium. In: *Encyclopedia of Reagents for Organic Synthesis.* John Wiley & Sons, Ltd. 2001
134. Berk SC, Grossman RB, Buchwald SL. Development of a titanocene-catalyzed enyne cyclization/isocyanide insertion reaction. *J Am Chem Soc.* 1994;116(19):8593–8601.
135. Aguilar-Calderón JR, Metta-Magaña AJ, Noll B, Fortier S. C(sp3)−H oxidative addition and transfer hydrogenation chemistry of a titanium(II) synthon: mimicry of late-metal type reactivity. *Angew Chem Int Ed.* 2016;55(45):14101–14105.
136. Aguilar-Calderón JR, Murillo J, Gomez-Torres A, et al. Redox character and small molecule reactivity of a masked titanium(II) synthon. *Organometallics.* 2020;39(2):295–311.
137. Saito T, Nishiyama H, Tanahashi H, Kawakita K, Tsurugi H, Mashima K. 1,4-Bis(trimethylsilyl)-1,4-diaza-2,5-cyclohexadienes as strong salt-free reductants for generating low-valent early transition metals with electron-donating ligands. *J Am Chem Soc.* 2014;136(13):5161–5170.
138. Frey G, Hausmann JN, Streuff J. Titanium-catalyzed reductive umpolung reactions with a metal-free terminal reducing agent. *Chem A Eur J.* 2015;21(15):5693–5696.
139. Urabe H, Narita M, Sato F. A titanacycle-to-carbocycle relay leading to an expedient synthesis of cyclopentadienols. *Angew Chem Int Ed.* 1999;38(23):3516–3518.
140. Sato F, Urabe H, Okamoto S. Synthesis of organotitanium complexes from alkenes and alkynes and their synthetic applications. *Chem Rev.* 2000;100(8):2835–2886.
141. Sato F, Okamoto S. The divalent titanium complex Ti(O-i-Pr)4/2 i-PrMgX as an efficient and practical reagent for fine chemical synthesis. *Adv Synth Catal.* 2001;343(8):759–784.
142. Takahashi M, Micalizio GC. Regio- and stereoselective cross-coupling of substituted olefins and imines. A convergent stereoselective synthesis of saturated 1,5-aminoalcohols and substituted piperidines. *J Am Chem Soc.* 2007;129(24):7514–7516.
143. Tarselli MA, Micalizio GC. Aliphatic imines in titanium-mediated reductive cross-coupling: unique reactivity of Ti(O-i-Pr)4/n-BuLi. *Org Lett.* 2009;11(20):4596–4599.

144. Chen MZ, McLaughlin M, Takahashi M, et al. Preparation of stereodefined homoallylic amines from the reductive cross-coupling of allylic alcohols with imines. *J Org Chem.* 2010;75(23):8048–8059.
145. Reichard HA, McLaughlin M, Chen MZ, Micalizio GC. Regioselective reductive cross-coupling reactions of unsymmetrical alkynes. *Eur J Org Chem.* 2010;2010(3):391–409.
146. Yang D, Micalizio GC. Convergent and stereodivergent synthesis of complex 1-aza-7-oxabicyclo[2.2.1]heptanes. *J Am Chem Soc.* 2011;133(24):9216–9219.
147. 5.34 Early transition metal mediated reductive coupling reactions. Micalizio GC, Knochel P, eds. *Comprehensive Organic Synthesis II.* 2nd ed. Amsterdam: Elsevier; 2014:1660–1737.
148. Kulinkovich OG. The chemistry of cyclopropanols. *Chem Rev.* 2003;103(7):2597–2632.
149. Hill JE, Balaich G, Fanwick PE, Rothwell IP. The chemistry of titanacyclopentadiene rings supported by 2,6-diphenylphenoxide ligation: stoichiometric and catalytic reactivity. *Organometallics.* 1993;12(8):2911–2924.
150. Ozerov OV, Ladipo FT, Patrick BO. Highly regioselective alkyne cyclotrimerization catalyzed by titanium complexes supported by proximally bridged p-tert-butylcalix[4]arene ligands. *J Am Chem Soc.* 1999;121(34):7941–7942.
151. Ozerov OV, Patrick BO, Ladipo FT. Highly regioselective [2 + 2 + 2] cycloaddition of terminal alkynes catalyzed by η6-arene complexes of titanium supported by dimethylsilyl-bridged p-tert-butyl calix[4]arene ligand. *J Am Chem Soc.* 2000;122(27):6423–6431.
152. Johnson ES, Balaich GJ, Rothwell IP. Regio- and stereoselective synthesis of the 1,3-cyclohexadiene nucleus by [2 + 2 + 2] cycloaddition reactions catalyzed by titanium aryloxide compounds. *J Am Chem Soc.* 1997;119(33):7685–7693.
153. Balaich GJ, Rothwell IP. Regio- and stereoselective formation and isomerization of 1,3-cyclohexadienes catalyzed by titanium aryloxide compounds. *J Am Chem Soc.* 1993;115(4):1581–1583.
154. Streuff J. A titanium(III)-catalyzed redox umpolung reaction for the reductive cross-coupling of enones with acrylonitriles. *Chem A Eur J.* 2011;17(20):5507–5510.
155. Streuff J, Feurer M, Bichovski P, Frey G, Gellrich U. Enantioselective titanium(III)-catalyzed reductive cyclization of ketonitriles. *Angew Chem Int Ed.* 2012;51(34):8661–8664.
156. Frey G, Luu H-T, Bichovski P, Feurer M, Streuff J. Convenient titanium(III)-catalyzed synthesis of cyclic aminoketones and pyrrolidinones—development of a formal [4 + 1] cycloaddition. *Angew Chem Int Ed.* 2013;52(28):7131–7134.
157. Feurer M, Frey G, Luu H-T, Kratzert D, Streuff J. The cross-selective titanium(iii)-catalysed acyloin reaction. *Chem Commun.* 2014;50(40):5370–5372.
158. Luu H-T, Wiesler S, Frey G, Streuff J. A titanium(III)-catalyzed reductive umpolung reaction for the synthesis of 1,1-disubstituted tetrahydroisoquinolines. *Org Lett.* 2015;17(10):2478–2481.
159. Streuff J, Feurer M, Frey G, et al. Mechanism of the TiIII-catalyzed acyloin-type umpolung: a catalyst-controlled radical reaction. *J Am Chem Soc.* 2015;137(45):14396–14405.
160. Bichovski P, Haas TM, Keller M, Streuff J. Direct conjugate alkylation of α,β-unsaturated carbonyls by TiIII-catalysed reductive umpolung of simple activated alkenes. *Org Biomol Chem.* 2016;14(24):5673–5682.
161. Leijendekker LH, Weweler J, Leuther TM, Kratzert D, Streuff J. Development, scope, and applications of titanium(III)-catalyzed cyclizations to aminated N-heterocycles. *Chem A Eur J.* 2019;25(13):3382–3390.

162. Luu H-T, Streuff J. Development of an efficient Synthesis of rac-3-demethoxyerythratidinone via a titanium(III) catalyzed imine-nitrile coupling. *Eur J Org Chem.* 2019;2019(1):139–149.
163. Weweler J, Younas SL, Streuff J. Titanium(III)-catalyzed reductive decyanation of geminal dinitriles by a non-free-radical mechanism. *Angew Chem Int Ed.* 2019;58(49): 17700–17703.
164. Chong E, Xue W, Storr T, Kennepohl P, Schafer LL. Pyridonate-supported titanium(III). Benzylamine as an easy-to-use reductant. *Organometallics.* 2015;34(20): 4941–4945.
165. Chaplinski V, de Meijere A. A versatile new preparation of cyclopropylamines from acid dialkylamides. *Angew Chem Int Ed Engl.* 1996;35(4):413–414.
166. Meijere AD, Kozhushkov SI, Savchenko AI. Titanium-mediated syntheses of cyclopropylamines. *J Organomet Chem.* 2004;689(12):2033–2055.
167. Tonks IA, Meier JC, Bercaw JE. Alkyne hydroamination and trimerization with titanium Bis(phenolate)pyridine complexes: evidence for low-valent titanium intermediates and synthesis of an ethylene adduct of titanium(II). *Organometallics.* 2013;32(12): 3451–3457.
168. Gilbert ZW, Hue RJ, Tonks IA. Catalytic formal [2+2+1] synthesis of pyrroles from alkynes and diazenes via TiII/TiIV redox catalysis. *Nat Chem.* 2016;8(1): 63–68.
169. Blanco-Urgoiti J, Añorbe L, Pérez-Serrano L, Domínguez G, Pérez-Castells J. The Pauson–Khand reaction, a powerful synthetic tool for the synthesis of complex molecules. *Chem Soc Rev.* 2004;33(1):32–42.
170. Davis-Gilbert ZW, Wen X, Goodpaster JD, Tonks IA. Mechanism of Ti-catalyzed oxidative nitrene transfer in [2 + 2 + 1] pyrrole synthesis from alkynes and azobenzene. *J Am Chem Soc.* 2018;140(23):7267–7281.
171. Chiu H-C, See XY, Tonks IA. Dative directing group effects in Ti-catalyzed [2+2+1] pyrrole synthesis: chemo- and regioselective alkyne heterocoupling. *ACS Catal.* 2019;9(1):216–223.
172. Chiu H-C, Tonks IA. Trimethylsilyl-protected alkynes as selective cross-coupling partners in titanium-catalyzed [2+2+1] pyrrole synthesis. *Angew Chem Int Ed:* 2018;57(21):6090–6094.
173. Pearce AJ, See XY, Tonks IA. Oxidative nitrene transfer from azides to alkynes via Ti(ii)/Ti(iv) redox catalysis: formal [2+2+1] synthesis of pyrroles. *Chem Commun.* 2018;54(50):6891–6894.
174. Davis-Gilbert ZW, Kawakita K, Blechschmidt DR, Tsurugi H, Mashima K, Tonks IA. In situ catalyst generation and benchtop-compatible entry points for TiII/TiIV redox catalytic reactions. *Organometallics.* 2018;37(23):4439–4445.
175. See XY, Beaumier EP, Davis-Gilbert ZW, et al. Generation of TiII alkyne trimerization catalysts in the absence of strong metal reductants. *Organometallics.* 2017;36(7):1383–1390.
176. Davis-Gilbert ZW, Yao LJ, Tonks IA. Ti-catalyzed multicomponent oxidative carboamination of alkynes with alkenes and diazenes. *J Am Chem Soc.* 2016;138(44):14570–14573.
177. Polse JL, Andersen RA, Bergman RG. Reactivity of a terminal Ti(IV) imido complex toward alkenes and alkynes: cycloaddition vs C−H activation. *J Am Chem Soc.* 1998;120(51):13405–13414.
178. Zhao G, Basuli F, Kilgore UJ, et al. Neutral and Zwitterionic low-coordinate titanium complexes bearing the terminal phosphinidene functionality. Structural, spectroscopic, theoretical, and catalytic studies addressing the Ti−P multiple bond. *J Am Chem Soc.* 2006;128(41):13575–13585.

179. Basuli F, Aneetha H, Huffman JC, Mindiola DJ. A fluorobenzene adduct of Ti(IV), and catalytic carboamination to prepare α,β-unsaturated imines and triaryl-substituted quinolines. *J Am Chem Soc.* 2005;127(51):17992–17993.
180. Botubol-Ares JM, Durán-Peña MJ, Hanson JR, Hernández-Galán R, Collado IG. Cp2Ti(III)cl and analogues as sustainable templates in organic synthesis. *Synthesis.* 2018;50(11):2163–2180.
181. Leijendekker LH, Weweler J, Leuther TM, Streuff J. Catalytic reductive synthesis and direct derivatization of unprotected aminoindoles, aminopyrroles, and iminoindolines. *Angew Chem Int Ed.* 2017;56(22):6103–6106.
182. Hao W, Wu X, Sun JZ, Siu JC, MacMillan SN, Lin S. Radical redox-relay catalysis: formal [3+2] cycloaddition of N-acylaziridines and alkenes. *J Am Chem Soc.* 2017;139(35):12141–12144.
183. Liedtke T, Spannring P, Riccardi L, Gansäuer A. Mechanism-based condition screening for sustainable catalysis in single-electron steps by cyclic voltammetry. *Angew Chem Int Ed.* 2018;57(18):5006–5010.
184. Zhang Z, Richrath RB, Gansäuer A. Merging catalysis in single electron steps with photoredox catalysis—efficient and sustainable radical chemistry. *ACS Catal.* 2019;9(4):3208–3212.
185. Rossi B, Prosperini S, Pastori N, Clerici A, Punta C. New advances in titanium-mediated free radical reactions. *Molecules.* 2012;17(12):14700–14732.
186. Knochel P, Molander GA. *Comprehensive Organic Synthesis.* Elsevier Science; 2014.